BRYOPHYTES OF THE PLEISTOCENE

BRYOPHYTES OF THE PLEISTOCENE

THE BRITISH RECORD AND ITS CHOROLOGICAL AND ECOLOGICAL IMPLICATIONS

J. H. DICKSON

Lecturer in Botany, The University of Glasgow

Formerly Senior Assistant in Research, The University of Cambridge and

Fellow of Clare College, Cambridge

CAMBRIDGE

AT THE UNIVERSITY PRESS

1973

Published by the Syndics of the Cambridge University Press
Bentley House, 200 Euston Road, London NW1 2DB
American Branch: 32 East 57th Street, New York, N.Y. 10022

Library of Congress Catalogue Card Number: 72-190419

ISBN: 0 521 08576 4

Printed in Great Britain
by Alden & Mowbray Ltd, Oxford

For

CAMILLA

CONTENTS

PREFACE

In writing this book I have been conscious of two categories of readers, firstly pollen analysts and students of related disciplines who may know little about bryophytes, and secondly bryologists who may know little about the many recent advances in Pleistocene research. I hope both groups may find the book of interest, even if of necessity parts of it must be familiar to some extent to one group or the other.

The bulk of the book is composed of a species-by-species compilation of the British Pleistocene records of bryophytes. There are abundant records of British mosses from the last 15,000 years - the period of most significance in understanding the present British bryoflora. However, where the British evidence is deficient I have not hesitated to use cogent evidence from other regions. Perhaps the most important in this respect are the Neogene assemblages from Russia and Poland.

My curiosity about bryophytes was first aroused by Mr A. C. Crundwell to whom I am grateful for his continued encouragement. Without my sojourn in Cambridge this book would not have been written. First I must extend my thanks to Professor Sir Harry Godwin for his support. To him and to all other members of the Sub-Department of Quaternary Research I owe a great debt for intellectual stimulation. In a smaller but equally pleasant way I have enjoyed co-operation with Miss A. P. Connolly and Dr W. Pennington (Mrs T. G. Tutin) of the University of Leicester and Professors G. F. Mitchell and W. A. Watts of Trinity College, Dublin. For seven years of my stay in Cambridge I was first a Research Fellow and later an Official Fellow of Clare College. This was an enlightening experience.

I have travelled widely outside the British Isles to my bryological benefit. I must mention for their support the Nature Conservancy (Sweden), the British Council and the Polish Academy of Sciences (Poland), Clare College and the University of Cambridge (Norway) and the National Museum of Canada (Arctic Canada).

I have sought the opinions of several bryologists at various times. Dr D. Briggs, Mr A. C. Crundwell, Miss U. K. Duncan, Mr S. W. Greene, Dr T. Koponen, Dr M. C. F. Proctor, Mrs J. Paton, Dr E. V. Watson and Mr E. C. Wallace, and no doubt others, deserve thanks for their forbearance in examining the all-too-often miserable scraps I proffered.

For permission to use their unpublished material I must thank Miss R. Andrew, Dr H. J. B. Birks, Miss A. P. Conolly, Dr W. Dickinson, Dr B. H. Green, Dr I. G. Simmons, Dr R. Squires, Mr D. M. Synott, Dr K. Thompson and Mr E. C. Wallace.

The scanning electromicrographs were taken by Cambridge Scientific Instruments Ltd. Plate 21*a* is published with permission of the National Museum of Ireland. The plates of macroscopic remains of mosses were taken by B. V. D. Goddard, F. T. N. Elborn and T. N. Tait.

I have reproduced or published, with additions, many distribution maps originally published in the *Transactions of the British Bryological Society* by various authors. I must thank Dr H. H. Birks, Dr H. J. B. Birks, Mr A. C. Crundwell, Miss U. K. Duncan, Mrs J. Paton, Professor C. D. Pigott, Dr M. C. F. Proctor, Dr D. A. Ratcliffe, Mr G. Rodway and Mr. A. McG. Stirling.

Figures 11, 12, 13, 14 and 16 have been reproduced from R. G. West's book *Pleistocene Geology and Biology* and figures 51, 54, 59, 71, 72, 76, 78 and 79 from P. Störmer's book *Mosses with a Western and Southern Distribution in Norway.*

January, 1972 J. H. DICKSON

1 INTRODUCTION

1.1 GEOGRAPHICAL PATTERNS IN THE PRESENT BRYOFLORA OF THE BRITISH ISLES

The primary aim of this book is to improve the understanding of the present bryoflora of the British Isles in historical terms. Can the many geographical patterns shown by bryophytes be explained as the results of the marked climatic-vegetational changes of the Pleistocene? By necessity the discussion is almost restricted to the mosses. The Pleistocene history of the liverworts is scanty, while that of the mosses is abundant. However, there is no overriding reason why conclusions reached for the mosses should not pertain to the liverworts.

The bryoflora of the British Isles, renowned for its richness, consists at present of some 960 species of which more than two-thirds are mosses. Table 1 reveals the richness compared to four other areas of Europe. Area for area it is clear that the British Isles support a large bryoflora, particularly of liverworts. With an area almost four times larger, Fennoscandia supports no markedly richer bryoflora. Poland, with an almost identical area but primarily a lowland country and with a continental climate, supports nearly 100 fewer species. Though the area of the British Isles is small, the great diversity of topography, edaphic and climatic conditions permits a varied bryoflora.

Table 1 *Numbers of bryophytes in four regions of Europe*

	Hepatics	Mosses	Total	Area (square miles)
Fennoscandia (Arnell 1956; Nyholm 1954-69)	300	725	1025	454,665
British Isles (Paton 1965; Warburg 1963; etc.)	285	675	960	121,362
Yugoslavia (Pavletic 1955)	200	715	915	98,760
France (Augier 1966)	250	660	910	212,209
Poland (Szafran 1957, 1961; Szweykowski 1958)	225	630	855	120,348

As Proctor (1964, 1967) has emphasised, for bryophytes no less than vascular plants the most significant geographical boundary in the British Isles is the Highland Line which demarks the mountainous region of the north and west from the lowland region of the south and east (Tansley 1939, p. 17). The highland zone supports by far the richer bryoflora. Angus (vice-county 90; p. 65), in the eastern Highlands of Scotland, is a small county with edaphically favourable mountains; some 615 species, including 164 liverworts, are listed by Duncan (1966). County Kerry (vice-counties H1 and H2) in southwestern Ireland, with great precipitation (over 220 wet days) and mild equable climate, supports some 550 species, including 175 liverworts. In contrast stands Cambridgeshire (vice-county 29), a lowland county of subdued relief, almost uniformly calcareous soils, low precipitation (less than 120 wet days) and intense cultivation, which supports only 260 species, including a mere 40 liverworts.

Table 1 gives 960 as the number of species in the British bryoflora. Perhaps a round figure of about 1000 species will prove to be a truer figure. Almost every year there are additions, and the following are by no means all of those that have been recorded for the first time in the British Isles during the last ten years.

Aongstroemia longipes. In Perthshire, an arctic-alpine species of Holarctic range (Crundwell 1965).

Fissidens celticus. Confined to western and southern coasts of the British Isles (Paton 1965*a*).

Fossombronia incurva. In Perthshire, a European species best known from the North German-Polish plain (Crundwell 1965).

Grimmia agassizii. In Perthshire, an arctic-alpine species of Holarctic range (Birks and Birks 1967).

G. borealis. In Angus, a species of Fennoscandia and Novaya Zemlya (Warburg 1965*a*).

Orthotrichum gymnostomum. In Easterness, a species of central and northern Europe, Japan and Newfoundland (Perry and Dransfield 1967).

Pohlia pulchella. Scattered from southern England to Argyll, also in the rest of Europe and eastern North America (Warburg 1965*b*).

Seligeria oelandica. In Co. Sligo, also in Öland, Gotland and Swedish Lapland (Crundwell and Warburg 1963).

Tortula vectensis. In the Isle of Wight and Louisiana (Warburg and Crundwell 1965; Reese 1967).

These discoveries emphasise not only the diversity of the bryoflora but also that much remains to be revealed of bryophyte distribution patterns. They reflect, moreover, the skill of numerous British bryologists and the intensity of bryological recording.

No comprehensive geographical analysis of the British bryoflora has ever been published, either with reference to the purely national distribution or to the European or world ranges. There is a partial analysis by Cardot (1930), and the oceanic species have been categorised by such authors as Gams (1952), Greig-Smith (1950) and Ratcliffe (1968), but for the great bulk of species there is no analysis comparable to those of Amann for Switzerland (1928), Szafran for Poland (1952) or Boros for Hungary (1968). For the purposes of this book this lack is an inconvenience but no more. Geographical categories are far from absolute, especially for bryophytes, which have been studied much less than flowering plants. Many bryophytes have enormous ranges, the details of which are little understood.

Species with strongly northern ranges in the British Isles are mostly *arctic-alpines.* The maps of *Arctoa fulvella* (fig. 1), *Conostomum tetragonum* (fig. 2), *Encalypta alpina* (fig. 3) and *Moerkia blyttii* (fig. 4) exemplify this type of distribution. *Arctoa fulvella*, widely scattered in the Scottish Highlands and the Hebrides, also occurs in the Southern Uplands of Scotland, the English Lake District and the hills of North Wales; there is a solitary Irish locality. *Conostomum tetragonum* is very similar in its distribution except that there are no localities in the Outer Hebrides, the Southern Uplands or Ireland. *Encalypta alpina* has a more limited range, though there is a solitary Irish occurrence, and the liverwort, *Moerkia blyttii*, is confined to the Scottish Highlands.

Some arctic-alpine species have highly restricted ranges. As examples mention may be made of *Cirriphyllum cirrosum, Saelania glaucescens*, and *Timmia austriaca,* all restricted to only three vice-counties in the Scottish Highlands, *Blindia caespiticia* and *Ctenidium procerrimum*, restricted to two, and *Brachythecium erythrorrhizon, Bryum arcticum* and *Mnium lycopodioides*, restricted to one. The arctic-alpine element of the British moss flora is composed of about 110 species which occur mostly at moderate or high altitudes in the highland zone. There is a close chorological parallel between these species and the vascular plants to which the term arctic-alpine primarily refers.

High-arctic species as defined by Steere (1953, 1965) have their greatest abundance at high northern latitudes where they have circumpolar ranges. This type of pattern can be exemplified by *Haplodon wormskjoldii*, almost the sole British species in this category (fig. 44).

Another group of species with a northern bias in the British Isles is here designated *northern-continental*, a term applying to the total European distributions which are strongly northern and eastern. In the present British moss flora there are *Calliergon trifarium, Catoscopium nigritum* (fig. 5), *Cinclidium stygium, Helodium blandowii, Homalothecium nitens, Paludella squarrosa, Meesia tristicha* and *Sphagnum warnstorfii.* They are species of mires, predominantly but not exclusively of the highland zone. In Europe they reach their greatest abundance in central-northern Fennoscandia where they inhabit fens, usually rich fens. In discussing the ecology and history of these mosses the Scandinavian concepts of mire ecology are followed (Du Rietz 1949, Sjörs 1950, Mårtensson 1956). *Mire* covers all types of fen, *minerotrophic* vegetation, and bog, *ombrotrophic* vegetation. *Rich* fens are rich in species, with water of circumneutral reaction and high electrical conductivity. *Poor* fens are poorer in species, with distinctly acid water of low conductivity. *Transitional* fens refer to intermediate types.

Species with strongly western or southwestern ranges in Europe can be broadly categorised as *oceanic*. The great richness of oceanic (or Atlantic) bryophytes, especially liverworts, in the British Isles has caused much comment; there are about 100 mosses and 100 liverworts. Major contributions towards the understanding of the distribution and ecology of oceanic species have recently been made by Ratcliffe (1968) and Störmer (1969), whose works should be consulted for data on the environmental requirements of these species. As has been stressed by various authors, oceanic species comprise a chorologically and ecologically very diverse group.

Within and without the British Isles oceanic species exhibit many distribution patterns. *Plagiochila carringtonii* (fig. 6) is strongly northern. It falls in the Northern Atlantic group of Ratcliffe (1968). Outside the British Isles it occurs in the Faeroe Islands and in Nepal—an enormous disjunction.

Myurium hebridarum (fig. 7) is strongly northwestern. Designated Widespread Atlantic by Ratcliffe (1968), the species is known elsewhere only in Macaronesia.

Dumortiera hirsuta (fig. 8) is strongly southwestern. Designated Southern Atlantic by Ratcliffe (1968), this species has a wide distribution in tropical and oceanic areas.

Leptodon smithii (fig. 9), strongly southern in the British Isles, has a European distribution centred on the Mediterranean; Mediterranean-Atlantic of Ratcliffe (1968).

Oedipodium griffithianum (fig. 10) falls in the Sub-Atlantic category of Ratcliffe (1968). Its British range is scattered but primarily northern. The extra-British distribution comprises Fennoscandia, Greenland, Alaska, Japan and the Falkland Islands.

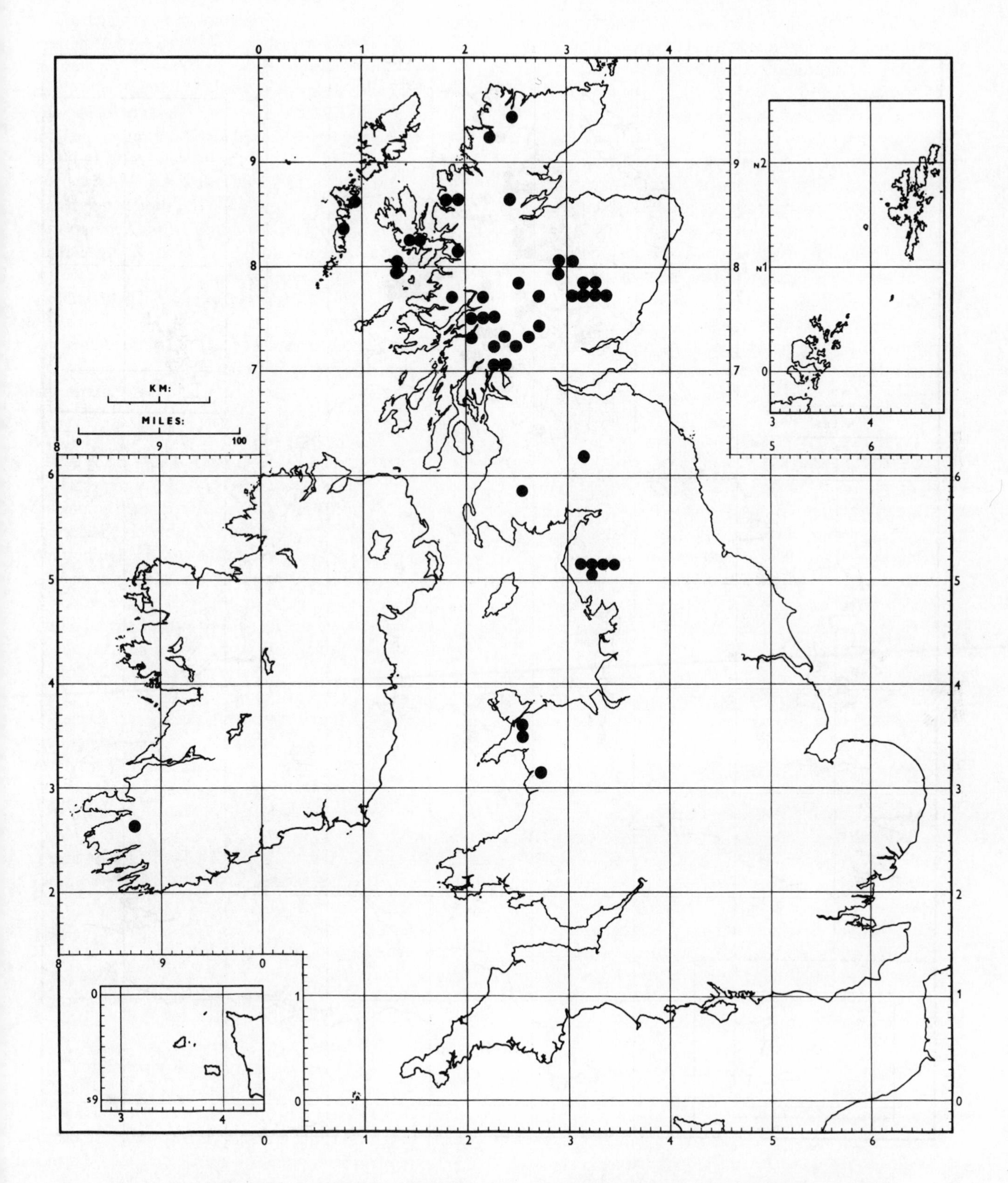

Fig. 1. Distribution of *Arctoa fulvella* in the British Isles. From Crundwell and Stirling, *Trans. Br. bryol. Soc.* 1970.

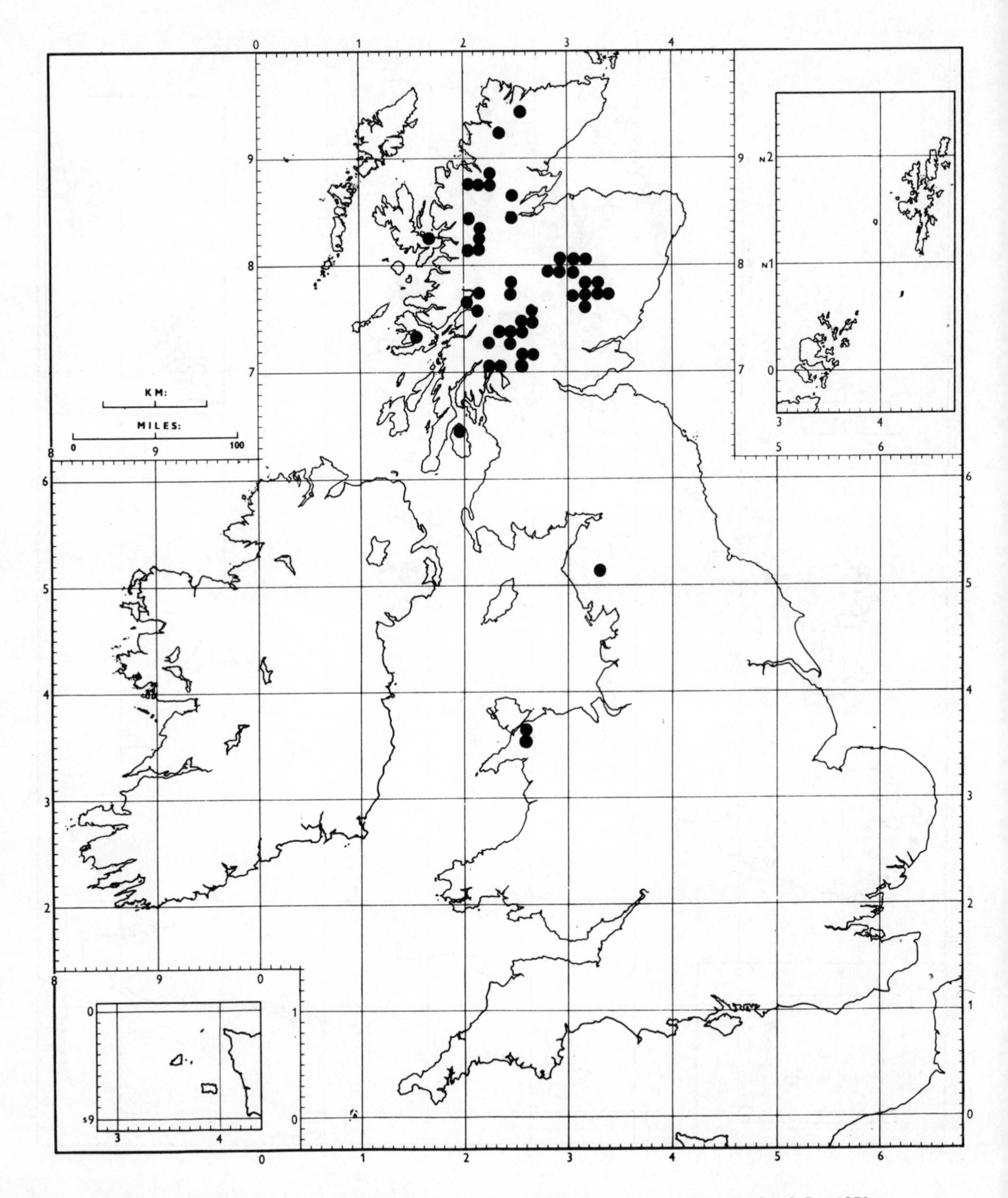

Fig. 2. Distribution of *Conostomum tetragonum* in the British Isles. From Duncan, *Trans. Br. bryol. Soc.* 1970.

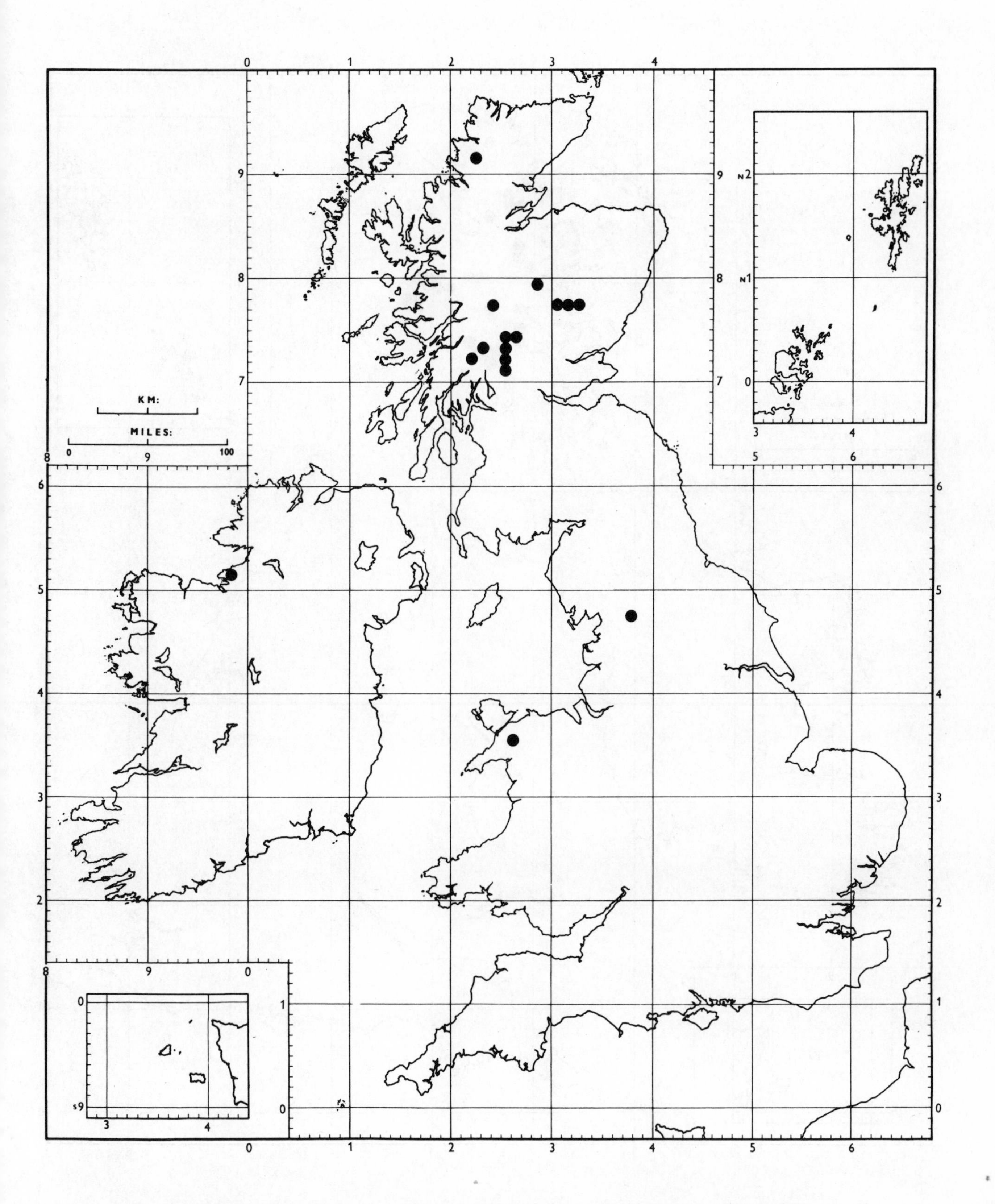

Fig. 3. Distribution of *Encalypta alpina* in the British Isles. From H. H. Birks, *Trans. Br. bryol. Soc.* 1969.

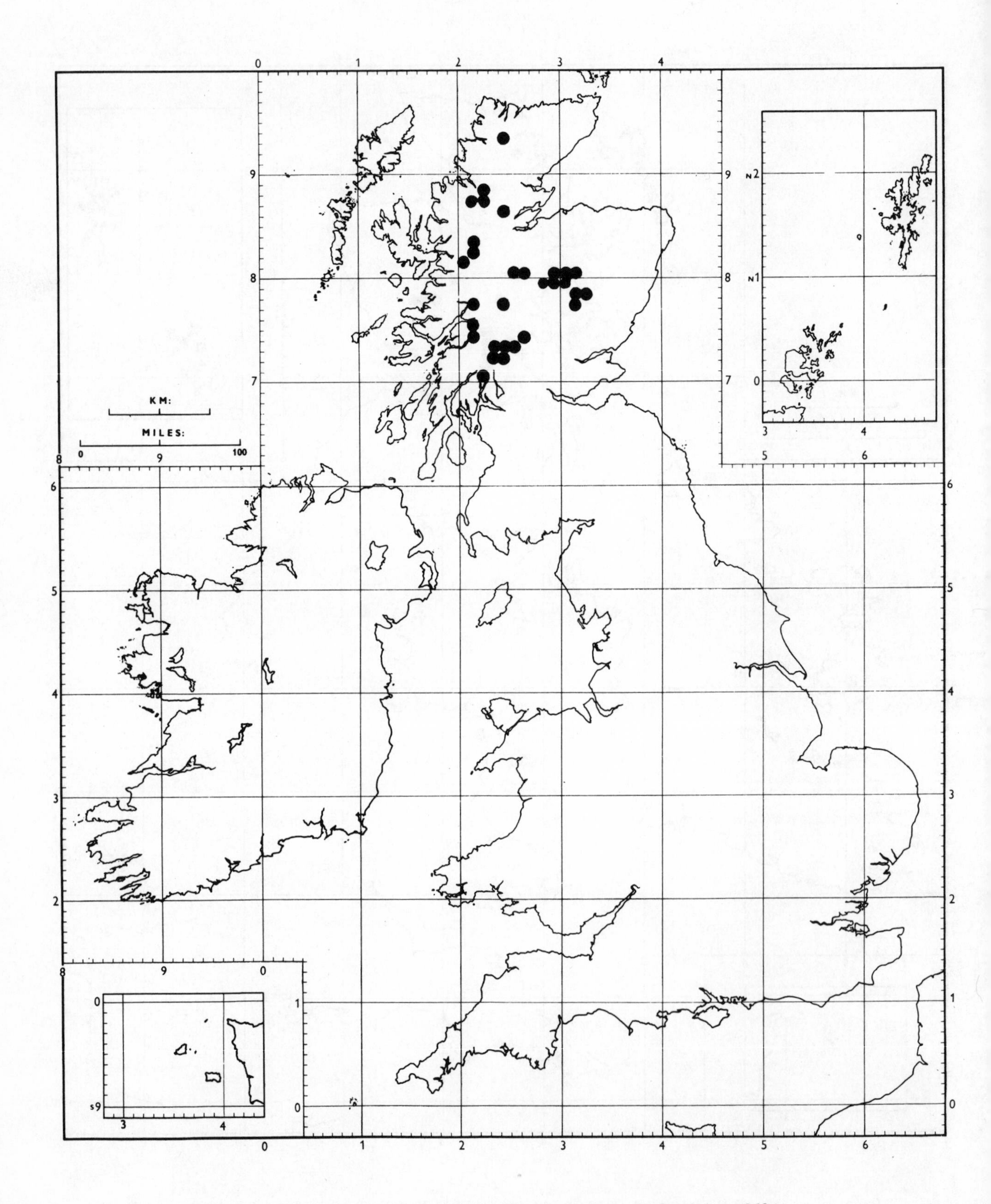

Fig. 4. Distribution of *Moerkia blyttii* in the British Isles. From Paton, *Trans. Br. bryol. Soc.* 1963.

Plate 1. (*a*) Ben Lawers, Perthshire. View from near the summit, 3984 ft, into the northeast corrie which was occupied by ice in zone III. The mica-schist rocks of the Ben Lawers range support a rich arctic-alpine bryoflora including *Calliergon turgescens, Ctenidium procerrimum, Hypnum revolutum, H. vaucheri* and *Ptychodium plicatum,* and many others.

Plate 1 (*b*) Ben Bulben, Co. Sligo. The limestone massif of the Dartry Mountains, commonly known as Ben Bulben lies partly in Co. Sligo, partly in Co. Leitrim. The cliffs and plateau, which reach only to 2100 ft, support one of the most notable arctic-alpine bryofloras in Ireland. Disjunct species include *Amblyodon dealbatus, Encalypta alpina, Mnium orthorhynchum, Orthothecium rufescens* and outstandingly *Seligeria oelandica. Timmia norvegica* occurs on the cliffs shown here. Ben Bulben may have been free of ice during the last glaciation. *Breutelia chrysocoma* is abundant. Note the ombrogenous bog (on the right) extending up the slopes.

Plate 2. (*a*) The Upper Lake, Killarney, Co. Kerry. In and around the *Quercus petraea* woodlands of Killarney the abundance of oceanic species such as *Cyclodictyon laetevirens, Dumortiera hirsuta, Sematophyllum* spp, *Telaranea sejuncta, Radula voluta* and species of Lejuneaceae, makes the area bryogeographically one of the most celebrated in Europe.

Plate 2. (*b*) *Quercus petraea* woodland of poorly grown trees at Keskadale in the English Lake District. *Antitrichia curtipendula* is locally abundant exclusively on rock. The oceanic species *Hedwigia integrifolia* also occurs.

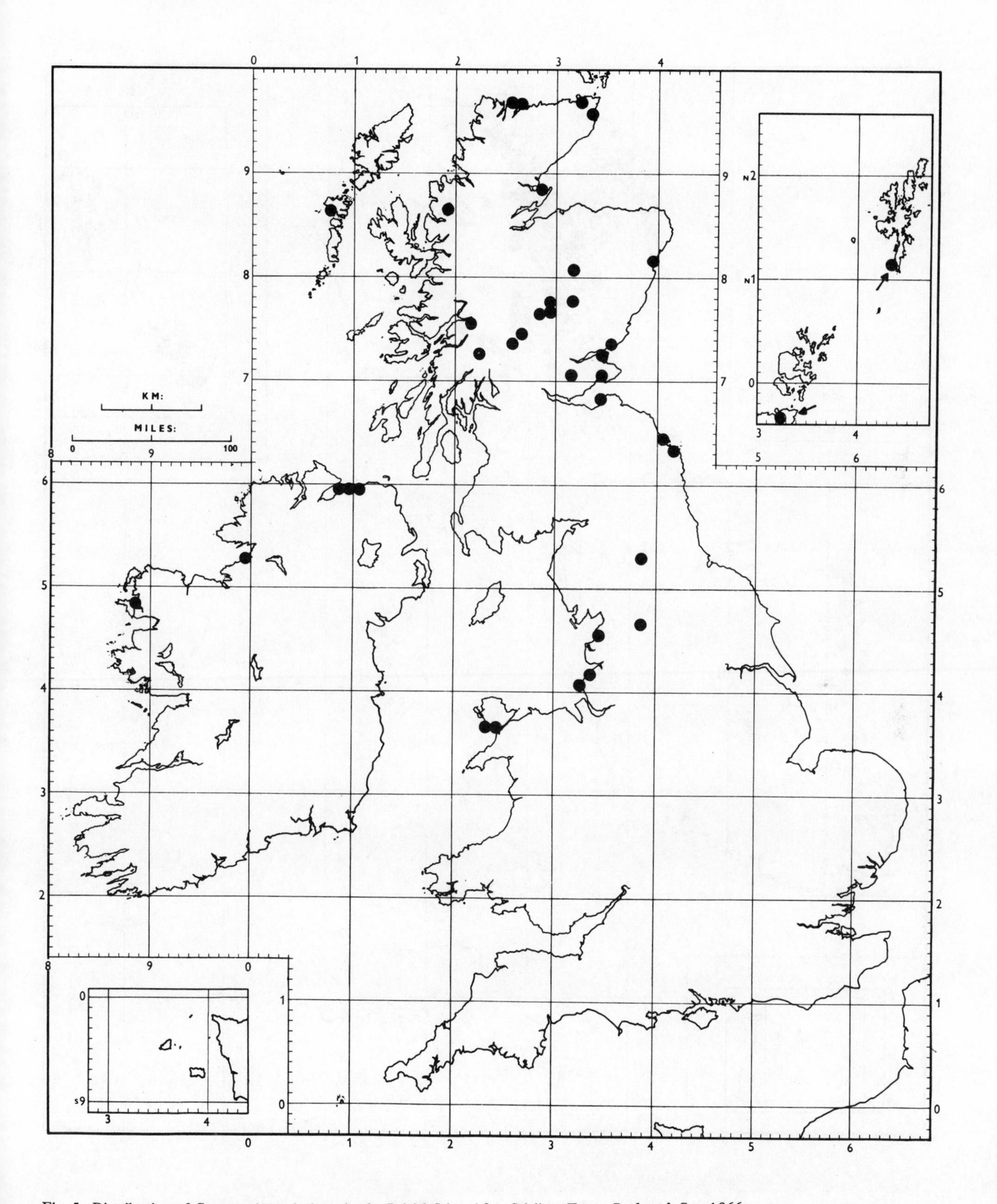

Fig. 5. Distribution of *Catoscopium nigritum* in the British Isles. After Stirling, *Trans. Br. bryol. Soc.* 1966.

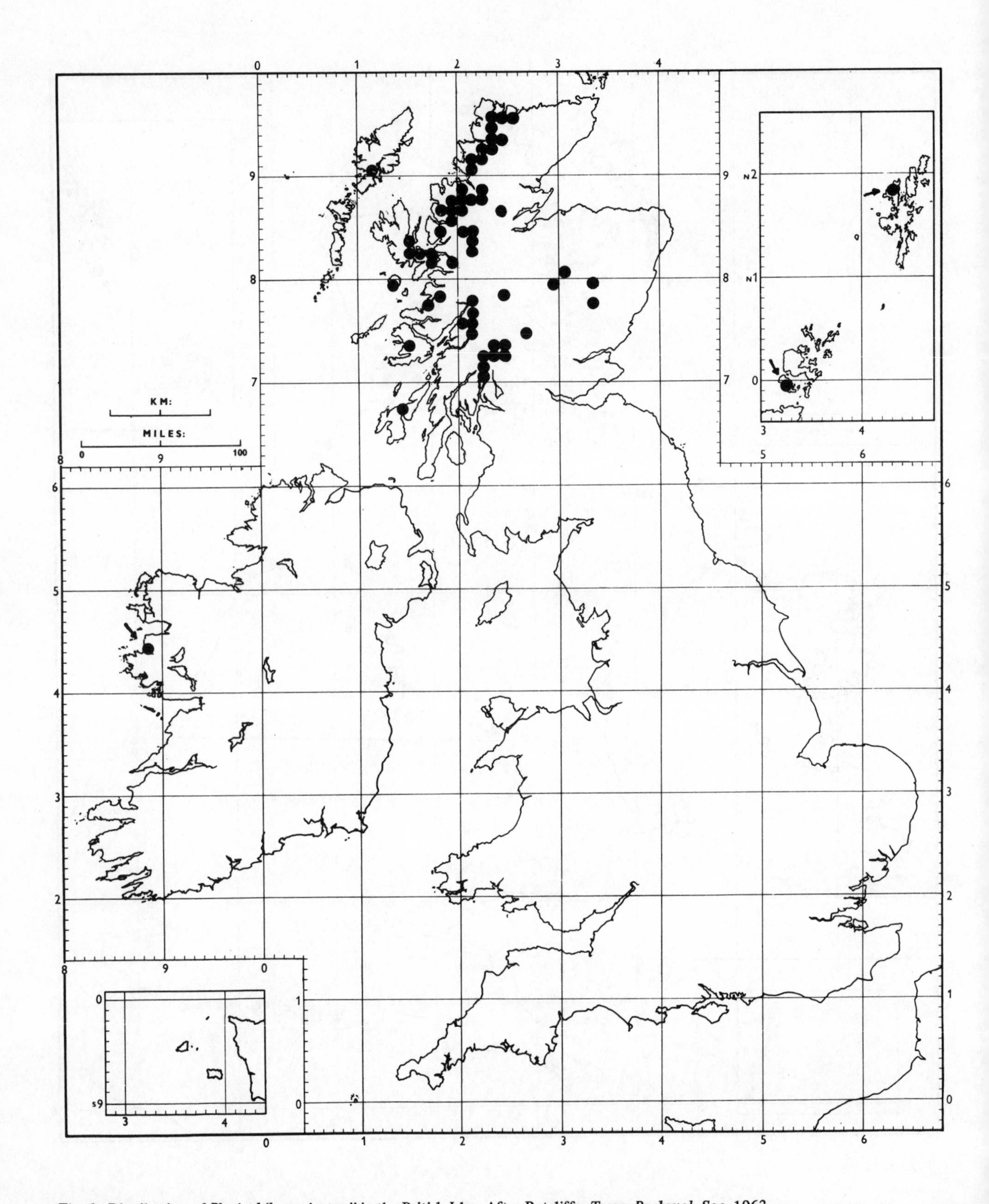

Fig. 6. Distribution of *Plagiochila carringtonii* in the British Isles. After Ratcliffe, *Trans. Br. bryol. Soc.* 1963.

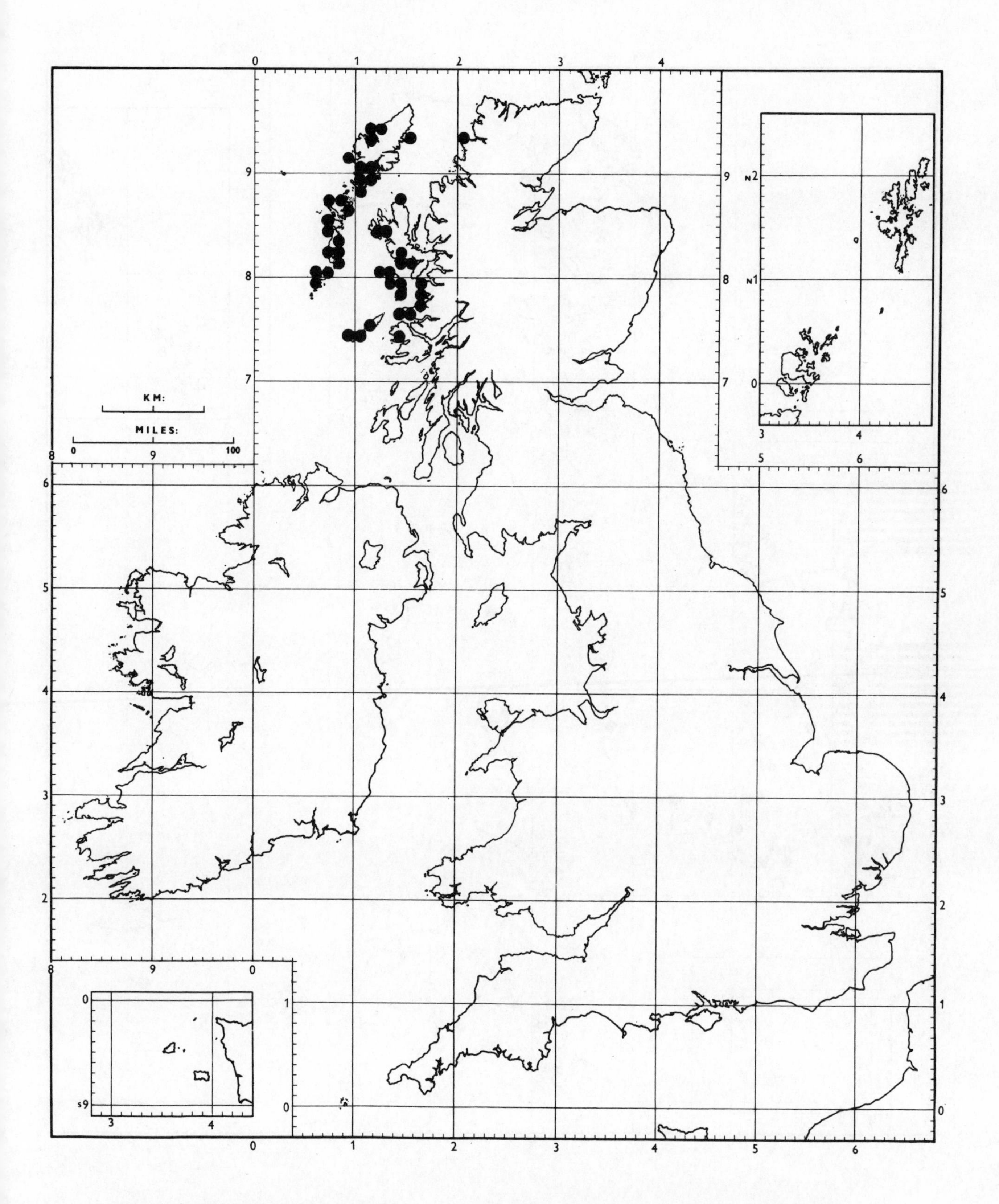

Fig. 7. Distribution of *Myurium hebridarum* in the British Isles. From Stirling and Rodway, *Trans. Br. bryol. Soc.* 1966.

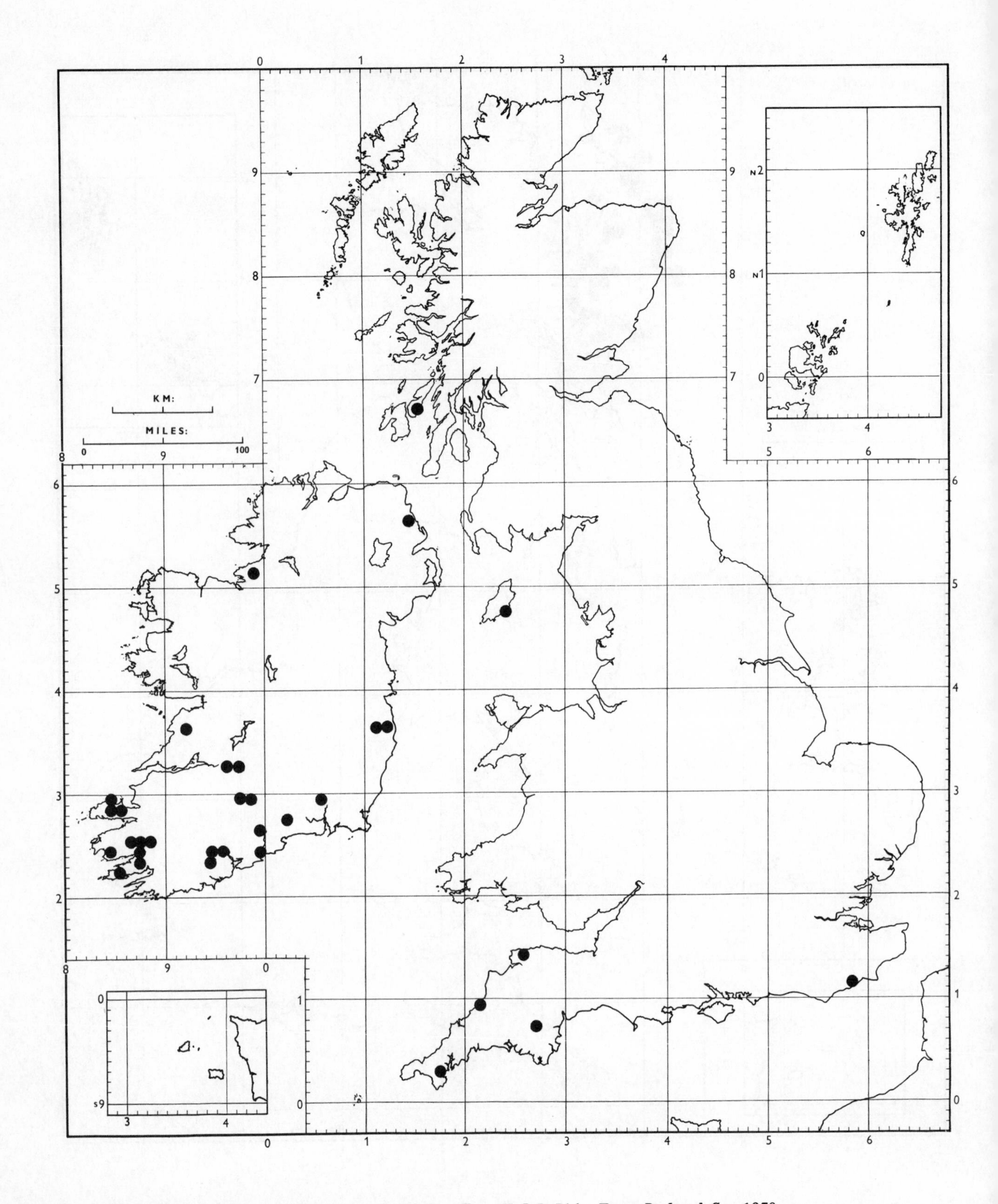

Fig. 8. Distribution of *Dumortiera hirsuta* in the British Isles. From H. J. B. Birks, *Trans. Br. bryol. Soc.* 1970.

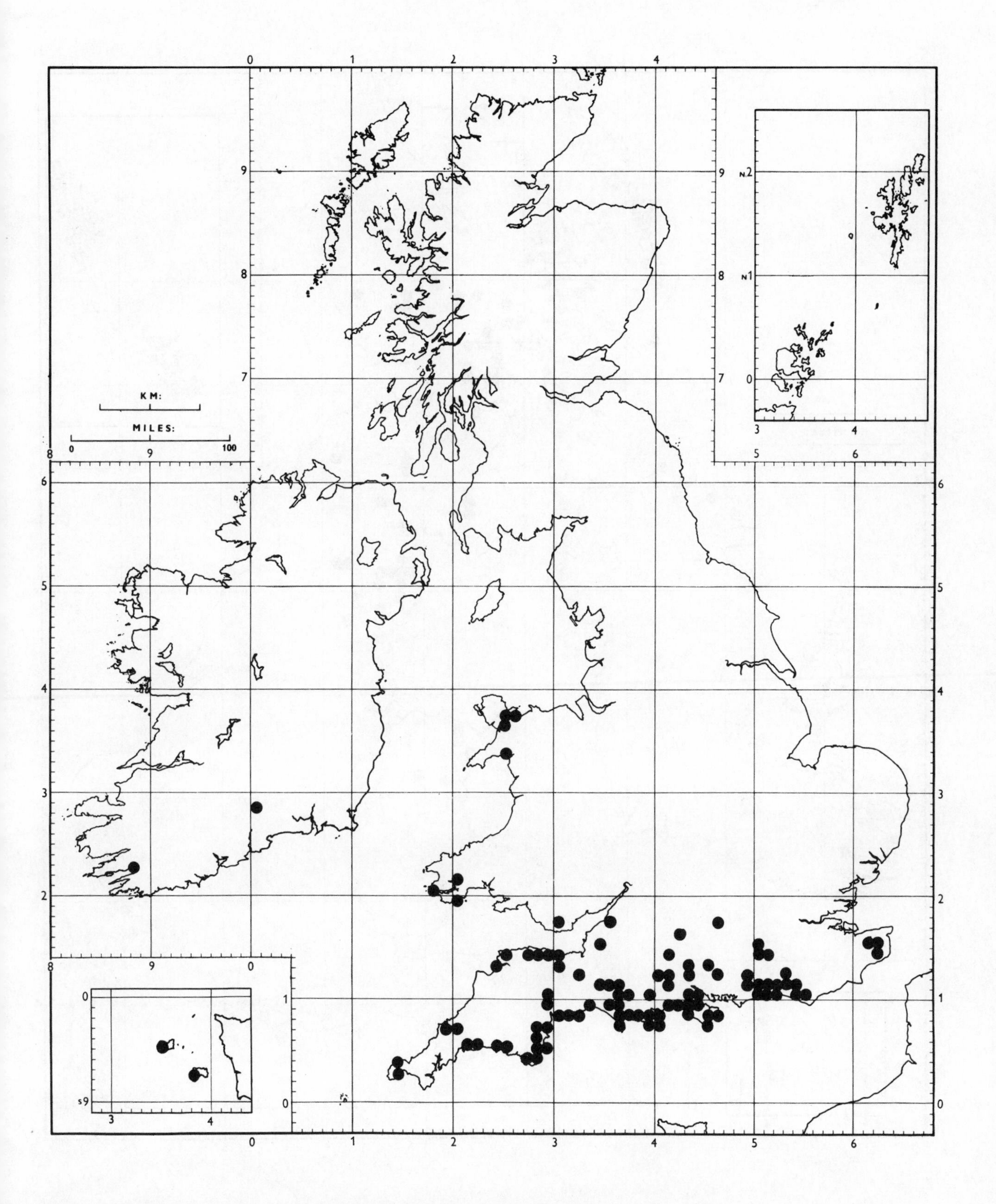

Fig. 9. Distribution of *Leptodon smithii* in the British Isles. From Proctor, *Trans. Br. bryol. Soc.* 1964.

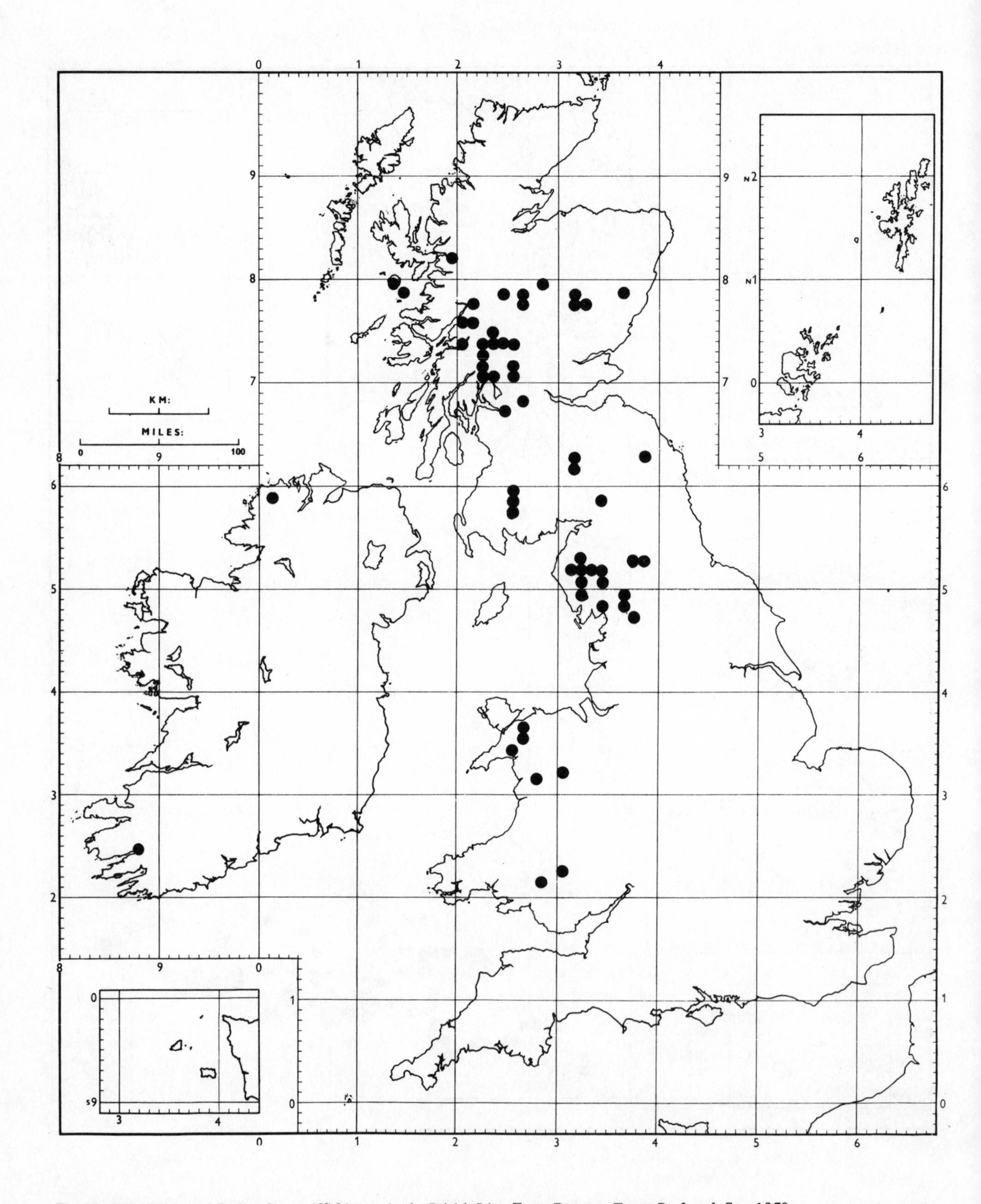

Fig. 10. Distribution of *Oedipodium griffithianum* in the British Isles. From Duncan, *Trans. Br. bryol. Soc.* 1970.

About twenty taxa of bryophytes have a claim to be considered as British *endemics.* However, of these, several may not be worthy of taxonomic recognition and several may be discovered elsewhere, especially on the nearby mainland of Europe. There are left six taxa which seem both clearly defined (according to Mr A. C. Crundwell) and less likely to be discovered outside the British Isles. They are as follows:

Barbula reflexa var. *robusta*
Campylopus shawii
Grimmia retracta
Gymnostomum recurvirostrum var. *insigne*
Radula voluta
Trichostomum hibernicum.

There being no subfossils, the histories of these taxa are totally obscure. Three are oceanic species; *Radula voluta* falls into the Southern Atlantic group of Ratcliffe (1968) and *Trichostomum hibernicum* and *Campylopus shawii* into the Northern Atlantic group. The last-named has a range very like that of *Myurium hebridarum* (fig. 7) except that it occurs in southwestern Ireland.

Many of the distribution patterns discussed so far are restricted or disjunct types. However, there are as many distribution patterns as there are species. In contrast to the species of limited occurrence in the British Isles stand those species which are very widely spread throughout these islands. *Ubiquitous* species, as here defined, are those which occur in all 152 vice-counties of the British Isles or in not less than 140. In this heterogeneous group there are about 85 species of mosses of varied ecology and varied extra-British range. Many of the species have wide ecological amplitudes.

Lastly in this discussion of bryogeographical categories mention must be made of *southern* species, which in the context of this book are those which have a limited northern extension in Fennoscandia.

It might be considered that because bryophytes can reproduce by spores and other microscopic or nearly microscopic propagules, gemmae, they have great facility in spreading abruptly over large distances. The high representations of bryophytes on some remote oceanic islands may be taken to support such a contention. However, in recent decades, many authors such as Herzog (1926), Irmscher (1929), Du Rietz (1940), Lazarenko (1958), Fulford (1951, 1963), Steere (1965), Crum (1966), Schofield (1969*b*) and Schuster (1969), who have discussed the disjunct distributions of bryophytes, deny to a greater or lesser extent the efficiency of long range dispersal. Historical factors are seen as the causes, including continental drift – an event not relevant to the patterns discussed here.

Two main arguments militate against long-range dispersal as having produced or producing the relevant disjunct patterns.

(1) *Groups of rare species.* It seems very improbable that long-range dispersal can account for groups of rare, disjunct species such as one found on Ben Lawers (plate 1*a*) and Ben Bulben (plate 1*b*) and other areas familiar to British botanists.

(2) *Reproductive capacity.* Many species with disjunct ranges have little or no power of spread by spores or gemmae. This has often been remarked for the oceanic species. The same is true for many of the arctic-alpine, northern-continental and other species discussed here, among which *Calliergon trifarium, C. turgescens, Hypnum bambergeri* and *Rhytidium rugosum* are among the best examples. However, reproductive capacity may have been greater in the past.

The reproductive capacity and dispersal of bryophytes are subjects about which there is little knowledge, especially critical knowledge based on experimentation. The striking experimental results obtained by Pettersson (1940) stand alone; using rain filter traps in southern Finland he caught and cultivated many spores, nearly 200 of which proved to be *Aloina*, a genus then unknown in Finland. Whether the spores came as a cloud from eastern Siberia, as claimed by Pettersson, or from some nearer area of Europe, as claimed by Persson (1944) and Bergeron (1944), the results remain cogent as a demonstration of long-range dispersal. *Aloina* has now been found growing in southwestern Finland by Koponen and Oittinen (1967) who believe that their discovery shows that long-range dispersal as demonstrated by Pettersson is not an extremely rare happening. More experimentation along the lines of Pettersson or, better, in an aircraft flying far from continental masses is greatly to be desired.

There are other recent claims in the Fennoscandian literature that disjunctions are the products of long-range dispersal. *Sphagnum lindbergii* in southern Sweden, having spread from the north (Sjörs 1949), *Anomodon viticulosus* and *Porella cordeana* in northernmost Sweden, having spread from the south (Mårtensson 1956), *Rhytidium* in southwestern Sweden, having spread from southern Norway (Hallberg 1959), *Polytrichum capillare* in Finland, having spread from northern Russia (Vaarama 1967), *Diphysicium foliosum* in southwestern Finland, having spread from Sweden (Karenlampi 1968), and *Oedipodium* in western Norway, having spread from the British Isles (Störmer 1969), are doubtless only a few examples which happen to be known to the author.

Whether all or none of these and similar claims are accepted, the fact of long-range dispersal cannot be denied. The problem is to gauge its frequency and importance. However, it may well have been insignificant in the formation of the disjunct patterns discussed in the book.

1.2 BRYOGEOGRAPHY AND PLEISTOCENE SUBFOSSILS

The analysis of subfossils, both microscopic and macroscopic, has proved of great value in elucidating many of the phytogeographical problems posed by the British vascular plant flora. Of all the numerous recent studies dealing with subfossil assemblages, Godwin's volume *The History of the British Flora* (1956) stands alone in its wide survey and detailed elaboration of the large quantity of modern British evidence. Many of Godwin's far-reaching conclusions regarding present distribution patterns, especially those of disjunct type, are based on macroscopic remains. The study of Post-glacial and Late glacial deposits, which are often rich in subfossils of leaves, seeds and fruits, proved particularly rewarding. Since 1956 many more investigations of the Late and Post-glacial have produced useful results. However, the greatest contributions towards the understanding of British phytogeography have stemmed from studies of the floras of the middle and early phases of the last glaciation and of the interglacial periods.

The potential bryogeographical value of the compilations of records of moss remains was shown as early as 1927 by Dixon's attempt at listing all known fossil mosses, and in 1932 Gams used the totality of European knowledge to great effect in drawing bryogeographical conclusions.

Modern knowledge of British subfossil mosses of the Pleistocene is assembled here for the first time. About 220 moss species have been recorded from the British Pleistocene. In other words, there are records of about 30 per cent of the present moss flora. This is a cogent body of data for the understanding of the present moss flora and its various distribution patterns. By contrast the Pleistocene record of the liverworts is very meagre with the exception of the Anthocerotales – easily recognised by spores (p. 62).

As stressed above, one of the outstanding features of the British bryoflora is the large representation of oceanic species. In recent years various papers have been published on the history of the oceanic element (e.g. Greig-Smith 1950, Gams 1952, Richards 1954, Proctor 1964 and Ratcliffe 1968). Of necessity the arguments rested solely on present distributions and ecology. It is to be much regretted that there is still little information based on subfossils. However, the few records of such species as *Breutelia chrysocoma, Dicranium scottianum* and *Hyocomium armoricum* are not without significance (p. 205).

Though subfossils of the most bryogeographically interesting oceanic species are scarce, a great many other species, including not a few of disjunct or otherwise remarkable ranges, have been found in the subfossil state. The history of the arctic-alpine element is well recorded. That the bryoflora of the last glaciation, especially of the last phases, was far from merely consisting of arctic-alpines and species of very broad ecological tolerances, is shown by the occurrences of species such as *Antitrichia curtipendula, Homalothecium lutescens, Plagiothecium undulatum* and *Sphagnum imbricatum*.

The northern-continental species, for the most part peat-forming species, have detailed Late Pleistocene histories. Also well known as subfossils are the ubiquitous and southern species.

The contribution of the renowned bryologist, H. N. Dixon, to bryological knowledge of the British Pleistocene was considerable. His long lists from the Middle Devensian deposits (table 2) in the Lea Valley (in Warren 1912) and from the Flandrian site at Fort William (1910) contain numerous species of much bryogeographical significance. The assemblage from the deposit of uncertain, perhaps Devensian, age at West Craigneuk, Lanarkshire (1907) consists of only eight species but is not without interest, as *Helodium blandowii* and *Cinclidium stygium* indicate.

Dixon's investigations covered periods older than the Devensian. He studied a large assemblage (in Bell 1904) from the site at Wolvercote, Oxfordshire (partly Hoxnian in age, partly Wolstonian) and a small assemblage (in Oakley 1939) from the deposit of Wolstonian age at Farnham, Surrey. Dixon identified three species from the Arctic Freshwater Bed (Cromerian) at Mundesley, Norfolk (1924, p. 548). Though only nine species were found, his list from the deposit at Castle Eden, Co. Durham (Early Pleistocene) represents one of the few British assemblages of such an age (in Reid 1920).

Apart from studying British material, Dixon identified subfossils from Icelandic (in Lewis 1911) and Russian peat (1925). However, his most important study of foreign material was the investigation of small assemblages, recovered by Reid and Reid (1915) from the Reuver (late Pliocene) and Tegelen (Early Pleistocene) clays of the Dutch-German border (Zagwijn 1960, 1963*a*).

In 1927 Dixon published a list of fossil mosses in *Fossilium Catalogus*. Pleistocene species, including those from British deposits, were recorded. Dixon had no intention of discussing the bryogeographical significance of the fossils. It is not surprising that he seems to have been unaware of several of the early British papers such as Mahony (1868*a* and *b*, 1869), Robertson (1881), Reid (1892), Bennie (1891, 1894*a* and *b*, 1896), Roeder (1899) and Sheppard (1910). It was a minor calamity that he mistakenly used a long list of glacial relicts postulated by Stark (1912) as though it was a list of Pleistocene records; many of these species have no subfossil history whatever.

Jovet-Ast (1967) has included Pleistocene records in her compilation of fossil bryophytes, the only attempt at such a difficult task since Dixon's list.

Many investigations of Neogene and Pleistocene mosses have been carried out in Europe; in North America fewer deposits have been studied. In the *Manual of Bryology*, Gams published a review article in which he ably summarised knowledge of Neogene and Pleistocene mosses and discussed many aspects of historical bryogeography, though some of his conclusions are now vitiated by geological reassessments of ages. It needs no stressing that knowledge of Neogene and particularly of the Pleistocene has been transformed since that time and numerous papers on subfossil mosses have been published. Gams (1932, pp. 297-9) has summarised the early history of investigation of subfossil mosses.

In the last twenty-five years the Polish botanist Professor Bronislaw Szafran carried out many investigations of Pleistocene mosses. He wrote summarising articles of earlier Polish results in 1948 and 1952. Co-operation with several members of the Kraków school of palynologists led to several bryological contributions (in Birkenmajer and Srodon 1960, Mamakowa 1962, 1968, Ralska-Jasiewiczowa 1966, Sobolewska *et al.* 1964, Srodon 1965, 1968, and Szczepanek 1960). Even more bryogeographical significance attaches to Szafran's work on Neogene mosses (1948*a*, 1949*a* and *b*, 1958, 1964 and in Szafer 1954). Szafran was not the only Polish botanist to investigate Pleistocene bryofloras. Jasnowski (1957*a* and *b*, 1959, and in Tolpa 1952, 1961) has produced significant results, as has Kuc (in Koperawa 1958).

Apart from Szafran's work, knowledge of Neogene mosses rests very largely on the important work of the Russian Dr I. I. Abramov and his wife A. L. Abramova. Of all their numerous papers, listed in the bibliography, probably the most significant is that of 1959*a* which deals with the Pliocene bryoflora of Duab in the Caucasus (p. 170). Much earlier Solonevicz (1935) produced a list of Late Pleistocene mosses, while Neujstadt has given many records in his book (1957) on the Post-glacial vegetational history of Russia. Bartosh (1963) has given a few Post-glacial records from Russia.

Apart from the Polish and Russian work, the author knows of the following European investigations which do not by any means comprise a complete list.

Fennoscandia: Störmer 1949, 1965; Persson 1960, and in Lundquist 1964; Ruuhijarvi 1963; Mornsjö 1969.
Germany: Koppe in Reich 1953, in Beug 1957, in Lesemann 1969, in Behre 1970; Grosse-Brauckmann 1962, 1963, 1968, 1969; Grüger 1968*a*; von Hübschmann in Körber-Grohne 1967.
The Netherlands: Landwehr 1949, 1951; Casparie 1969.
Belgium: Vanden Berghen 1950, 1951, and in Stockmans and Vanhoorne 1954.
France: Jovet-Ast and Huard 1966.
Italy: Koppe in Grüger 1968*b*.
Czechoslovakia: Pilous 1968; Rybnicek and Rybnicekova 1968.
Hungary: Boros 1952.

In North America, Steere has published two papers on Pleistocene mosses (1938, 1942) as well as a review paper of Cainozoic and Mesozoic bryophytes (1946) and a survey of Pleistocene mosses (1965). Numerous moss records are given in Rosendahl's paper (1948) on the Pleistocene of Minnesota. Culberson (1955) has studied the moss flora of the Two Creeks Forest Bed formed at the end of the last (Wisconsin) glaciation of North America. Small numbers of species are recorded in such papers as those by Argus and Davis (1962), Fries *et al.* (1961) and Muller (1964). Robinson (1959) has discussed *Calliergon subsarmentosum*, a species remarkable in having only one living locality, Vancouver Island, and one subfossil locality, a Late Wisconsin deposit in Minnesota. The most recent substantial results from North America are those of Welch (in Kapp & Gooding 1964, last interglacial and Wisconsin deposits in Indiana), De Vries and Bird (1965, Late Wisconsin deposit in Saskatchewan) and Miller (with Benninghoff 1966, a Late Wisconsin deposit in Michigan, and 1969, a Late Wisconsin deposit in New York) and Brassard and Steere 1968, Late Pleistocene deposits on Bathurst Island.

2 GEOLOGICAL AND PALYNOLOGICAL BACKGROUND

2.1 GEOLOGY

Since Dixon's catalogue of fossil mosses (1927) and Gams' review of Pleistocene bryophytes (1932) knowledge of Pleistocene stratigraphy and chronology has changed dramatically as a result of pollen analysis and radiometric dating. The records discussed by these authors and the great deal of new data accumulated since then can now be fitted into a highly detailed picture of geological and vegetational history. Much of the following discussion is taken from West's able survey (1968*a*) of this modern geological and palynological knowledge. Readers seeking further elaboration should consult West's book.

One of the significant advances is that the Pleistocene is now considered to have had up to twice or even three times the duration formerly commonly accepted as about one million years. Potassium-argon dating points to about two million years. Within this time span in the British Isles there have been three glacial periods preceded by at least four cold periods from which glacial deposits are unknown. The glacial periods have alternated with four interglacial periods, including the present warm period or Post-glacial, and the four preglacial cold periods have alternated with three warm periods. In all there have been at least seven major cold-warm fluctuations (table 2).

Table 2 *Stages of the British Pleistocene, adapted from West (1970)*

Stage	Climate
LATE	
Flandrian: the present interglacial period or Post-glacial	t
Devensian: the last glacial period or Weichselian	c g p
Ipswichian: the last interglacial period	t
Wolstonian: 2nd last glacial period or Gippingian	c g p
MIDDLE	
Hoxnian: 2nd last interglacial period	t
Anglian: 3rd last glacial period or Lowestoftian	c g p
Cromerian: 3rd last interglacial period	t
Beestonian	c p
Pastonian	t
EARLY	
Baventian	c
Antian	t
Thurnian	c
Ludhamian	t
Waltonian	c

t = temperate; c = cold; g = glacial deposits known; p = periglacial features known.

Very little is yet known about the time spans of these fluctuations. Radiocarbon dating has shown that the *Flandrian* (Post-glacial) began about 10,000 years ago and the *Devensian* (last glaciation) about 70,000 years ago or more; this last figure, resting on a very few radiocarbon dates around 60,000 years (Van der Hammen *et al.* 1967) beyond which the method is ineffective, is perhaps an underestimate (Shotton 1966; Shackleton 1969). Shackleton (1969) has given an estimate of around 120,000 years for the beginning of the last interglacial which may have lasted 11,000 years. Quoting the work of Frechen and Lippolt, West (1968*a*, tab. 9.5) has placed the second last interglacial about 150,000 years ago and the third last interglacial about 300,000 years ago. The *Hoxnian* (second last interglacial) may have had a duration of about 30,000 to 50,000 years (Shackleton and Turner 1967).

The term *Pleistocene* is here used in the sense of West (1968) as an *Epoch* of the *Cainozoic Era* which also comprises the *Pliocene* and *Miocene* Epochs, together composing the *Neogene Period*, and the *Oligocene, Eocene,* and *Palaeocene* Epochs (composing the *Palaeogene Period*). The nomenclature of the *stages* of the Pleistocene of the British Isles follows the *Recommendations on Stratigraphical Usage* (1969) of the Geological Society of London. The pollen zonation of the Flandrian and Late Devensian follows Godwin (1956); that of the Ipswichian, Hoxnian and Cromerian follows West (1970). The terms *interglacial* and *interstadial* are used in the sense of West (1968*a*). The radiocarbon dates quoted here are as published in the *Radiocarbon Supplement* of the *American Journal of Science* and in *Radiocarbon*.

For the purposes of this book the Devensian is split into three time units as recommended by Shotton and West (1969); the Early Devensian from the beginning of

Table 3 *Radiocarbon dates relevant to Devensian moss assemblages*

Flandrian zone IV	Greenock, Renfrewshire	9890 ± 160 Birm.-1210
Late Devensian zone III	Kirkmichael, Isle of Man	10,270 ± 170 Q-673
	Drumurcher, Co. Monaghan	10,515 ± 195 Birm.-239
	Greenock, Renfrewshire	10,560 ± 180 Birm.-121
Late Devensian zone II	Aby Grange, Lincolnshire	11,205 ± 120 Q-279
Alleröd interstadial	Low Wray Bay, Lake Windermere	11,878 ± 120 Q-284
Late Devensian zone I	Loch Droma, Ross and Cromarty	12,814 ± 155 Q-457
	Colney Heath, Hertfordshire	13,560 ± 210 Q-385
Late Devensian pre-zone I	Dimlington, Yorkshire	18,240 ± 250 Birm.-108
		18,500 ± 400 I-3372
	Barnwell Station, Cambridgeshire	19,500 ± 650 Q-590
Middle Devensian	Lea Valley Arctic Bed	28,000 ± 1500 Q-25
	Great Billing, Northamptonshire	28,225 ± 330 Birm.-75
	Derryvree, Co. Fermanagh	30,500 ± $^{1170}_{1030}$ Birm.-166
	Fladbury, Worcestershire	38,000 ± 700 GRO-1269
	Upton Warren, Worcestershire	41,900 ± 800 GRO-1245
Early Devensian Brörup interstadial	Chelford, Cheshire	60,800 ± 1500 GRO-1480

Last interglacial ends *c.* 70,000

the period until about 50,000 years ago, the Middle Devensian from then until about 26,000 years ago, and the Late Devensian from then until the final deglaciation at about 10,000 years ago. The beginning of Late Devensian zone I is here taken as about 13,500 years ago; hence the moss-rich deposit at Colney Heath, Hertfordshire is included in zone I. The period from about 26,000 years ago until 13,500 years ago, from which as yet there is little bryological information, is here referred to as Late Devensian pre-zone I.

Even at the maximum extent of the ice in the last glaciation large areas of southern England and southern Ireland remained free of ice (figs. 11 and 12); the same is true of the earlier glaciations. There is still considerable debate on the precise limits of the ice sheets. For instance, Mitchell (1967) claims that during the second last glaciation the ice abutted on the northern coasts of Cornwall and the Scilly Isles. Much has still to be learned about the limits and ages of the retreat stages of the last glaciation (fig. 11). That the major advance of the glaciers occurred in the Late Devensian is now well known; among the most dramatic British evidence is the radiocarbon dating of about 18,000 to 18,500 years obtained of a moss bed under boulder clays at Dimlington, Yorkshire (Penny *et al.* 1969).

Much attention has been given in recent years to the study of periglacial phenomena. The aspect stressed here is the formation of ice wedges (plate 3), which are known to have occurred widely in the British Isles during the last glaciation and certainly occurred in earlier glaciations and cold periods (West 1968*b*). There is a marked concentration of ice wedge casts in southeastern England but it is to

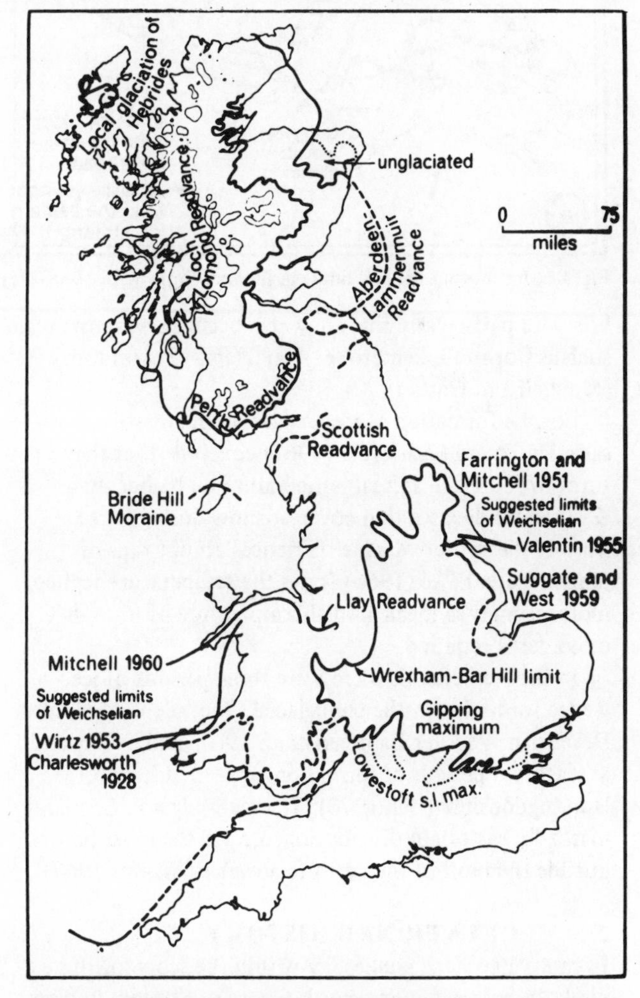

Fig. 11. Limits of main glacial advances and of Devensian retreat stages in Britain. From West 1968*a*. Weichselian = Devensian; Gipping = Wolstonian; Lowestoft = Anglian.

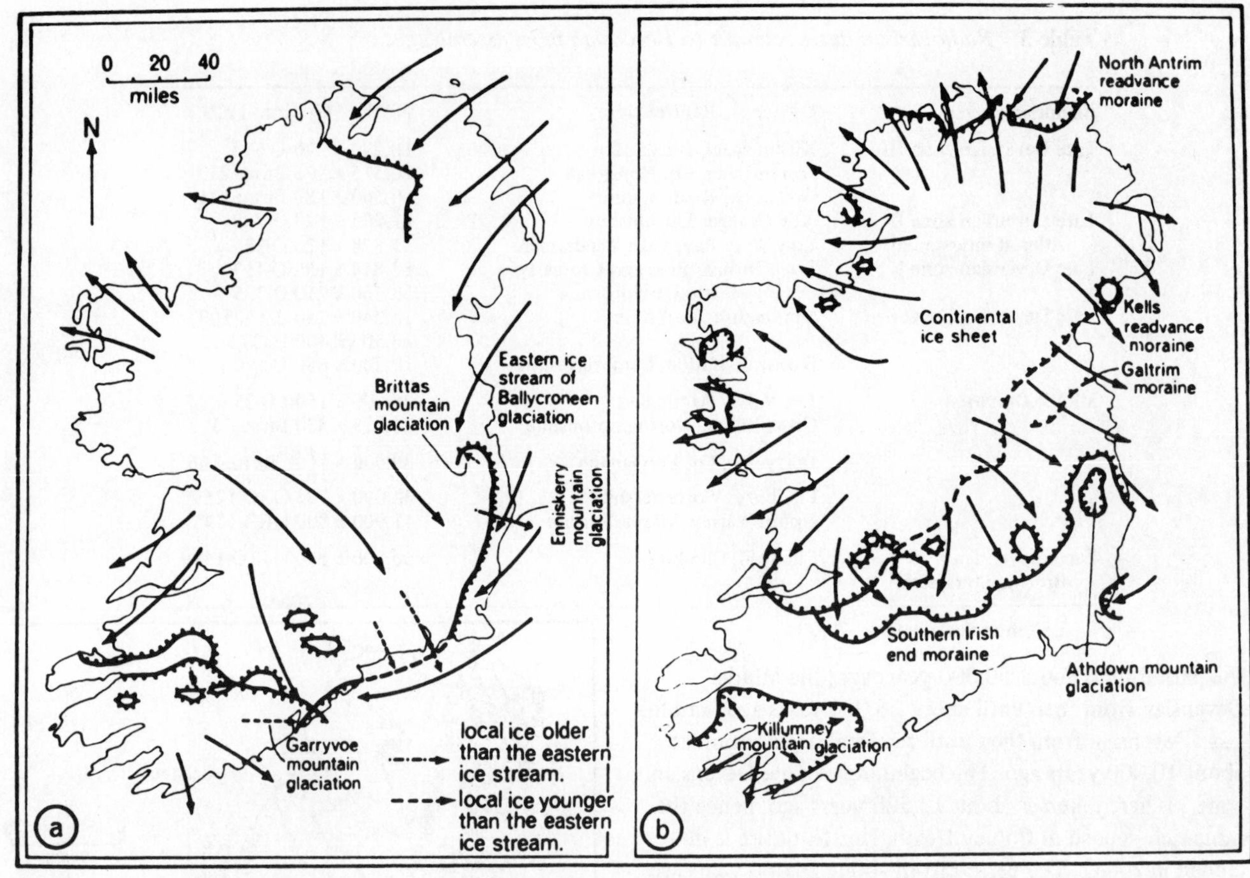

Fig. 12. Ice movements and limits in Ireland. From West 1968*a*. (*a*) shows second last glaciation, (*b*) shows last glaciation.

be noted particularly that they also occur in western areas such as Cornwall, Pembroke, Argyll (fig. 13) and Kerry (Mitchell and Watts 1970).

For the formation of ice wedges, permafrost is necessary. Dylik and Maarleveld (1967) conclude that they form with a mean annual temperature no higher than -2 °C and indicate a thin cover of snow during part at least of the winter which experiences abrupt falls of temperature. Péwé (1966) states the temperature regime more strongly; a mean annual temperature of -6 to -8°C or colder is required.

In the Netherlands there were three periods of ice wedge formation in the pleniglacial (equivalent to Middle Devensian; Van der Hammen *et al.* 1967) and in England at least two periods of formation, as seen at Earith, Huntingdonshire (Bell 1970). The ice wedges in Scotland, markedly eastern in distribution, are for the most part outside the limit of the Perth re-advance (Sissons 1967).

2.2 VEGETATIONAL HISTORY

Even if Palaeogene studies lay within the scope of this book the solitary moss records (those of *Sphagnum* from the Lough Neagh Clays, p. 173) would not justify a summary of the Irish floras reviewed by Watts (1970) or the renowned tropical flora of the London Clay, which, in any case, is apparently devoid of mosses. However, Boulter's (1971) fine palynological studies of Neogene deposits in sinkholes in Derbyshire demand mention. His records of Sphagnaceae and *Hypnodendron* are discussed on p. 170. From the Miocene-Pliocene transition a rich flora, with a great variety of trees, well known from Neogene deposits in other parts of Europe, is revealed (table 4).

Table 4 *Climatic groups of some members of the Derbyshire Neogene assemblage, after Boulter (1971)*

Warm element	Intermediate element	Temperate element	Facies element
Araliaceae	*Abies*	*Carpinus*	*Alnus*
Leiotriletes	*Carya*	Compositae	Ericaceae
Podocarpus	*Cedrus*	*Corsinipollenites*	*Ilex*
Sapotaceae	*Juglans*	*Corylus*	*Nyssa*
Symplocaceae	*Keteleeria*	Gramineae	*Pinus*
Verrucatosporites favus	Quercoid types	*Liquidambar*	Sphagnaceae
		Neogenisporis	Taxodiaceae
		Picea	
		Polypodiaceae	
		Sciadopitys	
		Tsuga	
		Ulmus	

Plate 3. (*a*) *Dryas*-Cyperaceae tundra at 72° N on Banks Island, Canadian Arctic Archipelago. *Calliergon turgescens* and *Scorpidium scorpioides* fill the channel above the ice wedge running from front to back of the photograph.

Plate 3. (*b*) An ice wedge seen in section, about 2 ft across at the top, on Garry Island, Mackenzie Delta, Canada. The dwarf shrub tundra above supports *Aulacomnium palustre*, *Rhytidium rugosum* and *Timmia* sp.

Plate 4. (*a*) Stannon, Cornwall, a Late Devensian allochthonous deposit. The moss-rich silts and fine sands, in which the lamination can be clearly seen, are overlain by unbanded silt and late Flandrian blanket peat. Kaolin gravel has been scattered by excavating machinery.

Plate 4. (*b*) Colney Heath, Hertfordshire. One of the large erratics of organic mud (close to spade), transported when frozen in Late Devensian Zone I, embedded in torrent gravels. Erratic *B* yielded a calcicolous bryoflora of rich fen and terricolous species. Photo by H. Godwin.

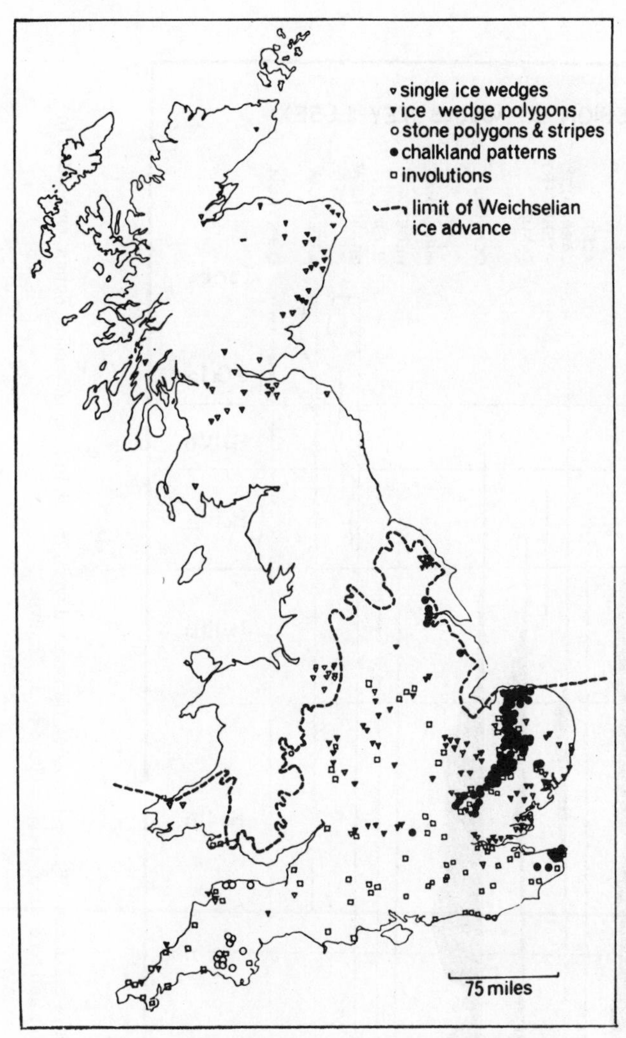

Fig. 13. Distribution of ice wedge casts and other periglacial features in Britain. From West 1968*a*. Weichselian = Devensian.

The elucidation of the Early Pleistocene vegetation of Britain is a result of West's work on the marine deposits in Norfolk. Fig. 14 shows the arboreal and some other components of the vegetation through a long time span of the Early Pleistocene. Amongst the exotic taxa, greatly reduced since the Neogene, *Tsuga* and *Pterocarya* are noteworthy; the latter last occurred in the Hoxnian (fig. 15). The high representation of Ericales is another striking feature.

Of the temperate stages associated with glacial deposits the earliest, Cromerian, is the least known. The Hoxnian is better understood; Turner's work at Marks Tey in Essex resulted in a pollen diagram outstanding in its span from the Late Anglian to Early Wolstonian (fig. 15). Also demanding mention here is the deposit from Gort, Co. Galway (Jessen *et al.* 1959) because of the great significance of the rich vascular moss plant and assemblages which include many oceanic and other plants of phytogeographical interest; *Brasenia purpurea* (tentative), *Daboecia cantabrica, Erica ciliaris, E. scoparia, Eriocaulon septangulare* (tentative), *Hymenophyllum* and *Rhododendron ponticum* are outstanding examples.

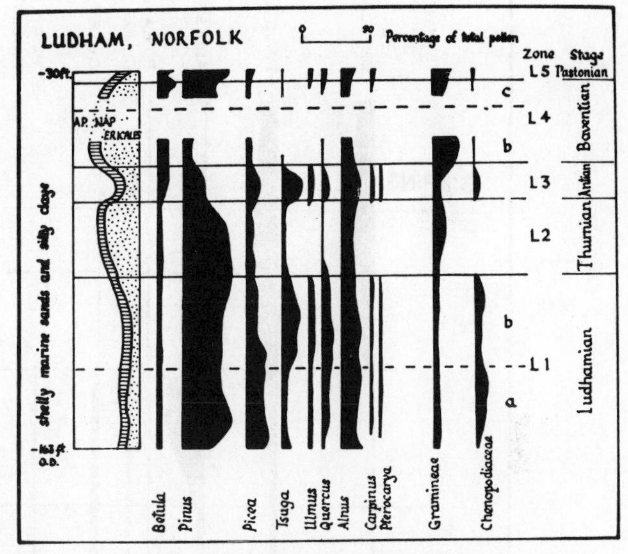

Fig. 14. Schematic pollen diagram through the Crag at Ludham, Norfolk. From West 1968*a*.

West's composite pollen diagram (fig. 16) shows the vegetational development of the Ipswichian interglacial period. Striking features are the high representations of *Acer, Carpinus* and *Corylus.* For bryogeographers no less than other students of the Pleistocene, it is a disappointment that Ipswichian deposits are mysteriously absent from Ireland despite informed searches (Watts 1970).

To discuss fully the extensive literature on the Devensian and Flandrian stages would take much more space than can be justified here. The vascular plant vegetation of the period of some 70,000 or more years can be discussed by reference to some important macroscopic and pollen-analytic assemblages in which significant moss remains were found.

The Early Devensian vegetation is as yet poorly known, with the marked exception of the interstadial deposits at Chelford, Cheshire (Simpson and West 1958) where some 60,000 years ago grew a boreal forest of *Pinus, Picea* and *Betula* with species-rich herbaceous and dwarf shrub vegetation in the vicinity (fig. 17).

This coniferous forest gave way to the treeless vegetation of the Middle Devensian. At Upton Warren, Worcestershire (Coope *et al.* 1961) about 42,000 years ago the variety of herbs included halophytes (such as *Glaux maritima*) and steppe elements (such as *Androsace septentrionalis*), both now firmly established features of Middle Devensian vegetation (Bell 1969).

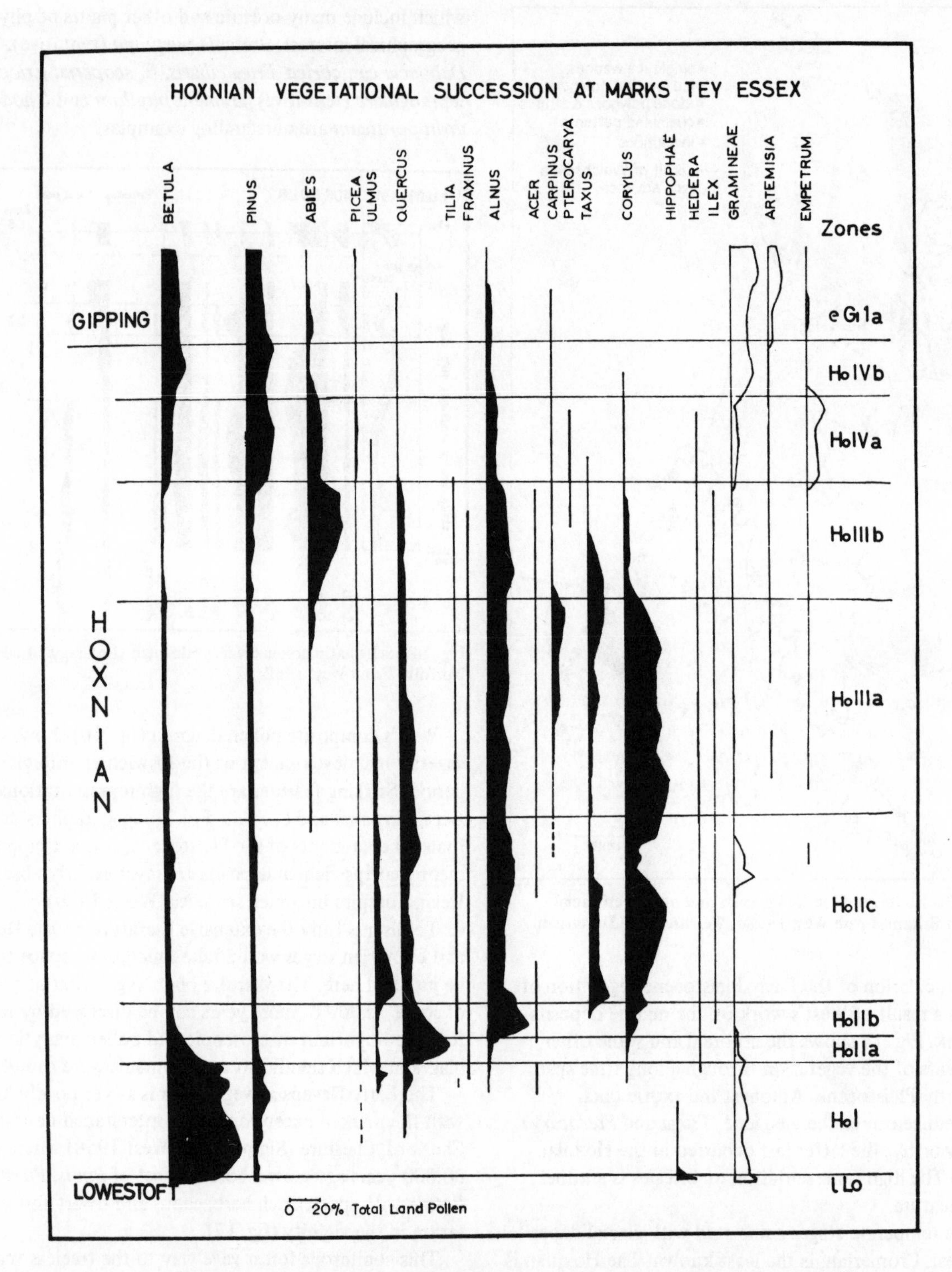

Fig. 15. Composite pollen diagram of the Hoxnian vegetational succession at Marks Tey, Essex. After Turner 1970.

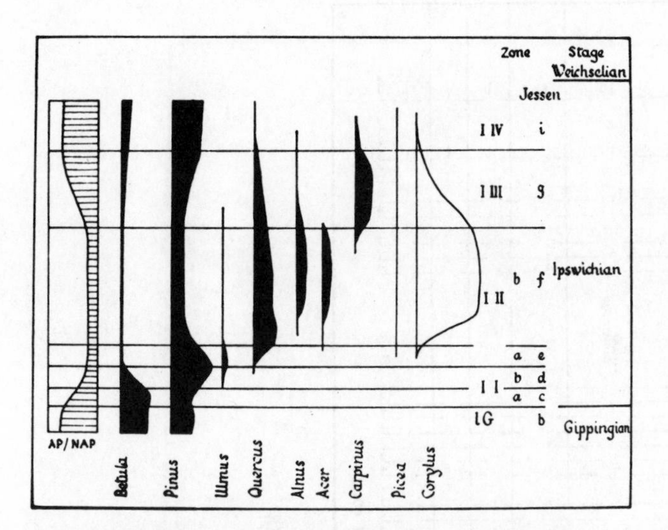

Fig. 16. Composite pollen diagram of the Ipswichian vegetational succession. After West 1968*a*. Weichselian = Devensian; Gippingian = Wolstonian.

Mosses may well have been important in the vegetation at Upton Warren as they certainly were at least locally at Derryvree, Co. Fermanagh, some 30,000 years ago (Colhoun *et al., Proc. Roy. Soc.* in press). *Dichodontium pellucidum, Drepanocladus fluitans, Philonotis fontana* and a species of *Brachythecium* were the commonest moss remains and *Ranunculus* subgenus *Batrachium, Arenaria ciliata* or *norvegica, Armeria maritima* and *Taraxacum* spp were the commonest macroscopic remains of flowering plants. Gramineae and Cyperaceae were the dominant pollen types.

At Dimlington, Yorkshire (Penny *et al.* 1969) some 18,500 years ago, very close in time to the major ice advance, there occurred the moss *Pohlia wahlenbergii* probably var. *glacialis* (plate 20) and perhaps little else but sparse grasses.

In Late Devensian zone I as seen at Colney Heath (fig. 18) and at Loch Droma, Ross and Cromarty (fig. 19), the vegetation still lacks megaphanerophytes. At Loch Droma, and less certainly at Colney Heath, mosses were an important component of the vegetation. Late snow bed communities occurred at Loch Droma.

The vegetation of the last few thousand years of the last glaciation can be further illustrated by reference to Low Wray Bay, Lake Windermere (fig. 20) where macroscopic remains of tree birches have been found in zone II (the Alleröd interstadial).

As a particular example of the Flandrian development of forests, shown for the regions of the British Isles in table 5, the pollen diagram from Seathwaite Tarn, in the English Lake District, is reproduced. The gross human interference with the forests in the last 5000 or more years (Smith 1970, Turner 1970) differentiates the Flandrian forest development from that of the earlier temperate stages.

The Flandrian development of ombrogenous mires is an aspect of vegetational history stressed in the later chapters of this book. Though oligotrophic mires are known from scattered areas of the British Isles in zones IV and V the extensive development of raised and blanket bogs, which are so characteristic a feature of the highland zone, came later in late zones VI, VII and VIII.

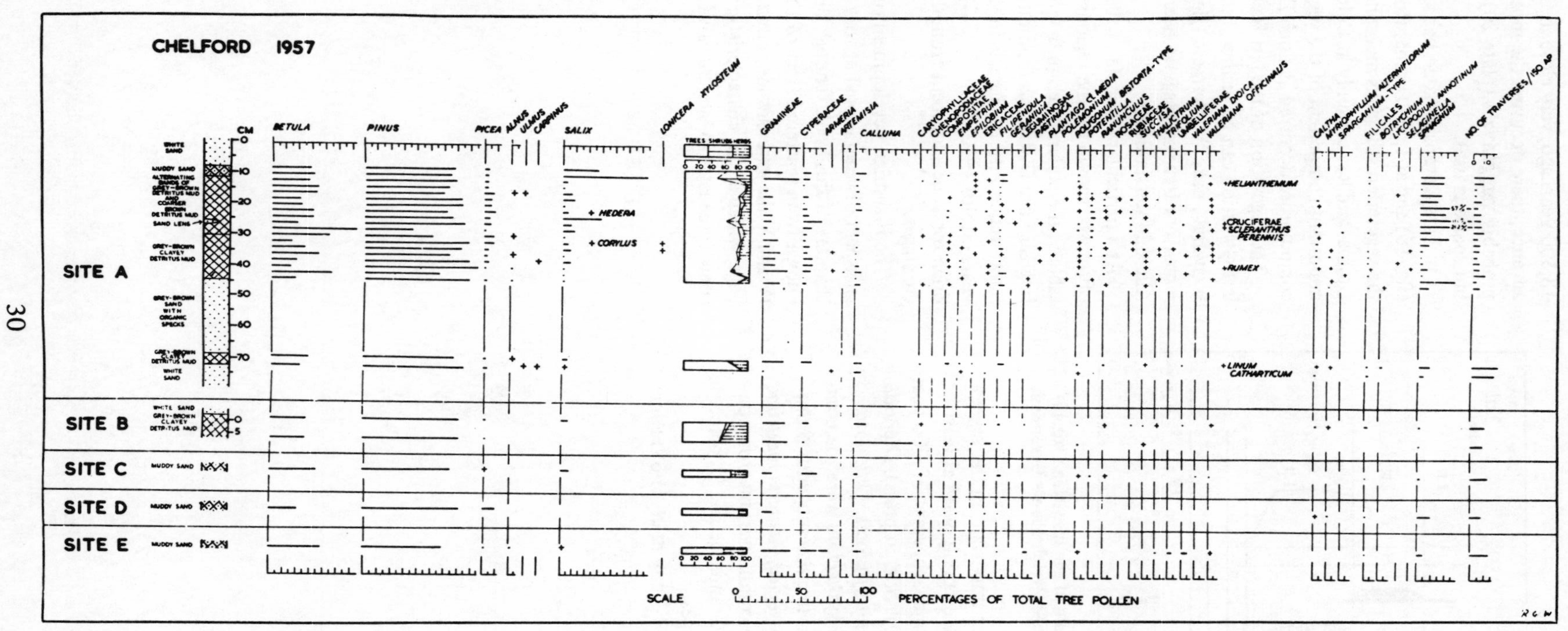

Fig. 17. Pollen diagrams from the Early Devensian interstadial deposits at Chelford, Cheshire. From Simpson and West 1958.

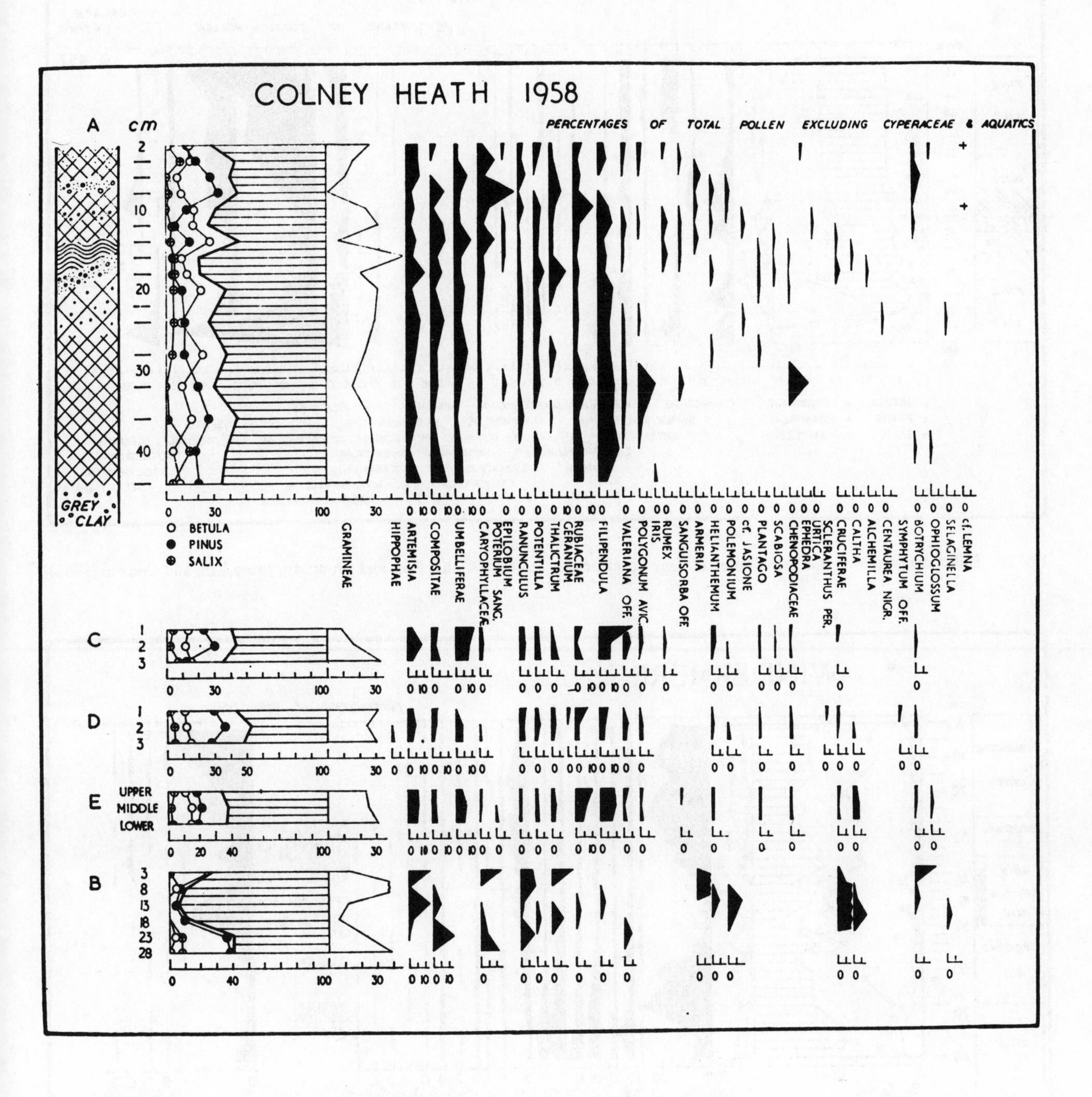

Fig. 18. Pollen diagrams from the Late Devensian zone I organic erratics at Colney Heath, Hertfordshire. From Godwin 1964.

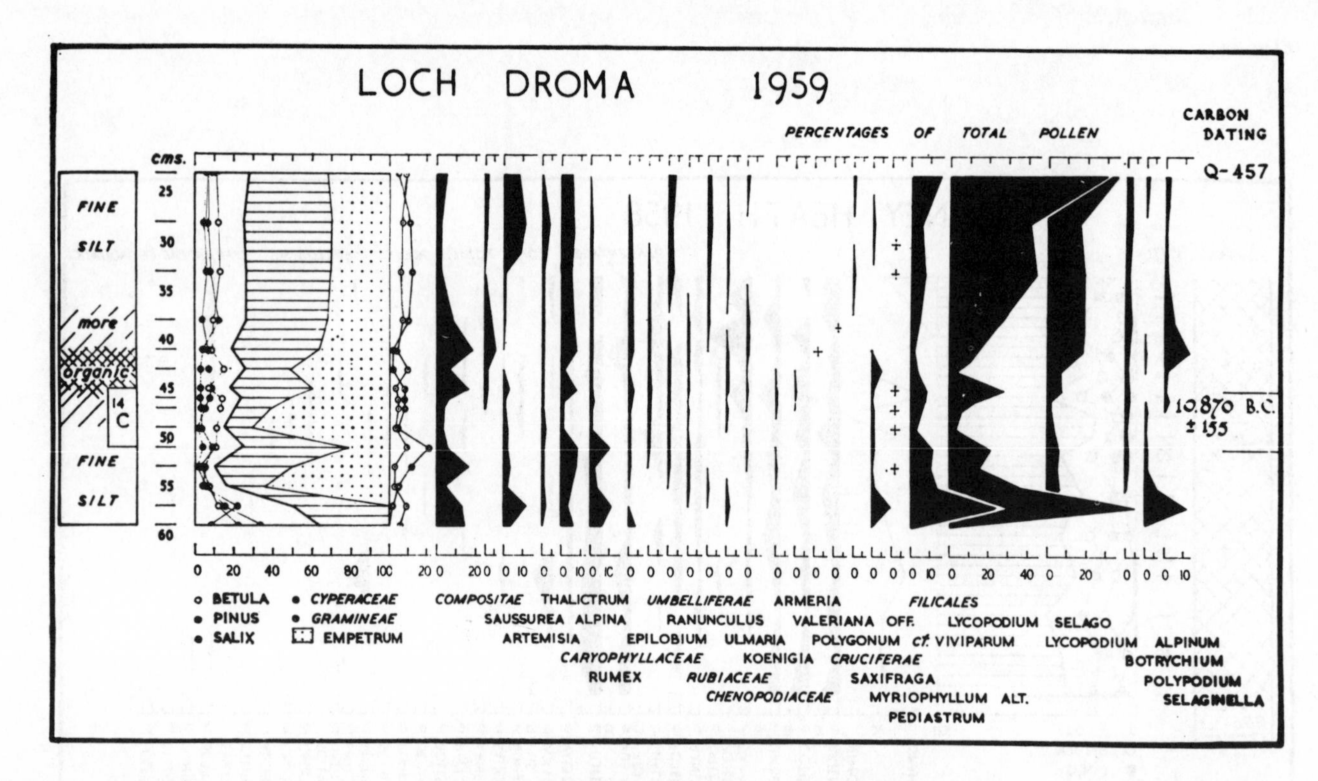

Fig. 19. Pollen diagram from the Late Devensian zone I organic silts at Loch Droma, Ross and Cromarty. From Kirk and Godwin, 1963.

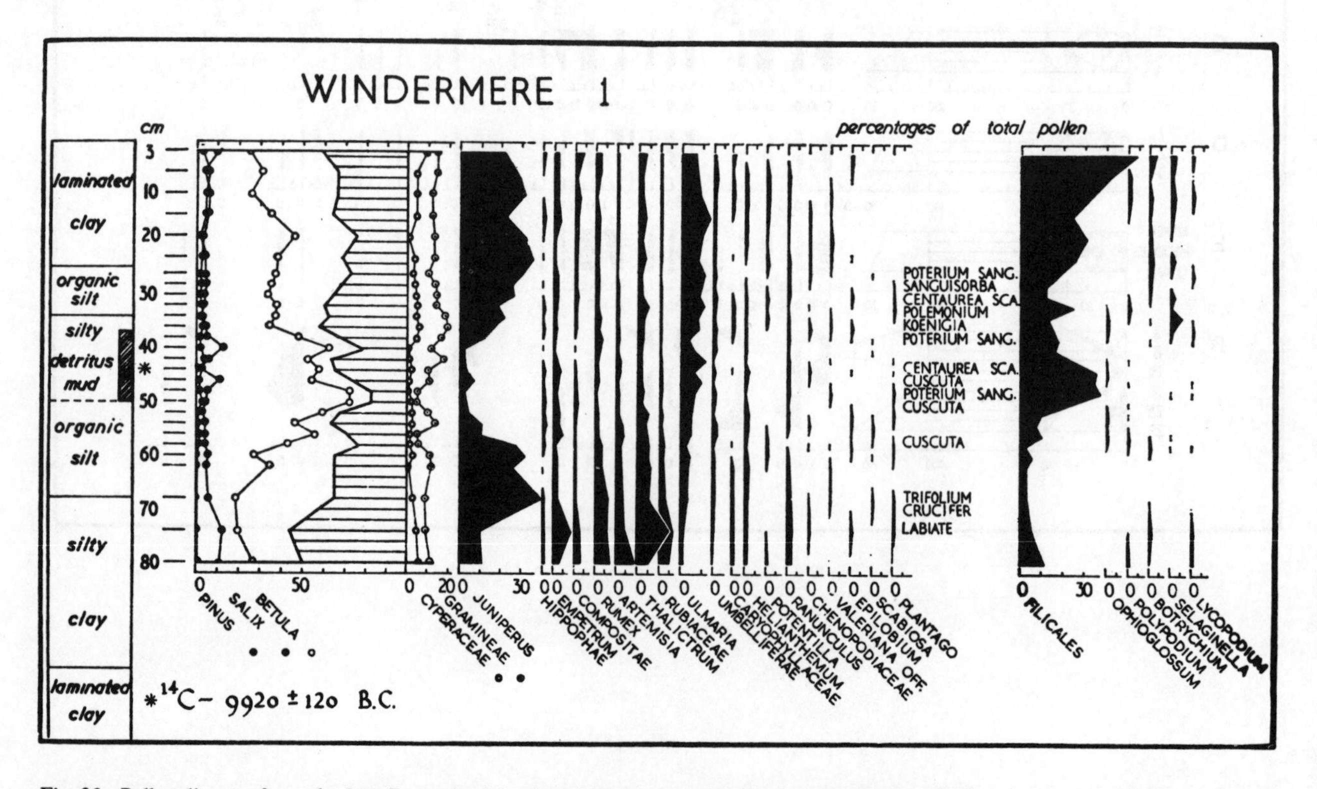

Fig. 20. Pollen diagram from the Late Devensian lake deposits from Low Wray Bay. From Godwin 1960.

Table 5 *Sequences of forest changes during the Late Devensian and Flandrian in the British Isles, modified from West (1968a)*

Stage	Time based on C^{14} dating	Blytt and Sernander periods	ENGLAND AND WALES Zone	ENGLAND AND WALES Characteristics of vegetation		IRELAND Zone	IRELAND Characteristics of vegetation		SCOTLAND Zone	SCOTLAND Characteristics of vegetation	
Flandrian		Sub-Atlantic	VIII modern		Afforestation	VIII modern	*Pinus-Fagus*	Afforestation	VIII modern		Afforestation
										Fagus	
	1000		VIII	*Alnus-Quercus -Betula (-Fagus -Carpinus)*		VIII	*Alnus-Quercus -Betula*				
								Ulmus decline			
	A.D. B.C.								VIII-VII*b*	*Alnus-Quercus -Betula*	
	1000	Sub-Boreal						*Quercus* maximum			
			VII*b*	*Alnus-Quercus -Tilia*		VII*b*	*Alnus-Quercus*				
	2000				Deforestation			Deforestation			Deforestation
	3000				*Ulmus* decline			*Ulmus* decline			*Ulmus* decline
	4000	Atlantic	VII*a*	*Alnus-Quercus -Ulmus-Tilia*		VII*a*	*Alnus-Quercus -Ulmus-Pinus*		VII*a*	*Alnus-Quercus-Ulmus*	
Flandrian	5000	Boreal									
					c *Quercus-Ulmus -Tilia*			c *Pinus* max.			
	6000		VI	*Pinus-Corylus*	b *Quercus-Ulmus*	VI	*Corylus-Pinus*	b *Ulmus* maximum	V-VI	*Betula-Pinus-Corylus*	
					a *Ulmus Corylus*			a *Corylus* maximum			
	7000		V	*Corylus-Betula-Pinus*		V	*Corylus-Betula*				
	8000	Pre-Boreal	IV	*Betula-Pinus*	AP rise	IV	*Betula*	AP rise	IV	*Betula*	AP rise
Late Devensian			III	Park tundra		III	Tundra (Younger *Salix herbacea* period)		III	(Highland Readvance)	
	9000		II	*Betula* with park tundra		II	Park tundra with *Betula* (Late glacial birch period)			The Zone II Interstadial is best indicated by biogenic sediments compared with the solifluction deposits of Zones I and III	
	10,000		I	Park tundra		I	Tundra (Older *Salix herbacea* period)		I	(Perth-Aberdeen Readvance)	

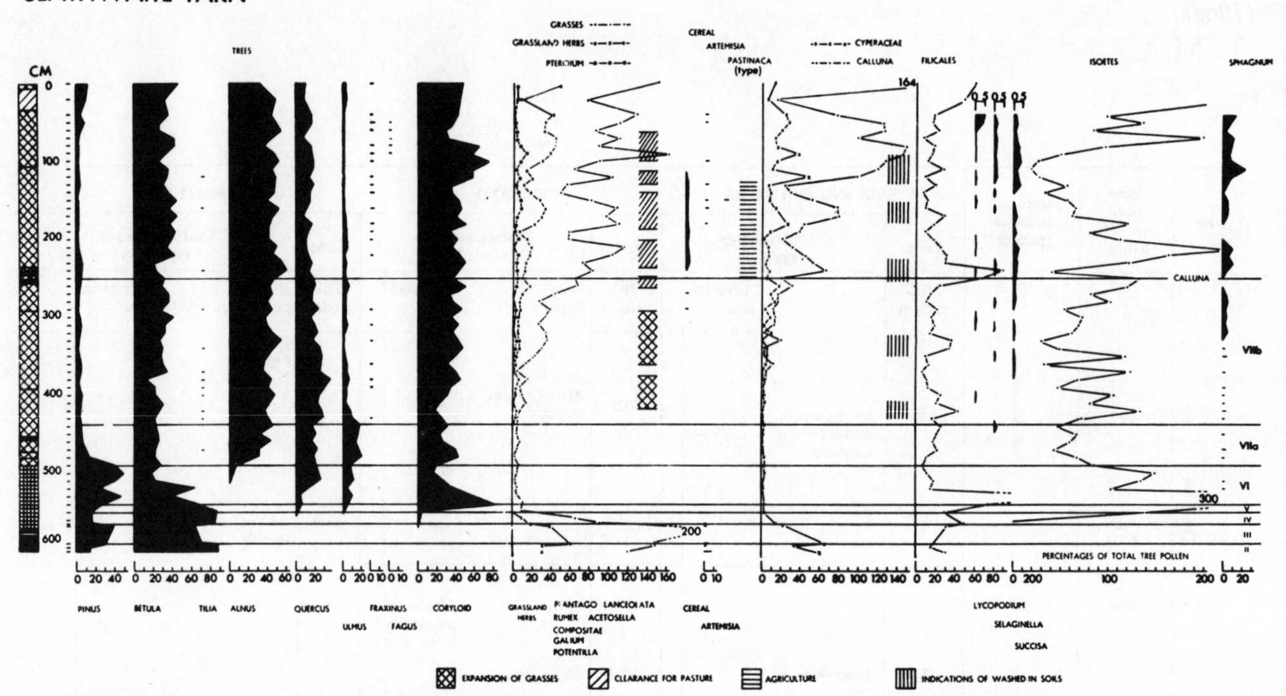

Fig. 21. Pollen diagram from the Late Devensian and Flandrian deposits of Seathwaite Tarn in the English Lake District. From Pennington 1964.

3 MATERIALS AND METHODS

3.1 SOURCES

The material on which this book is based consists almost entirely of fragments of gametophytes. In contrast to spores, leafy stems of mosses are readily recognised and have been reported a great many times. In Pleistocene and Neogene deposits moss remains sometimes occur abundantly and sufficiently well preserved to facilitate identification to the species level, but sometimes only small numbers are present and in such poor condition as to be recognisable only with difficulty.

Moss remains are found in a variety of Pleistocene deposits such as marine and lacustrine sediments, river terraces and peats. Details of the sediment types of deposits studied by the author are given in Appendix 2.

Sites yielding the largest numbers of species of greatest ecological diversity are allochthonous, i.e. lacustrine and fluviatile deposits (fig. 22). As pointed out by Reid (1899) and Godwin (1956), sediments of this nature afford the possibility of preservation of subfossils transported from wide catchment areas. Such assemblages as those of Fort William and Seathwaite Tarn (plate 5) with representatives of rupestral, woodland, aquatic and mire communities contrast strongly with autochthonous assemblages, i.e. peats such as Burtree Lane, Co. Durham (plate 6), and Holme Fen, Huntingdonshire, where the comparatively small numbers of species have been preserved *in situ* and only mire communities can be represented (fig. 23). Nevertheless, such peats may be of considerable bryogeographical interest, as *Paludella squarrosa* and *Helodium blandowii* from Burtree Lane and *Meesia longiseta* and *Dicranum undulatum* from Holme Fen testify.

It is noteworthy that the British deposits richest in moss remains are the estuarine-marine beds at Greenock (p. 226), which have yielded about sixty species so far, and the largest moss flora ever extracted from a Pleistocene site is that from Skaerumhede, Denmark, where about sixty-five species were recovered from the marine layers of the deposit (Hesselbo 1910; Anderson 1961).

The great variety of species in the Greenock beds may be accounted for in the same way as the richness of freshwater sediments. The mollusca and seaweeds indicate shallow water (Robertson 1881; Brett and Norton 1969). The mosses and angiosperms could have been washed into a sheltered bay from the nearest source, the north-facing slopes on which Greenock now stands. However, the possibility of transport from more distant areas, perhaps by floating ice, cannot be excluded.

Archaeological excavations are yielding an increasing number of ecologically and ethnobotanically interesting records (fig. 24). Prehistoric boats, the seams of which have been plugged with bryophytes, have proved particularly rich. Eleven of the twenty-one hepatics known from the British Pleistocene were the caulking material of a late Bronze Age canoe (Sheppard 1910; p. 187).

These are the major sources of subfossil mosses not just from the British Isles but also from the rest of Europe and North America, the only other areas where substantial research on Pleistocene bryophytes has been carried out.

Other, minor sources may be mentioned briefly. Tufa deposits have yielded a small number of species (for example Douin 1923, 1927; Gams, 1932; Boros 1952). I know of no tufa records from the British Isles, at any rate none of any great antiquity or bryogeographical significance.

Polytrichum norvegicum, Drepanocladus revolvens and *Campylium stellatum,* identified by Camus (1915), came from an unique source, the stomach of a mammoth, one of the better preserved of many from Siberia (Farrand 1961). Savicz-Ljubitzkaja and Abramova (1954) list some thirty-nine bryophytes preserved with a mammoth skeleton. It is remarkable that this assemblage includes three liverworts, previously unknown in the subfossil state, namely *Blepharostoma trichophyllum, Cephaloziella divaricata* and *Sphenolobus minutus*, which came from under the beast's right hind leg. In 1934 Szafran recorded six mosses found in the loam surrounding the body of a woolly rhinoceros, preserved by oil-bearing strata in what was then part of south-eastern Poland.

Lastly there are a small number of bryophytes from the Baltic amber (Czeczott 1961). For the most part, however, these occurrences, being mostly Eocene and Oligocene, are outside the scope of this book. No bryophytes have been recorded from amber picked up from East Anglian beaches (Reid 1884), as far as I know.

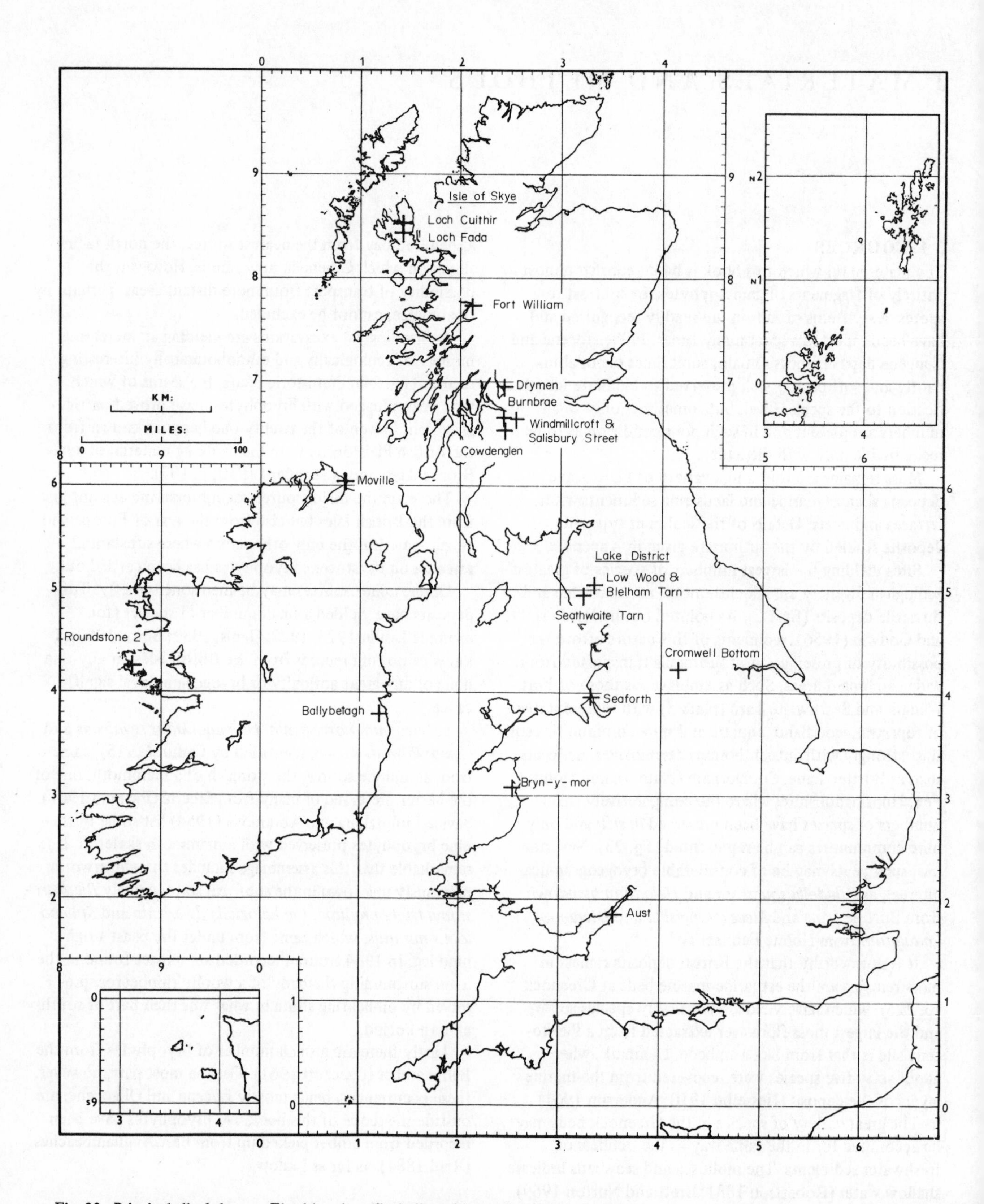

Fig. 22. Principal allochthonous Flandrian sites (fluviatile and lacustrine deposits).

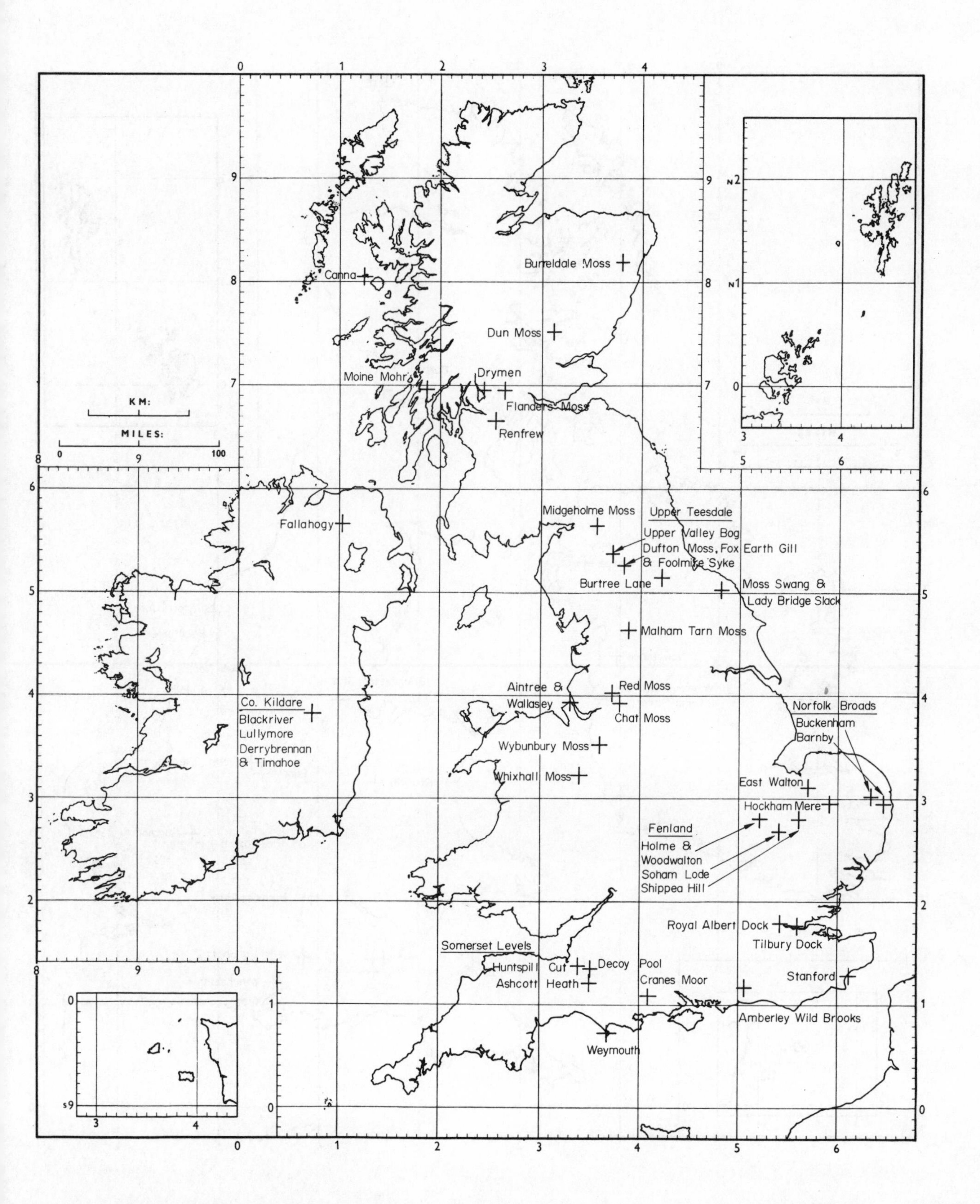

Fig. 23. Principal autochthonous Flandrian sites (mires).

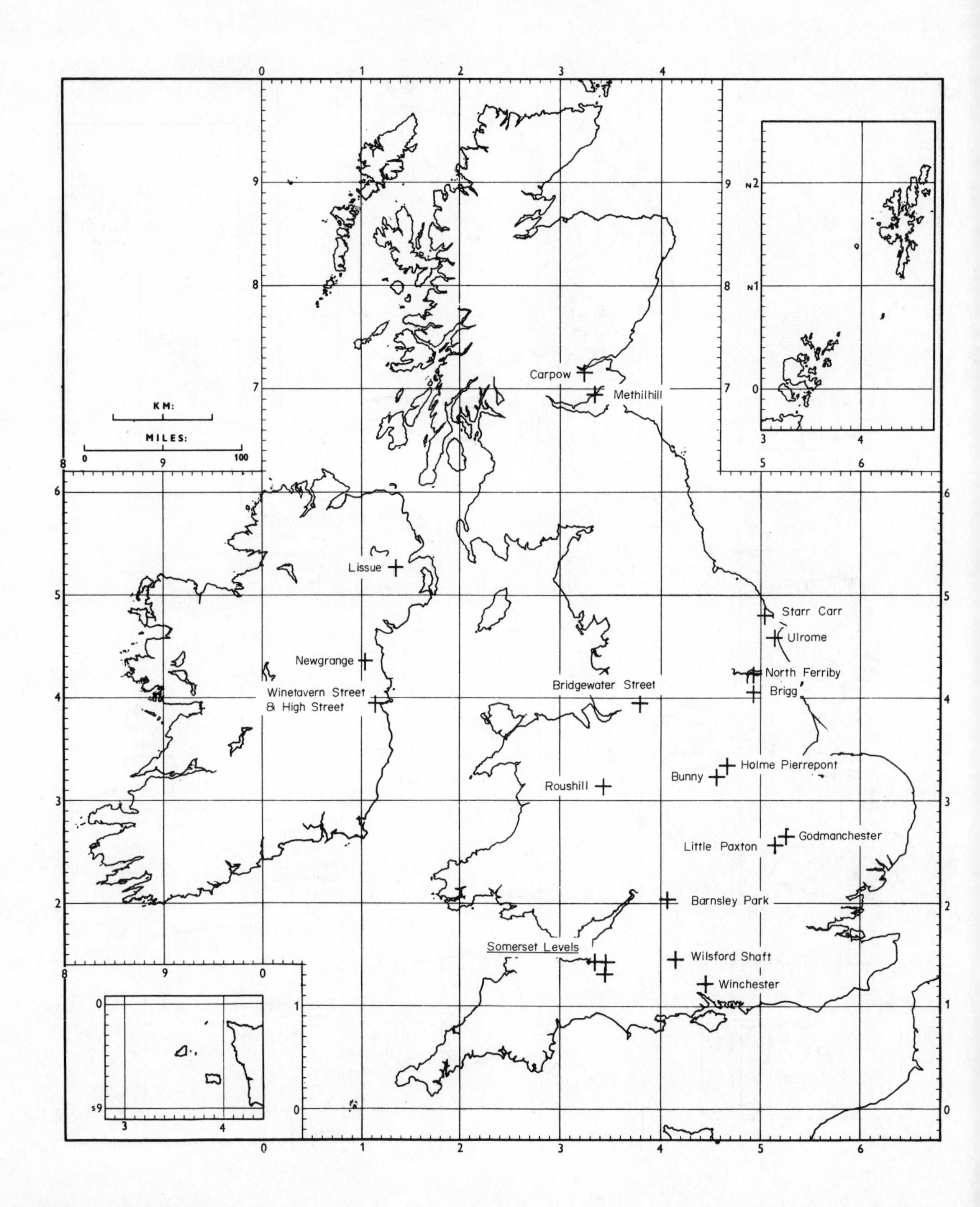

Fig. 24. Principal Flandrian archaeological sites.

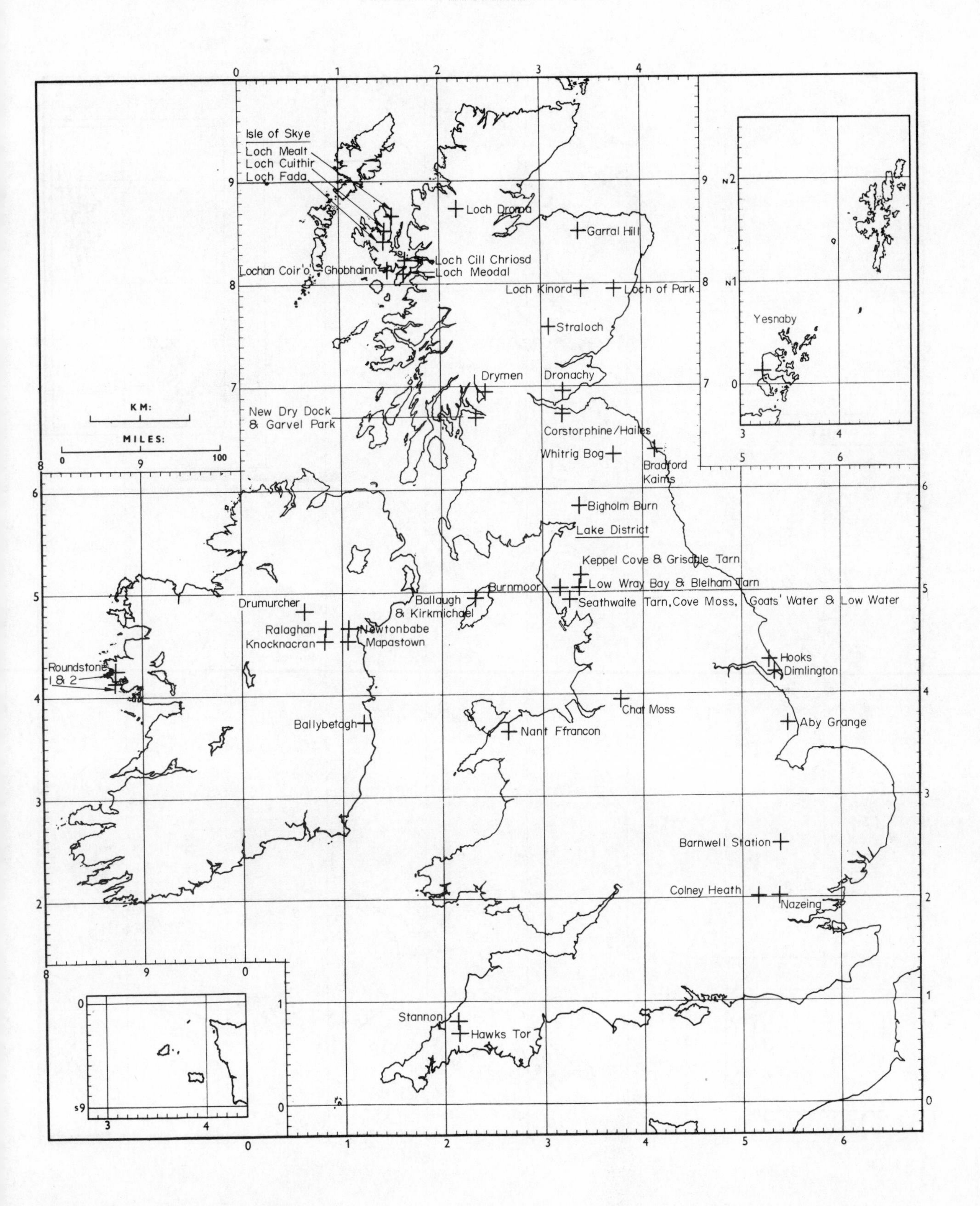

Fig. 25. Principal Late Devensian sites.

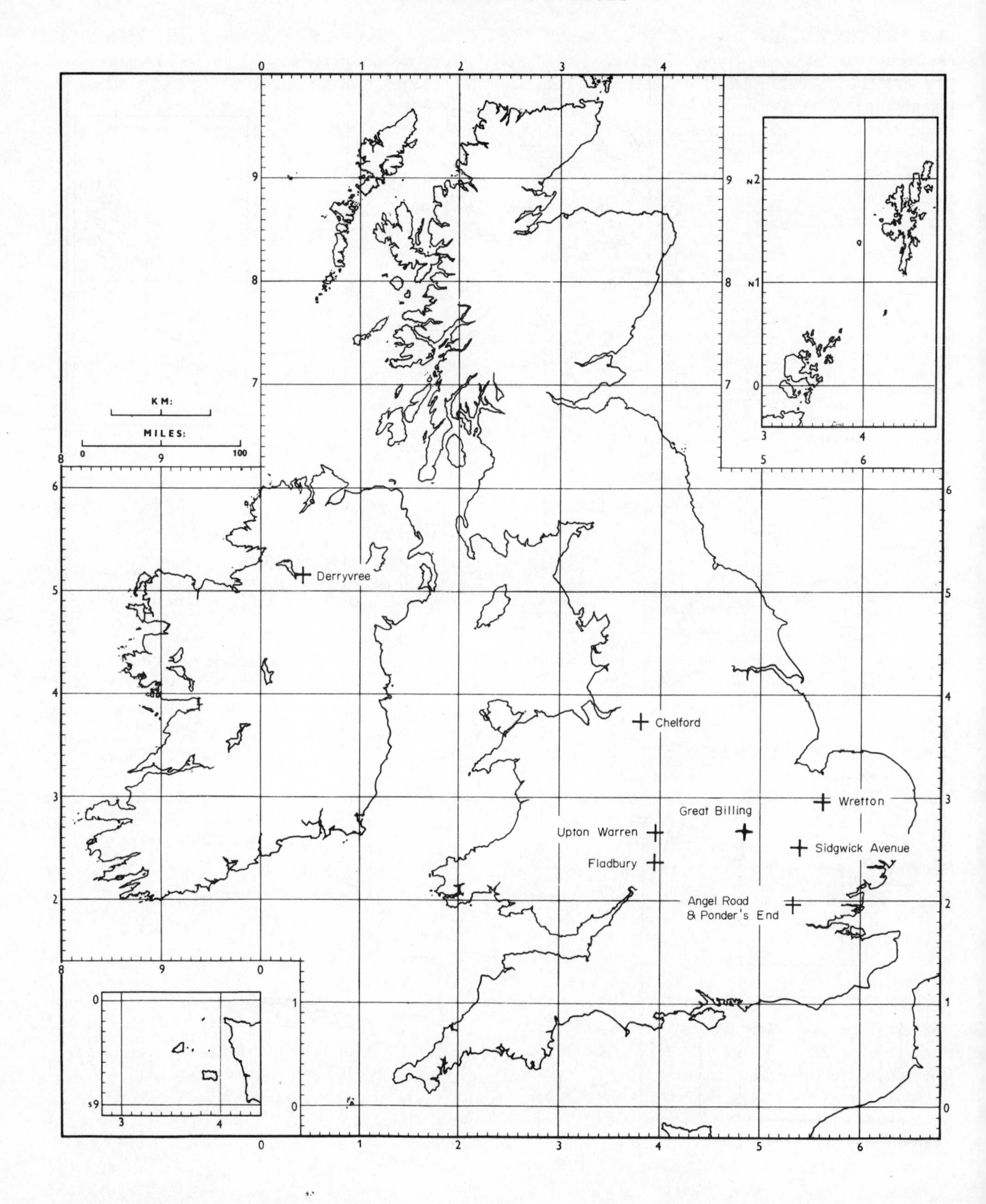

Fig. 26. Middle and Early Devensian sites.

3.2 PRESERVATION AND EXTRACTION

As discussed by Gams (1932, pp. 300-2), the habitat of a species influences the chances of fossilisation. More than twenty of the thirty species, most often recorded as subfossils from British Pleistocene deposits (table 6), grow in such habitats as mires and margins of lakes where the chances of preservation must be good. Ten of the thirty species belong to the Amblystegiaceae, a family which accounts for about 25 per cent of all the British subfossil records.

Table 6 *Species with fifteen or more Pleistocene localities in the British Isles*

Antitrichia curtipendula	*Hylocomium splendens*
Aulacomnium palustre	*Hypnum cupressiforme*
Calliergon giganteum	*Neckera complanata*
C. stramineum	*Paludella squarrosa*
Calliergonella cuspidata	*Pleurozium schreberi*
Campylium stellatum	*Polytrichum alpinum*
Climacium dendroides	*P. commune*
Cratoneuron commutatum	*Rhacomitrium lanuginosum*
C. filicinum	*Rhytidiadelphus squarrosus*
Dicranum scoparium	*Scorpidium scorpioides*
Drepanocladus exannulatus	*Sphagnum cuspidatum*
D. fluitans	*S. imbricatum*
D. revolvens	*S. palustre*
Eurhynchium praelongum	*S. papillosum*
Homalothecium nitens	*Thuidium tamariscinum*

Many rupestral and epiphytic families and genera are very poorly represented or unknown as subfossils, for example Andreaeaceae, Encalyptaceae, Orthotrichaceae, Ptychomitraceae, Hedwigiaceae, Cryphaeaceae and *Seligeria, Cynodontium, Grimmia* and *Bartramia*. The author is responsible for the sole records of *Plagiopus, Bartramia, Grimmia apocarpa* and *Ulota* from the British Pleistocene. The record of *Grimmia apocarpa* is especially remarkable since the material bears a capsule (plate 17). A highly exceptional occurrence is that of both *Grimmia apocarpa* and *G. ovalis* in early Flandrian clay-mud from Frankental in the Vosges mountains (det. F. Koppe in Firbas *et al.* 1948).

However, not all rupestral and corticolous species are scarce as subfossils. Both *Antitrichia curtipendula* and *Neckera complanata* have been recorded on numerous occasions (table 6). This may possibly be a genuine indication that these species were abundant. It may also be that the large size of the fragments and ease of identification have led to frequent recovery and recognition.

Small plants which grow on soil, e.g. species of *Ephemerum, Phascum, Pottia* and *Weissia*, are unknown as subfossils from Britain. However, Abramova and Abramov (1962) have recorded '*Pottia* sp' from an upper Tertiary deposit in the central Volga region.

Of the eighty-seven species ubiquitous in the British Isles at present (recorded from 140 or more vice-counties) only fifteen are unknown as Pleistocene subfossils (table 7); they are composed of species from terricolous, saxicolous and corticolous situations, all habitats unconducive to fossilisation.Differential destruction, the next topic, is not operating here.

Table 7 *Ubiquitous species unknown from British Pleistocene deposits*

Barbula convoluta	*Fissidens taxifolius*
B. cylindrica	*Grimmia pulvinata*
B. revoluta	*Orthotrichum affine*
B. rigidula	*O. anomalum*
B. tophacea	*Tortula muralis*
B. unguiculata	*Ulota crispa*
Bryum argenteum	*Weissia crispa*
Encalypta streptocarpa	

In view of present abundance and habitats favourable to preservation, the lack or poor representation of various species in the fossil record is remarkable. As Gams (1932) and Steere (1942) point out, many species of common occurrence in bogs, e.g. species of *Cephalozia, Mylia, Odontoschisma* and *Campylopus*, have seldom or never been recovered from peats. The scarcity of such widespread semi-aquatic plants as *Cinclidotus fontinaloides* and *Leptodictyum riparium* from the British Flandrian is so striking that one may wonder if differential preservation is involved. Another species giving rise to such thoughts is *Leucobryum glaucum*, which despite its wide range and common occurrence on ombrogenous mires, barely qualifies as a subfossil (p. 88).

One of the most striking absentees from Pleistocene deposits is *Dicranella palustris*, a common species of flushes and streamsides in the hills at present. The frequent occurrence of subfossil *Philonotis fontana* in Devensian deposits shows no lack of suitable habitats in the past.

Other noteworthy absentees are streamside species such as *Tortula latifolia, Rhacomitrium aciculare, Leskea polycarpa, Grimmia alpicola* and lastly *Hygohypnum* spp, of which there are very few subfossil records.

The major instance of under-representation is the small total of hepatics as a whole. Only eighteen liverworts are known as macroscopic subfossils from the British Devensian and Flandrian. This contrasts with some 179 mosses. Gams (1932), Steere (1942) and lately Culberson (1955) have discussed differential preservation in relation to cell-wall chemistry. Little experimental work has been carried out on this problem. Clymo (1965) has shown that some cymbifolian *Sphagnum* species preserve more readily than other members of the genus.

The bryophytes found in amber throw an indirect light on this matter. In her collation of records of amber plants Czeczott (1961) lists seventeen taxa of mosses and eighteen taxa of liverworts, almost all described as new species, principally by Casparie. Whatever one thinks of the worth of these taxa, it is clear that liverworts were at least as likely as mosses to be encased in amber, the difference from preservation in Pleistocene peats and sediments being that in amber preservation was so swift that the liverworts did not decay.

Apart from the destructive methods of extraction discussed below, very often the state of preservation is poor. In many cases only remnants of leaves remain. Commonly leaves are badly torn or have lost their tips and margins. The latter condition is characteristic of *Rhacomitrium* subfossils. This accounts for the Late Devensian records made by the author merely as '*Rhacomitrium* undetermined species'. Even the very large subfossils of *R. lanuginosum* from Loch Droma (fig. 41), well enough preserved to show the habit of the species clearly, have lost the tips and margins of the distal parts of the leaves.

On the other hand, the excellent state of preservation of some specimens may be surprising (plates 9, 18 and 19). Thus, the numerous long stems of *Meesia longiseta* from Holme Fen are in perfect condition (plate 18). Rarely cells may even retain some remnants of contents. I have seen plastid-like structures in the subfossils of *Eurhynchium speciosum* from Shippea Hill (zone VII*b*) and *Fontinalis* sp from Seathwaite Tarn (zone VIII). Rosendahl (1948, p. 295) records 'plastids' in a *Mnium* sp from Bronson station I, Minnesota (an Early Wisconsin (last glacial) deposit; R. O. Kapp personal communication) and Wilson (1932, p. 41) records 'remnants of chloroplasts' in species from the Two Creeks Forest Bed, Wisconsin (Late Wisconsin).

It appears that only very rarely do subfossils retain anything of their original colour. Dixon (1907, p. 99), referring to *Cinclidium stygium* from the deposit of uncertain, possibly Devensian, age at West Craigneuk, Lanarkshire, states 'The red colour is perfectly preserved.' Most subfossils are brownish in colour. However, fragments of a few species such as *Antitrichia curtipendula* and *Polytrichum* spp often have acquired a dark, very opaque colour which renders the specimens easily discernible.

The technique for extraction of moss remains depends on the nature of the substrate and is similar to the methods described by Godwin (1956, pp. 6-8) for recovery of vascular plant macrofossils. Where the sediment type permits it is most satisfactory to pick fragments from the faces of cleavage planes. Particularly fine specimens, especially the large fragments of *Rhacomitrium lanuginosum* (fig. 41) from Loch Droma, were obtained in this way.

Most often, however, the substrate has to be broken up into small lumps and loosened by soaking in dilute sodium hydroxide or nitric acid and then sieved. This drastic treatment is undesirable because it renders moss fragments even more fragmentary. In comparison with tough, compact fruits and seeds – which are capable of surviving such extraction more or less intact – many moss subfossils are vulnerable, finely branched structures. It can be readily imagined that the detached moss leaves frequently found in subfossil assemblages have been torn from their stems by the extraction. Such large-leaved mosses as species of Mniaceae are often in very bad condition and may well have been damaged by these crude but unavoidable methods of recovery.

Where possible it is preferable to break sediments up with water rather than with the chemical agents which make moss fragments mucilaginous if soaking is prolonged. Some little-humified peats need little treatment. For example, excellently preserved specimens of *Helodium blandowii* were recovered from zone II peat from Hooks, Holderness and of *Paludella squarrosa* from zone IV peat from Burtree Lane (plate 6).

3.3 IDENTIFICATION

After sieving, the fragments are transferred to a large petri dish, examined and segregated under a stereomicroscope before final identification by means of high-powered microscopy. The specimens, permanently mounted in gum choral on microscope slides, can be easily stored.

Unlike the determination of fruits and seeds, the identification of moss fragments is facilitated by the modern standard floras (Dixon 1924, Nyholm 1954-69, Watson 1968) before the necessary comparison with herbarium material begins. Subfossil fragments can be handled like herbarium specimens, except that they may become brittle if allowed to dry out. The author regularly cuts sections, e.g. of *Polytrichum* leaves, in order to make identifications to the species level (plate 19).

The fragmentary nature of moss subfossils complicates the determinations. Even if the preservation is perfect, information of an important, if not a decisive, kind may be denied the investigator. Sporophyte and inflorescence characters are almost always unavailable. Often habit is unknown, as is habitat unless the specimens were preserved *in situ*.

It is well known that species of such genera as *Calliergon*, *Cratoneuron* and *Drepanocladus*, which are abundantly represented as subfossils, are very variable. The pronounced variation of *Drepanocladus exannulatus* and *D. fluitans* according to light intensity, concentration of the medium and aquatic or subaerial growth has been described by Lodge (1959, 1960). It is not surprising if tiny fragments of these genera prove difficult or impossible to identify to the species level. Almost every major site yields specimens which the author sets aside as unidentifiable. These residues consist very largely or exclusively of badly preserved or minute fragments of hypnoid genera.

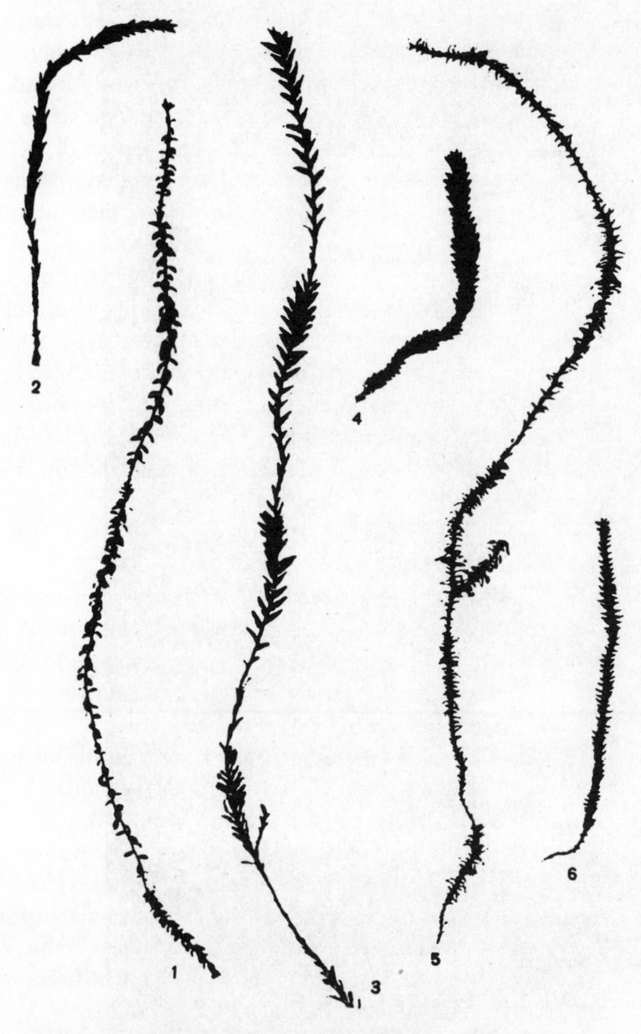

Fig. 27. Submerged and terrestrial forms of some Swedish mosses. From Persson 1942. 1, 3 and 5: Submerged forms of *Bryum pseudotriquetrum, Fissidens adianthoides*, and *Meesia tristicha*. 2, 4 and 6: Terrestrial forms of same species.

Not only hypnoid mosses are modified by submergence. Eighteen species, including in addition to aquatics such as *Fontinalis* spp and *Eurhynchium riparoides*, the more unexpected *Cinclidium stygium, Fissidens adianthoides* and *Meesia tristicha*, were dredged from depths down to 6·5 m in northern Swedish lakes (Persson 1942). Examples of the greatly altered shoots found by Persson are shown in fig. 27.

Miss R. Peck gave the author bryophytes brought up from a depth of almost 5 m in Oakdale Reservoir, near Northallerton, Yorkshire. There were two large fragments, one being *Scapania undulata* and the other *Polytrichum commune* (plate 5). The genus of the former was at once recognisable but that of the latter was obvious only when one of the few remaining leaves which had been produced subaerially was examined microscopically. The brown, washed-in shoot of *Polytrichum*, several centimetres long and showing no green colour whatever, bore four green shoots which had obviously grown at the bottom of the lake. The grossly modified leaves were short, broad in proportion to length, abruptly tapered, undulate and markedly squarrose. The sheaths were not clearly marked from the blades and the margins and teeth less distinct than in the typical state. However, the striking modification was the reduction of lamellae from the usual sixty to less than ten. The elamellose margin was sixteen cells wide on each side. The apical cells of the lamellae, not having developed a sinus, appeared rounded in transverse section. In appearance the modified shoot was reminiscent of *Atrichum* rather than *Polytrichum*. It is worth recalling that Dixon (1924, p. 45) found a *Polytrichum* with few lamellae which was first described as *Catharinea* (*Atrichum*) *dixoni* by Braithwaite but which proved merely to be a form of *P. longisetum*.

The greatly altered *P. commune* parallels the modifications found by Persson and previously by Leach (1931*a*) who showed that to grow a wide range of mosses such as *Polytrichum, Fissidens, Dicranum* and *Aulacomnium* under conditions of darkness and moisture saturation was to promote less of distinguishing features. This is a phenomenon familiar to all bryologists since the advent of polythene. Live, moist specimens abandoned in a polythene bag quickly produce grossly simplified shoots.

This phenotypic plasticity of mosses cautions the investigator of subfossil assemblages. As pointed out by Culberson (1955), subfossil fragments extracted from deposits of shallow lakes may have grown before fossilisation and the investigator unfamiliar with the resulting atypical shoots is baffled.

The fragmentary nature of subfossils, poor state of preservation and other difficulties discussed above account for the many tentative identifications given in chapter 5. About 220 species have been recovered from British Pleistocene deposits. Of these eight are based solely on tentative records, as follows:

Bryum caespiticium	*F. cristatus*
B. intermedium	*Isopterygium pulchellum*
Eurhynchium confertum	*Orthotrichum diaphanum*
Fissidens bryoides	*Polytrichum nanum*

A further twelve have only one or a few records in need of checking, for the most part in view of taxonomic advance. They are as follows:

Plate 5. (*a*) Seathwaite Tarn in the English Lake District. The moss remains were obtained from cores taken from the bed of the tarn towards the far bank as shown here. The landscape is devoid of phanerophytes except for very sparse *Sorbus aucuparia* and *Juniperus communis* on the crags. The oceanic species *Campylopus atrovirens, Hyocomium armoricum* and *Rhacomitrium aquaticum* are abundant around the tarn; *Breutelia chrysocoma* and *Leptodontium flexifoluim* occur in small quantity.

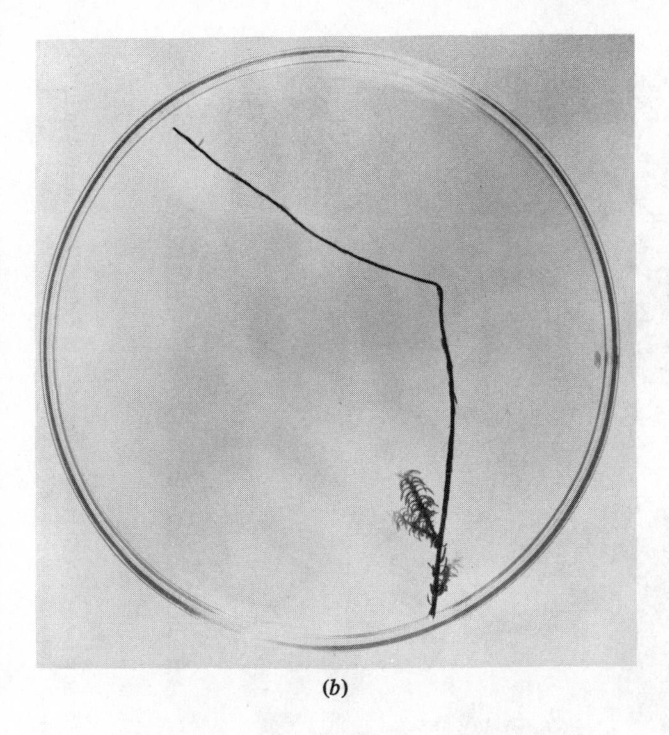

(*b*)

(*c*)

Plate 5. (*b*) and (*c*). Shoot of *Polytrichum commune* dredged from a depth of almost 5 m at Oakdale Reservoir, Yorkshire. (*b*) The shoot, several cm long, has lost almost all the subaerially grown leaves. The top of the shoot is at the base of the photograph. (*c*) Close-up of the top of the shoot showing three new shoots grown under water. The grossly modified leaves can be seen clearly.

Brachythecium reflexum	*Gymnostomum sp*
Bryum creberrimum	*Rhacomitrium ellipticum*
B. erthyrocarpum	*Sphagnum russowii*
Calliergon richardsonii	*Tortella inclinata*
Campylopus pyriformis	*Tortula subulata*
Distichium inclinatum	*Trichostomum sp*

In addition, the early records of such species as *Plagiothecium denticulatum* and *Tortella fragilis* require checking.

Because of the minimal attention given to description of the material and of the criteria on which the determinations rest, some doubts must arise about various others of the early identifications, e.g. some of those by Mahony (1868*a*, 1869), by Gepp (in Bennie 1894*a*), by Hobkirk (in Bennie 1894*b*) and by Fergusson (in Robertson 1881). It is certainly desirable that as many as possible of these records be checked. However, the species listed above are the ones based exclusively on one or a few records which the author considers to be in particular need of revision. In order to distinguish them in the following chapters and tables a bold **T** is placed against the names.

Because the majority, if not all, of the records must have been made from material lacking capsules, it is perhaps best to regard as tentative records of species of such genera as *Bryum, Distichium, Encalypta* and *Pohlia* in which sporophyte characters are important or decisive taxonomically.

Though such records as var. *densum* of *Ditrichum flexicaule*, var. *compactum* of *Oncophorus virens* and var. *brachydictyon* of *Drepanocladus exannulatus* are listed in chapter 5, little or no weight can be given to them in the light of modern taxonomic opinion. Records of varieties given by Warburg (1963) are listed and accepted.

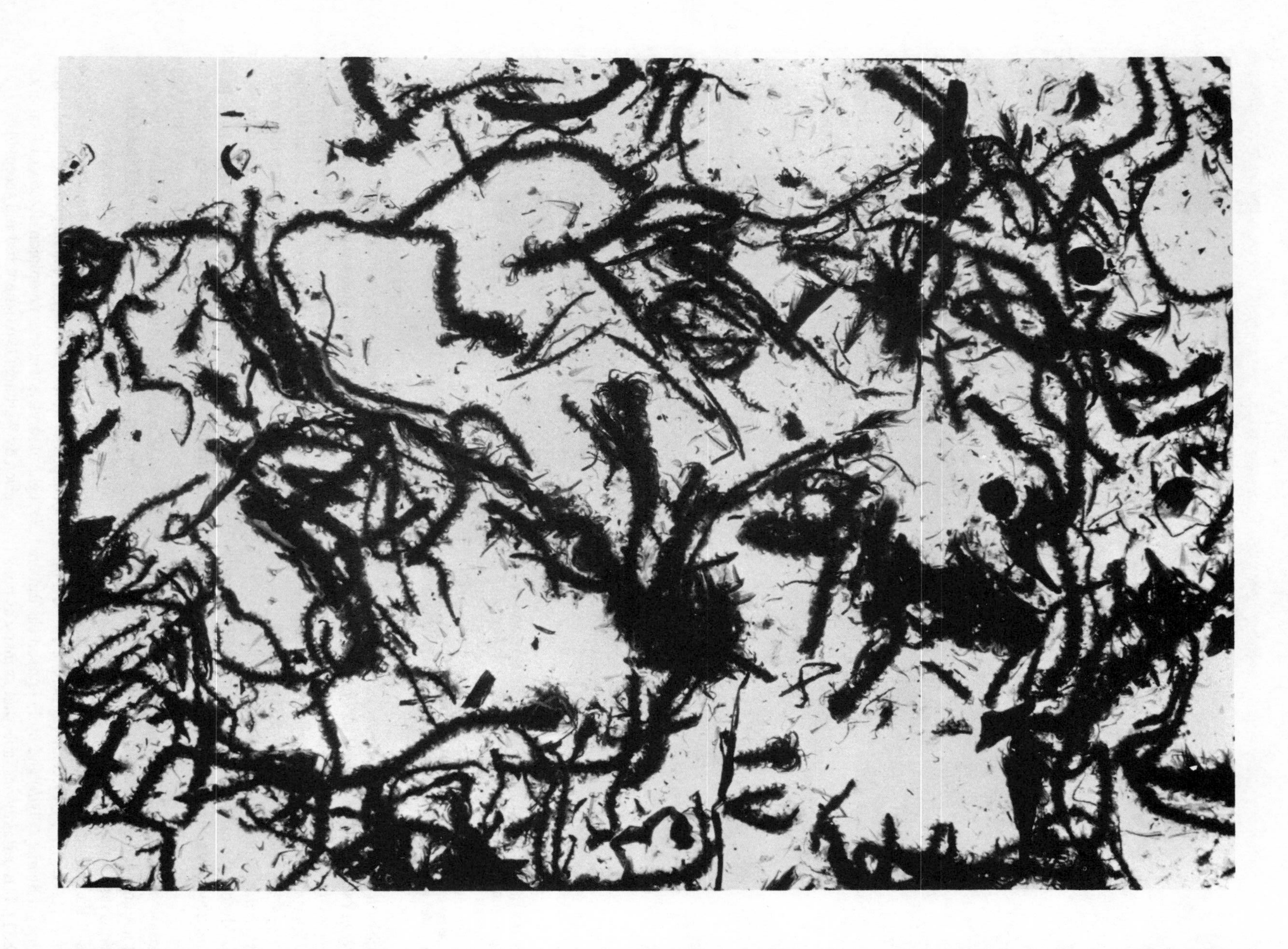

Plate 6. (*a*) Peat of *Paludella squarrosa*. The zone IV peat from Burtree Lane has been teased out to show the very well preserved remains of *Paludella*, distinctive in its squarrose leaves. Also to be seen are a few shoots and many detached leaves of *Homalothecium nitens*. The stems of *Paludella* are about 1 to 3 cm long.

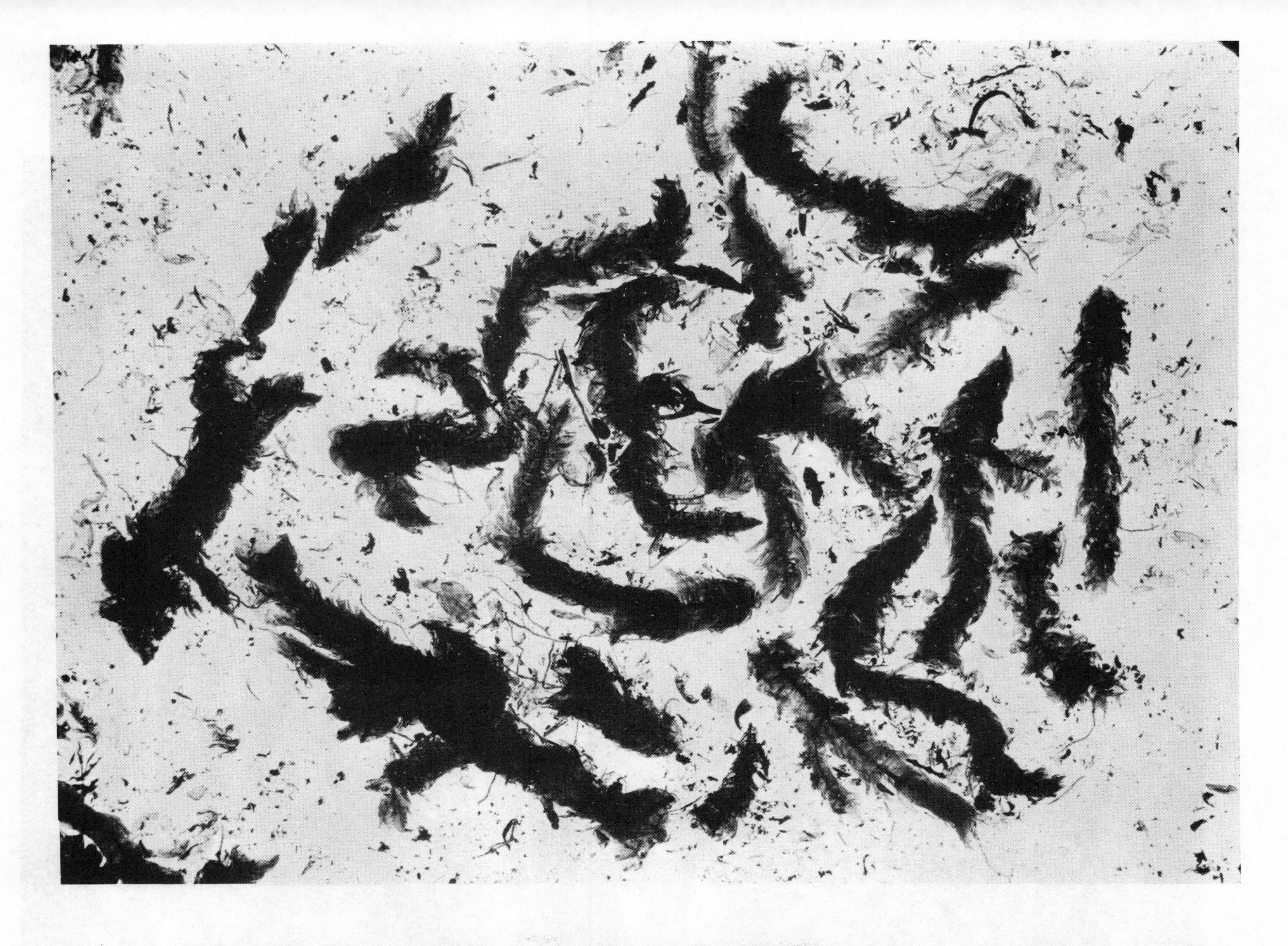

Plate 6. (*b*) Peat of *Scorpidium scorpioides* from the Somerset Levels. The well preserved remains, about 1 to 3 cm long, have been teased out. A few detached leaves show the broad, asymmetric shape and lack of a nerve.

Plate 7. (*a*) *Sphagnum* peat from Flanders Moss. The peat has been teased out to show the acuminate branch leaves characteristic of sub-genus *Litophloea*. The leaves are about 1 to 2 mm long.

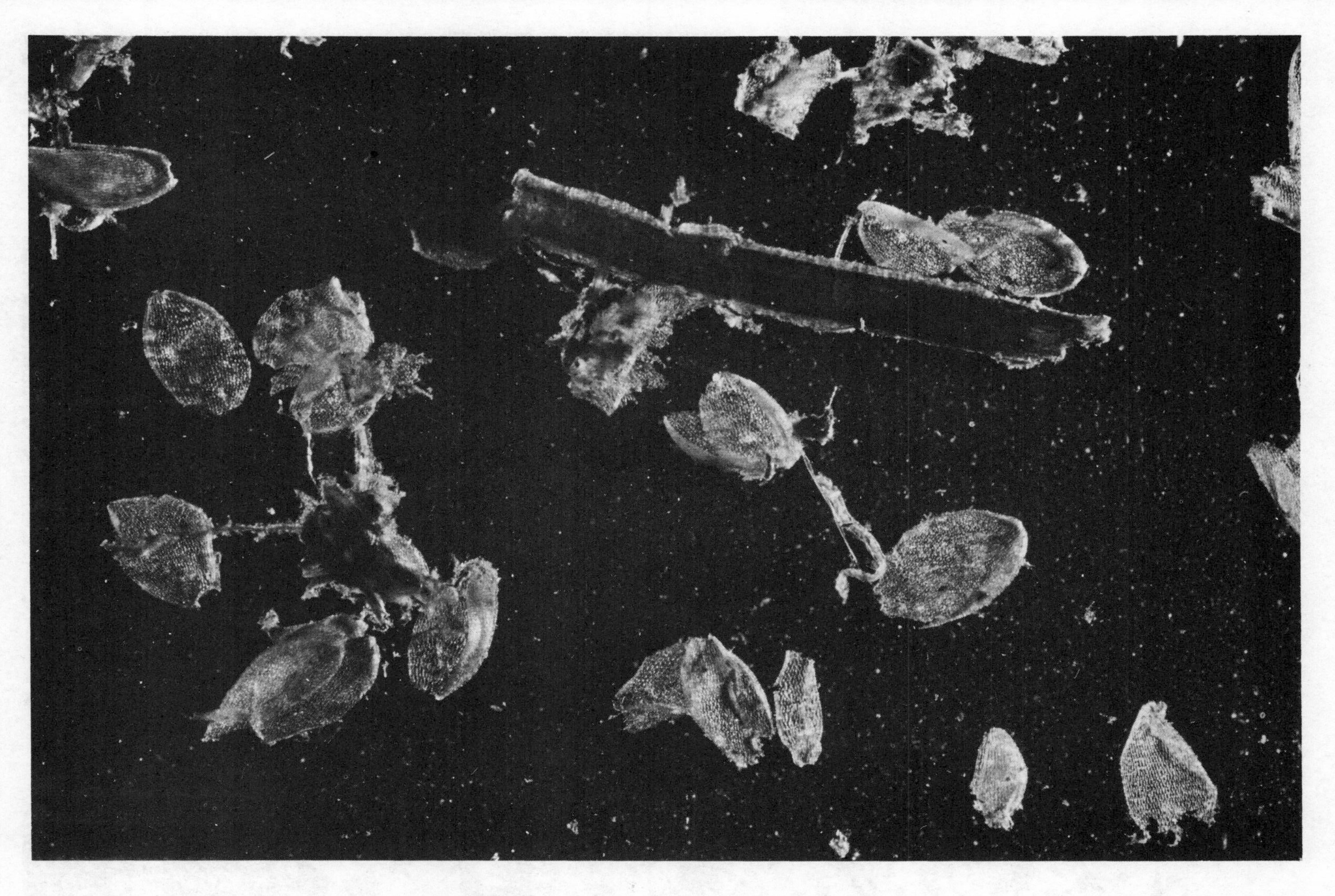

Plate 7. (*b*) *Sphagnum* peat from Holme Fen. Again the peat has been teased out, this time to show the broad very obtuse branch leaves characteristic of subgenus *Inophloea*. The leaves are about 2 mm long.

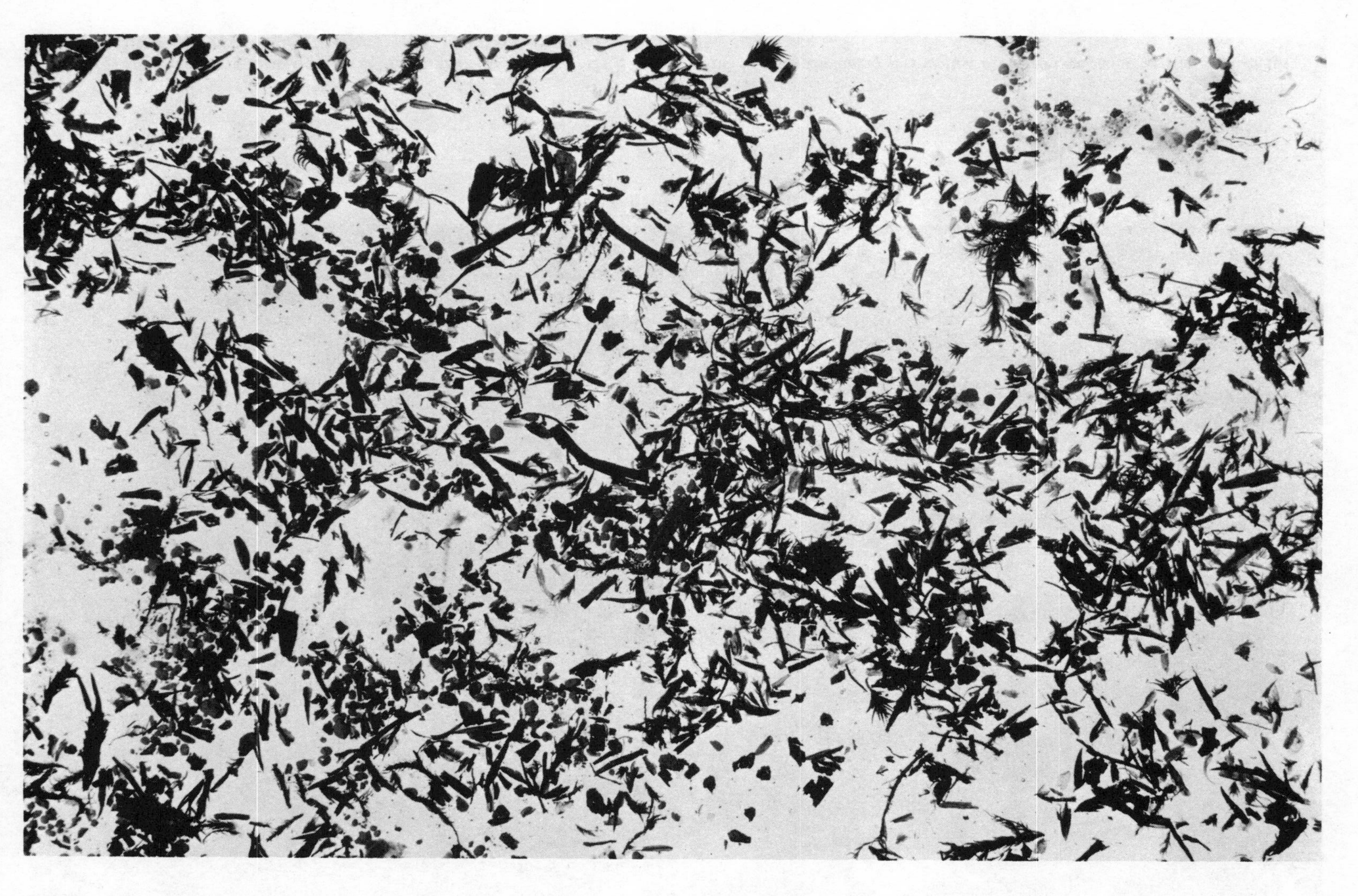

Plate 8. (*a*) An allochthonous assemblage from the Late Devensian layers at Nant Ffrancon. The photograph shows an abundance of moss remains including *Drepanocladus*, *Polytrichum* and *Rhacomitrium*, after washing and sieving but before sorting.

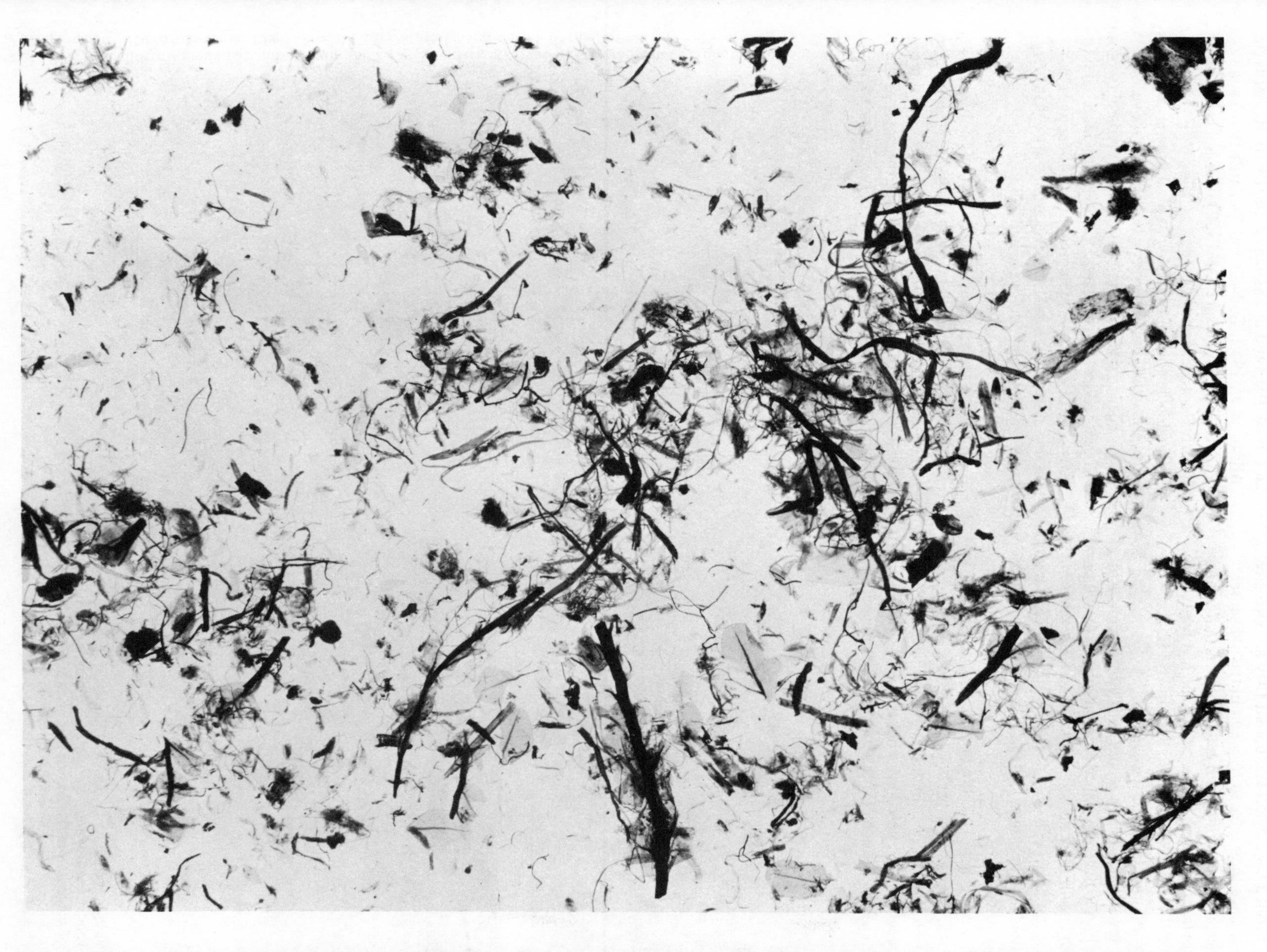

Plate 8. (*b*) Moss remains from the interstadial deposit at Chelford. The large stems and large detached leaves are *Pseudobryum cinclidioides*. There are also small detached leaves of *Sphagnum* subg *Inophloea*.

Various records of '*Hypnum*' peats occur in the literature. From a bryological standpoint they are virtually useless since they refer to *Hypnum* in the old sense. Numerous species of several genera such as *Calliergon*, *Campylium* and *Drepanocladus* may be involved. Only a few of these records are given in chapter 5 under '*Hypnum* undetermined species'.

A few of the modern records made by botanists of little bryological experience are spurious. *Aulacomnium androgynum*, *Eurhynchium murale* and *Rhacomitrium aquaticum* have been deleted from the Pleistocene record as a result of checking by the author. Other changes are purely minor.

However, there is no doubt that the vast majority of subfossil records are trustworthy. The very considerable contribution of H. N. Dixon has already been outlined and recently such noted bryologists as Hesselbo (in Jessen 1949), Holmen (in Mitchell 1953; in Jessen *et al.* 1959), Proctor (in West 1956; in Duigan 1955; in Sparks and West 1959 and in Lambert *et al.* 1960), Richards (in Allison *et al.* 1952; Godwin and Richards 1946) and Tuomikoski (in Allison *et al.* 1952) have published on British Pleistocene mosses.

No conclusions of bryogeographical import depend exclusively on the doubtful records discussed above.

In order to give some indication of the reliability of my determinations, I have classified leafy stem material into four categories, albeit subjective intergrading ones, describing state of preservation. The groups are called 'very good', 'good', 'bad' and 'very bad'.

The first is illustrated by plate 9*a*. The size of the subfossil and its preservation are such that not only are all or almost all of the leaves intact but the habit of the species can be seen. The second includes fragments as well preserved, or almost so, as in the first but the habit of the species is not shown (plate 9*b*). The 'bad' category includes

(*a*)

(*b*)

Plate 9. Preservation categories. (*a*) *Thamnobryum alopecurum*. Very good. Superbly preserved subfossil, 3 cm long, showing habit of species, from the Roman well, at Barnsley Park. (*b*) *Rhytidiadelphus squarrosus*. Good. Well preserved fragment, 2 cm long, all the leaves intact but habit not shown. From New Dry Dock.

(*c*)

(*d*)

Plate 9. Preservation categories. (*c*) *Dicranum undulatum*. Bad. Leaves mostly badly preserved. Fragment, 1.5 cm long, from Holme Fen. (*d*) *Aulacomnium palustre*. Very bad. Preservation is very poor, not a single intact leaf. Fragment about 1 cm long, from New Dry Dock.

fragments a stage worse in preservation. Few leaves are intact (plate 9*c*). Specimens in the fourth group are in such bad condition as to be unrecognisable to the species level in many cases (plate 9*d*). Two categories have been adopted for detached leaves; 'good' and 'bad'. Specimens in the first group have intact margins and apices. 'Bad' leaves are damaged to some extent. In appendix 2, against each species in the subfossil assemblages is given the number of fragments, the length of the longest fragment and the preservation category of the best preserved fragment.

My collection of Pleistocene bryophytes, the vast bulk being mounted on microscope slides, is housed in a cabinet lodged in the Botany Department of Glasgow University. A much smaller set of slides is retained at Cambridge in the Botany School.

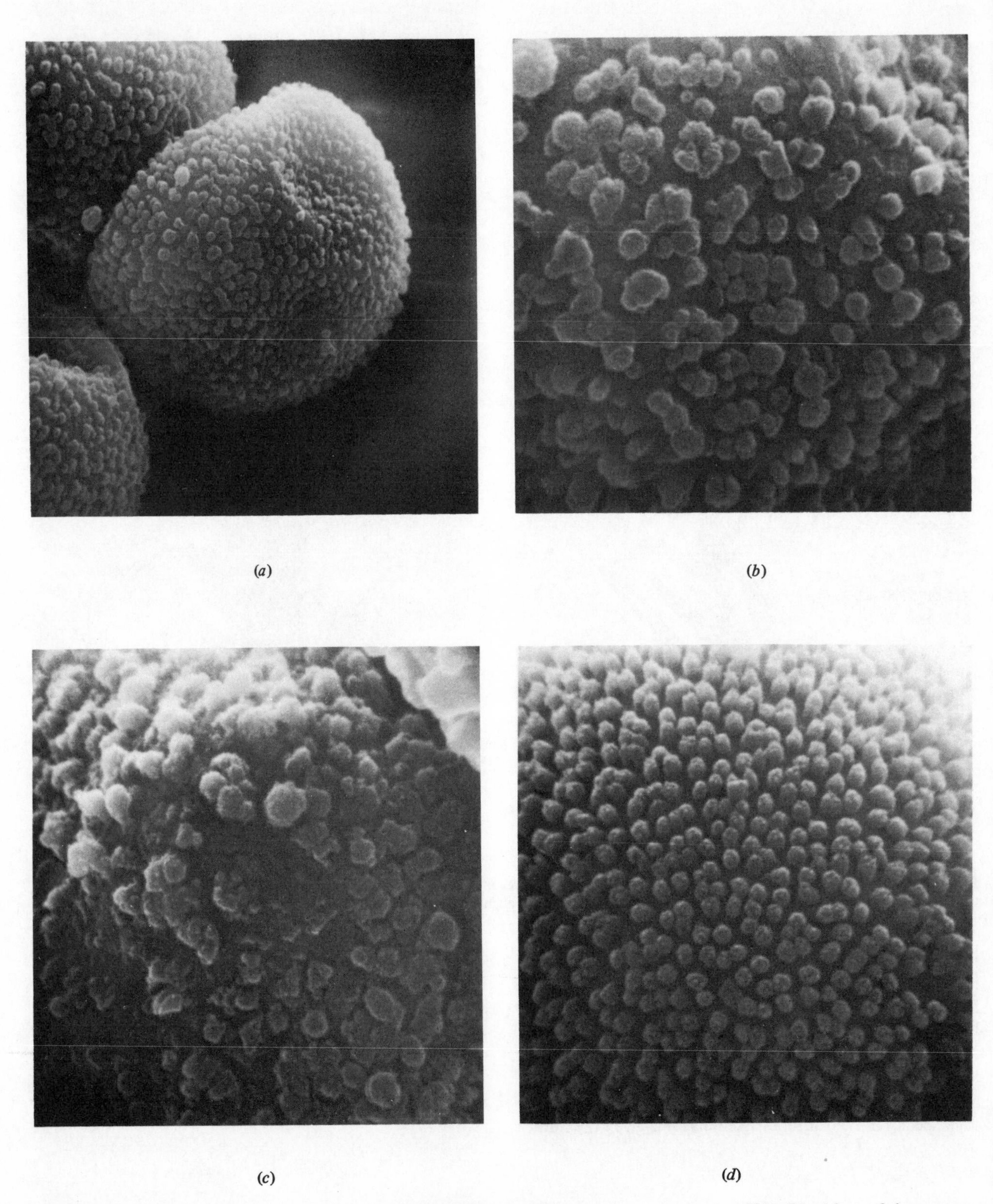

Plate 10. Scanning electronmicrographs of moss spores. (*a*) *Distichium capillaceum*, Angus, Scotland. × 3300. Distal faces of three spores. (*b*) The same. × 10,950. Detail of sculpturing. (*c*) *Homalothecium nitens*, Abisko, Sweden. × 11,200. Detail of sculpturing. (*d*) *Funaria hygrometrica*, Glasgow, Scotland. × 10,800. Detail of sculpturing.

4 SPORES AND SPOROPHYTES

4.1 SPORES

The number of taxa of bryophytes recorded by pollen analysts on the basis of spores is very small. *Sphagnum* spores are widely familiar but, for example in Britain, there is only one other genus of mosses authentically determined from spores; this is *Meesia*.

Many pollen diagrams give curves for 'Bryales' spores or similar vague assignments which, for the most part, are not well substantiated. Spores of *Tilletia sphagni* described below have been a considerable source of error. However, the main cause of confusion is lack of familiarity with bryophyte spores.

The great majority of moss spores have diameters in the range 10 to 20 μm and are therefore small compared to pollen grains. Moreover, the sculpturing is so tiny as to be only indistinctly seen with the light microscope. McClymont (1955) in a survey of the spores of some 653 North American species found that '60 to 80% were morphologically similar', though he did stress that the remainder were very diverse. Plate 10 shows how similar the spores of widely unrelated mosses can be. The roughly granular pattern is very widespread in the mosses.

Apart from McClymont's paper, much can be learned of bryophyte spore morphology and wall structure from the following publications: Knox (1939), Udar (1964), McClymont and Larson (1964) and Erdtman (1957, 1965).

If the condition of subfossil *Meesia* spores is typical, moss spores show even fewer distinguishing characters after fossilisation and extraction than before. *Meesia tristicha* spores, which have now been recovered from two Flandrian deposits in England (p. 109), are about 30 μm in diameter and bear granules coarse enough to be readily visible with the light microscope (plate 11). The subfossil spores show that the granules present on the ripe spores are lost during fossilisation or during the chemical digestion of the peat. Spores from both localities exhibit a range from complete covering of granules to total lack, and there are intermediate states with bare patches. In this context it is worth mentioning that Tallis (1964) used potassium hydroxide preparations because acetolysis destroys the perine (outer layer) of *Sphagnum* spores.

Therefore it may be that many moss spores are overlooked in the subfossil state not only because they are small but also because they are smooth, if only as a result of fossilisation or chemical preparation.

Spores of *Polytrichum alpestre* (*c*. 6-9 μm diameter) and *P. commune* (*c*. 8-10 μm diameter) are worthy of mention. Seen with the light microscope, these tiny spores appear as almost smooth spheres. This is perhaps why, despite the abundant representation of fruiting plants in mire vegetation, there are no properly authenticated records of subfossil spores. Seen with the scanning reflection electron microscope, *Polytrichum* spores exhibit diagnostic features as yet unexploited by taxonomists (plates 12 and 13).

Somewhat different but also instructive is the case of *Aulacomnium palustre*, the macroscopic remains of which make it the commonest Pleistocene moss. This is also an abundant mire moss, but, unlike the two *Polytrichum* species, it bears sporophytes infrequently in the British Isles today. If this was also the case in the past, the lack of spore records will not be remarkable because, even if all the spores were preserved and recognisable, their occurrence would be very local and possibly their numbers small.

Some authors have claimed to have found spores of *Funaria hygrometrica*, the hemerophilous species often inhabiting sites of former bonfires. The problem here again is that the spores are small (plate 10) and the sculpturing, though distinctive, is fine (see plate IV in Erdtman 1957). A charcoal horizon indicating clearance by burning is the obvious place to search for *Funaria* spores, but even here precise recognition will be difficult.

There are some striking exceptions to the quantitative and qualitative paucity of moss spores. *Sphagnum* is the most obvious example. Abundantly present in oligotrophic peats, spores of this genus have long been recognised. Despite the possibilities of specific determinations as shown by various authors, especially Tallis (1962) (fig. 28), few workers have attempted more than generic identification. In any case, several of the main peat-forming species are easily recognisable as macroscopic remains.

Encalypta, a genus of several species in Europe, has long been known to have spores widely different from species to species. It is well known from macroscopic remains. However, few spore records have been published and none as yet from Britain. *E. rhabdocarpa*, which has verrucose spores, should turn up in European glacial

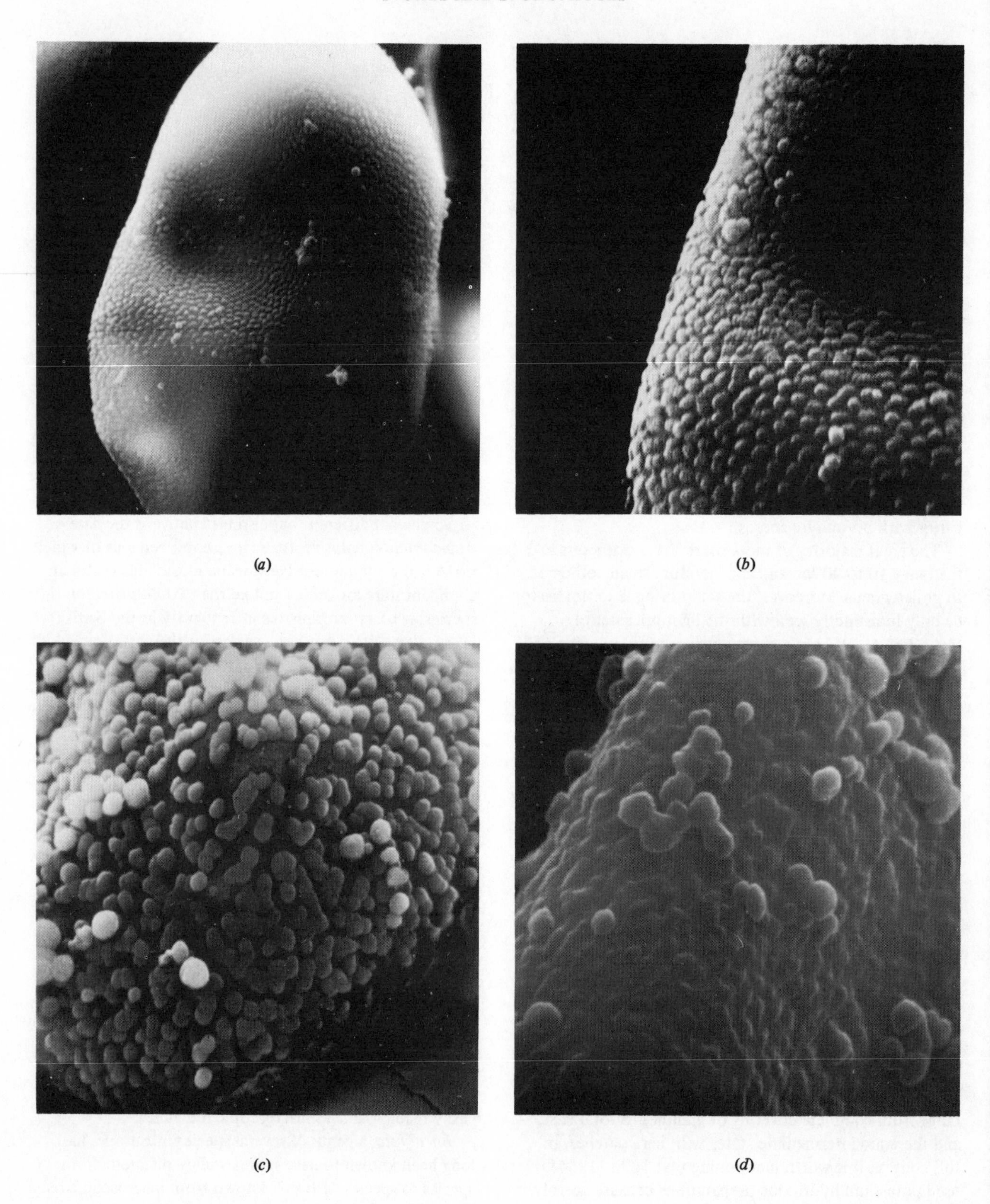

Plate 11. Scanning electronmicrographs of spores of Meesiaceae. (*a*) *Meesia tristicha*, Åbisko, Sweden. × 2300. (*b*) The same. × 5800. Detail of sculpturing. (*c*) *Meesia uliginosa*, Perthshire, Scotland. × 2700. Detail of sculpturing. (*d*) *Amblyodon dealbatus*, Lancashire, England. × 10,800. Detail of sculpturing.

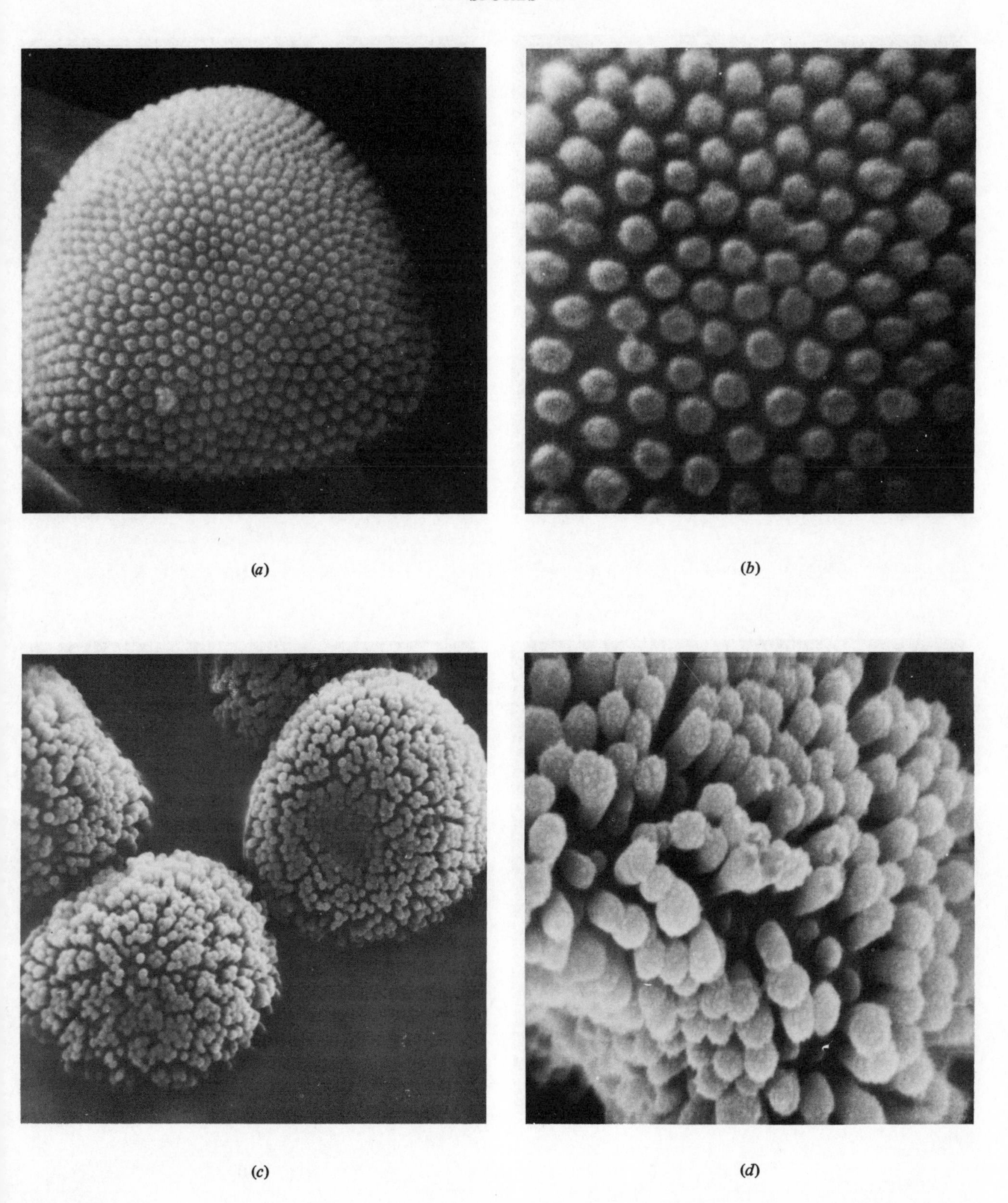

(a) (b) (c) (d)

Plate 12. Scanning electronmicrographs of *Polytrichum* spores. (*a*) *Polytrichum commune*, Norfolk, England. × 11,200. (*b*) The same. × 33,200. Detail of sculpturing. (*c*) *Polytrichum formosum*, Argyllshire, Scotland. Four spores. × 4650. (*d*) The same. × 14,100. Detail of sculpturing.

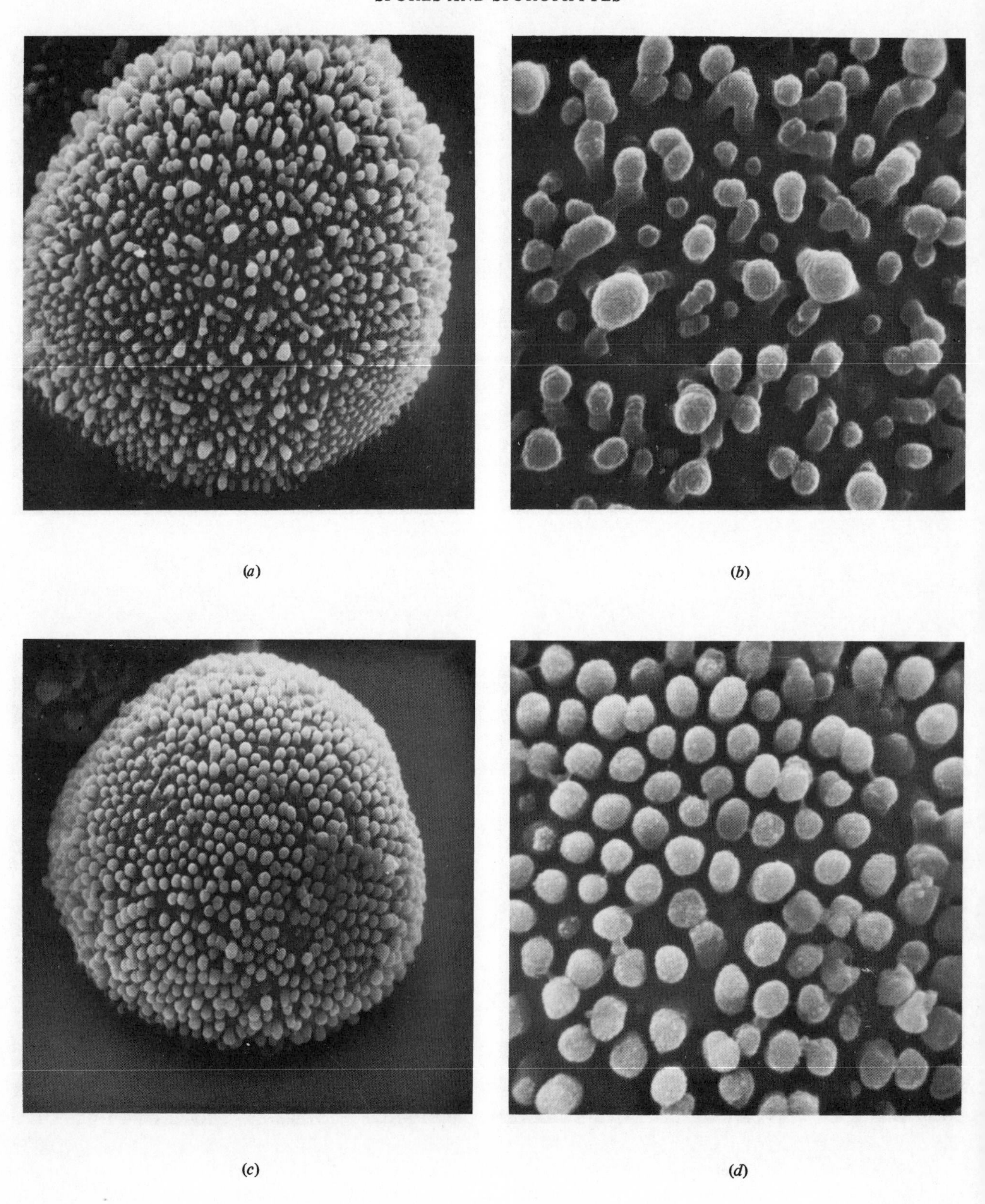

Plate 13. Scanning electronmicrographs of *Polytrichum* spores. (*a*) *Polytrichum longisetum*, Yorkshire, England. × 4050. (*b*) The same. × 14,100. Detail of sculpturing. (*c*) *Polytrichum alpinum*, Hordaland, Norway. × 4350. (*d*) The same. × 14,450. Detail of sculpturing.

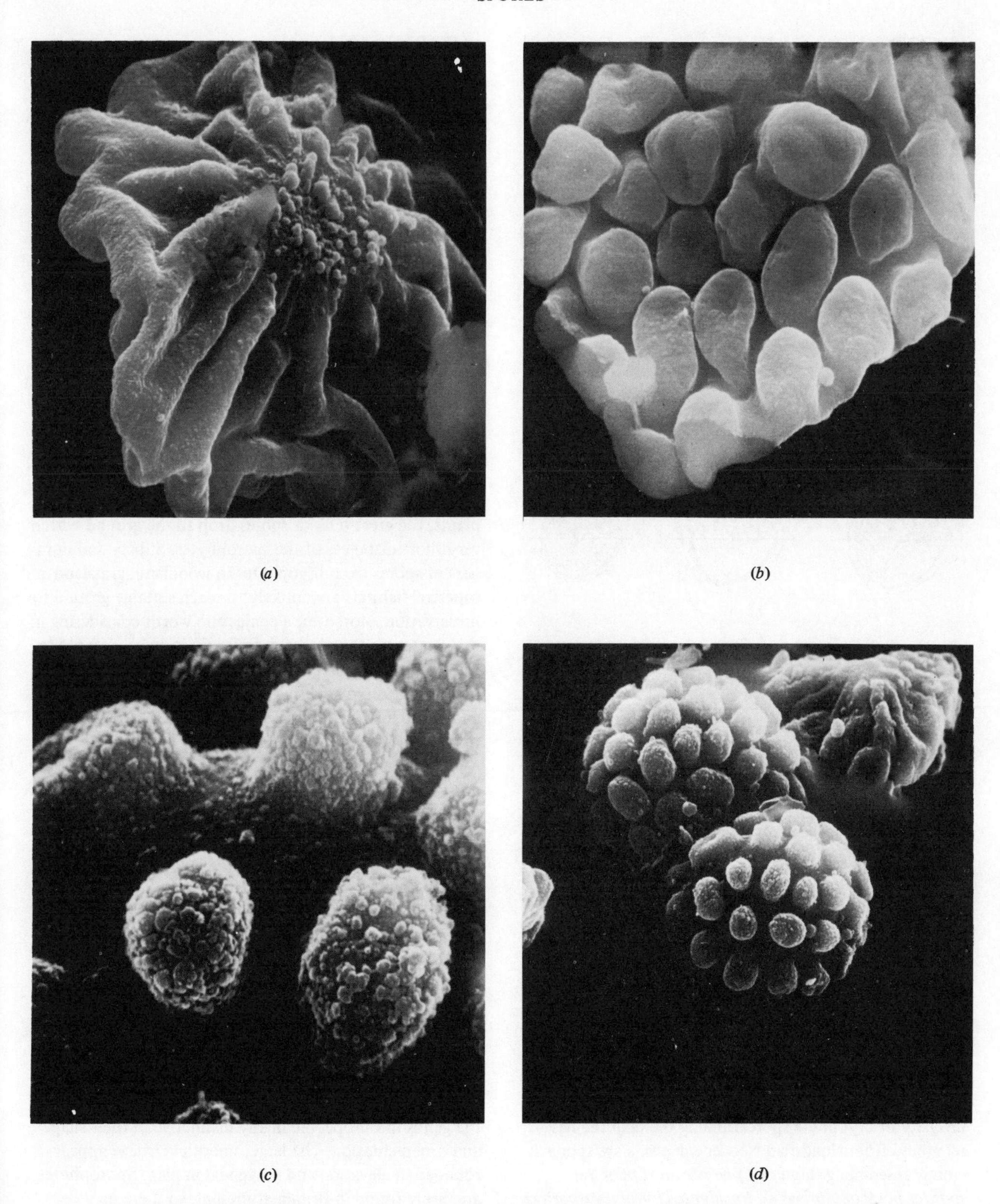

Plate 14. Scanning electronmicrographs of *Encalypta* spores. (*a*) *Encalypta vulgaris*, Tyrol, Austria. × 3800. Proximal face seen obliquely. (*b*) The same. × 3800. Part of distal face. (*c*) *Encalypta rhabdocarpa*, Valais, Switzerland. × 8700. Part of distal face. (*d*) The same. × 1700. Two spores showing distal faces, one showing proximal face very obliquely.

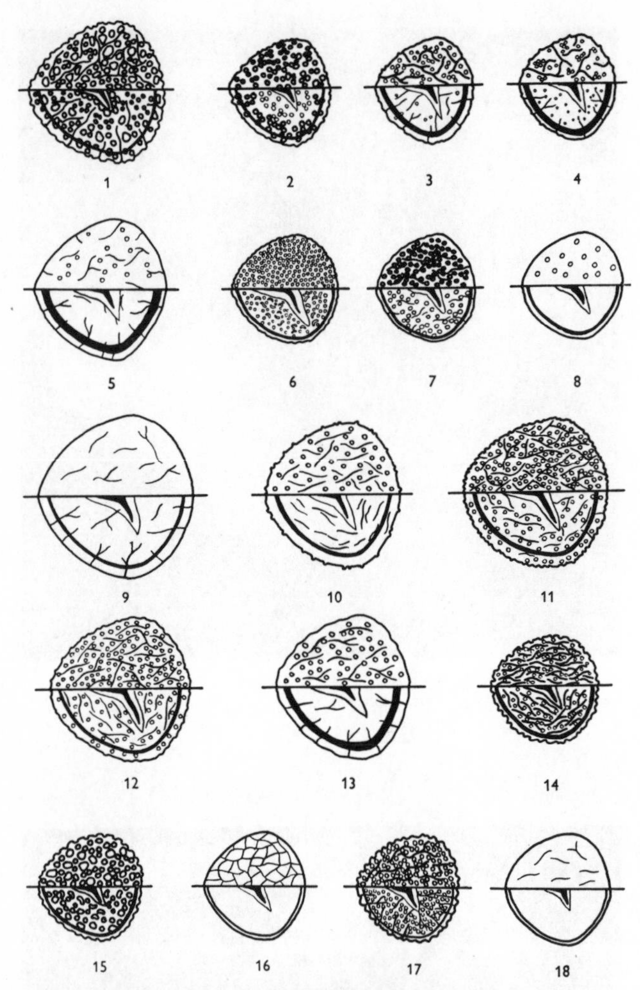

Fig. 28. Illustrations of *Sphagnum* spores. From Tallis 1962. The upper part of each drawing shows the distal face, the lower part the proximal face. 1. *S. papillosum* (spores *c.* 34-40 μm diam.); 2. *S. palustre* (*c.* 28-30 μm); 3. *S. imbricatum* (*c.* 28-30 μm); 4. *S. magellanicum* (*c.* 28-30 μm); 5. *S. compactum* (*c.* 38-40 μm); 6. *S. squarrosum* (*c.* 28-30 μm); 7. *S. teres* (*c.* 28-30 μm); 8. *S. flexuosum* (*c.* 28-30 μm); 9. *S. tenellum* (*c.* 45 μm); 10. *S. pulchrum* (*c.* 38-40 μm); 11. *S. cuspidatum* (*c.* 38-40 μm); 12. *S. subsecundum* (*c.* 38-40 μm); 13. *S. molle* (*c.* 34-40 μm); 14. *S. fimbriatum* (*c.* 28-30 μm); 15. *S. subnitens* (*c.* 28-30 μm); 16. *S. rubellum* (*c.* 34-36 μm); 17. *S. nemoreum* (*c.* 28-30 μm); 18. *S. fuscum* (*c.* 28-30 μm).

deposits (plate 14); Dr P. Colinvaux (personal communication) has recorded them from a Late Pleistocene deposit in Alaska. Plate 14 shows how strikingly varied are the spores of this genus.

The work of Nagy (1968) has revealed an exceptional diversity of bryophyte spores; among the taxa she has recovered from Hungarian Neogene deposits are spores closely resembling those of *Ephemerum* (*Ephemerisporites borsodensis* Nagy), *Encalypta* (*Encalyptasporites pliocenicus* Nagy) and *Sphagnum* (*Stereisporites*, a genus often recovered from Neogene deposits) and four species referred to the Anthocerotaceae and two species referred to the Ricciaceae.

It is no accident that the liverwort spores so far known as Pleistocene subfossils are both large and distinctive, just like those of the few moss taxa. *Anthoceros* and *Riccia* are both genera with large spores showing striking variation from species to species. The former is particularly well known from the Neogene and Pleistocene (interglacial and Flandrian). The distal and proximal faces of *Anthoceros* spores are markedly different in most species, including all the British taxa (plate 16). *Riccia* spores have so far eluded British palynologists but not the Dutch workers Koelbloed and Kroeze (1965), who found them in a late Flandrian deposit from the same horizons as spores of both *Anthoceros punctatus* and *A. laevis*.

Only a handful of genera have been mentioned above as being certainly recorded from Pleistocene deposits. However, too much of a mystery must not be made of this absence of bryophyte spores. It has to be remembered that the rain of bryophyte spores is light compared to that of pollen and, merely because of the diminutiveness of the plants, the great bulk of spores drop to the ground within very short distances of the sporophytes. Substantial numbers of spores from bryophytes in woodland, grassland and rupestral habitats are unlikely to reach suitable ground for preservation. Moreover, a conjecture worth considering in this context is that perhaps only a fraction of bryophyte spores are ever preserved because of ease of germination; on reaching a moist surface spores rupture in germinating and so are lost, or at least greatly altered, before fossilisation.

There are several bryophytes which on ecological grounds must have a considerable chance of fossilisation and which in addition have large, distinctive spores, as yet unknown as subfossils. For example, the species of *Physcomitrium*, inhabiting the muddy shores of lakes and ponds, have largely densely spinous spores (plate 15).

Less likely to be found as subfossils are *Oedipodium griffithanum* and *Breutelia chrysocoma*, although the spores are most remarkable (plate 15). On gross morphology the monotypic *Oedipodium* can be considered far removed from its allies; the spore serves to confirm this isolation. *Oedipodium* grows in earthy crevices of montane cliffs and some fortunate pollen analyst working on tarn deposits might find this spore.

The spores of *Breutelia chrysocoma*, the sole species of the genus in Europe, are highly distinctive in their shape and ornamentation. The large, smooth verrucae appear rounded in elevation and polygonal in plan. Sporophytes are rarely found in Britain at present, so there may be little chance of subfossil spores being discovered.

The spores of the fungus *Tilletia sphagni* Naw. have frequently been mistaken for spores of 'Bryales' or 'moss'

Plate 15. Scanning electronmicrographs of moss spores. (*a*) *Breutelia chrysocoma*, Argyllshire, Scotland. × 1200. Several spores in various views. (*b*) *Physcomitrium pyriforme*, Bedfordshire, England. × 2950. The spore has split. (*c*) *Oedipodium griffithianum*, Aberdeenshire, Scotland. × 1100. Tetrad showing three distal faces. (*d*) The same. × 2300. Proximal face seen very obliquely.

Plate 16. Scanning electronmicrographs of *Anthoceros* spores. (*a*) *Anthoceros laevis*, Sussex, England. × 1100. Proximal faces of two spores. (*b*) The same. × 1100. Distal face. (*c*) *Anthoceros punctatus*, Surrey. England. × 1200. Proximal faces of three spores. (*d*) The same. × 3300. Part of distal face.

and have even been referred to *'Polytrichum'* and *'Dicranum'*. *T. sphagni* (*Bryophytomyces sphagni* (Naw.) Ciferri) is a taxon of uncertain position despite the usually given generic name; probably it is a smut fungus but until spores can be germinated its precise relationships will remain obscure. In the living state the spores (plate 17) are found in *Sphagnum* capsules where they replace the spores of the moss; evidently this is a better known phenomenon on the European mainland than in Britain where the fungus has been reported only once (Walker 1948). In the subfossil state *T. sphagni* spores are commonplace in *Sphagnum* peats from a wide range of British localities; the frequency curves of the spores parallel those of *Sphagnum*. As yet the taxon is unknown from British interglacial deposits but Jessen and Milthers (1928) recovered it from three Eemian deposits in Denmark.

4.2 SPOROPHYTES

No liverwort sporophytes are known from Pleistocene deposits; this is much as would be expected from the delicate tissues and ephemeral life of the mature sporophyte. In contrast, moss sporophytes, much more rigid and longer lived structures, are frequently found.

Sphagnum capsules are commonplace in acid Flandrian peats and easily recognisable by their almost spherical shape; detached opercula are also frequent. Detached calyptras and opercula of many other mosses may be recovered and are often determinable at least to the genus. The shape of *Encalypta* calyptras is diagnostic (fig. 40), as is the particular hairiness of those of *Polytrichum.*

The range of form of moss sporophytes is considerable; some capsules are at once recognisable to the genus, if not to the species. The Late Devensian capsule of *Splachnum sphaericum* described by Conolly and Dickson (1969) is an outstanding example (plate 17).

Sporophytes of the following taxa are known from the British Pleistocene:

Bryum spp	*Grimmia apocarpa*
Distichium?	*Lescuraea patens*
Encalypta rhabdocarpa	*Neckera complanata*
Eurhynchium speciosum	*Polytrichum commune*
Fissidens bryoides	*Rhacomitrium* sp
Fontinalis antipyretica	*Sphagnum* spp
Funaria hygrometrica	*Splachnum sphaericum*

The list will increase in variety as work continues.

Abramova *et al.* (1965) have described sporophytes of nine genera from the Russian Pleistocene; *Encalypta vulgaris* and *Stegonia latifolia*, both tentatively identified, are two of the most noteworthy discoveries. Three years previously Abramova and Abramov published a record of subfossil hepatic elaters; seemingly this is an unique discovery.

The importance of recognising bryophyte spores and sporophytes is not merely to strive for completeness. Information regarding past reproductive biology may be gained, and thus light may be shed on the factors responsible for limitation of ranges. By far the best example so far is the discovery of spores of *Meesia tristicha* (p. 108).

5 THE BRITISH PLEISTOCENE RECORD

5.1 INTRODUCTION

This chapter is a compilation of all the discoveries of British Pleistocene bryophytes known to the author. It is certain that some records are overlooked. An exhaustive search of every journal, however obscure, devoted to local archaeology, geology and natural history would yield more, but not necessarily significant, information.

Advisedly, some types of records have not been collated. There would be little if any gain in gathering the vast number of *Sphagnum* spore records, or the records such as *'Hypnum'* as used by peat stratigraphers, which has scarcely any meaning in the modern context.

For each species the records are arranged by locality in ascending order of vice-counties, which are shown in Fig. 29. Next the age of the subfossil is given, followed by an expression of tentativeness if necessary, reference to the original publication, and finally the nomenclature of the original publication if different from that of this book.

The conventions adopted are as follows:

! specimen identified or confirmed by J. H. Dickson

An Anglian	Fl Flandrian
Ba Baventian	Ho Hoxnian
Be Beestonian	Ip Ipswichian
Cr Cromerian	Pa Pastonian
De Devensian	Pl Pleistocene
	Wo Wolstonian
E Early M Middle	L Late
ar archaeology	t tentative identification
pc personal communication	**T** all records need verification

* pollen zone assigned on archaeological/stratigraphical criteria

The discussion of each species includes points regarding identification, present British range and ecology, past range and its significance for present range, world distribution and non-British subfossils. Extended treatment is given to those species with many Pleistocene records which elucidate present chorology. Conversely there is little discussion of some species for which subfossils have little significance.

The nomenclature and generic concepts are those of the British Bryological Society Census Catalogues (Warburg 1963; Paton 1965*b*) except in the following cases:

Abietinella abietina (Hedw.) Fleisch. for *Thuidium abietinum* (Hedw.) B., S. & G.
Amblystegium includes *Leptodictyum* and *Hygroamblystegium.*
Calliergon for *Acrocladium.*
Calliergonella for *Acrocladium cuspidatum* (Hedw.) Lindb.
Homalia for *Omalia.*
Homalothecium for *Camptothecium.*
Hyocomium armoricum (Brid.) Wijk & Marg. for *H. flagellare* B., S. & G.
Mniaceae after Koponen (1968).
Lescuraea includes *Pseudoleskea patens* (Lindb.) Limpr.
Polytrichum longisetum Brid. for *P. aurantiacum* Sw.
Sphagnum after Isoviita (1966).
Thamnobryum for *Thamnium.*

The arrangement of genera and species follows the Census Catalogues except where species now extinct in the British Isles are interpolated and except for *Pseudoscleropodium* which is placed in the Brachytheciaceae, and *Isothecium*, placed in the Lembophyllaceae, following the Neckeraceae.

5.2 LIVERWORTS

Anthocerotales

Spores referred to various taxa of the Anthocerotales are well known from the Neogene of Germany and eastern Europe (Krutzsch 1963*a*, Stuchlik 1964, Nagy 1968) and there are many younger fossils, particularly from the British Isles and the Netherlands, if not elsewhere.

The taxonomic value of the spores of *Anthoceros* has been realised for many years. Of the three commonly accepted British species *A. punctatus* has spores easily distinguishable from *A. laevis* but not from the closely related *A. husnotii*, a taxon which is regarded as conspecific with *A. punctatus* by Proskauer (1958). These species have markedly ornamented spores (plate 16).

Anthoceros unidentified species
Barra (110); Fl VI; Blackburn 1946.

Fig. 29. Vice-counties of the British Isles: 1. West Cornwall with Scilly; 2. East Cornwall; 3. South Devon; 4. North Devon; 5. South Somerset; 6. North Somerset; 7. North Wiltshire; 8. South Wiltshire; 9. Dorset; 10. Isle of Wight; 11. South Hampshire; 12. North Hampshire; 13. West Sussex; 14. East Sussex; 15. East Kent; 16. West Kent; 17. Surrey; 18. South Essex; 19. North Essex; 20. Hertford; 21. Middlesex; 22. Berkshire; 23. Oxford; 24. Buckingham; 25. East Suffolk; 26. West Suffolk; 27. East Norfolk; 28. West Norfolk; 29. Cambridge; 30. Bedford; 31. Huntingdon; 32. Northampton; 33. East Gloucester; 34. West Gloucester; 35. Monmouth; 36. Hereford; 37. Worcester; 38. Warwick; 39. Stafford; 40. Shropshire; 41. Glamorgan; 42. Brecon (Brecknock); 43. Radnor; 44. Carmarthen; 45. Pembroke; 46. Cardigan; 47. Montgomery; 48. Merioneth; 49. Caernarvon; 50. Denbigh; 51. Flint; 52. Anglesey; 53. South Lincoln; 54. North Lincoln; 55. Leicester with Rutland; 56. Nottingham; 57. Derby; 58. Cheshire; 59. South Lancashire; 60. West Lancashire; 61. Southeast Yorkshire; 62. Northeast Yorkshire; 63. Southwest Yorkshire; 64. Midwest Yorkshire; 65. Northwest Yorkshire; 66. Durham; 67. South Northumberland; 68. North Northumberland (Cheviotland); 69. Westmorland with North Lancashire; 70. Cumberland; 71. Isle of Man; 72. Dumfries; 73. Kirkcudbright; 74. Wigtown; 75. Ayr; 76. Renfrew; 77. Lanark; 78. Peebles; 79. Selkirk; 80. Roxburgh; 81. Berwick; 82. Haddington (East Lothian); 83. Edinburgh (Midlothian); 84. Linlithgow (West Lothian); 85. Fife with Kinross; 86. Stirling; 87. West Perth with Clackmannan; 88. Mid Perth; 89. East Perth; 90. Forfar (Angus); 91. Kincardine; 92. South Aberdeen; 93. North Aberdeen; 94. Banff; 95. Elginshire (Moray); 96. Easterness; 97. Westerness; 98. (Main) Argyll; 99. Dumbarton; 100. Clyde Isles; 101. Kintyre (Cantyre); 102. South Ebudes; 103. Mid Ebudes; 104. North Ebudes; 105. West Ross; 106. East Ross; 107. East Sutherland; 108. West Sutherland; 109. Caithness; 110. Outer Hebrides; 111. Orkney; 112. Zetland (Shetland); H1. South Kerry; H2. North Kerry; H3. West Cork; H4. Mid Cork; H5. East Cork; H6. Waterford; H7. South Tipperary; H8. Limerick; H9. Clare (with Aran Islands); H10. North Tipperary; H11. Kilkenny; H12. Wexford; H13. Carlow; H14. Laois (Queen's County); H15. Southeast Galway; H16. West Galway; H17. Northeast Galway; H18. Offaly (King's County); H19. Kildare; H20. Wicklow; H21. Dublin; H22. Meath; H23. Westmeath; H24. Longford; H25. Roscommon; H26. East Mayo; H27. West Mayo; H28. Sligo; H29. Leitrim; H30. Cavan; H31. Louth; H32. Monaghan; H33. Fermanagh; H34. East Donegal; H35. West Donegal; H36. Tyrone; H37. Armagh; H38. Down; H39. Antrim; H40. Londonderry.

As Blackburn realised, the genus is unknown in the Outer Hebrides, therefore it is a pity that there was no greater precision.

Anthoceros punctatus L.

Amberley Wild Brooks (13); Fl VIII. *Frogholt* (15); Fl VII*b*; Godwin 1962. *West Runton* (27); Cr I and II. *Beetley* (28); Ip II. *Flagrass* (29); Fl VII*a*. *Godmanchester* (31); Fl VIII. *Fugla Ness* (112); Ho IV; Birks and Ransom 1969; *Baggotstown* (H9); Ho III; t; Dickson 1964*a*. *Gort* (H15); Ho III and III or IV; Jessen *et al.* 1959. *Vazon Bay* (C); Fl VII. *Grande Mare* (C); Fl VII and VII*b*; Churchill 1970 pc.

Seven of the eleven records were made by Miss R. Andrew, to whom I am indebted for the information.

The records must be viewed in the light of Proskauer's opinion of the status of *A. husnotii.* Otherwise they would best be given as '*A. husnotii* or *A. punctatus*'. *A. punctatus* stands alone in the Bryophyta in the detail of its history based on spore records which extend back to the Cromerian and include the two later interglacials as well as the Flandrian.

Several Flandrian records of *Anthoceros* from the Netherlands are given by Koelbloed and Kroeze (1965) who found spores not only of *A. punctatus* but also of *A. laevis* and *Riccia* together in the same post-Atlantic horizons at Spoolde-Zuid (fig. 30). Rightly in this context they regard the spores as indicators of agriculture. Quite apart from the association with pollen of such species as *Centaurea cyanus, Plantago* and cereals, the present-day habitats justify such an assessment. The *Anthoceros* species discussed here often occur in stubble fields. The pollen analysis and radiocarbon assay of detritus mud from Frogholt in Kent (Godwin 1962) revealed *A. punctatus* as present continuously for 500 years (fig. 31). Pollen and macroscopic fossils are consistent with the habitat having been marshy pasture.

Interglacial occurrences can hardly pertain to anthropogenic vegetation; clay banks, marshy ground, streamsides and even dunes are some of the natural habitats.

In the British Isles *Anthoceros* species have scattered distributions of southern tendency. None is present in the Outer Hebrides, Orkneys or Shetlands; Irish records are sparse.

It is possible that the ranges were less patchy, at least in the south, in the last 5000 years, because some of the fossils show occupation of areas now abandoned such as East Kent and Huntingdonshire. Both *A. punctatus* and *A. husnotii* are absent from the Channel Isles, hence two Flandrian occurrences are of some interest.

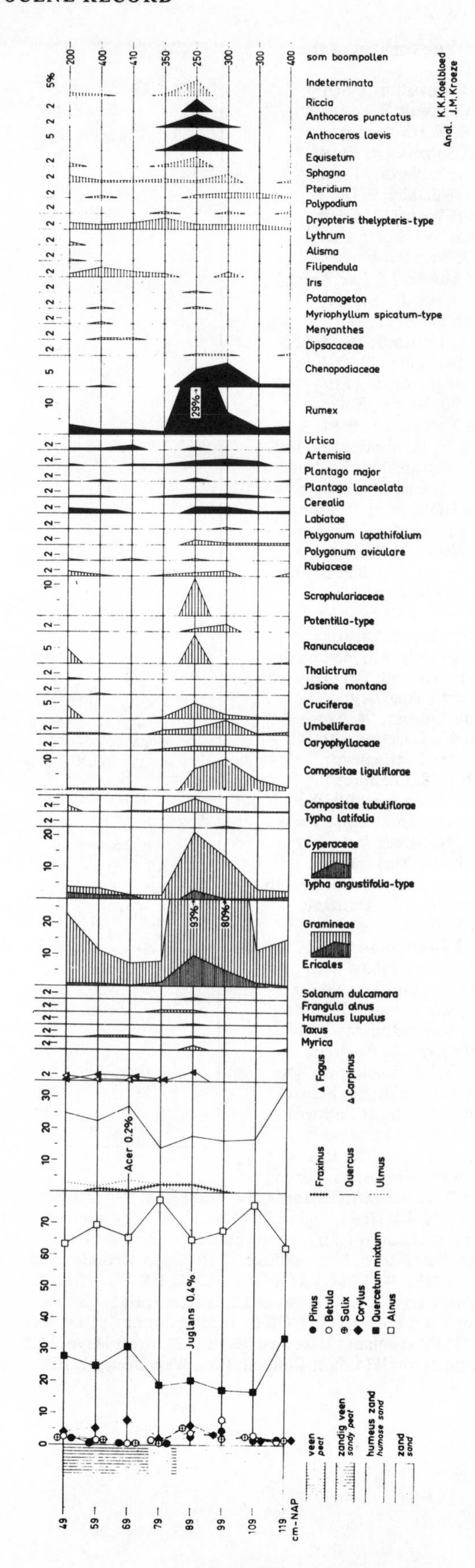

Fig. 30. Pollen diagram from Spoolde-Zuid, the Netherlands. From Koelbloed and Kroeze 1965.

Marchantiales

Lunularia cruciata (L.) Dum.

Brigg (54): Fl VII*b**; Bronze Age ar; Sheppard 1910.

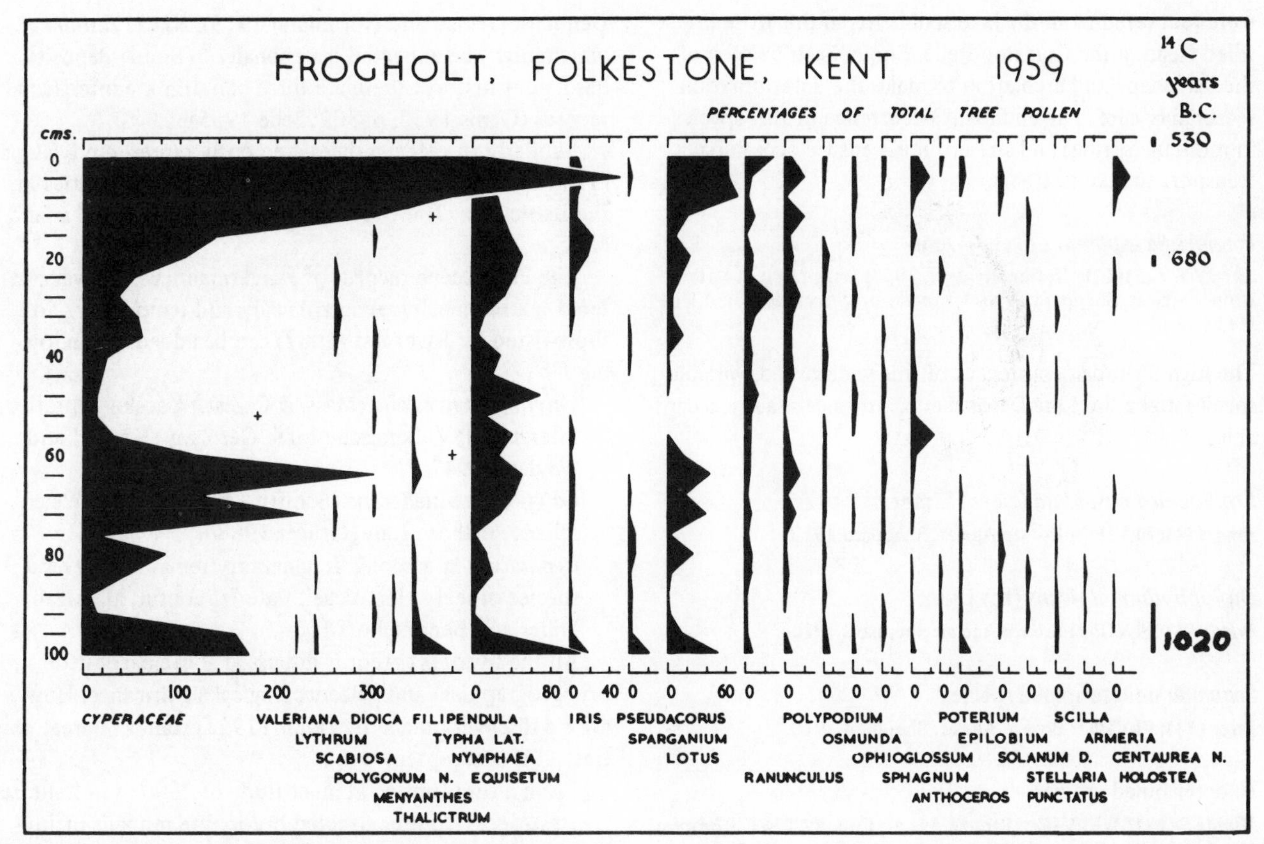

Fig. 31. Pollen diagram from Frogholt, Kent. From Godwin 1962.

Lunularia was one of ten species of liverworts and twelve of mosses used as caulking material for a monoxylous boat found at Brigg in Lincolnshire (p. 187). Because of its marked abundance in artificial habitats such as greenhouses and paths *Lunularia* has been called in question as a native species. This Bronze Age record puts paid to such a notion, unless, of course, the canoe was caulked outside the British Isles; nothing in the material suggests this, though *Distichum capillaceum* is unknown in Lincolnshire at present.

Metzgeriales

Riccardia undetermined species

Brigg (54); Fl VII*b**; Bronze Age ar; Sheppard 1910 as *Aneura* sp.
Seaforth (59); Fl VII/VIII*; Travis 1913.

Riccardia latifrons (Lindb.) Lindb.

West Craigneuk (77); De only; t; Dixon 1907 as *Aneura latifrons.*

Pellia epiphylla (L.) Corda

Brigg (54); Fl VII*b**; Bronze Age ar; Sheppard 1910.

Metzgeria furcata (L.) Dum.

Brigg (54); Fl VII*b**; Bronze Age ar; Sheppard 1910.

Jungermanniales

Lepidozia reptans (L.) Dum.

Brigg (54); Fl VII*b**; Bronze Age ar; Sheppard 1910.

T *Calypogeia trichomanis* (L.) Corda emend. Buch

Brigg (54); Fl VII*b**; Bronze Age ar; Sheppard 1910 as *Kanta trichomanis* (L.) G. & B.

The specific identity of this subfossil must be called in question in view of the confusion revealed by Paton's study of the British material of the genus (1962).

Leiocolea badensis (Gottsche) Jorg.

Seaforth (59); Fl VII/VIII*; Travis 1913 as *Lophozia badensis.*

Solenostoma triste (Nees) K. Müll.

! *New Dry Dock* (76); LDe; t.

Three small fragments are referred tentatively to this species on grounds of leaf shape and areolation. There are no perianths.

Solenostoma cordifolium (Hook.) Steph.

! *Cove Moss* (69); Fl only; t.

Abundant but poorly preserved fragments of a liverwort

were recovered from the lake sediments in this little infilled basin in the Coniston Fells. Enough can be seen of the leaf shape and areolation to make the determination reasonably sure. The abundant occurrence of this species in montane springs and streams must ensure occasional transport to lake bottoms.

Plagiochila asplenioides (L.) Dum.

Hawks Tor (2); LDe II; Conolly *et al.* 1950. *Brigg* (54); Fl VII*b**; Bronze Age ar; Sheppard 1910. ! *Winetavern Street* (H21); Fl VIII*; Medieval ar.

The high altitude occurrences of this widespread, variable species make the Late Devensian occurrence readily acceptable.

Lophocolea cuspidata (Nees) Limpr.

Brigg (54); Fl VII*b**; Bronze Age ar; Sheppard 1910.

Diplophyllum albicans (L.) Dum.

Brigg (54); Fl VII*b**; Bronze Age ar; Sheppard 1910.

Scapania undetermined species

Brigg (54); Fl VII*b**; Bronze Age ar; Sheppard 1910.

Undetermined species

Minnis Bay (15); Fl VII*b**; Bronze Age ar; Conolly 1941. ! *Broome Heath* (27); Wo. *Paisley* (76); LDe only; '*Jungermannia*'; Mahony 1868*b*.

The subfossil from Broome Heath is minute (*c*. 1 mm) and in a very bad state of preservation. Those very few cells of the remnants of leaves which can be seen at all clearly have the appearance of hepatic rather than moss cells.

As mentioned in chapter 3, it is clear that there is no real hope that the Neogene and Pleistocene history of the liverworts, in contrast to that of the mosses, will ever be well known. The tiny representation of some fifteen species listed above gives scarcely any guide to the distribution, numbers, and abundance of British hepatics in the Pleistocene. This is much to be regretted in view of the many, varied geographical patterns shown by the liverworts.

The occurrence of Pleistocene remains of liverworts is just as sparse and erratic in other countries as in the British Isles. However, various species, unknown from the British Pleistocene, deserve brief comment.

Amongst the few members of the Marchantiales which have been found in the subfossil state *Marchantia* itself has several records. There is a Pliocene record of the genus from Frankfurt (Gams 1932, p. 303) and at least three of *M. polymorpha*; the oldest is from the penultimate interglacial at Kalisz, Poland (Tolpa 1952), the next from the last interglacial at Ejstrup, Denmark (table 20) and the youngest from a Late Weichselian site at Norre Lynby, Denmark (Jessen and Nordmann 1915). *Riccia fluitans*, which might be expected occasionally in limnic deposits, has a poor history: there are three penultimate interglacial records (Gams 1932, p. 309; Selle 1955).

Similarly the Metzgeriales are poorly represented. Gams (1932) states that *Riccardia pinguis* has three localities in the Baltic area. There seem to be few records of *Pellia* and *Metzgeria*.

The Pleistocene records of Jungermanniales are very scattered taxonomically, geographically and temporally. To those listed by Jovet-Ast (1967) can be added the following:

Calypogeia neesiana (Mass. & Carest.) Loeske; tentative; interglacial; Vulkanische Eifel, Germany; Krausel and Weyland 1942.

Calypogeia muellerana (Schiffn.) K. Mull.; Late Weichselian; northern Italy; Grüger 1968*b*.

Cephalozia sp; among 'fragmentary remains of several species of leafy Hepaticae'; Late Wisconsin; Michigan; Miller and Benninghoff 1966.

All the liverwort records discussed above carry little bryogeographical and palaeoecological significance. However, a discovery made by Douin (1923) stands in great contrast, if it is accepted.

From a tufa formed at an altitude of 2000 m in Lautaret, Hautes Alpes, Douin extracted bryophyte remains including a single small fragment which he claimed as *Plagiochila carringtonii* (*Jamesoniella carringtonii*). This is an oceanic species known only from the western parts of the British Isles (fig. 6), the Faeroe Islands and Nepal (Grolle 1964).

The subfossil was regarded as a new variety (var. *alpina*) by Douin, who also recognised five taxa of mosses and four other liverworts, the latter being *Lophozia hornschuchiana, Plagiochila asplenioides, Pedinophyllum interruptum* and *Solenostoma triste.* Though the fragment shows no cell structure because of the calcareous encrustation Douin confidently asserts that it must be *Plagiochila carringtonii*. Not only because the leaves are less decurrent and less reniform than those of the typical plant but also because of the calcareous conditions the new variety was erected.

Until the specimen is re-examined, it is best to regard Douin's identification as doubtful; the morphological and ecological differences from *Plagiochila carringtonii* serve to make one seek another species before accepting an identification which holds great bryogeographical import (p. 205).

5.3 MOSSES

Sphagnales

Sphagnaceae

The large genus *Sphagnum*, with forty-three species in Europe according to Isoviita (1966), has thirty British species including the presumed extinct *S. obtusum* (Warburg

1963) and also *S. majus* (*S. dusenii*), recognised a few years ago by Maass (1965). It differs from all other bryophyte genera in the vast profusion of spores recorded from Pleistocene deposits, particularly oligotrophic peats in which macroscopic remains of *Sphagnum* are often major components.

About thirty of the European species have Pleistocene records; eighteen or nineteen of those from Britain are known in the subfossil state, some abundantly such as *S. imbricatum* and *S. cuspidatum*. Not all the records have been listed here. Even if they were they would represent a gross underestimate of the occurrences in the Flandrian ombrogenous peats of northern and western areas. Other species have been identified only a few times or merely once, such as *S. russowii* and *S. molle.*

The few cymbifolian species (subgenus *Inophloea*, section *Sphagnum* or section *Palustria*) are readily recognisable in peats by their robustness and obtuse, cucullate leaves (plate 7), just as they are in the living state. If preservation is good, microscopically they present few difficulties, especially *S. imbricatum* with its comb fibrils.

Some of the numerous remaining species (subgenus *Litophloea*, all other sections) are also easily determinable if the subfossil material is good enough. However, often all one has to deal with are small branches or detached branch leaves which may only allow assignment to a section such as *Cuspidata* or *Acutifolia*, both of which contain several species of considerable ecological diversity. Hence, the reward may not match the effort.

The taxonomic treatment of the Fennoscandian species by Nyholm (1969) includes all the British species. Also readily accessible are the keys by Fearnsides (1938), Proctor (1955) and Duncan (1962). The recent key by Cöster and Pankow (1968) is abundantly furnished with photomicrographs.

Several of the British species unknown as yet in the subfossil state have restricted distributions. For example, *S. riparium, S. warnstorfii* and particularly *S. lindbergii* have markedly northern ranges which may prove to be relict.

Section *Sphagnum*

Sphagnum palustre L.

Dogger Bank; Fl IV-V; t; Whitehead and Goodchild 1909 as *S. cymbifolium.* ! *Honeygore Track* (6); Fl VII*b*, ! *Chilton* (6); Fl VII*a*, VII*b*. ! *Bell Track* (6); Fl VII*b*. *Weymouth* (9); Fl only; Gepp 1895 as *S. cymbifolium* Ehrh. *Cranes Moor* (11); Fl V, VI; Seagrief 1955. ! *Hockham Mere* (28); Fl VI and VII*a*. *Wood Fen* (29); Fl VII*a*, VII*b*; Godwin *et al.* 1934. *Wybunbury Bog* (58); Fl VII; Poore and Walker 1958; Fl VII and VIII; Green 1963 pc. *Chat Moss* (59); Fl IV-V; Birks196 5*a*. *Bridgewater Street* (59); Fl VIII*; Roman ar; Roeder 1899 as *S. cymbifolium. Ladybridge Slack* (62); Fl VII*b*; Simmons 1968 pc. *Moss Swang* (62); Fl VII*b*; Simmons 1968 pc. *Tadcaster* (64); Fl VII*a*; Bartley 1962. *Fox Earth Gill* (65); Fl VII*a*; Squires 1968 pc. ! *Seathwaite Tarn* (69); Fl VIII. *Loch Dungeon* (73); Fl V or VI; H. H. Birks 1969 pc. *Snibe Bog* (73); Fl VII*b*; H. H. Birks 1969 pc. *Renfrew* (76); Fl IV or V? ; Thompson 1968 pc. *Methilhill* (85); Fl VII*b*; Bronze Age ar; Lambert 1964. *Canna* (104); Fl VII*b* or VIII; Flenley and Pearson 1967. *Loch Mealt* (104); LDe II; Birks 1969 pc. *Gort* (H15); Ho II and III; Jessen *et al.* 1959. *Roundstone 2* (H16); Fl VII*a*; Jessen 1949. *Derrybrennan* (H19); Fl VI? ; Synnott 1970 pc. ! *Newtonbabe* (H31); LDe II.

Unrecorded from very few vice-counties, *S. palustre* is one of the most widespread *Sphagnum* species in the British Isles where it occurs in wet woodland, mires and streamsides.

The great bulk of records belong to zone VII*a* and later, but there are Late Devensian records from the Isle of Skye and Co. Louth. *S. palustre* has often been recovered from Pleistocene deposits outside the British Isles; there is a Pliocene record from Poland (table 10).

S. palustre is a bipolar species widespread in the Northern Hemisphere, slightly suboceanic (Isoviita 1966), and southern in Fennoscandia.

Sphagnum magellanicum Brid.

Woodwalton Fen (31); Fl VII*b*; Godwin 1938. *Wybunbury Bog* (58); Fl VII or VIII; Green 1963 pc. *Chat Moss* (59); Fl IV-V; Birks 1965*a*. *Dean Head Hill* (63); Fl VIII; spores; Tallis 1964. *Hutton Henry* (66); Ip II; Beaumont *et al.* 1969. *Abbott Moss* (70); Fl VII*b* and/or VIII; Walker 1966. *Loch Dungeon* (73); Fl V or VI; H. H. Birks 1969 pc. *Gort* (H15); Ho II or III; t; Jessen *et al.* 1959. ! *Derrycunlaugh* (H16); Fl IV or V. *Fallahogy 2* (H40); Fl VIII; Walker and Walker 1961.

Though most abundant in the highland zone *S. magellanicum* is also known from southern and eastern England. It is a species of oligotrophic mires, usually ombrogenous bogs. Wider occupation of the lowland zone is indicated by the Huntingdonshire subfossil. This disappearance from the western Fenland margin may be anthropogenic.

Apart from the many records from Flandrian peats outside the British Isles, there is a Polish last interglacial one (Tolpa 1952). However, the oldest occurrence is the tentative one from Gort. *S. magellanicum* is a bipolar species widespread in the Northern Hemisphere.

Sphagnum papillosum Lindb.

! *Stannon* (2); LDe only and Fl VII*b* and/or VIII. *Cranes Moor* (11); Fl V and VI; Seagrief 1955. *Hurston Warren* (13); Fl VIII*; Rose 1953. *Ockley Bog* (17); Fl VIII*; Rose 1953. *Chartley Moss* (39); Fl VII and VIII; Green 1964 pc. *Cwm Idwal* (49); Fl IV-V; Godwin 1955. *Nant Ffrancon* (49); LDe only. *Bettisfield* (51); Fl VIII; Hardy 1939. *Wybunbury Bog* (58); Fl VII and VIII; Green 1963 pc. *Chat Moss* (59); Fl IV-V; Birks 1965*a*. *Dean Head Hill* (63); Fl VIII; spores: Tallis 1964. *Wissenden* II (63); Fl VIII; spores; Tallis 1964. ! *Seathwaite Tarn* (69); Fl VIII. *Foolmire Sike* (69); VI to VIII. ! *Russland Moss* (69); Fl only; Dickinson 1968 pc. ! *Blelham Tarn* (69); Fl IV-V. *Scaleby Moss* (70); Fl VII*b* and/or VIII; t; Walker 1966. *Abbot Moss* (70); Fl VII*b* and/or VIII; Walker 1966. *Snibe Bog* (73); Fl VII*b*; H. H. Birks 1969 pc. *Long Range* (H2); Fl VIII; Jessen 1949. *Gort* (H15); Ho II, III and III or IV; Jessen *et al.* 1959.

Castlelackan Desmesne (H27); Fl VIII; Jessen 1949. *Raheelin* (H29); Fl VII*b*; Jessen 1949. *Ardlow Inn* (H30); Fl VIII; Jessen 1949. *Fallahogy* (H39); Fl VII and VIII; Smith 1958*a*.

The only major absence of *S. papillosum* in the British Isles occurs in the eastern Midlands and East Anglia. Apart from its abundant occurrence in oligotrophic mires the species is also found in such habitats as rock ledges and wet heaths.

The two Late Devensian records from North Wales and Cornwall testify to a wide range, at least in western Britain, from the earliest Flandrian times. Many of the later records derive from raised bog peats but the Stannon, Cwm Idwal, Blelham Tarn and Seathwaite Tarn remains cannot have such an origin and probably relate to the rock ledge or poor fen habitats.

S. papillosum, a species with an extensive, slightly sub-oceanic Holarctic range, has often been extracted from Flandrian peats in the rest of Europe (e.g. Aletsee 1959; Grosse-Brauckmann 1962, 1963, 1968).

Sphagnum imbricatum Hornsch. ex Russ.

! *Stannon* (2); LDe only. ! *Meare Track* (6); Fl VII*b* or VIII. ! *Paston* (27); Pa. ! *Mundesley* (27); EAn. *Chartley Moss* (39); Fl VII and VIII; Green 1964 pc. ! *Nant Ffrancon* (49); LDe only. *Bettisfield* (51); Fl VIII; Hardy 1939. *Ringinglow* (57); Fl VIII; Conway 1954. *Danes Moss* (58); Fl VII and VIII; Birks 1967 pc. *Wybunbury Bog* (58); Fl VII and VIII; Green 1963 pc; Birks 1967 pc. *Flaxmere* (58); Fl VIII; Birks 1967 pc. *Hatchmere* (58); Fl VIII; Birks 1967 pc. *Carrington Moss* (58); Fl VII and VIII; Birks 1967 pc. *Lindow Moss* (58); Fl VII and VIII; Raistrick and Blackburn 1932; Birks 1965*b*. *Holcroft Moss* (59); Fl VII*b* and VIII; Birks 1965*b*. *Chat Moss* (59); Fl IV-V to VIII; Birks 1965*a*. *Risley Moss* (59); Fl VII and VIII; Birks 1967 pc. *Hatfield Moss* (63); Fl VIII; Smith 1958*c*. *Snake Pass* (63); Fl VII*b* and VIII; Tallis 1964. *Linton Mires* (63); Fl VII*a*; Raistrick and Blackburn 1938. *Malham Tarn Moss* (64); Fl VII*a*, VII*b* and VIII; Pigott and Pigott 1963. ! *Hutton Henry* (66); Ip III; forma *affine* (Ren. & Card.) Warnst.; Beaumont *et al.* 1969. *Foulshaw Moss* (69); Fl VIII; Smith 1959. *Shoulthwaite Moss* (69); Fl VIII; Pennington 1969 pc. *Rigg Beck Upper Bog* (69); Fl VII; Pennington 1969 pc. *Helsington Moss* (69); Fl VIII; Smith 1959. *Nichols Moss* (69); Fl VIII; Smith 1959. *Bowness Common* (70); Fl VIII; Walker 1966. *Snibe Bog* (73); Fl V or VI, VII*a* and VII*b*; H. H. Birks pc. *Moss of Cree* (74); Fl VII or VIII; Lewis 1905. ! *Bloak Moss* (75); Fl VIII. ! *Flanders Moss* (87); Fl VI, VII*a*, VII*b* and VIII. *Long Range* (H2); Fl VIII; Jessen 1949. *Dromsallagh* (H10); Fl VII*b* and VIII; Mitchell 1956. *Cloonmoylan 2* (H15); Fl VIII; Walker and Walker 1961. *Gort* (H15); Ho III; Jessen *et al.* 1959. *Roundstone 2* (H16); LDe II and Fl V and VII*a*; Jessen 1949. *Pollagh Bog* (H18); Fl VIII; Moore 1955. *Timahoe* (H19); Fl VII; Synnott 1970 pc. *Canbo* (H25); Fl VIII; Mitchell 1956. *Castlelackan Desmesne* (H27); Fl VIII; Jessen 1949. *Ardlow Inn* (H30); Fl VIII; Mitchell 1956. *Derrytagh North* (H37); Fl VIII; Jessen 1949. *Ballyscullion Bog* (H39); Fl VIII; Jessen 1949. *Cloughmills* (H39); Fl VIII; Jessen 1949. *Fallahogy* (H40); Fl VIII; Smith 1958*a*.

S. imbricatum is familiar to peat stratigraphers because of the readily observed comb fibrils of the hyaline cells. However, this diagnostic feature may not be developed (forma *affine*) as in the Hutton Henry material; the broadly triangular shape of the green cells is also diagnostic.

S. imbricatum has a long, if intermittent, history in the British Isles. Perhaps the Oligocene record discussed on p. 174 represents this species. The earliest certain record is Pastonian, followed by an Early Anglian one and that from the Hoxnian at Gort. The latter is paralleled by a discovery from the penultimate interglacial in Russia (Abramova and Abramov 1967*c*).

This species has occasioned more comment from British ecologists than any other species considered in this book. The interest stems from the abundance of remains in ombrogenous peats of zone VII*a* and particularly of zones VII*b* and VIII, and on the present scarcity or absence from the mires. It is clear that *S. imbricatum* had a high cover value on the mires, as it still has on some undamaged bogs of the north and west.

Figure 32, originally published by Pigott and Pigott (1963), shows the past and present distribution of *S. imbricatum* in the British Isles. Many more living localities could now be added (in the north and west including a single locality on Dartmoor, Devon). Numerous subfossils can be added as well (again mostly in the north and west but also in the south Lancashire-Shropshire-Staffordshire area). The picture would be augmented but not transformed; there has been a marked contraction towards the north and west.

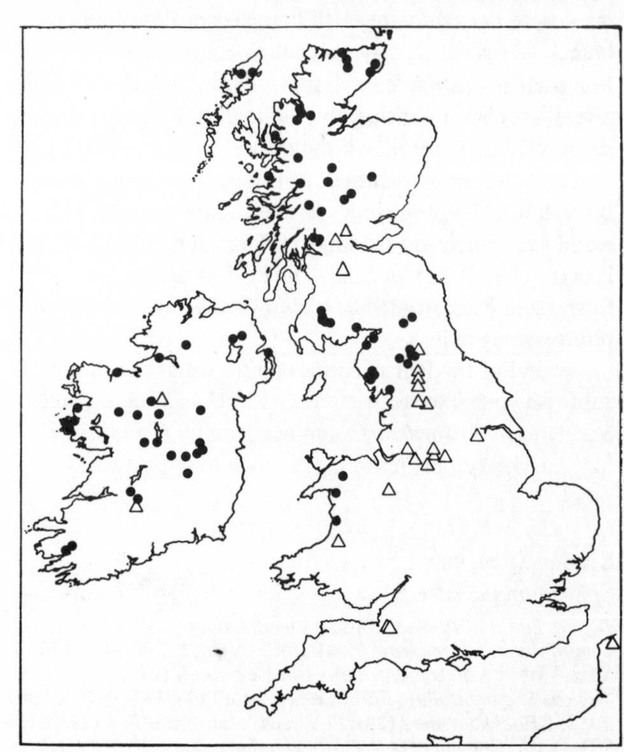

Fig. 32. British range and Flandrian subfossils of *Sphagnum imbricatum*. From Pigott and Pigott 1963. The open triangles show Flandrian subfossils.

The Flandrian history of the species is similar over much

of northwestern Europe, such as the Netherlands and Belgium where there are no living occurrences but remains are well known from Sub-Boreal and Sub-Atlantic peats, and also northwestern Germany where the range is markedly coastal (Schumacher 1958). In Norway the range is both strongly southern and coastal (map in Lid 1925).

At present in the British Isles, *S. imbricatum* is a locally abundant plant of blanket and raised bogs in the highland zone; in England it occurs only in the far north (fig. 32). It can form very large hummocks conspicuous not only by size but also by the rich brown colour, as in the bogs of central Ireland (Bellamy and Bellamy 1966). However, it is not merely a species of hummocks in ombrogenous bogs. As emphasised by Green (1968) its tolerance includes mesotrophic, even perhaps eutrophic conditions. It is capable of forming extensive lawns, as it did in the past (Green 1968) at Wybunbury Bog, Cheshire, and elsewhere in the British Isles and perhaps also at Draved Kongsmose, Denmark, where according to Hansen (1966) 'probably the surface was slightly undulating without significant hummock-hollow topography'. A 'lax low hummock form' was found to be abundant in a flush bog in Galloway (Ratcliffe and Walker 1958, p. 413). A similar low-growing form occurs at the celebrated locality of *Scheuchzeria palustris* on the Moor of Rannoch, at Loch Morlich in Rothiemurchus Forest, and no doubt elsewhere.

Apart from Early Pleistocene records, *S. imbricatum* has occurred in the British Isles during the last two interglacials. Much more interest attaches to the three Devensian records, one from western Ireland, one from Cornwall and one from North Wales. The Irish record, made by Jessen (1949), I have been tempted for long to consider a contaminant. The species grows on the surface of the blanket peat which was sampled by a Hiller borer and also was a component of the zone V peat. However, no doubts can attach to the Welsh record. Mr C. Burrows recovered several leaves from the Late Devensian layers of a carefully sampled 4-inch core. The Late Devensian occurrence in western Britain is thus established. And, of course, there is nothing inherently improbable in a west of Ireland population at those times.

What the Late Devensian habitats may have been is an interesting speculation. There is little evidence of bog development (Dickson *et al.* 1970), though there may well have been poor and intermediate fens. One would like to know more of its high latitude occurrences in Russia and North America; from Umiat, Alaska (69° 20′ N) it was found with *S. balticum* in a 'marsh' (Persson 1949). Another possibility is occurrence on rock by waterfalls as in the Southern Appalachians at present (Billings and Anderson 1966). This might have happened at Nant Ffrancon, though I know of no such occurrences in Britain at present.

There is confusion in the British ecological literature about the range of *S. imbricatum.* Neither is the distribution 'very oceanic' nor 'more or less cosmopolitan' as two contrasting opinions have it. *S. imbricatum* is both amphi-atlantic and amphipacific as well as bipolar (in Chile). The isolated localities at the River Lena region (Isoviita 1966), the mouth of the River Ob (Tyuremnov 1963) and the Arctic coast of North America (as at the Mackenzie Delta, K. Holmen 1967 pc) have to be borne in mind when the range is assessed as suboceanic (Isoviita 1966), as it clearly is in Europe at present. The map reproduced here (fig. 33, from Tyuremnov 1963) is deficient and inaccurate in detail but gives a general impression of the range.

The most southerly subfossils in the British Isles are from the derelict mires of the Somerset Leveis where the species is abundant in the oligotrophic peats. The species is unrecorded from the blanket peats of Cornwall. Perhaps it never has occurred there. The same may be said of the eastern Fenland margin; the ombrogenous peats of Holme Fen and Woodwalton Fen have yielded various *Sphagnum* species but not *S. imbricatum.* Perhaps this observation can be extended to the whole of southeastern and south-central England where the species has never been recorded living in valley bogs, nor in the subfossil state, as the admittedly sparse stratigraphic records show.

A climatic interpretation of the decline of *S. imbricatum* was reiterated by Godwin (1956) who envisaged increasing dryness as the responsible factor; earlier Godwin and Conway (1939) had discussed in similar terms the disappearance from Tregaron Bog, where remains come within a few centimetres of the surface but no living plant could be found.

Most authors in recent years have favoured an anthropogenic interpretation. Pigott and Pigott (1963), primarily considering Malham Tarn Moss, have summed up the north of England disappearance as follows:

> *Sphagnum imbricatum* became abundant at the opening of Zone VII at Malham and is the main constituent of the peat of the Tarn Moss until Zone VIII, when the area it occupied began to diminish. Its decline began on the marginal sloping parts of the moss where early in Zone VIII it was replaced by *Eriophorum vaginatum*, but it must have survived until quite recently on the central part as remains reach within a few centimetres of the present surface. Most of the deeper blanket bogs in the north Pennines show a similar abundance of *Sphagnum imbricatum* almost to their present surface and on Widdy Bank in Teesdale, for example, a few hummocks still survive associated with seven other species of *Sphagnum*, including *S. fuscum.* The bog surface in this locality is

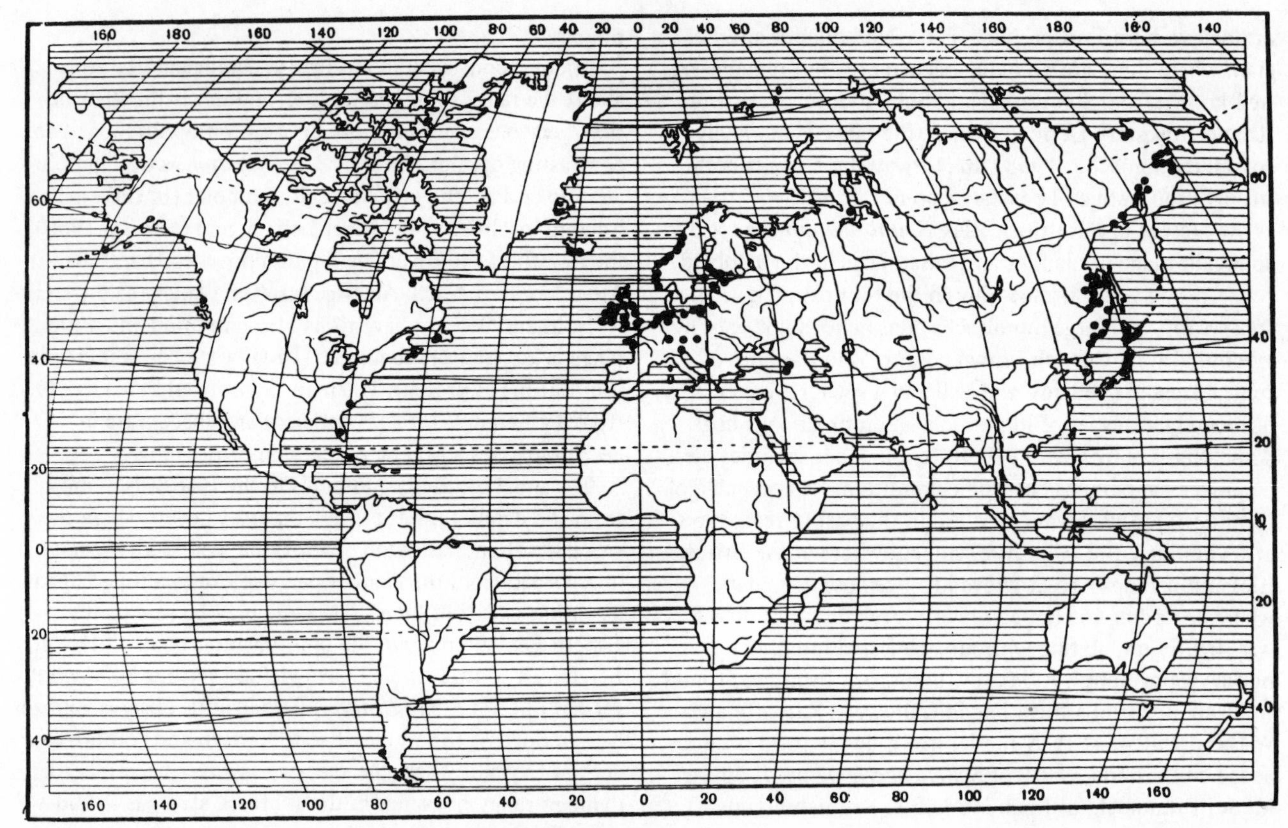

Fig. 33. World range of *Sphagnum imbricatum*. From Tyuremnov 1963.

unbroken, and very wet but it is clear that *S. imbricatum* is the first species to disappear around the heads of peat haggs which are cutting back into peat composed of its remains. There seems little doubt that the recent extinction of this moss over most of the Pennines is to be attributed to the intensive modification of bogs by grazing, burning and trampling, rather than to any direct climatic change.

Pearsall (1956) considered *S. imbricatum* and *S. fuscum* to be the species of the genus most damaged by a light fire which occurred on a blanket bog in Sutherland. McVean and Ratcliffe (1962, p. 135) cite the occurrence of *S. imbricatum* and also the rare *Dicranum undulatum* on tiny islands in a mixed mire which were isolated from fires by small streams. Landwehr (1951) believes that but for human interference *S. imbricatum* would still exist in the Netherlands. However, there are also supporters of climatic change. Hansen (1966), describing a past occurrence on Draved Kongsmose similar to that at Tregaron Bog, found only three small tufts alive on the surface. He sees the disappearance in climatic terms and refers to unfavourable annual distribution of precipitation.

There are still other theories. Competition from other *Sphagnum* species and autogenic bog processes are invoked by King and Morrison (1956) and Morrison (1959). Green (1968) tentatively considers changing trophic state to be significant; the mires may be poorer now than in the past when *S. imbricatum* flourished.

Nonetheless, the abundance of *S. imbricatum* on seemingly undamaged bogs, contrasted with its scarcity or absence from obviously damaged bogs, alone makes one favour an anthropogenic hypothesis. Whatever the details of its decrease, which may have been complex, the final extirpation over much of the English range, if not elsewhere in the British Isles, can reasonably be ascribed to man because the mires are, in many cases, grossly altered to the extent that the entire genus *Sphagnum* is reduced or vanished

However, amidst all the speculations there has been little autecological and, in particular, field study of *S. imbricatum* which is not a very rare plant as some authors make out, at any rate in Ireland. Few, if any, mosses are more worthy of autecological study.

Section *Rigida*

Sphagnum compactum DC.

Cranes Moor (11); Fl V; det. Proctor; Seagrief 1955.

S. compactum occurs extensively in the British Isles but is absent from much of central and southeastern England.

Distinctive in its branch leaves, and occurring in a range of habitats including oligotrophic mires and wet rocks, this species might well have been expected to be a common subfossil. On the contrary, however, the above is the only British one known to the author. Ruuhijarvi (1963) has recorded the species from ombrogenous peat in northern Finland and Jovet-Ast (1967) gives a record from the Polish Pleistocene.

The distribution is wide in the Holarctic; in Fennoscandia, where the habitats are diverse, *S. compactum* is a common species (Lange 1969).

Section *Squarrosa*

Sphagnum teres (Schimp.) Angstr.

Buckenham Broad (27); Fl VII-VIII; Lambert *et al.* 1960. *Dufton Moss* (65); Fl VII*a*; Squires 1968 pc. *Fox Earth Gill* (65); Fl VII*a*; Squires 1968 pc. *Burtree Lane* (66); Fl V; Bellamy *et al.* 1966.

S. teres is primarily a species of the highland zone but also occurs in East Anglia and along the southern coast of England. There are a few records from Ireland. It is a species of mesotrophic and eutrophic fens, carr and flushes.

There are several Pleistocene occurrences of this widespread Holarctic species including a Middle Pleistocene one and a second last interglacial one from Russia (Abramova and Abramov 1962, 1967*c*), and also a Late Weichselian one from Germany.

Sphagnum squarrosum Crome

Cranes Moor (11); Fl V or VI; Newbould 1953. *Hockham Mere* (28); Fl VI, VIII; t; Godwin and Tallantire 1951. *Elan Valley* (36); Fl VI; Moore and Chater 1969. *Moss Swang* (62); Fl VII*b*; Simmons 1969 pc. *Fox Earth Gill* (65); Fl VII*a*; Squires 1968 pc. *Abbot Moss* (70); Fl VII*a* and/or VII*b*; Walker 1966.

S. squarrosum is unrecorded from a few scattered vice-counties in Britain and more in Ireland, particularly in central and western areas. It very often occurs in wet woodlands, particularly carr, but also occurs in mires and flushes.

There are few other subfossils, including an Aftonian one from Iowa (Steere 1942), of this widespread Holarctic species. According to Rönning (1961) this species is common in Spitzbergen, where it reaches 80° 30′N, the northernmost locality of any *Sphagnum* species.

Section *Cuspidata*

Sphagnum flexuosum Dozy & Molk.
[*S. recurvum* P. Beauv.]

Woodwalton Fen (31); Fl VII-VIII; Godwin and Clifford 1939. *Chartley Moss* (39); Fl VII and VIII; Green 1964 pc. *Sweat Mere* (40); Fl VIII*; Sinker 1962. *Wybunbury Bog* (58); Fl VII and VIII; Green 1963 pc. *Moss Swang* (62); Fl VII*b*; Simmons 1969 pc. *Dufton Moss* (65); Fl VII*a*; Squires 1969 pc. *Fox Earth Gill* (65); Fl VII*a*; Squires 1963 pc. *Hutton Henry* (66); Ip III and IV; t; Beaumont *et al.* 1969.

All these records were made under the name *S. recurvum* P. Beauv. See Isoviita (1966) for the reason why it should be discarded. The three species *S. flexuosum* Dozy & Molk., *S. fallax* (Klinggr.) Klinggr. and *S. angustifolium* (Russ.) C. Jens., often recognised by Scandinavian bryologists, have been regarded as one species by British bryologists. In that case (Crundwell 1971) the name with priority is *S. flexuosum* Dozy & Molk.

S. flexuosum, extensively distributed in the Holarctic and widespread in the British Isles, is a species of a variety of habitats including wet woodland, mires and flushes. There are interglacial records (from Poland, Tolpa 1952, 1961) as well as younger discoveries from Europe.

Sphagnum tenellum (Brid.) Pers. ex Brid.

Cranes Moor (11); Fl V; Seagrief 1955. *Fenn's Moss* (50); Fl VIII; Hardy 1939. *Chat Moss* (58); Fl IV-V; Birks 1965*a*. *Snake Pass* (63); Fl VIII; Tallis 1964. *Dean Head Hill* (63); Fl VIII; spores; Tallis 1964.

S. tenellum, absent from much of the lowland zone, occurs in such habitats as wet heaths, ombrogenous bog, flushes and rock ledges. There appear to be few other subfossils (e.g. Bartosh 1963; Casparie 1969), of this suboceanic species which is extensively distributed in the Holarctic and also occurs in South America.

Sphagnum cuspidatum Ehrh. ex Hoffm.

Chartley Moss (39); Fl VII and VIII; Green 1964 pc. *Whixall Moss* (40); Fl VII*b* and VIII; Hardy 1939. *Bettisfield* (51); Fl VIII; Hardy 1939. *Wybunbury Bog* (58); Fl VII*a*; Poore and Walker 1958. *Holcroft Moss* (58); Fl VII*b*; Birks 1965*b*. *Chat Moss* (59); Fl IV-V; Birks 1965*a*. *Snake Pass* (63); Fl VIII; Tallis 1964. *Dean Head Hill* (63); Fl VIII; Tallis 1964. *Wissenden 2* (63); Fl VII*b* and VIII; spores, Tallis 1964. *Tadcaster* (64); Fl VI; Bartley 1962. *Helsington Moss* (69); Fl VII*b*; Smith 1959. *Abbot Moss* (70); Fl VII and/or VIII; Walker 1966. *Bowness Common* (70); Fl VIII; Walker 1966. *Timahoe* (H19); Fl VII; Synnott 1970 pc. *Agher B* (H22); Fl VII*b*; Mitchell 1956. *Canbo* (H25); Fl VIII; Mitchell 1956. *Cloughmills* (H39); Fl VIII; Jessen 1949.

Unrecorded from only a few vice-counties, *S. cuspidatum* is an abundant species of pools in oligotrophic mires. There appear to be no records younger than zone IV-V and, of course, the great bulk of records derive from the ombrogenous peats of zones VII and VIII.

Similarly there are many subfossils of this age from the rest of Europe but older material seems scarce; there is a second last interglacial record from Russia (Abramova and Abramov 1967*c*). *S. cuspidatum* is a bipolar, amphipacific, amphiatlantic and suboceanic species (Isoviita 1966).

Section *Subsecunda*

Ugg Mere (31); Fl VII*b*; Godwin and Clifford 1939.

Sphagnum subsecundum Nees

Weymouth (9); Fl only; t; Gepp 1895. *Elan Valley* (46); Fl VII*b* and VIII; Moore and Chater 1969. *Malham Tarn Moss* (64); Fl VI; Proctor and Birks 1963 pc. *Cloonmoylan 2* (H16); Fl VIII; var. *auriculatum*; Walker and Walker 1961.

This polymorphic species (in the wide sense) is unrecorded from about a dozen scattered vice-counties. British sphagnologists have yet to recognise the specific status of *S. auriculatum* Schimp. and *S. inundatum* Russ. accepted widely elsewhere. It is not clear precisely what taxa are intended by Gepp (1895) or Moore and Chater (1969). Var. *auriculatum*, recorded by Walker and Walker (1961), occurs in a variety of habitats including mires, irrigated rocks and flushes.

Remains of section *Subsecunda* are well known from Flandrian peats of north and central Europe (Grosse-Brauckmann 1968, Mornsjö 1969, Rybnicek and Rybnicekova 1968). There are Late Weichselian records from Belgium (Vanhoorne 1954) and northern Italy (Grüger 1968*b*) and pre-last glacial records from Russia (Abramova and Abramov 1967*c*). *S. subsecundum sensu lato* is widespread in the Holarctic.

Section *Acutifolia*

Hoxne (25); EWo; West 1956. *Chartley Moss* (39); Fl VII and VIII; Green 1964 pc. *Wybunbury Bog* (58); Fl VII and VIII; Green 1963 pc.

Sphagnum fimbriatum Wils.

Buckenham Broad (27); Fl VII-VIII; Lambert *et al.* 1960. *Moss Swang* (62); Fl VII*b*; Simmons 1969 pc. *Moine Mohr B* (98); Fl VII and VIII; Chesters 1931.

S. fimbriatum is widely distributed in the British Isles, though unrecorded from much of Ireland. It often occurs in wet woodland and more rarely in treeless habitats such as dune slacks and mires. The bipolar, extensive Holarctic range reaches to Spitzbergen. There are few subfossils from outside the British Isles (Neujstadt 1957; Nichols 1967).

Sphagnum fimbriatum Wils. or *Sphagnum girgensohnii* Russ.

Fort William (97); Fl V or VI*; Dixon 1910.

S. girgensohnii has a markedly northern range in the British Isles where it often grows in wet woodland, flush-bogs, by stream-sides and on mountain rock. Its range is extensive in the Holarctic.

T *Sphagnum russowii* Warnst.
[*S. robustum* (Russ.) Roll.]

Woodwalton Fen (31); Fl VII-VIII; det. W. R. Sherrin; Godwin and Clifford 1938.

S. russowii hardly occurs in the lowland zone and is mainly found in the north and west where it is lacking from many vice-counties including the great majority in Ireland. Duncan (1966) describes the occurrence in Angus as 'flushes and among wet rocks, rather common on the mountains, much rarer elsewhere'.

The zone VII-VIII record from Huntingdonshire, if correct, implies wider lowland distribution. *S. russowii* is close to *S. girgensohnii*; Nyholm (1969, p. 720) considers colour the only reliable distinction. However, even if the subfossil is *S. girgensohnii* the chorological comment still applies with little diminished force.

There appear to be few other subfossil occurrences of this species (last interglacial, Poland; t; Tolpa 1952, and Flandrian, Canada; t; Nichols 1967) which is widespread in the Holarctic.

Sphagnum fuscum (Schimp.) Klinggr.

Woodwalton Fen (31); Fl VII-VIII; t; det. W. R. Sherrin; Godwin and Clifford 1938. *Dean Head Hill* (63); Fl VIII; spores; t; Tallis 1964. *Timahoe* (H19); Fl VII; Synnott 1970 pc. *Edera Bog* (H24); Fl VIII; King and Morrison 1956.

With the exception of West Sussex, *S. fuscum* is confined to the highland zone where it is primarily a plant of blanket bog. Hence the zone VII-VIII record from Huntingdonshire, if correct, shows past occurrence well outside the present range. This widespread Holarctic species is well known from Flandrian peats of Fennoscandia (e.g. Ruuhijarvi 1963; Mornjsö 1969), Russia (Neujstadt 1957) and Canada (Nichols 1967).

Sphagnum rubellum Wils.

Hoxne (25); EAn; t; West 1956. *Dean Head Hill* (63); Fl VIII; spores; Tallis 1964. *Malham Tarn Moss* (64); Fl VII*a* and VII*b*; Pigott and Pigott 1963. *Hutton Henry* (66); Ip III; t; Beaumont *et al.* 1969. *Cloonmoylan 2* (H16); Fl VIII; Walker and Walker 1961. *Derrybrennan* (H19); Fl VII; Synnott 1970 pc. *Fallahogy 2* (H40); Fl VIII; Walker and Walker 1961.

S. rubellum, absent from much of central and eastern England, is an abundant species of ombrogenous bog of the north and west of the British Isles. All but the tentative Hoxne subfossil certainly relate to this type of habitat. There are several other subfossils (e.g. Aletsee 1959; Mornjsö 1969; Casparie 1969) of this suboceanic widespread Holarctic species which also occurs in the Azores and South America.

Sphagnum nemoreum Scop.
[*S. capillaceum* (Weiss) Schrank]

Dean Head Hill (63); Fl VIII; spores; Tallis 1964.

Growing in such habitats as ombrogenous bogs and wet heaths, *S. nemoreum* is unrecorded from only a few vice-counties, mainly in south-central England. This may be the only subfossil record. The range is extensive in the Holarctic and includes South America.

Sphagnum subnitens Russ & Warnst.
[*S. plumulosum* Roll.]

Cranes Moor (11); Fl V; Seagrief 1955. *Sweat Mere* (40); Fl VIII*; Sinker 1962. *Dean Head Hill* (63); Fl VIII; spores; Tallis 1964. *Tadcaster* (64); Fl VI; Bartley 1962. *Burtree Lane* (66); Fl V; Bellamy *et al.* 1966. *Fallahogy 2* (H40); Fl VIII; Walker and Walker 1961.

All these records were given under *S. plumulosum* Roll. *S. subnitens* is common in the blanket bogs of the north and west of the British Isles and also occurs on wet heaths and raised bogs. There may be no other subfossils of this species, which is suboceanic, amphiatlantic and amphipacific.

Sphagnum molle Sull.

Gort (H15); Ho III or IV; Jessen *et al.* 1959.

This is probably the only subfossil of *S. molle* ever found, though the denticulate branch leaves render determination easy. The range is scattered in the British Isles but is mainly northern and western. Habitats include wet heaths, mires and lake shores.

This suboceanic species is known only from Europe and eastern North America.

Andreaeales

Andreaeaceae

There are very few records of *Andreaea* from Pleistocene deposits. *A. rothii*, *A. rupestris* and *A. blyttii* have been recorded from the Polish Pleistocene. Jovet-Ast (1967) gives a record of *A. huntii* from the 'glaciation of Bavaria'.

Andreaea rupestris Hedw.

Loch Fada (104); LDe; Birks 1969 pc.

Andreaea rothii Web. & Mohr

Fort William (97); Fl V or VI*; Dixon 1910.

Both these rupestral, calcifuge species are common in the highland zone of the British Isles. *A. rupestris* is bipolar while *A. rothii* is confined to Europe and North America. It is hard to believe that they were not abundant in the Late Devensian; the habitats afford little scope for fossilisation. *A. nivalis*, a species with a limited British range (mainly the northern Scottish Highlands), may turn up in Late Devensian deposits in company with *Polytrichum norvegicum.*

Bryales

Polytrichales

Polytrichaceae

Species of *Polytrichum* are often recovered from Late Pleistocene deposits. All but one of the eleven British species (*P. aloides*) are recorded as subfossils. The fragments have an opaque dark brown or blackish colour. The two other British genera, *Atrichum* and *Oligotrichum*, have only the scantiest histories. *Psilopilum*, the remaining European genus, with two species in Scandinavia, is unknown as yet in the subfossil state but might occur in Scottish Devensian deposits.

Atrichum undulatum (Hedw.) P. Beauv.

Leeds (64); Fl VI/VII/VIII*; Raistrick and Woodhead 1930.

A. undulatum, which is extensively spread in the Holarctic, occurs throughout the British Isles and is frequently abundant in woodland. There are very few other subfossils of the genus.

Oligotrichum hercynicum (Hedw.) Lam. & DC.

! *Nant Ffrancon* (49); LDe only.

That this is the sole subfossil not just from the British Isles but from the rest of Europe is surprising. *O. hercynicum*, a calcifuge species, is a colonist of unstable substrata such as screes and gravel by streams. Therefore a widespread glacial occurrence in the north and west of Britain seems likely. The Late Devensian deposits of the Lake District Mountains will yield *Oligotrichum*. The fragments from Nant Ffrancon are finely preserved, clearly exhibiting the characteristic leaf shape and flexuous lamellae. One of them bears a perigonium and the succeeding shoot (plate 19). *O. hercynicum* is known from Europe, North America and Greenland.

Polytrichum undetermined species

Ashcott Heath (6); Fl VII*b*; Dewar and Godwin 1963. *Elstead* (17); Fl VI; Seagrief and Godwin 1960. *Woodwalton Fen* and *Trundle Mere* (31); Fl VII*b*; Godwin 1938. *Kirby Thore* (69); Fl VI; Walker and Lambert 1955. ! *Blelham Tarn Kettle Hole* (69); LDe III. *Swath Beck and Fall Crag* (69); Fl VI/VII; Johnson and Dunham 1963. *Abbot Moss* (70); LDe I; Walker 1966. *Cross Fell* (70); Fl VI and VII; Godwin and Clapham 1951. *Hailes* or *Corstorphine* (83); LDe; both *Polytrichum* sp and *Pogonatum* sp; Bennie 1894*a*. ! *Flanders Moss* (87); Fl VII*a*. ! *Loch Kinord* (92); LDe II; Vasari and Vasari 1968. ! *Loch of Park* (92); LDe I; Vasari and Vasari 1968. *Leigh* (H11); Fl VII; t; Mitchell 1956. *Gort* (H15); Ho III; Jessen *et al.* 1959. *Ballynakill* (H23); Fl VI; Mitchell 1956. *Ardlow Inn* (H30); Fl VII; Jessen 1949. *Mapastown* (H31); LDe II; Mitchell 1953. *Roddans Port* (H38); LDe III; Morrison and Stephens 1965.

The bulk of these records, particularly those of Flandrian age, in all probability refer to *P. commune* or *P. alpestre.* Bad preservation accounts for much of the uncertainty. The spore record by Woodhead and Hodgson (1935) needs confirmation.

Subfossil fragments usually present little difficulty in determination; the necessary leaf characters (sheaths, margins and lamellae) are very often easily observed and sections readily cut (plate 19).

In pre-Late Devensian zones I to III deposits *Polytrichum* appears to be poorly represented; this is certainly true of the British Isles where even Middle Devensian sites have yielded few remains of the genus. In the case of such deposits as Barnwell Station, the Lea Valley Arctic Bed and Sidgwick Avenue this need cause little surprise, despite the long lists of taxa represented. They are calcicolous floras indicating edaphic conditions unsuitable for *Polytrichum* (p. 209), possibly with the exception of *P. alpinum*. Numerous fragments of *Polytrichum* were recovered from Upton Warren and fewer from Fladbury but only one species was represented (*P. juniperinum*).

Late Devensian deposits often yield three or more species. The granitic uplands at Stannon supported at least five species, *P. alpinum, P. commune, P. juniperinum, P. norvegicum* and *P. urnigerum*, and the glaciated terrain at Loch Droma at least four (as Stannon but not *P. commune*) and that at Nant Ffrancon at least four also (as Stannon but not *P. norvegicum*).

All the species discussed here have very extensive Holarctic ranges. Several are bipolar and a few may be considered cosmopolitan. *Polytrichum* species are known from areas of great climatic rigour. *P. piliferum, P. juniperinum* and *P. alpinum* occur in Peary Land, the northernmost land in the world (Holmen 1960). The last-named is one of the few mosses recorded from floating ice islands in the Arctic ocean (Polunin 1958; Hulten 1962).

The earliest British *Polytrichum* is of Late Anglian age (Hoxne). There are Middle Pleistocene records from Russia (Abramova and Abramov 1962) including *P. fragile* and *P. jensenii*. The sole Neogene record is that of an undetermined species from the Polish Pliocene (table 10).

Polytrichum nanum Hedw.

Cowdenglen (76); Fl? only; t; Mahony 1869 as *Pogonatum nanum*; Bennie 1891 as *Polytrichum subrotundum* Huds.

Mahony's record would carry more conviction if the criteria for the determination were explained. Sterile *P. nanum* and *P. aloides* can be difficult to separate. This probably accounts for the doubt attached to Bennie's record which is given as '*Polytrichum subrotundum* Huds., or *P. aloides* but probably the former'. There are few if any other subfossil records of *P. aloides* or *P. nanum.*

Polytrichum urnigerum Hedw.

!*Stannon* (2); LDe only. *Cwm Idwal* (49); Fl IV-V; Godwin 1955. !*Nant Ffrancon* (49); LDe only. !*Burnbrae* (76); Fl VII/VIII*. !*New Dry Dock* (76); Fl IV*. !*Loch Droma* (105); LDe I; Kirk and Godwin 1963. !*Drumurcher* (H32); LDe III.

In the author's experience detached, toothed leaves of *Polytrichum* which are short and broad often prove to be *P. urnigerum*; the apical cells of the lamellae as seen in transverse section in all the material listed above are rounded (plate 19) rather than truncate as in *P. capillare*, a non-British species which might occur in Devensian deposits.

Though recorded from many lowland vice-counties, *P. urnigerum* is chiefly a species of the uplands of the British Isles, where it inhabits montane ledges and gravelly and sandy soil. Dune-slacks in Aberdeenshire are an unusual habitat (Gimingham 1964).

The subfossils come from areas well within the present British range and are predominantly Devensian, as one would expect. *P. urnigerum* is well known from glacial deposits in Denmark, Germany and Poland.

Polytrichum alpinum Hedw.

!*Stannon* (2); LDe only. *Hawks Tor* (2); LDe I and II; also cf. var. *septentrionale* Brid.; Conolly *et al.* 1950. !*Hoxne* (25); LAn; Turner 1968. !*Nant Ffrancon* (49); LDe only. *Aby Grange* (54); LDe II; Suggate and West 1959. *Chat Moss* (59); LDe III and Fl IV-V; Birks 1965*a*. !*Kersall Moss* (59); LDe III. !*Low Wray Bay* (69); LDe II. *Low Wray Bay* and *High Wray Bay* (69); LDe I, II and III; Pennington 1947 and 1962 as *P. alpinum* L. !*Blelham Tarn* (69); LDe II. !*Blelham Tarn Kettle Hole* (69); LDe II. !*Keppel Cove* (69); LDe III-IV. !*Kirkmichael 3a* (71); LDe I, II and III; Dickson *et al.* 1970. !*Bigholm Burn* (72); LDe II. *Garvel Park* (76); LDe/Fl IV*; Robertson 1881 as *Pogonatum alpinum.* !*New Dry Dock* (76); Fl IV*. *Faskine* (77); De only; Bennie 1894*b* as *Pogonatum alpinum* L. *Whitrig Bog* (81); LDe III; Conolly 1963 pc. !*Straloch* (89); LDe. !*Garral Hill* (94); LDe II. *Loch Fada* (104); LDe I; Birks 1969 pc. *Loch Mealt* (104); LDe I; Birks 1969 pc. *Loch Gill Chriosd* (104); LDe I; Birks 1969 pc. *Lochan Coir 'a' Ghobhainn* (104); LDe II and III; Birks 1969 pc. !*Loch Droma* (105); LDe I; Kirk and Godwin 1963. !*Yesnaby* (111); Fl IV. !*Spa* (H2); LAn. *Ralaghan* (H30); LDe III; Jessen 1949. !*Knocknacran* (H32); LDe III. !*Drumurcher* (H32); LDe III.

A Devensian *Polytrichum* with serrate leaves is most likely to be *P. alpinum.* However, examination of the lamellae is necessary for diagnosis.

Various authors have stressed the wide ecological amplitude of *P. alpinum*, a bipolar species. Typical is Mårtensson's assessment (1956, p. 33) that 'It is the only member of the genus which may occur scattered among definitely calcicolous species (e.g. in *Dryas* heath rich in bryophytes). In the subalpine belt it occurs in fens, in wet birch forest, in moist meadows, etc. . . . In the alpine belt it is scattered almost everywhere and may be abundant in late or latish snow areas, in heath, etc.'

Gjaerevoll (1956) and McVean and Ratcliffe (1962) have demonstrated the commonness of this species in various Scandinavian and British snow-bed communities.

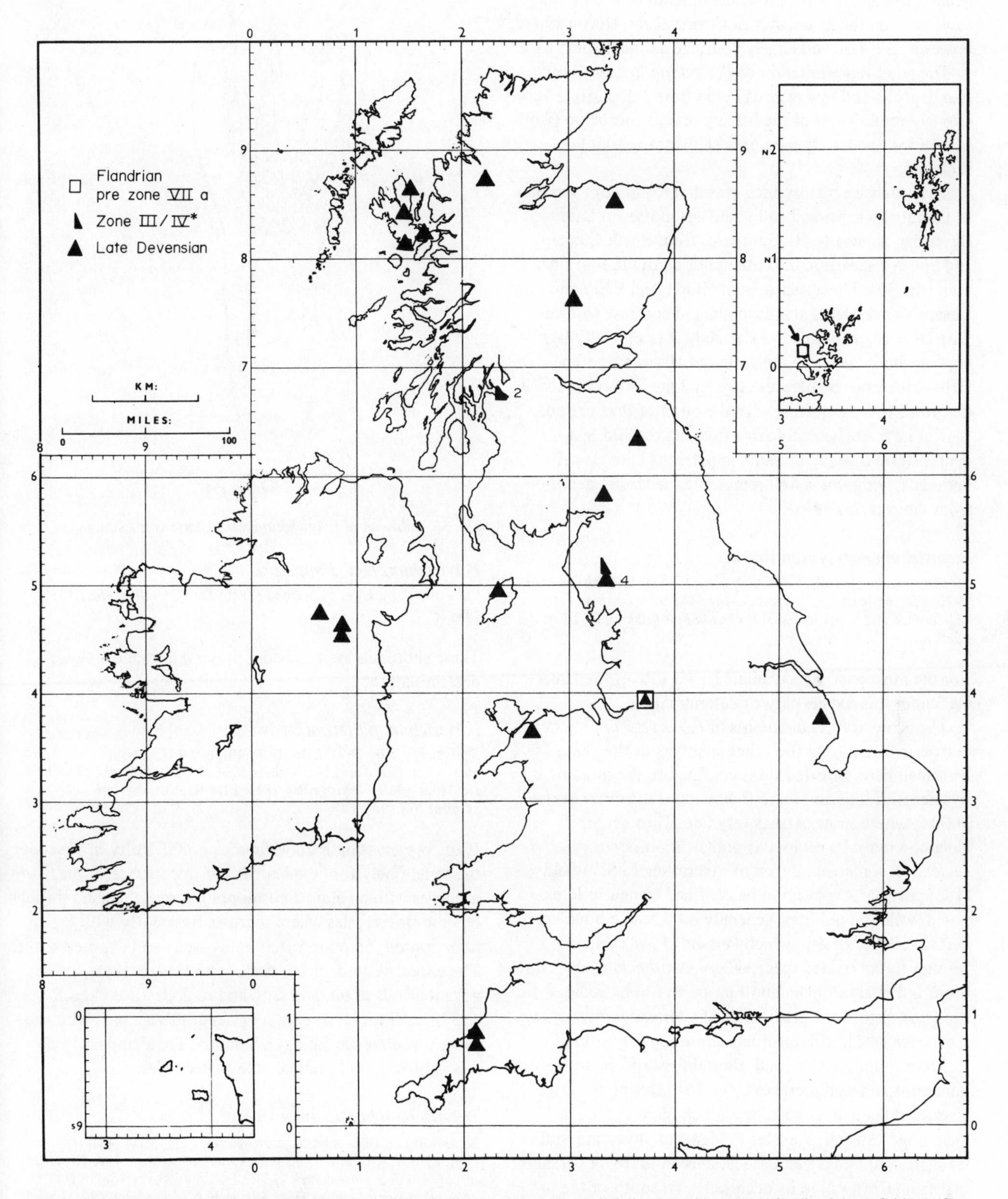

Fig. 34. Late Pleistocene subfossils of *Polytrichum alpinum*. In this grid map and all following grid maps subfossils of doubtful identification or very vague age have been omitted. Apart from Barnwell Station (LDe pre-zone I) the black triangles indicate derivation from LDe zone I, II or III.

P. alpinum occurs throughout the uplands of the British Isles, even on the moorlands of Cornwall and Devon where, however, it is rare and largely confined to the highest tors.

The good representation of *P. alpinum* in Late Devensian deposits and few records in Pre-Boreal deposits is strongly reminiscent of the history of *Salix herbacea* (Godwin 1956); the two species have similar ranges and ecologies.

The extensive history indicates that *P. alpinum* was one of the most widespread and abundant mosses in Late Devensian zones I to III. Subfossils from North Lincoln and South Lancashire indicate minor contradiction since then (fig. 34). The absence from Middle and Early and Late Devensian pre-zone I stands in marked contrast with the Late Devensian abundance. Certainly it is a paucity lacking conviction as a true indication of absence or even rarity. However, perhaps the twenty Late Devensian localities point to a flourishing with the onset of that period.

The Late Anglian subfossils from Hoxne and Spa demonstrate lowland invasion in previous Late glacial periods. *P. alpinum* is well represented in glacial deposits from the rest of Europe.

Polytrichum norvegicum Hedw.

! *Stannon* (2); LDe only. ! *Ballaugh* (71); LDe II-III; Dickson *et al.* 1970. ! *Loch Mealt* (104); pre-LDe zone I. ! *Loch Droma* (105); LDe I; Kirk and Godwin 1963. *Inchnadamph* (108); LDe III; Birks 1968 pc.

For the most part, the cucullate leaves with entire lamellae render this species easy of determination.

The ecological requirements of *P. norvegicum* are much narrower than any of the other members of the genus described here. In Britain it is confined to the more massive Scottish hills and there to base-poor substrata over 3000 ft where snow persists very late. Throughout its Holarctic range its ecology is similar, though var. *vulcanicum* is somewhat different. According to Schofield (1966, p. 34) 'it appears to be confined to relatively exposed, well-drained sites, generally on volcanic boulders or cliffs . . .' This variety, which Persson (1968) believes belongs to the related species *P. sphaerothecium*, does not occur in Britain or mainland Europe and so its ecology need not disturb the acceptance of subfossil *P. norvegicum* as an excellent indicator of late snow bed vegetation.

From Stannon, Cornwall, the subfossils were both numerous and well preserved (fig. 35). During the last glaciation the British range was much more extensive, including Cornwall, the Isles of Man and Skye, and Sutherland (fig. 36). A Belgian Late Devensian subfossil parallels that from Cornwall in its occurrence far south of the present localities (Van der Flerk and Florschütz 1950). There are glacial records from Poland and Russia.

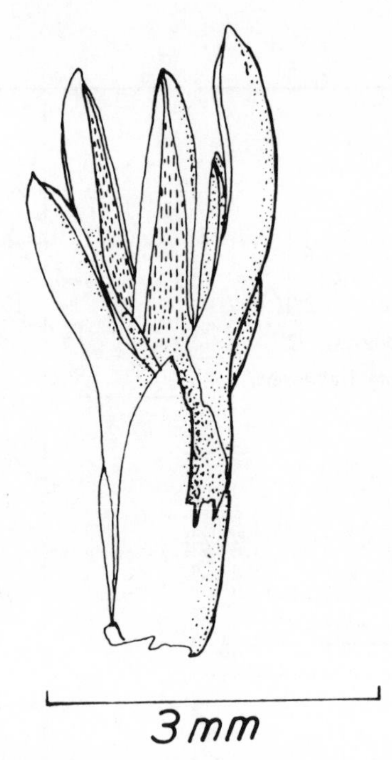

Fig. 35. Subfossil of *Polytrichum norvegicum* from Stannon.

Polytrichum sect. *Juniperina*

! *Wretton* (28); EDe. ! *Chelford* (58); EDe. ! *Newtonbabe* (H31); LDe II.

These subfossils are too badly preserved to allow closer determination.

Polytrichum piliferum Hedw.

Skelsmergh Tarn (69); LDe III; t; Walker 1955. *Abbot Moss* (70); LDe I and I-II; Walker 1966. ! *Little Lochans* (74); LDe III; Moar 1969*a*. *Whitrig Bog* (81); LDe III; Conolly 1963 pc. ! *Garral Hill* (94); LDe II; t.

If the preservation is good there is no difficulty in observing the long, finely toothed leaf tips of this species. *P. piliferum* a species with an almost cosmopolitan range, occurs throughout the British Isles where it grows frequently with *P. juniperinum*, on poor, often sandy, soils and exposed rocks. The extensive underground shoot and rhizoid system are advantageous in binding sand and rock detritus (Leach 1931*b*). All the subfossils are glacial: this accords with the present preference for open habitats. There appear to be few subfossils from outside the British Isles.

Polytrichum juniperinum Hedw.

! *Stannon* (2); LDe only. ! *Upton Warren* (37); MDe. *Abbot Moss* (70); LDe II; t; Walker 1966. ! *Kirkmichael 3a* (71); LDe III; Dickson *et al.* 1970. *Garvel Park* (76); LDe only; Robertson 1881. ! *New Dry Dock* (76); Fl IV*. *Allt na Feithe Sheillach* (96); Fl only; Lewis 1906. *Loch Meodal* (104); LDe I; Birks 1969 pc. ! *Loch Droma* (105); LDe I; Kirk and Godwin 1963. *Roundstone 2* (H16);

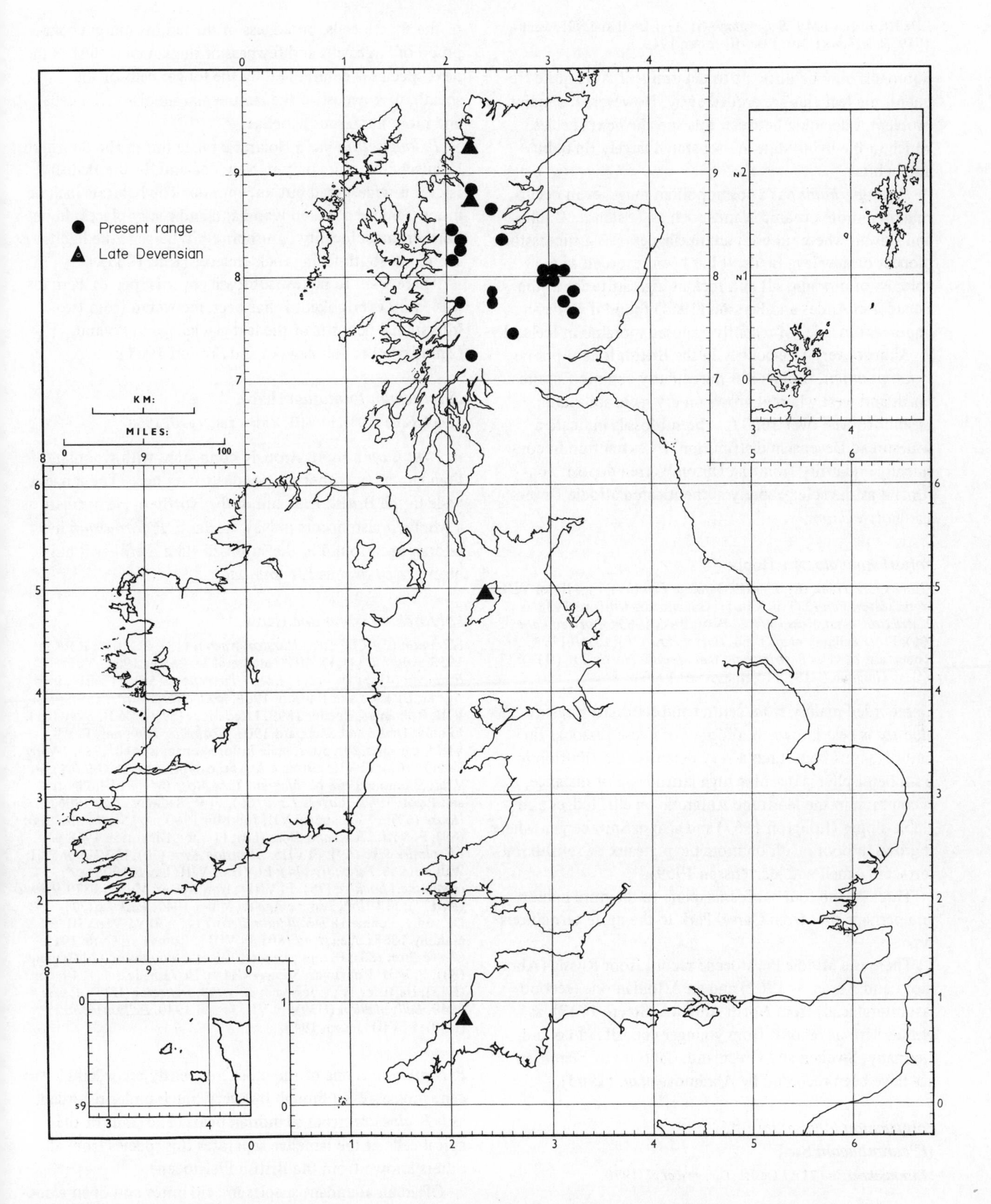

Fig. 36. Devensian subfossils and present range of *Polytrichum norvegicum*. Partly after H. J. B. Birks, *Trans. Br. bryol. Soc.* 1968.

LDe III; Jessen 1949. *Ballybetagh* (H21); LDe II and VI; Jessen 1949. *Ralaghan* (H30); LDe III; Jessen 1949.

Subfossils may be difficult to determine if, as is quite frequent, the leaf tips are eroded away. However, the major problem is deciding between this and the next species, which in the living state are separated largely on habit and habitat.

P. juniperinum has a cosmopolitan range, even occurring on remote oceanic islands such as Tristan da Cunha and Hawaii where, in both archipelagoes, it is a successful pioneer of new lava fields. It has been recorded as a colonist of moranic silt and rock at high altitudes in the Ecuadoran Andes and Persson (1964) found it to be an important invader of recently exposed moraine in Iceland.

Almost every vice-county in the British Isles supports *P. juniperinum*, though it is particularly common in the north and west where it grows on dry, acid soils and ascends to well over 3000 ft. The subfossils indicate a widespread Devensian distribution. Its extraction in considerable quantity from the Upton Warren deposit confirms it as the sole properly authenticated Middle Devensian *Polytrichum.*

Polytrichum alpestre Hoppe

! *Honeygore Track* (6); Fl VII*b*. ! *Holme Fen* (31); Fl VII*b* or VIII. *Woodwalton Fen* (31); Fl VII*b*; t; Godwin and Clifford 1938 as *P. strictum. Chat Moss* (59); Fl IV-V; Birks 1965*a*. *Burtree Lane* (66); Fl V; Bellamy *et al.* 1966. *Garvel Park* (76); LDe/Fl IV*; Robertson 1881 as *P. strictum.* ! *Ballydermot North Bog* (H19); Fl VII; t. *Timahoe* (H19); Fl VII; Synnott 1970 pc.

Unrecorded mainly from central and southern England, *P. alpestre* is best known as a former of dense tussocks on ombrogenous bog. It has a very extensive distribution in both hemispheres reaching high latitudes. For instance, it is common in the Maritime Antarctic on cliff ledges and stable slopes (Longton 1967) and also in Spitzbergen where it grows in poor heath on morainic plateaux or solifluxion terraces (Arnell and Mårtensson 1959).

Three of the four subfossils relate to the mire habitat and perhaps that from Garvel Park to the more terricolous type.

There is a Middle Pleistocene record from Russia (Abramova and Abramov 1962) and an Aftonian one (second last interglacial) from North America (Steere 1942), as well as various records from younger deposits in Poland, Germany, Sweden and Greenland. Capsules of Flandrian age have been recorded by Abramova *et al.* (1965).

Polytrichum longisetum Brid.
[*P. aurantiacum* Sw.]

! *Kirkmichael 3a* (71); LDe III; Dickson *et al.* 1970.

This solitary British subfossil is identified by the shortness of the sheath cells, broadness of the lamina margin composed of large cells and fewness of the lamellae, just as the next species is determined by the longer cells of the sheath, narrowness of the lamina margin of smaller cells and more numerous lamellae.

P. longisetum has a Holarctic range but in the Southern Hemisphere occurs only in New Zealand. In the British Isles it is widespread but uncommon. The habitats include mires, acid substrata in woodland and rocky places. In its northernmost locality (southern Spitzbergen) the habitat is sheltered, south-facing rock crevices (Kuc 1963).

There seem to be few subfossils of this species. Perhaps it is easily overlooked. It has been recovered from two interstadial deposits of the last glacial age in Poland (Zmuda 1914; Sobolewska and Srodon 1961).

Polytrichum formosum Hedw.

! *Loch Fada* (104); Fl VIII; Vasari and Vasari 1968.

P. formosum is more strongly connected with woodland than any of the other species dealt with here. The range is wide in the British Isles and in the Northern Hemisphere as a whole; it also occurs in New Zealand. *P. formosum* has seldom been found in the subfossil state. Perhaps it has been passed over as *P. commune.*

Polytrichum commune Hedw.

! *Stannon* (2); LDe only. *Hurston Warren* (13); Fl VIII*; Rose 1953. *Roushill* (4); Fl VIII*; Medieval ar; Barker 1961. *Nant Ffrancon* (49); LDe only. ! *Holme Pierrepont* (56); Fl VIII*; Iron Age ar; Dickson and Ransom 1968. *Bridgewater Street* (59); Fl VIII; Roman ar; Roeder 1899. ! *Red Moss* (59); LDe II. *Filey* (61); Fl only; Drake and Sheppard 1909. *Yorkshire Pennines*; Fl VII/VIII*; capsules, calyptras, male inflorescence; Burrell 1924. ! *North Ferriby* (61); Fl VII*b*; Bronze Age ar; calyptra. *North Gill* (62); Fl VII*a*; Simmons 1968 pc. *Malham Tarn Moss* (64); Fl VI; Pigott and Pigott 1963. *Burtree Lane* (66); Fl V; Bellamy *et al.* 1966. *Moor House* (69); Fl VII and/or VII; Johnson 1960 pc. ! *Seathwaite Tarn* (69); Fl VIII. ! *Russland Moss* (69); Fl only; Dickinson 1968 pc. ! *Foolmire Sike* (69); Fl VII*b*. *Ehenside Tarn* (70); Fl VII*b* or VIII; Walker 1966. *Palwhilly* (74); Fl VII or VIII; Lewis 1905 as *P. commune. Lochlee* (75); Fl VIII*; Iron Age ar; Munro 1879. *Ferniegair* (77); Fl VII*b**; Bronze Age ar; Miller 1947. *Faskine* (77); De? only; Bennie 1894*b*. *Windmillcroft* (77); Fl VI/VII/VIII*; Mahony 1868*a*. *Newstead* (80); Fl VIII*; Roman ar; Curle 1911. *Cleave Burn* (83); Fl only; Lewis 1905 as *P. commune* L. ! *Carpow* (88); Fl VIII; Roman ar. *Shower* (H10); Fl VII*b*; Jessen 1949. *Gort* (H15); He III or IV; t; Jessen *et al.* 1959. *Raheelin* (H29); Jessen 1949. *Ballyshallion* (H39); Fl VII; Jessen 1949. *Ballymacombs* (H40); Fl VIII; Jessen 1949.

P. commune is one of the most frequently recorded Pleistocene mosses, even though it is very much under-recorded, as is *P. alpestre*, from Flandrian peats. The sinus of the apical cells of the lamellae separates this species from all others known from the British Pleistocene.

Often an abundant species in acid mires and open woodland, *P. commune* occurs throughout the Holarctic, reaching southern Spitsbergen (Kuc 1963), and is widely scattered in the Southern Hemisphere. Almost all the subfossils are of Atlantic or later age. There are only three certainly glacial subfossils (Stannon, Nant Ffrancon and Red Moss) and the single interglacial record is a tentative one from Gort.

The tough, pliable stems of *P. commune*, the largest moss in our flora, have given the species economic significance right up to the modern period, if only in a very minor way. Curle (1911) listed nineteenth-century examples of brooms, baskets and hassocks and Dixon (1924) mentioned its use for pillows and beds by the Lapps and for stuffing mattresses in this country. Undoubtedly its importance was much greater in earlier, particularly Roman and Prehistoric, times (p. 192).

The species has been recovered from several archaeological sites ranging from Bronze Age to Medieval. There is an unfinished Roman basket of complex design (Curle 1911), a twist rope used as caulking in a Bronze Age canoe (plate 21) and a 'shroud' (Miller 1947). The uses of the 'fringes' and 'pigtail-like' structure from the Lochlee Crannog (Munro 1879) are not at once obvious.

The species is well known from Pleistocene deposits in the rest of Europe.

Fissidentales

Fissidentaceae

The Fissidentaceae is poorly represented in Pleistocene deposits. Apart from the species listed below, only *F. taxifolius* (from Poland, Czechoslovakia and Switzerland) and *Octodiceras fontanum* (from Finland, p. 202) are known as subfossils. In the British Isles the genus *Fissidens* is represented by twenty species, as many as any other European country except France (twenty-nine species, Augier 1966). Species such as *F. polyphyllus, F. serrulatus* and *F. celticus* have bryogeographically interesting ranges. The last-named is unknown outside the western areas of the British Isles.

Of the four species discussed below, three occur throughout the British Isles while the fourth, *F. osmundoides*, is restricted to the highland zone but there it is widespread. *F. bryoides* and *F. adianthoides* are bipolar species with widespread ranges in the Northern Hemisphere, as have *F. cristatus* and *F. osmundoides.*

Fissidens undetermined species

!*Nant Ffrancon* (49); LDe only. !*New Dry Dock* (76); LDe only.

As a genus, *Fissidens* with the boat-shaped sheathing laminae is unmistakable. However, determination of small sterile fragments to the species level may be difficult.

Fissidens bryoides Hedw.

!*Barnsley Park* (33); Fl VIII*; Roman ar; t. *Brigg* (54); Fl VII*b**; Bronze Age ar; with sporophytes; Sheppard 1910. !*Carpow* (88); Fl VIII*; Roman ar; t.

Fissidens osmundoides Hedw.

!*Nant Ffrancon* (49); LDe only; t. *Abbot Moss* (70); LDe I; Walker 1966. *Fort William* (97); Fl V or VI*; Dixon 1910.

Fissidens cristatus Wils.

!*Nant Ffrancon* (49); LDe only; t.

Fissidens adianthoides Hedw.

!*Nant Ffrancon* (49); LDe only. !*Bunny* (56); Fl VIII*; Roman ar. *Loch Droma* (105); LDe I; Durno 1960 pc.

The sediments of Late Devensian age in Nant Ffrancon in Snowdonia are remarkable in having yielded three species of the genus. There is nothing unexpected in their occurrence in glacial times. All three can occur on rocks by streams and so may be washed into lacustrine deposits. *F. cristatus* is given as an exacting calcicole, and *F. adianthoides* and *F. osmundoides* as calcicoles by McVean and Ratcliffe (1962).

Dicranales

Ditrichaceae

Few subfossils of the Ditrichaceae have been found apart from the species discussed here. *Ditrichum pusillium* (as *D. tortile*) was recovered from the Early Pleistocene of the Dutch-German frontier (Dixon in Reid and Reid 1915).

Ditrichum flexicaule (Schwaegr.) Hampe

Ponder's End (21); MDe; Warren 1912 as var. *densum.* ! *Upton Warren* (37); MDe. !*Fladbury* (37); MDe. !*Low Wray Bay* (69); LDe II. !*Kirkmichael 3a* (71); LDe I; Dickson *et al.* 1970. !*New Dry Dock* (76); Fl IV*. !*Loch Droma* (105); LDe I; Kirk and Godwin 1963. !*Drumurcher* (H32); LDe III.

Most of the subfossils determined by the author are short-leaved forms perhaps referable to var. *densum*, as was the Ponder's End material by Dixon. On the other hand the Loch Droma specimens belong to the typical plant.

In the British Isles *D. flexicaule* is as strictly calcicolous as any species considered in this book. It appears to show such a preference throughout the entire range which is extensive in the Northern Hemisphere and also includes New Zealand.

In the Arctic *D. flexicaule*, occurring in a variety of plant communities including rich fens, is often an abundant species. Habitats in the British Isles include calcicolous grassland, dunes, rock ledges and the margins of *Fagus* woodland. *D. flexicaule* is absent from fourteen scattered vice-counties. The eight subfossils point to widespread occurrence and imply an abundance greater than at

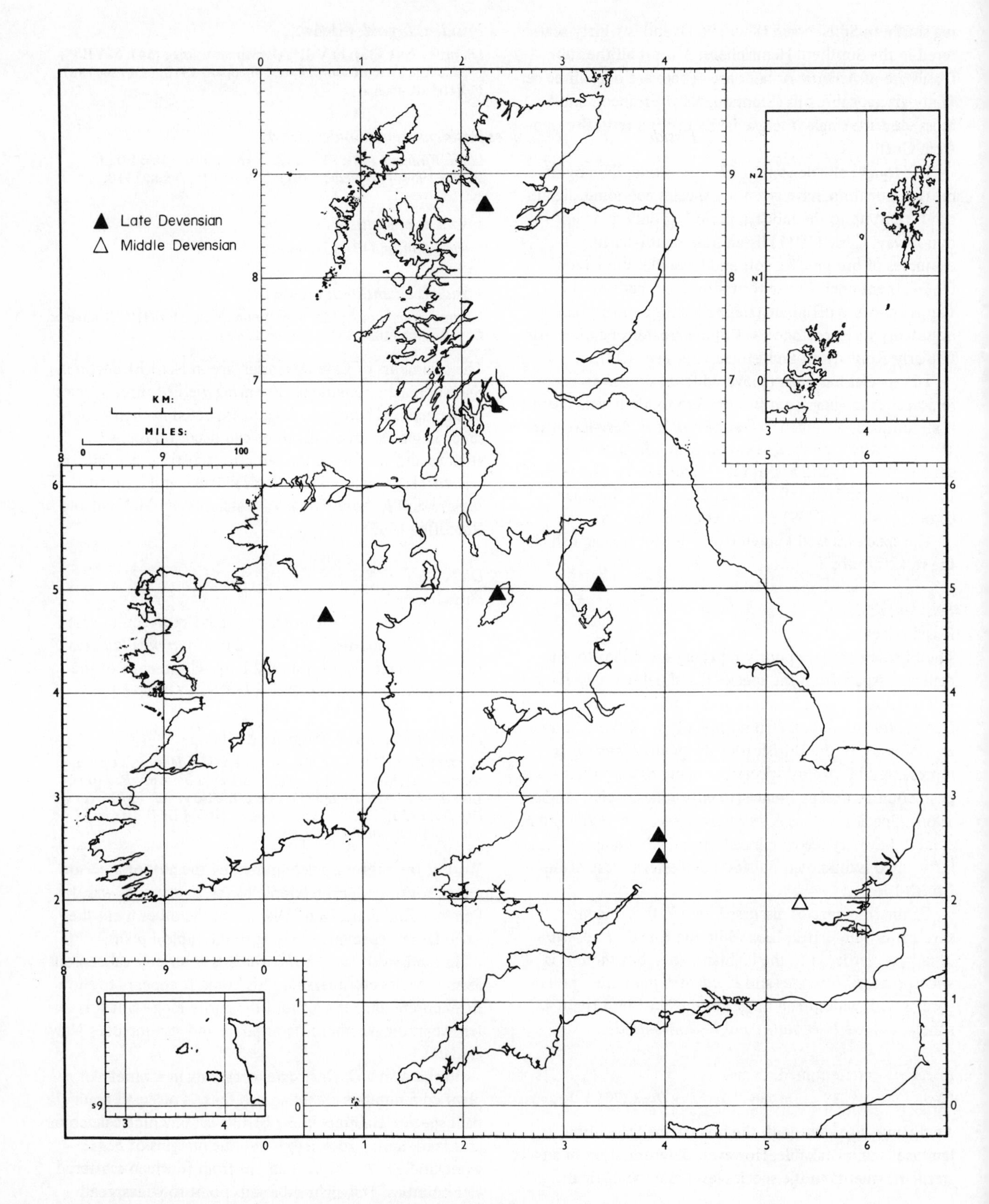

Fig. 37. Devensian subfossils of *Ditrichum flexicaule*.

present for much, if not all, of the last glaciation (fig. 37).

The species is well known from Pleistocene glacial deposits in the rest of Europe and North America.

Distichium capillaceum (Hedw.) B., S. & G.

Ponder's End (21); MDe; Warren 1912. ! *Broome Heath* (27); Wo; t. ! *Upton Warren* (37); MDe; t. *Brigg* (54); Fl VII*b**; Bronze Age ar; Sheppard 1910 as *Swartzia montana* Lindb. ! *New Dry Dock* (76); Fl IV*; t. *Whitrig Bog* (81); LDe III; Conolly 1963 pc. *Dronachy* (85); LDe only; Bennie 1896. ! *Loch Droma* (105); LDe I; Kirk and Godwin 1963. ! *Baggotstown* (H8); LAn and Ho III; t; Dickson 1964*a*. ! *Kildromin* (H8); Ho I; t. *Ballybetagh* (H21); Fl VI; Jessen 1949. *Mapastown* (H31); LDe III; Mitchell 1953. ! *Drumurcher* (H32); LDe III; t.

The leaf insertion and morphology make subfossils easily recognisable at the genus level. In the absence of capsules separation of this and the next species (and also *D. hagenii*, the northern latitude species) can be very difficult; leaf stance is then the only and perhaps not reliable criterion. Hence the tentative records given above. The very short leaves of the material from Upton Warren and Broome Heath make one think of var. *compactum*, the form from unfavourable habitats.

In most texts, *D. capillaceum* is considered cosmopolitan while *D. inclinatum* has a very wide range in the Northern Hemisphere. Both reach the highest northern latitudes. *D. capillaceum* is by far the commonest moss in Peary Land (Holmen 1960) where it occurs under almost all ecological conditions including, in contrast to the British habitats, springs.

In the British Isles, *D. capillaceum* is calcicolous in tendency and is restricted to the highland zone with the striking exception of Surrey, where the habitat was a wall, a surprising habitat so far from the main area (Wallace 1941). It is known from walls elsewhere, but in the highland zone as, for instance, Selkirkshire (Simpson 1924). Base-rich rock ledges are the usual habitat.

D. capillaceum is tolerant of prolonged snow lie, and its abundant occurrence in such habitats led Gjaerevoll (1956) to state '*Distichium capillaceum* is a very good characteristic species of snow-bed communities rich in calciphiles (*Salix reticulata-S. polaris*-and *Saxifraga oppositifolia*-socs) and may grow abundantly in places devoid of vasculars.' Communities in the latter situation are grouped under Alliance *Distichion capillacei.*

In the last glacial times the species was widespread over much of the British Isles including the lowland zone. Two records are unexpected in their Flandrian age. That from Brigg is the more surprising; the material was part of the caulking of a Bronze Age canoe; *D. capillaceum* no longer grows in Lincolnshire. There are numerous Pleistocene records from the rest of Europe.

T *Distichium inclinatum* (Hedw.) B., S. & G.

Garvel Park (76); LDe/Fl IV*; Robertson 1881.

The species determination must be held in some doubt. *D. inclinatum* is a calcicole, confined to the highland zone where it is more local than *D. capillaceum.* There is a last glacial record from Poland.

Ceratodon purpureus (Hedw.) Brid.

! *Stannon* (2); LDe only. *Ponder's End* (21); MDe; t; Warren 1912. ! *Wretton* (28); Ip II. ! *Godmanchester* (31); Fl VIII*; Roman ar; t. ! *Upton Warren* (37); MDe. ! *Kirkmichael 3a* (71); LDe I and III; Dickson *et al.* 1970. ! *Baggotstown* (H8); Ho III; Dickson 1964.

A cosmopolitan species, *C. purpureus* occurs throughout the British Isles in a great variety of habitats from the lowlands to high in the mountains. It reaches the highest northern latitudes.

The subfossils from Cornwall and the Isle of Man point to an extensive Late Devensian occurrence which may also be true of the Middle Devensian period if the Ponder's End tentative identification is taken to support the Upton Warren discovery.

There are several glacial occurrences from the rest of Europe, particularly Poland, and North America; Abramova *et al.* (1965) record capsules from the Middle Pleistocene of eastern Siberia.

Seligeriaceae

Blindia acuta (Hedw.) B., S. & G.

Hawks Tor (2); LDe II; Conolly *et al.* 1950 as *B. acuta* B. & S. *Faskine* (77); De? only; Bennie 1894*b* as *B. acuta* Dicks. *Fort William* (97); Fl V or VI*; Dixon 1910 as var. *trichoides* Braithw. *Loch Droma* (105); LDe I; Durno 1960 pc.

The Hawks Tor record is in some doubt; Conolly *et al.* (1950) describe the leaf apical cells but the alar cells, characteristic of the species, are left unmentioned.

The range of *B. acuta* is widespread in the Northern Hemisphere, where the highest latitudes are reached, and includes Australia and South America. The habitats are similar throughout, wet rock on cliffs, or rocks by or in streams. In the British Isles the species does not extend into the lowland zone. There are few subfossils from the rest of the world. *Seligeria* is unknown in the subfossil state.

Dicranaceae

With the principal exception of *Dicranum* this family is poorly represented in Pleistocene deposits. Genera such as *Rhabdoweisia, Cynodontium, Arctoa, Dicranodontium* are unknown in the subfossil state.

Dicranella undetermined species

! *Ilford* (18); LWo; t; Dickson 1964*b*. ! *Loch Fada* (104); Fl VIII;

Vasari and Vasari 1968.

One of the outstanding absentees from the British Pleistocene is *D. palustris*, a species which occurs abundantly by springs and flushes in the highland zone. The chances of fragments being incorporated in lacustrine sediments seem high. The lack of subfossils is made all the more strange by the frequent recovery of often associated species such as *Philonotis fontana.* There appear to be only two non-British discoveries of subfossils, one from Poland (Szafran 1952) and one from Italy (Clerici 1892).

D. crispa is known from the Danish Pleistocene (Hesselbo 1910), *D. schreberiana* from the Late Wisconsin of New York (Miller 1969) and *D. cerviculata* from the Flandrian of Northwestern Germany (Grosse-Brauckmann 1968) and from the Late Pleistocene of Poland (Zmuda 1914).

Dicranella varia (Hedw.) Schimp.
Holme Fen and *Trundle Mere* (31); Fl VII*b*; Vishnu-Mittre 1959.

D. varia is absent from only a few vice-counties in Scotland and Ireland. It is a calcicole often found on clayey ground. With the addition of North Africa the range is like that of *D. heteromalla.* There are two records from the Polish Weichselian (Szafran 1952).

Dicranella heteromalla (Hedw.) Schimp.
Moine Mohr A and B (98); Fl VII and/or VIII; Chesters 1931.

The above are the only subfossils of this common calcifuge species of rotten wood, soil and rock. In Fennoscandia the range, which includes Eurasia, Macaronesia and North America, is markedly southern.

Oncophorus virens (Hedw.) Brid.
! *Broome Heath* (27); Wo. *Garvel Park* (76); LDe/Fl IV*; Robertson 1881 as *Cynodontium virens* var. *compactum.*

Apart from the Cumbrian Mountains, in the British Isles *O. virens* is confined to the Grampian Highlands where it grows on wet rocks and stony ground. McVean and Ratcliffe (1962) regard it as an exacting calcicole. The Garvel Park subfossils prove an extension of range in Late Devensian times while those from Broome Heath, which are numerous and well preserved, prove a much greater expansion in the preceding glaciation.

The range, extending north to Peary Land and Spitzbergen is extensive in the Holarctic. There is a glacial subfossil from Skaerumhede, Denmark (Hesselbo 1910).

Dichodontium pellucidum (Hedw.) Schimp.
Wolvercote (23); Wo?; Bell 1904. ! *Nant Ffrancon* (49); LDe only. ! *Seathwaite Tarn* (69); Fl VIII. *Garvel Park* (76); LDe only; Robertson 1881. ! *New Dry Dock* (76); LDe only. *Fort William* (97); Fl V or VI*; Dixon 1910 as *D. pellucidum* B. & S. !*Loch Fada* (104); III-IV; Vasari and Vasari 1968; LDe I; Birks 1969 pc. *Loch Mealt* (104); LDe III; Birks 1969 pc. ! *Derryvree* (H31); MDe. ! *Drumurcher* (H32); LDe III.

D. pellucidum is largely a species of the highland zone but it extends along the southernmost counties to Kent and there are records from East Norfolk and Bedfordshire. It is a species of calcicolous tendency; wet rocks and gravel by streams are the usual habitats. Hence there is no surprise in extracting subfossils from lacustrine sediments; as, for example, at Seathwaite Tarn and Loch Fada. *D. pellucidum* was widespread during Late Devensian zones I to III and perhaps also in the Middle Devensian (fig. 38).

The distribution is extensive in Eurasia and North America. There are few subfossils from outside the British Isles.

Dicranum undetermined species
Hawks Tor (2); LDe II; Conolly *et al.* 1950. ! *Beeston* (27); Pa. ! *Hockham Mere* (28); Fl VI. *Woodwalton Fen* A (31); Fl VII or VIII; Godwin and Clifford 1939. ! *Bryn-y-Mor* (48); Fl VII*a*; ! *Kirkmichael 1* (71); LDe; Dickson *et al.* 1970. *Gort* (H15); Ho III or IV; Jessen *et al.* 1959.

The Beeston, Bryn-y-Mor, Hockham Mere and Kirkmichael specimens are too fragmentary for further determination but certainly belong to section *Dicrana scoparia* as does the Hawks Tor material.

The seventeen species of the genus *Dicranum* in the British Isles exhibit considerable diversity of geographical pattern. Of the species known from the British Pleistocene, *D. scoparium, D. majus* and *D. bonjeanii* are widespread, *D. scottianum* is markedly western, *D. undulatum* and *D. spurium* are scattered and markedly disjunct and *D. elongatum* has an arctic-alpine range.

Of those unknown from the British Pleistocene, *D. blyttii, D. starkei, D. falcatum* and *D. glaciale* have arctic-alpine patterns; they should be sought in Devensian deposits. *D. leioneuron*, only recently recognised as British (Ahti *et al.* 1965), has a solitary Perthshire locality. *D. montanum, D. flagellare* and *D. strictum* are southern and eastern. *D. fuscescens* is widespread but absent from much of southern and eastern England and central Ireland, and *D. polysetum* is scattered and disjunct.

D. fuscescens, D. montanum, D. muelenbeckii, D. spadiceum and *D. congestum*, unknown as British subfossils, have been recovered from the European and Siberian Pleistocene.

Dicranum scottianum Turn.
Fort William (97); Fl V or VI*; Dixon 1910.

A rupestral species, *D. scottianum* has a scattered, markedly northern and western range in the British Isles (Widespread Atlantic; Ratcliffe 1968). It is absent from all of south-

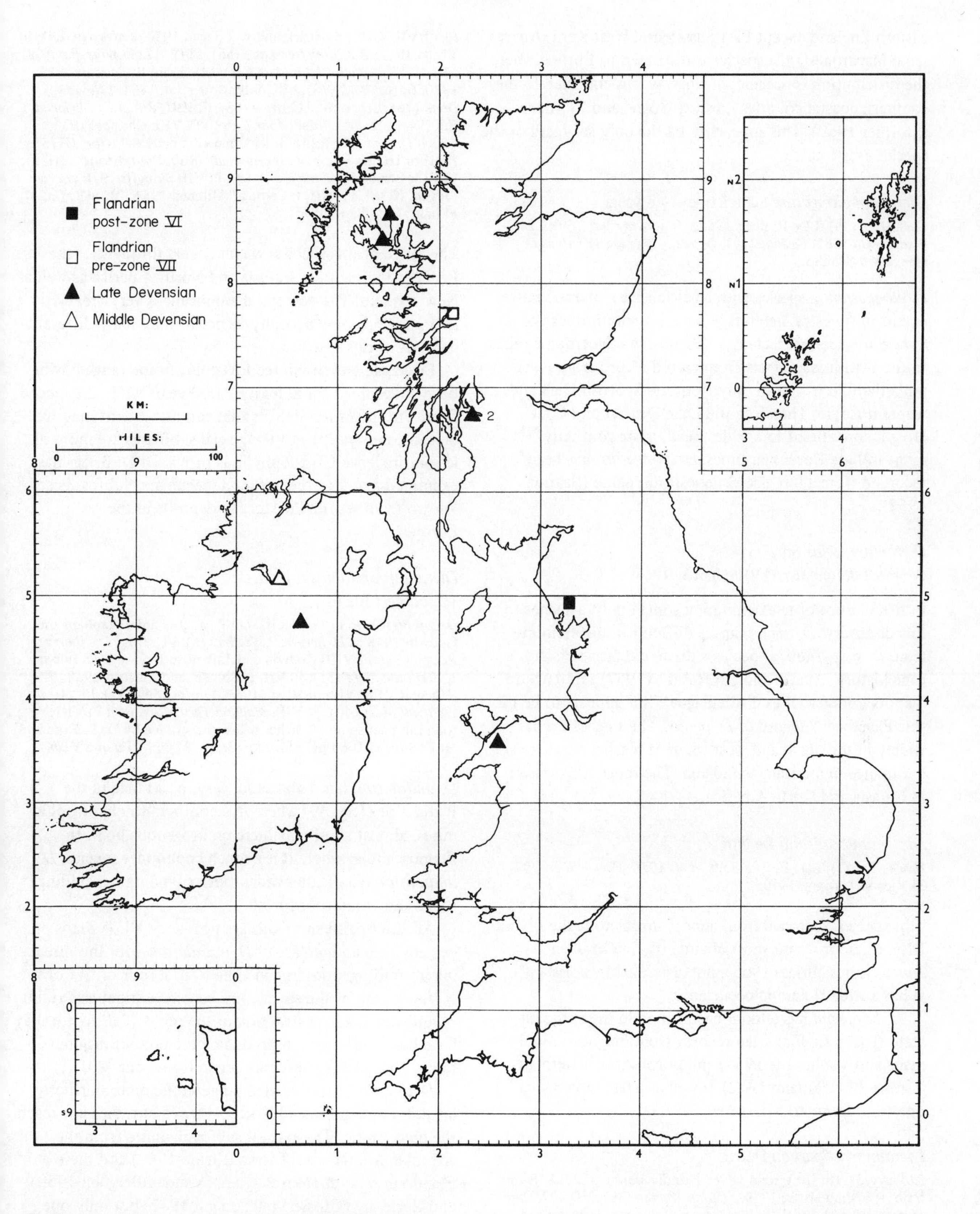

Fig. 38. Late Pleistocene subfossils of *Dichodontium pellucidum*.

eastern England except East Sussex and West Kent. Apart from Macaronesia the species is restricted to Europe where the distribution is oceanic; in Norway it is confined to the southern coastal counties north to Sogne and Fjordane (Störmer 1969). This appears to be the only subfossil of the species ever found.

Dicranum elongatum Schleich. ex Schwaegr.

Chat Moss (59); LDe II; Birks 1965*a*. *Low Wray Bay* (69); LDe II; Pennington 1962. *Faskine* (77); De? only; Bennie 1894*b* as *D. elongatum* Schimp.

D. elongatum, a species with an extensive Holarctic range, occurs in rupestral habitats as well as on hummocks of *Sphagnum* species, such as *S. fuscum*, in oligotrophic mires. In the British Isles, apart from an old record from Northumberland, it is known only from the Scottish Highlands where it is rare. Therefore the Lake District and Chat Moss records point to a widespread, more southerly range in Late Devensian times. *D. elongatum* has been recorded from a last glacial deposit in Poland (Szafran 1952).

Dicranum majus Sm.

Cromwell Bottom (6); Fl VII*b*; Bartley 1964.

In the absence of re-examination some doubt attaches to this discovery; *D. majus* can be difficult to differentiate from *D. scoparium*, a species with an extensive Pleistocene history. Apart from Jovet-Ast's (1967) reference to 'dépôts interglaciaires d'Allemagne', this appears to be the sole Pleistocene record of *D. majus.* The species occurs widely in the north and west where it reaches its greatest luxuriance in montane woodland. The range is extensive in Eurasia and North America.

Dicranum bonjeanii De Not.

Hawks Tor (2); LDe II; t; Conolly *et al.* 1950. *Fort William* (97); Fl V or VI*; Dixon 1910.

This species, recorded from almost throughout the British Isles, where it occurs most abundantly in *Carex* and *Juncus* mires (Briggs 1965), but also inhabits grasslands. It has a wide Holarctic occurrence.

D. bonjeanii has seldom been found in the subfossil state. There are Flandrian records from Belgium (Stockmans and Vanhoorne 1954) and northwestern Germany (Grosse-Brauckmann 1962) as well as a last interglacial one from Denmark (Hartz 1909).

Dicranum scoparium Hedw.

Hoxne (25); Ho III; t; Reid 1896. *Trundle Mere* (31); Fl VII*a* and VII*b*; t; Vishnu-Mittre 1959. ! *Upton Warren* (39); MDe. ! *Holme Pierrepont* (56); Fl VIII*; Iron Age or Roman ar; Dickson and Ransom 1968. *Chat Moss* (59); Fl IV-V; Birks 1965*a*. *Leeds* (64); Fl VI/VII/VIII*; Raistrick and Woodhead 1930. *Tadcaster* (64); Fl VI, Bartley 1962. ! *Burtree Lane* (66); Fl IV. ! *Low Wray Bay* (69); LDe II. ! *Kirkmichael 1 and 3a* (71); LDe I and II; Dickson *et al.* 1970. *Garvel Park* (76); LDe only; Robertson 1881. ! *New Dry Dock* (76); LDe only. ! *Carpow* (88); Fl VIII; Roman ar. ! *Garral Hill* (94); LDe III. *Moine Mohr B* (98); Fl VIII; Chesters 1931. ! *Loch Droma* (105); LDe I; Kirk and Godwin 1963. *Gort* (H15); Ho II or III and III or IV; Jessen *et al.* 1959. *Roundstone 2* (H16); Fl V; Jessen 1949. *Timahoe* (H19); Fl VII; Synnott 1970 pc. ! *Lissue* (H38); Fl VIII; t; Celtic ar; Mitchell 1951. ! *Moville* (H34); Fl only; McMillan 1957.

The range of this species, which occurs throughout the British Isles, includes Eurasia and North America as well as New Zealand. The ecological amplitude is very great (Briggs 1965); few bryophytes occur in as many different habitats as this species.

Five Late Devensian records point to the range having been widespread for at least 12,000 years while the record from Upton Warren implies that this assessment may well be too conservative. All the glacial subfossils are more or less entire leaved, orthophyllous forms. Dr D. Briggs has examined many of the author's specimens. Subfossils are known from several Pleistocene deposits in the rest of Europe.

Dicranum undulatum Brid.
[*D. bergeri* Bland.]

Decoy Pool Wood (6); Fl VIII; det P. W. Richards; Clapham and Godwin 1948 as *D. bergeri.* ! *Westhay* (6); Fl VII*b*. ! *Cranberry Rough* (28); Fl VIII; Godwin and Tallantire 1951 as *Campylopus* sp. *Trundle Mere* (31); Fl VII; t; Godwin and Clifford 1938; Fl VII*a* and VII*b*; Vishnu-Mittre 1959. ! *Holme Fen* (31); Fl VII*b*. ! *Whixall Moss* (40); Fl VIII. *Malham Tarn Moss* (64); Fl VIII; found in peat some 18 inches below the surface by Dr F. Rose and Mr C. Sinker, 1963 pc. ! *Flanders Moss* (87); Fl VII*a* and VII*b*.

D. undulatum has a scattered, very local range in the British Isles (fig. 39) where it is confined to oligotrophic mires, almost all of which are ombrogenous bogs. In Fennoscandia, where it is a much commoner species, *D. undulatum* occurs in a wider range of habitats including rocks and coniferous forest.

All the British subfossils are post-zone VI. Perhaps one can envisage an increase of *D. undulatum* with the spreading of ombrogenous bog in zone VII. In four or five of the British localities the species has become extinct as a result of anthropogenic destruction of the mires. This factor may have been important in producing the present disjunct range; four of the subfossils derive from zone VIII.

D. undulatum has a wide range in the more northerly latitudes of Eurasia and North America but does not reach the high Arctic. The earliest subfossil comes from the Aftonian interglacial of Iowa (Steere 1942) and there are Flandrian records from Belgium (Vanden Berghen 1950) and Germany (Grosse-Brauckmann 1962) but only one glacial subfossil is known, that from Reichermoos,

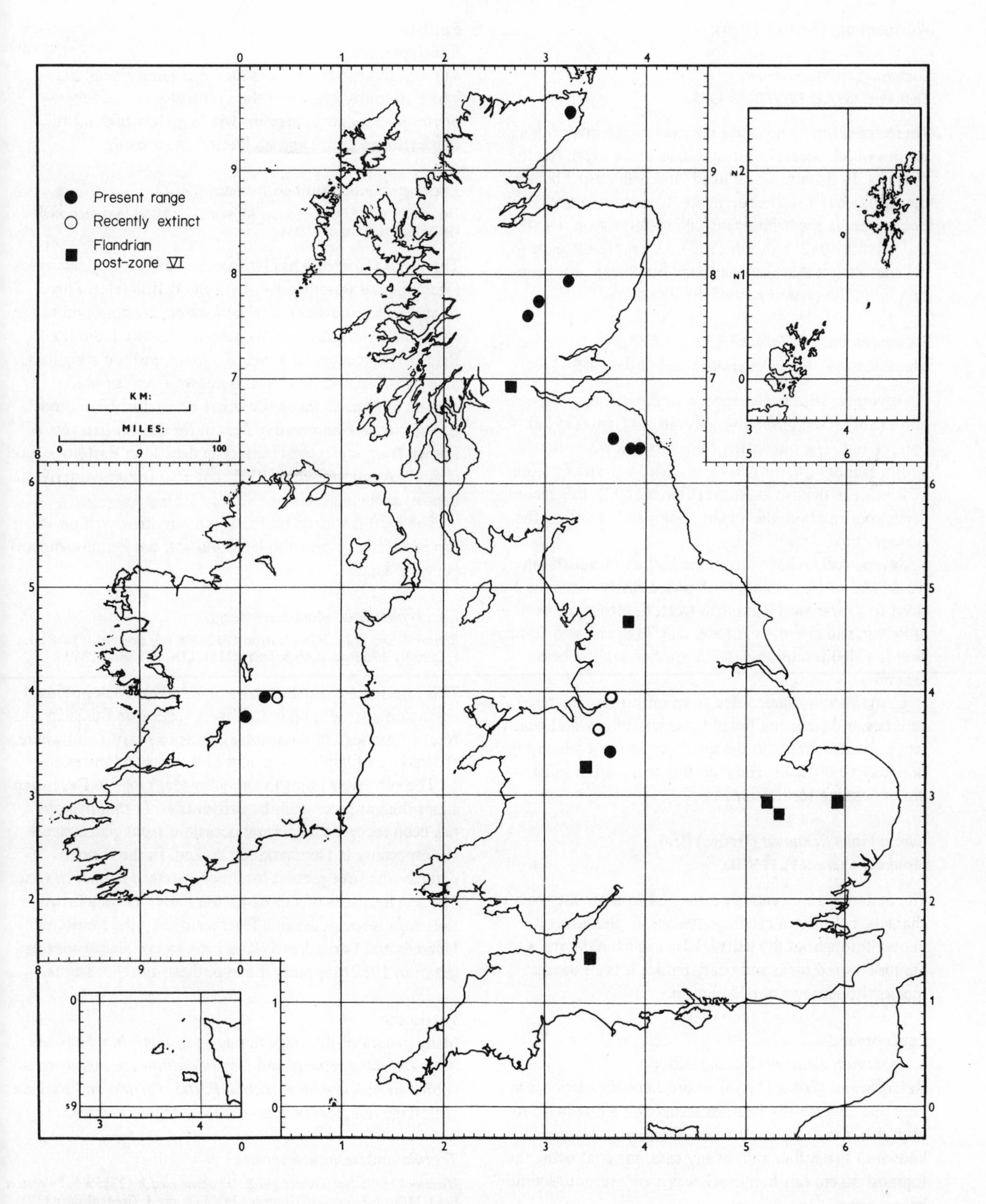

Fig. 39. Flandrian subfossils and present range of *Dicranum undulatum.*

Wurttemburg (Bertsch 1924).

Dicranum spurium Hedw.
Chat Moss (59); Fl IV-V: Birks 1965*a*.

There appear to be no other Pleistocene records of this species which occurs in Europe, Asia and eastern North America. In Britain, where heathland is the usual habitat, the species has a scattered, markedly eastern, range. In Fennoscandia the habitats include rock outcrops (Ahti and Isoviita 1962; Nyholm 1955). The subfossil leads to the speculation that the range may have been more continuous in the earliest part of the Flandrian.

T *Campylopus pyriformis* (Schultz) Brid.
Elan Valley Bog (46); Fl VII*a*; Moore and Chater 1969.

Campylopus, with twelve species in the British Isles, is remarkably rare in the subfossil state, as Gams (1932) pointed out. The only other discovery known to the author is that of *C. atrovirens* recorded tentatively from a Pleistocene deposit at Rome (Clerici 1892); this record needs confirmation, the locality being well south of the present range.

Several species have ranges the history of which one would like to understand. *C. shawii*, a British endemic confined to the far west and north west, *C. setifolius*, western in Britain and elsewhere known only in Spain, and *C. atrovirens*, a Mediterranean-Atlantic species, are the best examples.

C. pyriformis, unrecorded from only a few scattered vice-counties in Britain and Ireland, has an ecology and world range broadly similar to the next species from which separation may be difficult. The record is best held in some doubt pending verification.

Campylopus flexuosus (Hedw.) Brid.
!*Seathwaite Tarn* (69); Fl VIII.

The material is well enough preserved to show the specific characters clearly. A calcifuge species, *C. flexuosus* is known throughout the British Isles in such habitats as humus-covered rocks and peaty banks. It is an oceanic species little known outside Europe.

Leucobryaceae
Leucobryum glaucum (Hedw.) Schimp.
Ratcliffe and Walker (1958) record *Leucobryum* peat in only one of forty-five profiles across Snibe Bog in Galloway; this is the only occurrence of *Leucobryum* peat known to the author and, in any case, the peat being the topmost 30 cm of a hummock was a very recent accumulation. See p. 41.

Pottiales
Encalyptaceae
The Encalyptaceae includes the large genus *Encalypta*, which is poorly known in the subfossil state, and *Bryobrittonia*, a monotypic genus with a mainly high latitude range (Steere 1953) and no Pleistocene records.

Encalypta undetermined species
!*Wretton* (28); EDe; t. !*Upton Warren* (37); MDe. !*Baggotstown* (H8); Ho III; Dickson 1964*a*.

The genus *Encalypta* has five species, all exacting calcicoles (McVean and Ratcliffe 1962), in the British Isles. Three, *E. alpina, E. ciliata* and *E. rhabdocarpa*, are confined to the highland zone. Only the last-named is known from the British Pleistocene. However, *E. alpina*, easily distinguishable by its acuminate leaves, is known from the glacial deposit at Skaerumhede, Denmark (Hesselbo 1910) and *E. vulgaris*, a widespread species in the British Isles, is known from an Upper Pleistocene deposit in Eastern Siberia (fig. 40; Abramova *et al.* 1965) and also from an interglacial deposit at Weimar (Gams 1932, p. 309).

The Upton Warren subfossil, a diminutive, well preserved tuft, may prove to be *E. mutica*, the Fennoscandian endemic species.

Encalypta rhabdocarpa Schwaegr.
Ponder's End (21); MDe; Warren 1912. *Whitrig Bog* (81); LDe III; t; Conolly 1963 pc. *Ballybetagh* (H21); LDe II; Jessen 1949.

This rare species usually grows on montane rock but also occurs on coastal sand in Caithness. The wide Eurasian and North American distribution extends to Peary Land where the variety of habitats includes bird rocks and mires.

The subfossils point to a much wider range in Devensian times than at present in the British Isles. *E. rhabdocarpa* has been recorded on several occasions from glacial deposits especially in Denmark and Poland. In the former country the four present localities are clearly relict in view of three localities of subfossils. Similarly, in Poland, where the range is restricted to a few localities in the Southern Uplands and Carpathians (Kuc 1964) a last glacial subfossil (Szafran 1952) supports the hypothesis of relict status.

Pottiaceae
Many genera of this large family such as *Aloina, Phascum, Acaulon, Pleurochaete* and *Leptodontium* are unknown as subfossils. Even genera such as *Pottia, Tortula* and *Barbula* with large numbers of species have meagre histories.

Tortula undetermined species
!*Hoxne* (25); LAn; Turner 1968. !*Broome Heath* (27); Wo. !*Wretton* (28); MDe; t. !*Barnwell Station* (29); LDe pre-I. *Great Billing* (32); MDe; Morgan 1969. !*Barnsley Park* (33); Fl VIII*; Roman ar.

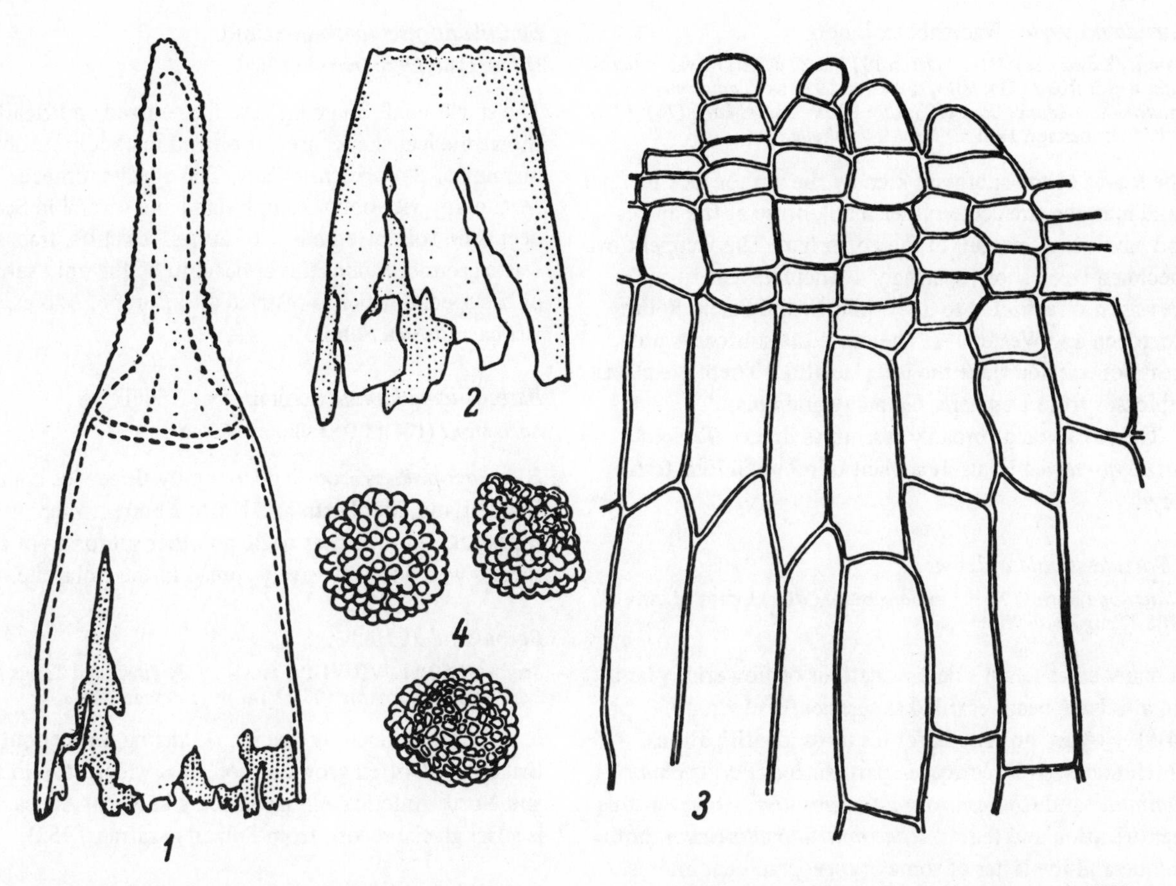

Fig. 40. Sporophyte remains and spores of *Encalypta vulgaris*? from the Upper Pleistocene of Siberia. From Abramova *et al.* 1965.

!*Low Wray Bay* (69); LDe II. *Goat's Water* (69); LDe III-IV; Pennington 1964. !*New Dry Dock* (76); LDe only. *Whitrig Bog* (81); LDe II; Conolly 1963 pc. *Loch Mealt* (104); LDe II; Birks 1969 pc. !*Baggotstown* (H8); Ho II, III, IV or EWo; Dickson 1964*a*. !*Kildromin* (H8); Ho I. !*Ardcavan* (H12); LDe or Fl. !*Drumurcher* (H32); LDe III.

Most of the above may be referrable to *T. ruralis* but firm or even tentative identification is precluded by the fragmentary and badly preserved state of the subfossils which in this genus usually lack the leaf tips.

Of the seventeen British species only two (*T. ruralis* and *T. norvegica*) are known with certainty from Pleistocene deposits. Indeed, the genus is poorly represented elsewhere in the subfossil state, apart from the two mentioned species. Schütrumpf (1937) recovered *T. mucronifolia* from a Late Weichselian deposit at Meiendorf, Germany, and De Vries and Bird (1965) recorded the same species from a Late Wisconsin deposit in Saskatchewan. *T. muralis* is recorded from an interglacial deposit at Weimar (Gams 1932, p. 309). However, there appear to be no other species.

T. vahliana, T. standfordensis (which may be an introduction), *T. vectensis, T. marginata* and *T. cuneifolia* are all markedly southern in the British Isles. Nothing is known of their Pleistocene history.

Tortula ruralis (Hedw.) Gartn., Meyer & Scherb.

!*Upton Warren* (37); MDe. *Kirkmichael 3a* (71); LDe I; Dickson *et al.* 1970.

Both these determinations rest on subfossils derived from robust plants without central strands and with leaves having margins distinct at the base and revolute close to the apices which are more or less obtuse.

T. ruralis is unrecorded from several Irish and Scottish vice-counties. Otherwise, it is a common species of a wide variety of habitats including soil, rock, tree trunks and thatch. Various authors mention a calcicolous tendency.

The wide Holarctic range reaches the highest latitudes. Brassard and Steere (1968, p. 379), referring to Bathurst Island, N.W.T., where they recovered the species from peat samples, state 'One of the most widespread and abundant species in the area, as in all arctic areas'.

Glacial subfossils are known from New York, Poland, Germany and Denmark and there is a Pliocene one from Russia.

Tortula norvegica Wahlenb. ex Lindb.

!*Ponder's End* (21); MDe; Warren 1912 as *T. aciphylla* var. *mucronata. Angel Road* (21); MDe; t; Warren 1912 as *T. aciphylla* var. *mucronata.* !*Keppel Cove* (69); LDe III-IV. *Garvel Park* (76); LDe/ Fl IV*; Robertson 1881 as *Barbula aciphylla.*

The leaves of the subfossils seen by the author lack distinct basal margins, are acute rather than obtuse at the apices and have plane margins in the upper half. The Keppel Cove specimen bears short, sparingly denticulate leaf tips. *T. norvegica* is restricted to a few hills in Mid Perth, South Aberdeen and Westerness. Therefore the subfossils prove great contraction since the last glaciation. There are glacial subfossils from Denmark, Germany and Poland.

The world range broadly resembles that of *T. ruralis* but *T. norvegica* is largely a plant of mountainous territory.

T *Tortula subulata* Hedw.

Winterhope Burn (79); Fl? var. *angustata* (Wils.) Limpr.; Lewis 1905 *T. angustata* Wils.

In many cases Lewis's determinations of flowering plant remains have been regarded as suspect (Godwin 1956, pp. 64-5). He gave no criteria for his moss identifications, which may well be correct in part, such as *Polytrichum commune* and *Rhacomitrium lanuginosum*. However, this identification and that of *Rhacomitrium ellipticum*, both unique and the latter of some bryogeographical interest, must be regarded with reservation.

Cinclidotus fontinaloides (Hedw.) P. Beauv.

!*Loch Cuithir* (104); Fl VII*a*; t; Vasari and Vasari 1968.

The generic determination is certain because of the characteristic thick border of the leaves. *C. fontinaloides*, widespread in the British Isles, is an aquatic species occurring in fluviatile and lacustrine positions liable to periodic emergence. There is a marked calcicolous preference; indeed Spence (1967) states that it is restricted to base-rich waters in Scottish lochs. That this is the only British subfossil is surprising in view of the habitat. The only other subfossil was found at Ripetta near Rome (Clerici 1892).

Barbula undetermined species

!*Broome Heath* (27); Wo. *Leeds* (64); Fl VI/VII/VIII*; Raistrick and Woodhead 1930. *Ralaghan* (H30); LDe III; Jessen 1949.

Only five of the twenty British species of this large genus have been recovered from the British Pleistocene. *B. convoluta* and *B. unguiculata* are known from glacial deposits in Poland and *B. rigidula* in Germany.

Barbula hornschuchiana Schultz

!*Ilford* (18); LWo; Dickson 1964*b*.

The single small fragment is well preserved and clearly shows the leaf shape, areolation and markedly revolute margins of *B. hornschuchiana.* This species, unrecorded from many vice-counties in Ireland and several in Scotland, occurs on soil, often basic, of dunes, sea cliffs, tracks and waste ground. Again, this appears to be the only subfossil of the species which is restricted to Europe, Azores, North Africa and Asia Minor.

Barbula icmadophila Schimp. ex. C. Müll.

Loch Mealt (104); LDe I; Birks 1969 pc.

B. icmadophila is recorded from only three vice-counties, Cumberland, Mid Perth and North Ebudes, where it is rupestral. There appear to be no other subfossils of this species which is extensively spread in the Holarctic.

Barbula fallax Hedw.

Seaforth (59); Fl VII/VIII*; Travis 1913; Travis and Travis 1913. *Hailes* or *Corstorphine* (83); LDe only; Bennie 1894*a*.

B. fallax, a calcicolous species, occurring throughout the British Isles, often grows on soil. It is widespread in Eurasia and North America and also occurs in North Africa. There is a last glacial record from Poland (Szafran 1952).

Barbula spadicea (Mitt.) Braithw.

Ormaig (98); 'Arctic Shell Bed'; LDe? only; Walton 1962 pc.

This species, also calcicolous, is almost entirely restricted to the highland zone apart from a few occurrences in the midlands, south and southeast of England. The habitats are rocks and sand by streams. There are no other subfossils.

Barbula recurvirostra (Hedw.) Dix.

Ponder's End (21); MDe; Warren 1910 as *B. rubella. Low Wray Bay* (69); LDe II; Pennington 1962.

B. recurvirostra, a species of calcicolous tendency, is known throughout the British Isles apart from a few Irish vice-counties and Huntingdonshire. Habitats, which may be shaded or open, include soil, rock (often stonework), and tree bases. A species with a wide Holarctic range, it extends south to Central Africa, New Guinea and New Zealand. The highest northern latitudes are reached. There are Pleistocene records from Denmark, Czechoslovakia, Saskatchewan and Michigan.

T *Gymnostomum* undetermined species

Garvel Park (76); LDe/Fl IV*; Robertson 1881.

This record means very little; the concept of the genus has narrowed since Fergusson identified Robertson's material.

This is also true of the record of *Trichostomum* sp.

Tortella fragilis (Hook. & Wils.) Limpr.

Garvel Park (76); LDe/Fl IV*; Robertson 1881 as *Barbula fragilis. Faskine* (77); De only; Bennie 1894*b* as *Mollia fragilis* L. *Hailes* or *Corstorphine* (83); LDe only; Bennie 1894*a* as *Barbula fragilis* Bruch & Sch. *Mapastown* (H31); LDe III; Mitchell 1953.

The species has often been misidentified in the past, as indicated by the large number of doubtful records in the Census Catalogue which gives only Fife, Mid Perth, Angus and Caithness with certainty. Perhaps all the nineteenth-century records of subfossils should be regarded very cautiously. But there is no need to question Holmen's record from Mapastown; this puts the species into the category which has vacated Ireland since Late Devensian times, if all the living specimens have been misidentified (Warburg 1963).

T. fragilis, a calcicole of rock and sand, has a wide Eurasian and North American distribution. The disjunct range in Denmark is shown on the map by Christensen (Holmen 1959).

The range, wide in Eurasia and northern North America, extends north to Peary Land. Subfossils have been found in Poland, Denmark, Germany and the Taimyr Peninsula.

Tortella tortuosa (Hedw.) Limpr.

!*Upton Warren* (37); MDe. !*Low Wray Bay* (69); LDe II. *Garvel Park* (76); LDe/Fl IV*; Robertson 1881 as *Barbula tortuosa. Faskine* (77); De? only; Bennie 1894*b* as *Mollia tortuosa* L. *Ballybetagh* (H21); Fl VI; Jessen 1949.

Widespread but rare or absent in some areas such as East Anglia, *T. tortuosa* is a calcicolous species of rock and soil usually in open situations. The range, extensive in the Northern Hemisphere, also includes Fuegia. *T. tortuosa* occurred in widely scattered areas of the British Isles during the last glaciation. There are glacial subfossils from Poland, Russia and North America.

T *Tortella inclinata* (Hedw. f.) Limpr.

Faskine (77); De? only; Bennie 1894*b* as *Mollia inclinata* Hedw.

Like the early records of *T. fragilis* this too must be treated with reserve. However, in defence of Hobkirk, who identified Bennie's material, he recognised three species of *Tortella* in the deposit. There are certain records of this species from only fifteen vice-counties scattered widely from Surrey to East Sutherland and Co. Down. *T. inclinata* has been recorded from a Late Wisconsin site at Lockport, New York State (Miller 1969).

T *Trichostomum* undetermined species

Garvel Park (76); LDe/Fl IV*; Robertson 1881.

See T *Gymnostomum* sp.

Grimmiales

Grimmiaceae

With the exception of *Bryum*, *Grimmia* with thirty-three species is the largest moss genus in the British Isles. Many are confined to the highland zone and several have greatly restricted ranges. *G. trichodon, G. atrofusca, G. anodon, G. elatior* and *G. agazissii* are good examples.

Grimmia appears to be very poorly represented in Pleistocene deposits; this is hardly surprising because the preference of many of the species for dry rocks must greatly curtail the chances of fossilisation. In contrast, the much smaller genus *Rhacomitrium* with eight species in the British Isles is very well represented in Pleistocene, especially Late Devensian deposits.

Grimmia undetermined species

Faskine (77); De? only; Bennie 1894*b*. *Hailes* or *Corstorphine* (83); LDe; Bennie 1894*a*. *Fort William* (97); Fl V or VI*; Dixon 1910.

Dixon (1910) states that the few fragments of the Fort William material may be *G. muelenbeckii* or *G. decipiens* var. *robusta* (as *G. robusta* Ferg.).

Grimmia apocarpa Hedw.

!*New Dry Dock* (76); LDe; with a sporophyte.

The single badly preserved fragment bears a scarcely exserted capsule (plate 17). *G. apocarpa*, a very variable species of calcicolous tendency, occurs throughout the British Isles on rock and walls. The range includes much of both the Northern and Southern Hemispheres.

There is a glacial record from Denmark (Hesselbo 1910) and an early Flandrian one from Frankental in Vosges where *G. ovalis* was also found (Firbas *et al.* 1948).

Rhacomitrium undetermined species

!*Stannon* (2); LDe. !*Broome Heath* (27); Wo. !*Elan Valley* (46); LDe III; given as *R. aquaticum* in Moore 1969. !*Low Wray Bay* (69); LDe II. *Blelham Tarn Kettle Hole* (69); LDe I and II. !*Kirkmichael 1 and 3a* (71); LDe I and III; Dickson *et al.* 1970. *Nick of Curleywee* (74); Fl IV; Moar 1969*a*. !*New Dry Dock* (76); Fl IV*. *Whitrig Bog* (81); LDe II and III; Conolly 1963 pc. *Corstorphine* or *Hailes* (83); LDe only; Bennie 1894*a*. !*Drymen* (86); LDe II and III; Vasari and Vasari 1968. *Morven 1, 2 and 3* (92); Fl V or VI; Romans *et al.* 1966. !*Loch of Park* (92); LDe II; Vasari and Vasari 1968. !*Garral Hill* (94); LDe. !*Loch Oich* (96); LDe. !*Loch Cuithir* (104); LDe III; Fl III-IV, IV, V and VIII; Vasari and Vasari 1968. !*Loch Fada* (104); LDe I, II and III; Birks 1969 pc; Fl III-IV; Vasari and Vasari 1968. *Loch Meodal* (104); LDe I, II and III; Fl IV; Birks 1969 pc. *Loch Mealt* (104); LDe I, II and III; Birks 1969 pc. *Loch Cill Chriosd* (104); LDe I and III; Birks 1969 pc. *Lochan Coir 'a' Ghobhainn* (104); LDe II and III; Fl IV; Birks 1969 pc. !*Duartbeg* (108); Fl VII*a*; Moar 1969 *b*. !*Yesnaby* (111); LDe only; Moar 1969*b*. !*Newtonbabe* (H31); LDe II. !*Drumurcher* (H32); LDe III. !*Derryvree* (H33); MDe.

Many of the above are very fragmentary and badly preserved. Often the remains lack leaf tips, the very feature most useful in determination. The loss may occur before fossilisation at least in the case of *R. lanuginosum.* Dead material from large cushions of this species often lacks the distinctive hyaline tips. It is likely that the majority of these undetermined subfossils belong to *R. lanuginosum.*

Both *R. aciculare* and *R. aquaticum* appear to be unknown from the Pleistocene. This is a somewhat unexpected lack in view of the aquatic or semi-aquatic habitats.

The recently discovered material from Derryvree in Co. Fermanagh, though merely a tiny battered scrap, is the only proof of the Middle Devensian presence of the genus; all the other numerous Devensian records are from zones I to III.

T *Rhacomitrium ellipticum* (Turn.) B. & S.

Yellow Tomach (73); Fl? only; Lewis 1905 as *Racomitrium ellipticum* B. & S.

This record must be regarded with suspicion for the same reasons as that of *Tortula subulata.*

Rhacomitrium fasciculare (Hedw.) Brid.

Lake Windermere (69); LDe I and II; Pennington 1962. *Low Wray Bay* (69); LDe II; Pennington 1962. *Skelsmergh Tarn* (69); LDe II; Walker 1955. *Faskine* (77); De? only; Bennie 1894.
Loch Meodal (104); LDe I; Birks 1969 pc. *Loch Fada* (104); LDe I and III; Birks 1969 pc. *Loch Mealt* (104); LDe I and II; Birks 1969 pc. *Loch Cill Chriosd* (104); LDe III; Birks 1969 pc.

R. fasciculare is rare or absent in much of southern and eastern England but elsewhere is abundant as a rupestral species in the mountains. It has a wide range in Europe, Asia and North America and has been recorded from Arctic regions such as southern Spitsbergen, Jan Mayen and Greenland.

The subfossils established a wide Late Devensian range at least in northern Britain. There are few records from outside the British Isles.

Rhacomitrium heterostichum (Hedw.) Brid.

!*Nant Ffrancon* (49); LDe only. !*New Dry Dock* (76); LDe only. *Fort William* (97); Fl V or VI*; Dixon 1910.

This species has a broadly similar British range and ecology to *R. fasciculare.* More Devensian discoveries are to be expected; there seem to be few subfossils from the rest of

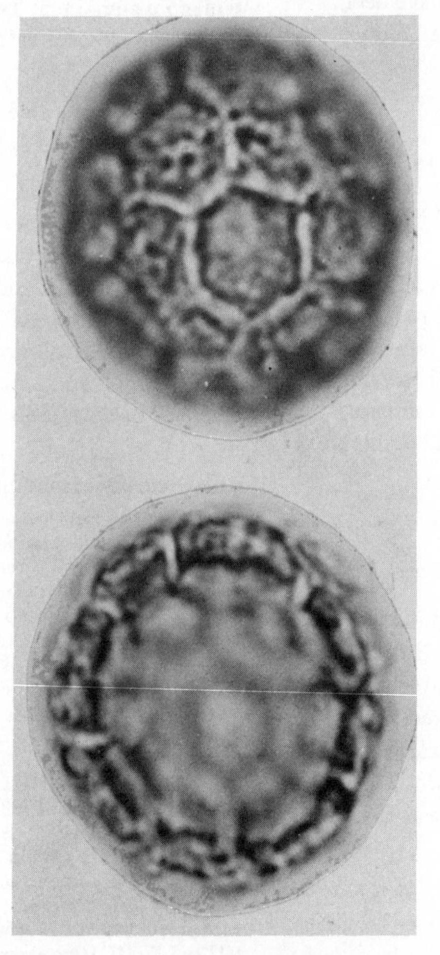

(*a*)

(*b*)

Plate 17. (*a*) Spores of *Tilletia sphagni.* c. 14 μm diam. Photos by A. M. M. Berrie. (*b*) *Grimmia apocarpa* from New Dry Dock. Fragment which bears a sporophyte is about 4 mm long.

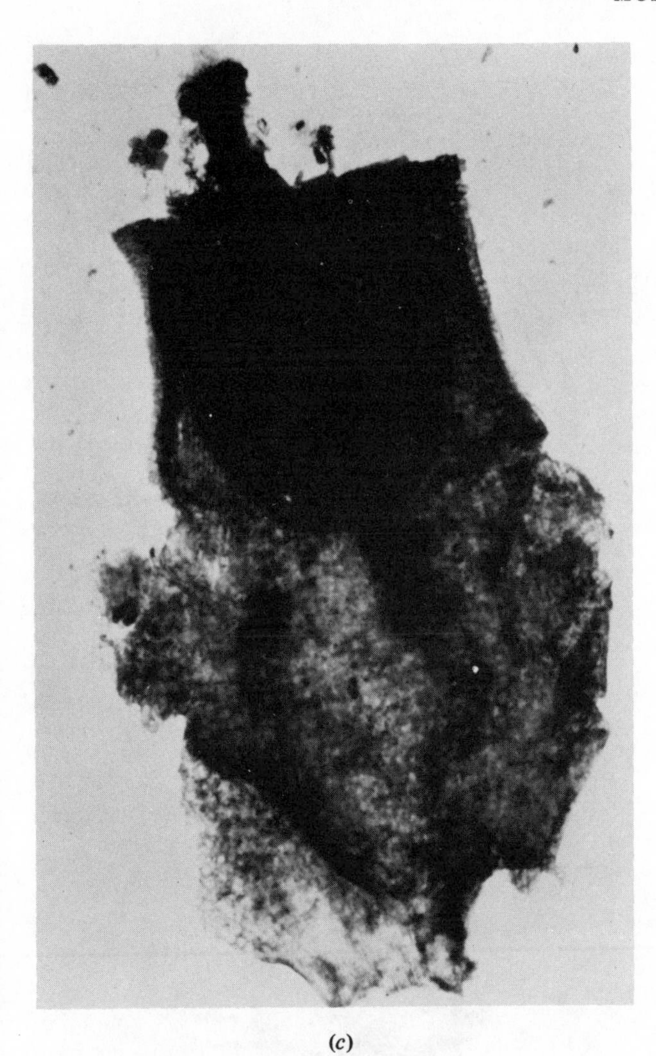

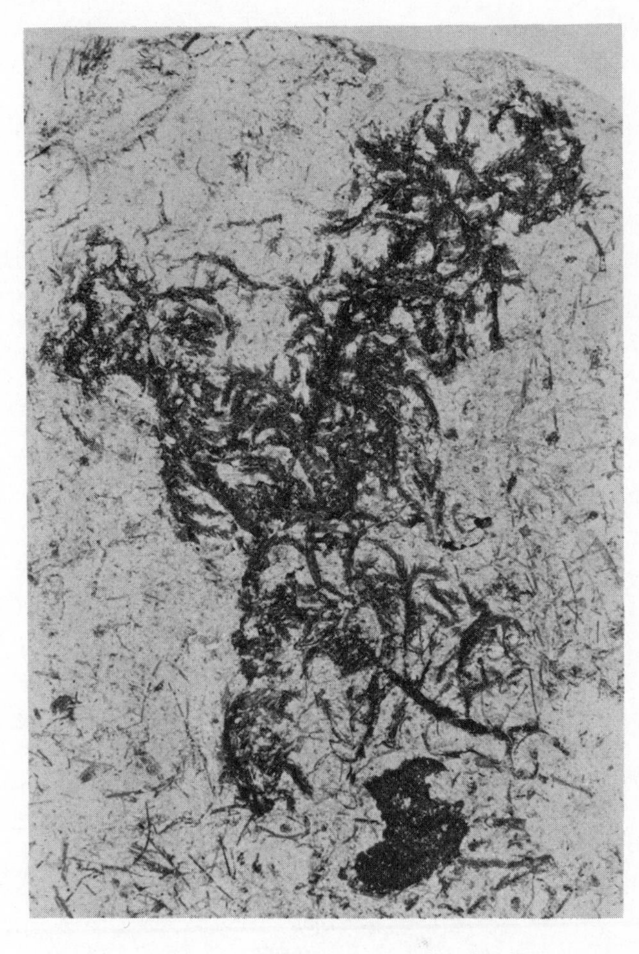

(*c*) (*d*)

Plate 17. (*c*) Sporophyte of *Splachnum sphaericum* from zone III of Whitrig Bog. (*d*) *Hylocomium splendens.* Well preserved, richly branched subfossil about 3 cm long, embedded in zone I silts from Loch Droma.

Europe where the species is widespread as it is in Asia and North America.

Rhacomitrium canescens (Hedw.) Brid.

Fort William (97); Fl V or VI*; Dixon 1910.

That this is the only record of such a common, easily recognised species is surprising. Perhaps it has been overlooked to some extent; indeed there are few records from the Pleistocene anywhere.

R. canescens occurs very widely in the British Isles, often on coarse-grained soils including alluvium in the flood zone of rivers. It is a frequent pioneer of glacial outwash and moraine as in central Norway (Faegri 1933) and Iceland (Persson 1964). Perhaps the subfossils from Skaerumhede grew in such a habitat. The range is very extensive in Eurasia and North America; the highest latitudes in Greenland are reached.

Rhacomitrium lanuginosum (Hedw.) Brid.

!*Nant Ffrancon* (49); LDe only. *Kinder 6* (57); Fl VII*b*–VIII; Conway 1954. *Bleaklow* (57); Fl VI, VII*b*-VIII and VIII; Conway 1954. *Malham Tarn Moss* (64); Fl VII*a* and VII*b*; Pigott and Pigott 1963. *Yorkshire Pennines*; Fl VII/VIII*; Burrell 1924. !*Seathwaite Tarn* (69); LDe II. *Swathbeck and Moss Flats* (69); Fl VIII; Johnson and Dunham 1963. *Garvel Park* (76); LDe/Fl VI*; Robertson 1881. !*New Dry Dock* (76); LDe only and Fl IV*. *Rhum* (104); Fl only; Eggeling 1965. *Loch Meodal* (104); LDe I; Birks 1969 pc. *Loch Fada* (104); LDe I and II; Birks 1969 pc. *Loch Mealt* (104); LDe I and II; Birks 1969 pc. *Loch Cill Chriosd* (104); LDe III and Fl IV; Birks 1969 pc. *Lochan Coir 'a' Ghobhainn* (104); LDe III; Birks 1969 pc. !*Loch Droma* (105); LDe I; Kirk and Godwin 1963. *Loanan Valley* (108); Fl only; Lewis 1907. *Loch Urigill* (108); Fl only; Lewis 1907. *Bragor* (110); Fl only; Lewis 1907. *Gort* (H15); LAn and Ho III; Jessen *et al.* 1959. *Roundstone 1* (H16); LDe II and III; Jessen 1949. *Roundstone 2* (H16); LDe I and II; Fl V; Jessen 1949. *Ballybetagh* (H21); LDe II; Fl VI; Jessen 1949. *Castlelackan Demesne* (H27); Fl VII-VIII; Jessen 1949. Walker

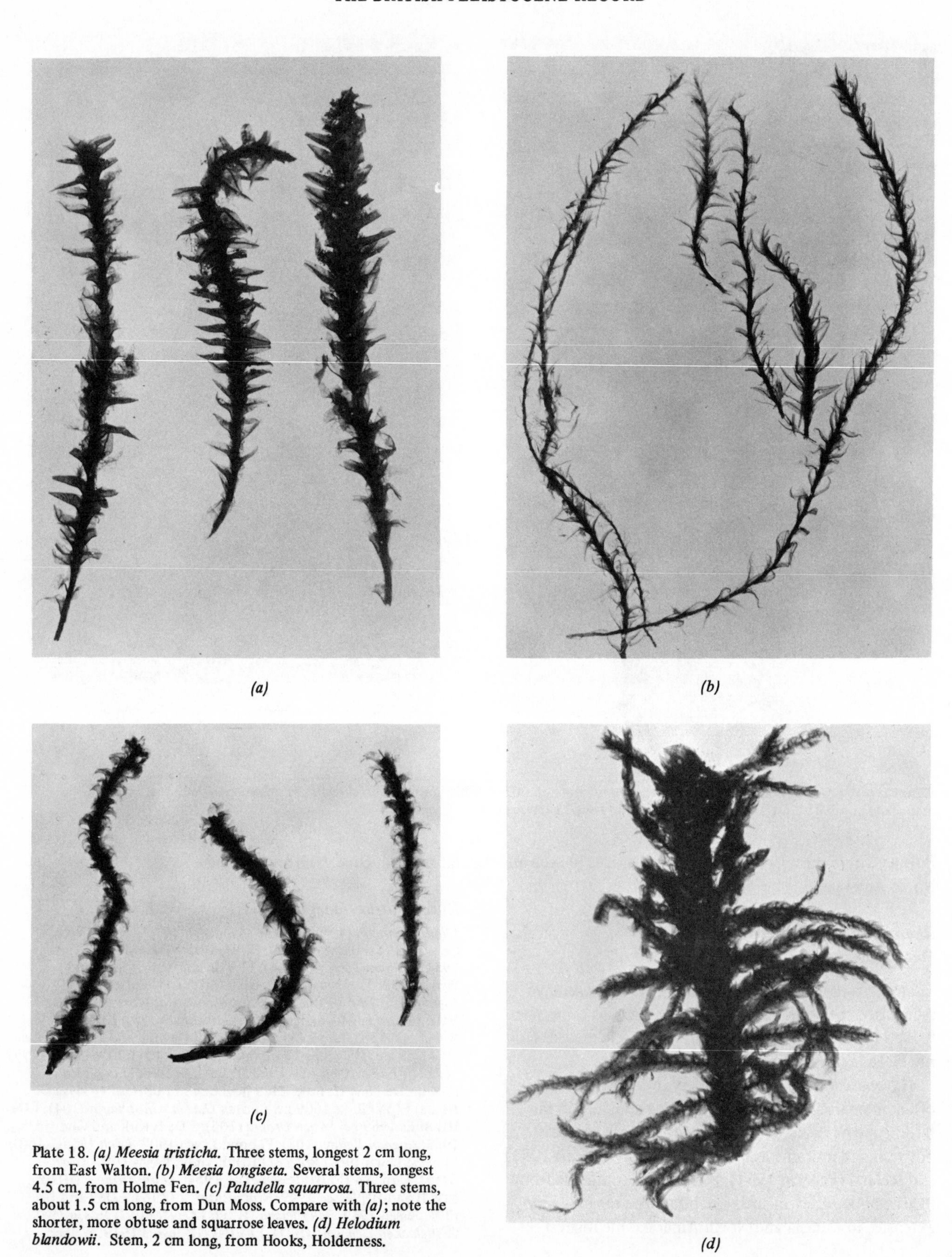

Plate 18. *(a) Meesia tristicha.* Three stems, longest 2 cm long, from East Walton. *(b) Meesia longiseta.* Several stems, longest 4.5 cm, from Holme Fen. *(c) Paludella squarrosa.* Three stems, about 1.5 cm long, from Dun Moss. Compare with *(a)*; note the shorter, more obtuse and squarrose leaves. *(d) Helodium blandowii.* Stem, 2 cm long, from Hooks, Holderness.

and Walker (1961, p. 170) mention 'occasional occurrences' in late Flandrian peats of the eight Irish bogs they studied.

R. lanuginosum is amongst the commonest of Pleistocene bryophytes. The great majority of records come from Flandrian peats, a few from Late Devensian deposits (but see p. 92) and only one from an interglacial.

The habit of this species is distinctive and can be well seen in some of the above including those of the Late Devensian localities, Nant Ffrancon, Seathwaite Tarn and Loch Droma (fig. 41).

The species is lacking in several vice-counties of southern and eastern England. Otherwise it is a common moss which in the highland zone gives its name, *Rhacomitrium* heath, to large stretches of the vegetation of mountain tops. It is a successful pioneer of freshly exposed substrata such as lava fields, and moraine and outwash in many parts of the world, Iceland, Hawaii, Patagonia and New Zealand.

Apart from growth on mineral soils, *R. lanuginosum* can be abundant as large cushions on the dry-growing stages of ombrogenous bog in Europe and elsewhere, for instance, British Columbia.

Many bryologists have considered *R. lanuginosum* to be one of the commonest mosses of the Arctic. However, greatest luxuriance is reached in the oceanic areas of temperate regions. This oceanic tendency has been noted by various authors including Jalas (1955) who plotted the markedly maritime range of the species in Finland.

R. lanuginosum is often considered cosmopolitan; the range in both the Northern and Southern Hemispheres is very great and includes many remote islands. However, there appear to be few records from Africa.

The subfossils point to widespread occurrence in western Britain in Late Devensian times (fig. 42); in this one is reminded of *Antitrichia curtipendula* and *Hylocomium splendens.*

Funariales

Funariaceae

Nine of the eleven British genera of the Funariales are unknown as subfossils. In most cases such as *Discelium, Physcomitrium, Physcomitrella, Nanomitrium, Ephemerum* and *Oedipodium* this is hardly surprising; minute habit or habitat or both are unlikely to have led to fossilisation and recovery. The monotypic *Oedipodium* has a unique disjunct range (fig. 10) discussed on p. 2.

Funaria undetermined species

Esthwaite (69); Fl VII*b*; spores; Franks and Pennington 1961.

When they made the record of *Funaria* spores no doubt Franks and Pennington referred to *F. hygrometrica*, the

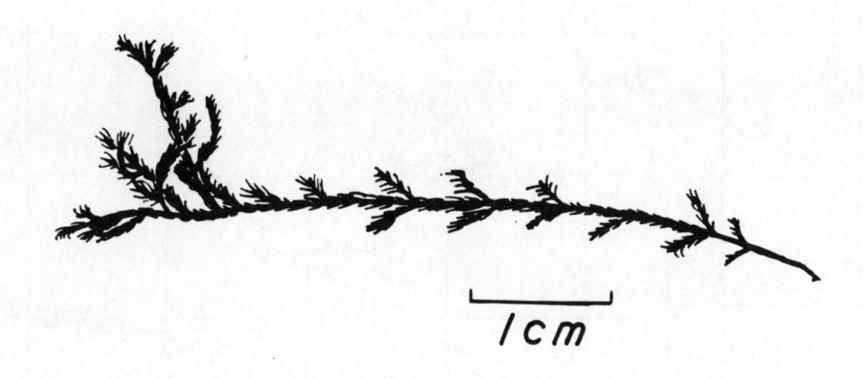

Fig. 41. Remains of *Rhacomitrium lanuginosum* from Loch Droma.

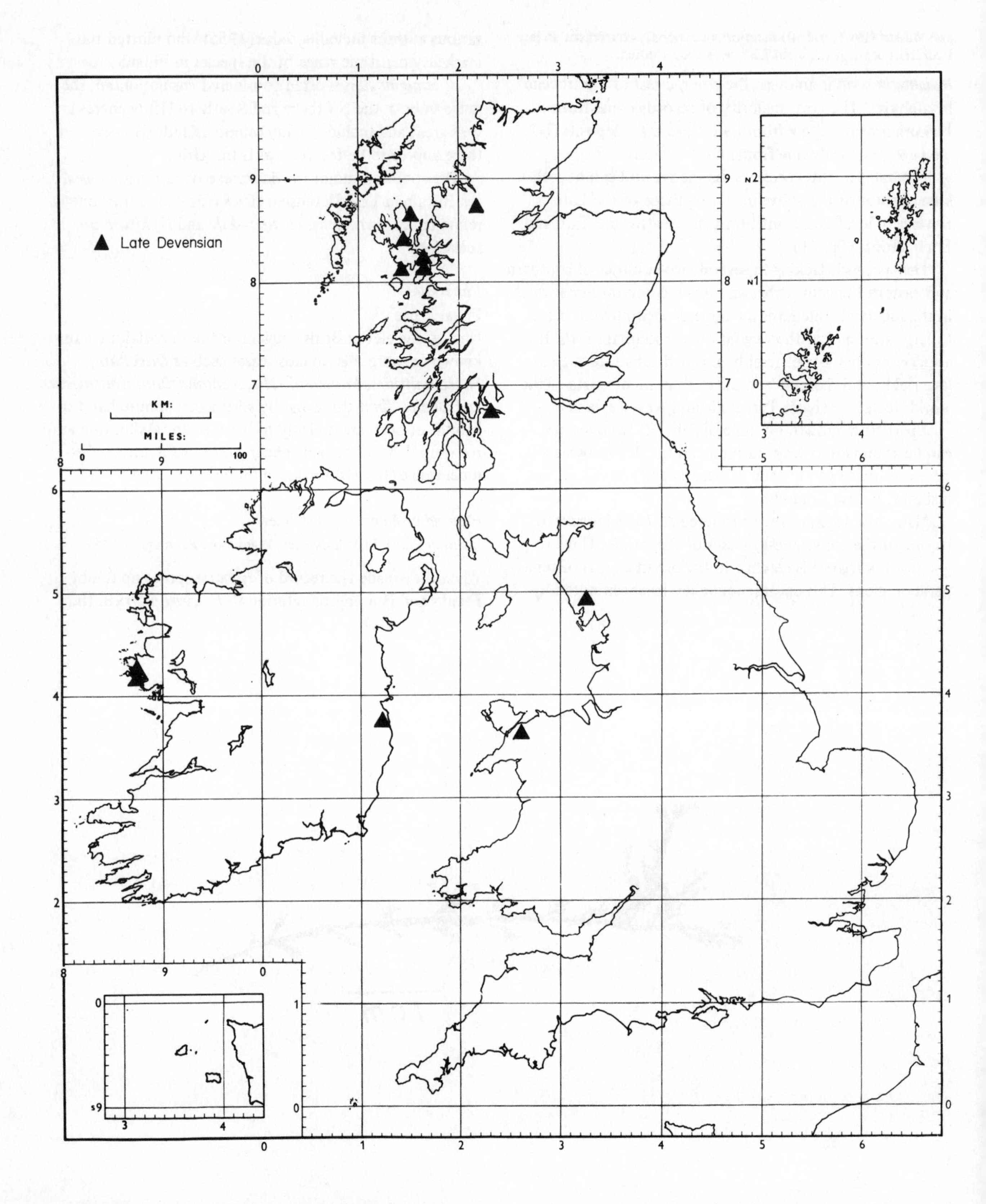

Fig. 42. Late Devensian subfossils of *Rhacomitrium lanuginosum*.

cosmopolitan hemerophilous species. The record must be regarded as tentative. Spores of *F. hygrometrica*, though bearing sculpturing which is very distinctive as seen by electron microscopy, can scarcely be determined with certainty by light microscopy.

That *F. hygrometrica* should occur in horizons representing forest clearance by prehistoric man is very likely. Ground enriched by burning is a favoured habitat of the species.

Funaria hygrometrica Hedw.
!*Crymlyn Bog* (41); Fl VI; capsule.

The discovery consists of a well preserved capsule, easily recognisable by shape and peristone teeth, and a leaf fragment. *F. hygrometrica* occurs throughout the British Isles where it grows in a variety of habitats, very often burnt ground. Capsules have been recovered from three Middle and one Upper Pleistocene deposits in Russia (Abramova *et al.* 1965).

Splachnaceae
Among the British Splachnaceae *Haplodon wormskjoldii* has an arctic as distinct from an arctic-alpine range; it does not occur in Central Europe. *Splachnum vasculosum* Hedw. is a widespread but rare species of the highland zone. The Splachnaceae are very poorly represented in Pleistocene deposits. Fluviatile deposits from northern Sweden yielded the only other record, namely *S. vasculosum* (Andersson 1896).

Mapastown (H31); LDe I and II; Mitchell 1953.

This record made by Dr K. Holmen is given merely as 'Splachnaceae' without qualification.

The majority of species in this family are coprophilous. Therefore it is noteworthy that at Mapastown and at Whitrig Bog skeletons of large mammals were found; the former yielded a giant Irish deer, the latter an elk. Dung of elk, often deposited in mires where the beasts graze, is a common habitat for species of *Splachnum*. Subfossil elk droppings have twice been discovered (the first is illustrated in Godwin 1956, plate VIII); however, when preserved the excreta were too fresh for growth of *Splachnum.*

Splachnum sphaericum Hedw.
!*Whitrig Bog* (81); LDe III; capsule; Conolly and Dickson 1969.

The genus *Splachnum* in particular, but also other genera of the family, exhibit a range of sporophytes by which the species can be determined. Identification can be made not only on colour and gross morphology (plate 17) but also on microscopic detail, especially the stomatal complexes. The latter character in particular enabled Conolly and Dickson (1969) to confirm the Whitrig Bog capsule as *S. sphaericum*, a widespread species of the highland zone.

Schistostegales
Schistostegaceae
Schistostega pennata (Hedw.) Web. & Mohr
Aylsham (27); VII*b**; Bronze Age ar; Clough 1971.

This discovery is of outstanding interest. The material was removed from the socket of an axe buried as part of a hoard in Bronze Age times, at Aylsham, north of Norwich.

Because the material is no longer extant the record must be held in some doubt. Mr Clough has published the following (1971, p. 165).

> The small spearhead contained a quantity of organic material. Fragments of a moss amongst this were identified by Mr C. C. Wilcock as *Schistostega pennata* (Hedw.) Hook & Taylor. The diagnostic features were the two-ranked nerveless leaves composed of large elongated cells with pointed ends. The remainder of the material was later examined by Dr J. H. and Mrs C. A. Dickson, who made the following identification of what at first seemed to be wood fragments: 'We think they are derived from a herbaceous plant rather than a tree or shrub. If this is correct a herb with some development of woody tissue is involved . . . Certainly the fragments are not from the heartwood of some tree, so it is misleading to talk of wood splinters.' Thus these fragments do not represent a wooden shaft.

No other specimen of *Schistostega* is yet known to have survived from an early period. *Schistostega* is a minute species with a unique habitat. It grows in areas which never receive direct sunlight and where the light intensity is very low, such as caves, mine shafts and deep fissures, but although it can tolerate such deep shade, it cannot exist in the total absence of light. The species is mainly restricted in its present-day distribution to southern and western Britain. The only East Anglian locality so far known was only recently discovered at Wolferton, West Norfolk; otherwise it has been recorded no nearer to Aylsham than Northamptonshire and other Midland counties. Its distribution is unlikely to have changed significantly in the last three thousand years because of its limited habitat, and it is therefore most unlikely that *Schistostega* was able to grow in the socket of this spearhead after its concealment in the ground at Aylsham. Neither was it introduced recently for this organic material adhered to the interior of the spearhead and was sealed in position by a quantity of earth. It is such a small and fragile moss that it can hardly have been placed deliberately in the socket as plugging or packing material. It therefore seems most likely that it was introduced by accident in antiquity. Possibly this spearhead had previously been concealed within an environment suitable for the moss, of which a fragment could have fallen into the socket and by chance have survived until the present day.

There remains the possibility that the *Schistostega* had magical significance. Another record in a similar context is greatly to be desired.

Eubryales

Bryaceae

With the marked exceptions of *Bryum, Pohlia* and *Leptobryum* there are no British Pleistocene subfossils of the six other genera of this family in the British flora.

Epipterygium tozeri, a Mediterranean-Atlantic species of strongly southwestern range in the British Isles, has been recovered from a Pliocene or Early Pleistocene deposit in the Caucasus (Abramova and Abramov 1959) where the species still occurs; the preservation is very fine (fig. 85).

Leptobryum pyriforme (Hedw.) Wils.

Brook (10); Fl VII*a*; Clifford 1936.

L. pyriforme is widely scattered in the British Isles, but is unrecorded from much of Ireland. Often it grows in man-made habitats such as burnt ground and pots in greenhouses. It has an almost cosmopolitan range and reaches the highest latitudes where it is unconnected with anthropogenic habitats. In Spitsbergen it grows in rock crevices and on morainic clay (Kuc 1963) and on soil below bird cliffs (Arnell and Mårtensson 1959). Favoured habitats in Peary Land are bones and old musk ox dung (Holmen 1960).

The Brook subfossil is the only one of Flandrian age. There are glacial subfossils from Denmark (Hesselbo 1910; Jessen and Nordmann 1915) and Saskatchewan (De Vries and Bird 1965). Abramova *et al.* (1965) have recorded tentatively capsules from Middle Pleistocene deposits in Russia.

Pohlia undetermined species

!*Stannon* (2); LDe only. !*Honeygore Track* (6); Fl VII*b*. *Dorchester* (23); Wo; Duigan 1955. !*Lowestoft* (25); An? West and Wilson 1968. !*Broome Heath* (27); Wo. !*Bradford Kaims* (68); LDe III; Bartley 1966. !*Kersall Moss* (69); LDe III. !*Moss of Cree* (74); Fl only. *Cowdenglen* (76); Fl?; Mahony 1869 as *Webera*. !*New Dry Dock* (67); Fl IV*. *Hailes* or *Corstorphine* (83); LDe only; Bennie 1894*a* as *Webera* sp. !*Loch Cuithir* (104); Fl IV; Vasari and Vasari 1968. !*Loch Droma* (105); LDe I; Kirk and Godwin 1963. !*Drumurcher* (H32); LDe.

The large genus *Pohlia* has sixteen species in the British Isles. Sporophytes and/or gemmae are often crucial for identification; hence the large number of indeterminate specimens. A majority of the species belong primarily to the highland zone.

Pohlia nutans (Hedw.) Lindb.

Holme Fen (31); Fl VII*b*; Vishnu-Mittre 1959. *Chat Moss* (59); Fl IV-V; Birks 1965*a*. *Seaforth* (59); Fl VII/VIII*; Travis 1913; Travis and Travis 1913. *Linton Mires* (64); Fl IV or V; Raistrick and Blackburn 1932 as *Webera nutans* Hedw. !*Bradford Kaims* (68); Fl VII*b*; t; Bartley 1966. *Timahoe* (H19); Fl VII; Synnott 1970 pc.

The recovery of this species mostly from ombrogenous peats is in accordance with the present ecology. *P. nutans* is a widespread calcifuge of a variety of habitats including oligotrophic mires. It is probable that subfossils have often been overlooked by peat stratigraphers.

The range is very wide in both the Northern and Southern Hemispheres. The numerous subfossils from outside the British Isles include a Pliocene one from Poland (table 10).

Pohlia wahlenbergii (Web. & Mohr) Andr.

Hoxne (25); Ho II or EWo; Reid 1896 as *P. albicans*. !*Wretton* (28); EDe; t. !*Upton Warren* (37); MDe. !*Dimlington* (61); LDe pre-I; var. *glacialis*; t. !*Kirkmichael 3a* (71); LDe III; Dickson *et al.* 1970. *Garvel Park* (76); LDe only; Robertson 1881 as *Webera albicans*. *Loch Fada* (104); LDe I; t; Birks 1969 pc. !*Derryvree* (H33); MDe; t.

The very lax areolation of this strongly variable species (cells *c.* 20 μm or more wide) is a character which applies to all the material seen by the author.

The Dimlington material consists of hundreds of badly preserved fragments. The largest exceeds 5 cm long. No other mosses were found, only a few grass caryopses. Apart from the size of the fragments, the large pure mass hints strongly at var. *glacialis* which often occurs in extensive patches free of other bryophytes or almost so, by springs and runnels fed by snowbeds in the mountains (plate 20).

Var. *glacialis*, in the British Isles, apart from Derbyshire, is recorded only from Caernarvonshire, northernmost England, the Scottish Highlands and Co. Sligo. *P. wahlenbergii* was widespread in last glacial Britain. Occurring widely in both the Northern and Southern Hemispheres, *P. wahlenbergii* is well known from Pleistocene deposits.

Pohlia delicatula (Hedw.) Grout

Trundle Mere (31); Fl VII*a* and VII*b*; Vishnu-Mittre 1959.

This species, alike the previous one but of smaller habit, grows mainly on clay banks by streams and rivers. Widespread in the British Isles and the Holarctic, *P. delicatula* has few, if any, other Pleistocene records.

Bryum undetermined species

!*Honeygore Track* (6); Fl VII*a* and VII*b*. *Minnis Bay* (15); Fl VII*b**; Bronze Age ar; Conolly 1941. !*Colney Heath* (20); LDe I; Godwin

1964. *Dorchester* (23); Wo; Duigan 1955. *Hoxne* (25); Ho III; Reid 1896; LAn; Turner 1968. !*Broome Heath* (27); Wo. !*Wretton* (28); EDe. *Soham Lode* (29); Fl VIIb. !*Barnwell Station* (29); LDe pre-I. *Great Billing* (32); MDe; Morgan 1969. !*Upton Warren* (37); MDe. !*Fladbury* (37); MDe. *Brigg* (54); Fl VII*b**; Bronze Age ar; Sheppard 1910. *Aby Grange* (54); LDe II; Suggate and West 1959. !*Bunny* (56); Fl VIII*; Roman ar. !*Russland* (69); Fl only; Dickinson 1968 pc. *Low Wray Bay* (69); LDe II; Pennington 1962. !*Kirkmichael 3a* (71); LDe III; Dickson *et al.* 1970. !*New Dry Dock* (76); LDe only. !*Burnbrae* (76); Fl VII/VIII*. *Whitrig Bog* (81); LDe III; Conolly 1963 pc. *Hailes* or *Corstorphine* (83); LDe only; Bennie 1894*a*. !*Loch of Park* (92); LDe I; Vasari and Vasari 1968. !*Loch Cuithir* (104); Fl III-IV and VII*a*; Vasari and Vasari 1968. !*Loch Fada* (104); Fl III-IV and VIII; Vasari and Vasari 1968; LDe I; Birks 1969 pc. *Loch Meodal* (104); LDe I and II; Birks 1969 pc. *Lochan Coir 'a' Ghobhain* (104); LDe II; Birks 1969 pc. !*Baggotstown* (H8); Ho III, IV or EWo; Dickson 1964*a*. !*Kildromin* (H8); Ho I. !*Drumurcher* (H32); LDe III. !*Derryvree* (H33); MDe.

As is evident from the above list, fragments of *Bryum* are often extracted from Pleistocene deposits. In the absence of sporophytes, which applies to all the specimens considered here, *Bryum* species are difficult or impossible to identify except in a few cases where leaf shape is diagnostic.

With over forty species *Bryum* is the largest genus in the British bryoflora. Many have disjunct or very restricted ranges. *B. arcticum, B. dixonii, B. lawersianum*, for instance, are markedly northern, being restricted to a few (or single) stations in the Scottish Highlands. In contrast *B. canariense* var. *provinciale*, a Mediterranean-Atlantic species, has a southern range.

A few species such as *B. tortifolium, B. wiegelii* and *B. neodamense* are known from the European Pleistocene but not that of Britain. *B. cryophilum*, a species primarily of high latitudes, was recovered from the excavation of a mammoth in the Taimyr peninsula and also from a Late Wisconsin deposit in Michigan (Miller and Benninghoff 1966), far to the south of the present North American range.

Bryum pallens Sw.

Ponder's End (21); MDe; Warren 1912. *Hoxne* (25); Ho III or EWo; Reid 1896. *Seaforth* (59); Fl VII or VIII*; Travis 1913; Travis and Travis 1913. *Garvel Park* (76); LDe/Fl IV*; Robertson 1881.

B. pallens, unrecorded from only eleven vice-counties, occurs in such habitats as calcicolous grassland, calcareous clay, flushes, stream-sides and sand-dunes. It has a wide Holarctic range and has been recorded from at least three Pleistocene deposits from the rest of Europe.

Bryum pseudotriquetrum (Hedw.) Gartn., Meyer & Schreb.

Ponder's End (21); MDe; Warren 1912. !*Upton Warren* (37); MDe; t. *Burtree Lane* (66); Fl VI; Bellamy *et al.* 1966. *Garvel Park* (76); LDe/Fl IV*; Robertson 1881. !*Drymen* (86); Fl V and VI; t; Vasari and Vasari 1968. *Fort William* (97); Fl V or VI*; female inflorescence; Dixon 1910. !*Loch Droma* (105); LDe I; t; Kirk and Godwin 1963.

This species has been recorded from almost throughout the British Isles where in the highland zone it is a conspicuous component of the bryophyte vegetation of springs and stream-sides. Other habitats include fens and dune slacks. The very extensive range in the Holarctic reaches the highest latitudes. There are many records of subfossils from the rest of Europe and North America.

T *Bryum creberrimum* Tayl.

Ponder's End (21); MDe; t; Warren 1912 as *B. cirratum*.

Dixon records this subfossil merely as *B. cirratum*. I have not re-examined the material and therefore it is with hesitation that the record is listed under *B. creberrimum* Tayl. *B. cirratum* is a poorly known taxon in need of further study.

Bryum intermedium (Brid.) Bland.

Ponder's End (21); MDe; t; Warren 1912.

B. intermedium, absent from much of Ireland and Scotland but commoner in England and Wales, occurs on moist soil and sand-dunes. The range, southern in Fennoscandia, includes much of the Holarctic and also Australia and New Zealand. There appear to be no other subfossils.

Bryum caespiticium Hedw.

Low Wray Bay (69); LDe II; t; Pennington 1962.

This species of soil and rock is known from almost all vice-counties of Britain and the great majority in Ireland. It is usually stated to be cosmopolitan. Again there are no other subfossils.

T *Bryum erythrocarpum* Schwaegr.

Wolvercote (23); Ho or EWo?; Bell 1904.

This record has little significance in the light of modern knowledge. The taxonomy of the *Bryum erythrocarpum* complex has been revolutionised by the study of rhizoid gemmae (Crundwell and Nyholm 1964).

Bryum rubens Mitt.

!*Bell Track* (6); Fl VII*a*; Andrews 1966 pc.

These subfossils are unique, as far as I am aware; they are the only detached bryophyte gemmae ever found. Identification is assured by the size of the tubers and the protuberant cells (Whitehouse 1966) which separate this species from all others of the *B. erythrocarpum* complex. *B. rubens* is widespread in the British Isles and Fennoscandia.

Bryum capillare Hedw.

Ponder's End (21); MDe; Warren 1912.

B. capillare, occurring extensively in both Northern and Southern Hemispheres, is recorded from all 152 vice-counties of the British Isles where its wide amplitude includes rock, soil and wood. This appears to be the only glacial locality; Pilous (1968) has recorded the species from the Post-glacial of Moravia.

Mniaceae

With a few exceptions, particularly the *Plagiomnium affine* group, subfossils of this family present little difficulty. Most species of the European Mniaceae are known from Pleistocene deposits in Britain and the rest of Europe. *Mnium lycopodioides, M. spinosum* and *Plagiomnium medium* have very restricted montane ranges in Scotland. They may well turn up in Devensian deposits as may the non-British *Cyrtomnium* species and also *Cinclidium arcticum* and *C. subrotundum*. These *Cinclidium* species are known from Pleistocene deposits in Germany, Denmark and Poland well south of their present European ranges which are restricted to Fennoscandia and those to the north. The Neogene records of *Trachycystis* have considerable bryogeographical importance (p. 170).

Mnium undetermined species

Hoxne (25); Ho III; Reid 1896. *Leeds* (64); Fl VI/VII/VIII*; Raistrick and Woodhead 1930. !*Loch of Park* (92); LDe II; Vasari and Vasari 1968.

Mnium hornum Hedw.

Weymouth (9); Fl only; Gepp 1895. *Woodwalton Fen* (31); Fl VII*b*; Godwin and Clifford 1938. *Brigg* (54); Fl VII*b**; Bronze Age ar; Sheppard 1910 as *M. hornum* L. *Cromwell Bottom* (63); Fl VII*b* and VII*b* or VIII; Bartley 1964. !*Seathwaite Tarn* (69); Fl VIII. *Fort William* (97); Fl V or VI*; Dixon 1910. !*Loch Maree* (105); Fl VII*b* or VIII; H. H. Birks 1969 pc.

The subfossils of this common species of acid substrata in woodlands are exclusively Flandrian. This conforms with the present range which is extensive in the temperate parts of the Holarctic. In Fennoscandia there is a southern and western tendency. *M. hornum* might turn up in Alleröd deposits of western Britain; it ascends above the tree line by inhabiting rock clefts. There are few subfossils from outside Britain. The absence from the Irish Pleistocene is striking and must be due to lack of investigation.

Mnium orthorhynchum Brid.

!*Mochras* (48); De? only. *Loch Fada* (104); LDe III; Birks 1969 pc.

The single dainty fragment from Mochras has the small and uniformly thickened leaf cells characteristic of this species.

M. orthorhynchum reaches the highest latitudes in its extensive Holarctic range. As exemplified by its occurrences in Ellesmere Island, Peary Land, Spitsbergen, Fennoscandia and the British Isles, the species shows a calcicolous tendency throughout its range. It is classed as an exacting calcicole by McVean and Ratcliffe (1962).

Calcareous rock ledges are the usual habitat in the British Isles where there are no localities south of Caernarvonshire. *M. orthorhynchum* is tolerant of late snow lie (Gjaerevoll 1956). The Mochras subfossil from Merioneth indicates a minor contraction of range.

M. orthorhynchum was found with a mammoth in the Taimyr peninsula (Savicz-Ljubitzkaja and Abramova 1954). Somewhat less expected are two Pliocene records, one from France (Boulay 1892) and one from Poland (table 10).

Mnium marginatum (With.) P. Beauv.

Hoxne (25); Ho III or EWo; t; Reid 1896 as *M. serratum. Windmillcroft* (77); Fl VI/VII/VIII*; Mahony 1868*a* as *M. serratum.*

This species has a broadly similar chorology and ecology to *M. orthorhynchum.* There are two Devensian records from Poland (Szafran 1952) and a Post-glacial one for Moravia (Pilous 1968).

Plagiomnium rostratum (Schrad.) Kop.
[*Mnium longirostrum* Brid.]

Ponder's End (21); MDe; Warren 1912 as *Mnium rostratum. Wolvercote* (23); Ho or EWo?; Bell 1904 as *M. rostratum.* !*Loch Fada* (104); IV; t; Vasari and Vasari 1968.

Confusion with the next two species is possible but there is no reason to doubt the above records. *P. longirostrum* occurs throughout the British Isles in such habitats as shaded banks and woodland. In Fennoscandia there is a southern tendency. The range covers much of both the Northern and Southern Hemispheres.

Plagiomnium affine (Funck) Kop.
[*Mnium affine* Bland.]

Hawks Tor (2); LDe II; Conolly *et al.* 1950. *Wolvercote* (23); Ho or EWo?; Bell 1904. *Bridgewater Street* (59); Fl VIII*; Roman ar; Roeder 1899. *Garvel Park* (76); LDe only; Robertson 1881. !*Winetavern Street* (H21); Fl VIII*; Medieval ar.

All these records must be considered as doubtful except the Irish one; there has been much confusion of this and the next species. Duckett and Little (1968) summarise the distinguishing features.

There are various non-British subfossils of the *P. affine* group including *P. seligeri* (Solonevicz 1935; Szafran 1952 and Körber-Grohne 1967) and *P. medium* (Störmer 1949), neither of which has yet been found in British deposits.

Plagiomnium rugicum (Laur.) Kop.
[*Mnium rugicum* Laur.]

!*Honiton* (3); Ip only. !*Broxbourne* (21); MDe; t. *Dorchester* (23);

Wo; t; Duigan 1955. !*Hoxne* (25); Ho III or EWo; t; Reid 1896; and !LAn; Turner 1968. !*Wretton* (28); EDe; t. !*Upton Warren* (37); MDe. !*Seathwaite Tarn* (69); Fl VIII. !*Cove Moss* (69); LDe III-IV. *Kirkmichael 3a* (71); LDe III; Dickson *et al.* 1970. !*New Dry Dock* (76); LDe only. !*Drumurcher* (H32); LDe III; t.

Dr T. Koponen has confirmed all the above subfossils with the exception of that from Dorchester. In Fennoscandia where the species is widespread he has described the ecology as follows (1968, p. 220).

> *Plagiomnium rugicum* grows in a wide variety of habitats, the only common feature of which seems to be that they are more or less wet. It is most abundant in eutrophic swamp forests and prefers the neighbourhood of springs and surface seepages. It is also to be found on the shores of lakes at the water's edge, on wet meadows by rivers, along brooksides, and in many kinds of localities with trickling water even if the trickling does not persist throughout the growing season. Like *P. elatum* it can grow with the old parts of the stem submerged.
>
> *Plagiomnium rugicum* suffers less from human influence than the other species under discussion. It can exist along ditches, in wet man-made meadows, and on regularly watered lawns. The writer even found it growing as a weed in an outdoor flowerpot.

Plagiomnium undulatum (Hedw.) Kop.
[*Mnium undulatum* Hedw.]

Amesbury (8); Fl VII*b**; Bronze Age ar; Newall 1931. *Burtree Lane* (66); Fl VI; Bellamy *et al.* 1966. !*Carpow* (88); Fl VIII; Roman ar. !*Newgrange* (H22); Fl VII*b*; Neolithic ar.

P. undulatum is a shade plant primarily of soil and wood in forests. However, the Newgrange fragments, and possibly others, did not derive from such a habitat; with *Brachythecium rutabulum* they grew in the turf of a damp, weedy grassland. Like *Mnium hornum* it ascends in the hills in shaded crevices. Again paralleling *M. hornum* the British subfossils are all Flandrian and there appear to be few discoveries outside the British Isles, as for instance at Ripetta, Italy (Clerici 1892); Mainz, Germany (Neuweiler 1905) and Ejstrup, Denmark (Hartz 1909).

Rhizomnium punctatum (Hedw.) Kop.
[*Mnium punctatum* Hedw.]

Hoxne (25); Ho III and Ho III or EWo; Reid 1896. *Brigg* (54); Fl VII*b**; Bronze Age ar; Sheppard 1910 as *M. punctatum* L. *Aby Grange* (54); LDe II; Suggate and West 1959. *Cromwell Bottom* (63); Fl VII*b* or VIII; Bartley 1964. *Neasham* (66); LDe II and III; Blackburn 1952. *Whitrig Bog* (81); LDe III; t; Conolly 1963 pc.

R. punctatum occurs throughout the British Isles in such habitats as damp rocks by streams, montane rock crevices and decaying wood in forests. Its extra-British range includes Eurasia and North America. Subfossils have been recorded by Hartz (1909), Störmer (1949), Szafran (1952) and Kapp and Gooding (1964).

R. punctatum or *R. pseudopunctatum*

Ponder's End (21); MDe; Warren 1912 as *Mnium punctatum* or *M. subglobosum.*

These two species may be difficult to separate, as pointed out by various authors, notably Mårtensson (1956).

Rhizomnium pseudopunctatum (B. & S.) Kop.
[*Mnium pseudopunctatum* B. & S.]

Garvel Park (76); LDe/Fl IV*; Robertson 1881 as *M. subglobosum. Altt na Feithe Sheilla* (96); Fl only; Lewis 1906. *Loch Einich* (96); Fl V or VI; H. H. Birks 1969 pc. *Loch Fada* (104); LDe III; Birks 1969 pc. !*Loch Droma* (105); LDe I; Kirk and Godwin 1963.

R. pseudopunctatum has a more scattered British range than *R. punctatum* and occurs principally in fens; McVean and Ratcliffe (1962) consider it calcicolous. In the Torneträsk area it may be abundant in rich fens with such species as *Aulacomnium palustre, Cinclidium stygium, Helodium* and *Paludella* (Mårtensson 1956). There is a Middle Pleistocene record from the Kama region of Russia (Abramova and Abramov 1962) and a last interglacial one from Denmark (Hartz 1909).

Pseudobryum cinclidioides (Huben) Kop.
[*Mnium cinclidioides* Huben]

!*Chelford* (58); EDe. *Upper Valley Bog* (69); Fl VII; Johnson 1960 pc.

The large leaves distinctive in shape and areolation render this species easy of identification. *P. cinclidioides* is widespread in Scotland but occurs only in a few northern counties of England and Wales and is absent from Ireland. In Fennoscandia also the range is northern.

The Chelford subfossils (plate 8) prove occurrence in a lowland area no longer occupied. This and other species show that the coniferous forest at Chelford supported a marshy floor. *P. cinclidioides* has been recovered from deposits in Germany (Beug 1957) and Poland (Szafran 1952).

Cinclidium stygium Sw.

Nazeing (18); LDe III; Allison *et al.* 1952. !*Colney Heath* (20); LDe I; Godwin 1964. !*Broome Heath* (27); Wo. *Gors Goch* (52); Fl VII*b*-VIII; Pigott 1963 pc. *Great Close Mire* (64); Fl VIII; Pigott and Pigott 1963. !*Kirkmichael 3a* (71); LDe I; Dickson *et al.* 1970. *West Craigneuk* (77); De? only; Dixon 1907.

C. stygium, a species of rich fens, springs, irrigated rocks and stream-sides, is rare in the British Isles where the distribution is disjunct. The map published by Proctor (1960) shows, apart from three localities in East Anglia, scattered occurrences in the highland zone. Since then the species has been found in several other localities, the most important addition being in Leitrim, Ireland.

The subfossils point to a greater abundance and a more continuous range in the Late Devensian; unlike many other species of similar ecology there are as yet no early Flandrian records. The distribution of *C. stygium* is extensive in the Holarctic and includes southern South America (Koponen 1969).

Cinclidium, a genus of four or five species, has several fossils of bryogeographical significance. Apart from the British discoveries there are subfossils of *C. stygium* from the late Pleistocene of Poland and Germany (Beug 1957). *C. subrotundum* is known from a last glacial site in Poland and subfossils of *C. arcticum* from Denmark and Germany are well south of the present range; both these species may turn up in the British Pleistocene.

However, the greatest interest attaches to a record of *C. latifolium* from the imprecisely dated but probably last glacial deposits at Skaerumkede, Denmark (Hesselbo 1910). An attempt to find the material for re-examination proved negative. However, there is no special need to question the determination. Hesselbo is explicit about the broad leaves and strikingly recurved margins diagnostic of this species which is one of the high-arctic species discussed by Steere (p. 212). Fig. 43 shows the range to be predominantly north of the Arctic Circle; many more localities would now be added as, for instance, in Spitsbergen (Arnell and Mårtensson 1959) and in western Greenland (Holmen 1957). Since its Danish occurrence, *C. latifolium* has retreated northwards some 20° of latitude (2500 km or 1060 miles).

Aulacomniaceae

Of the three European species *Aulacomnium palustre* is probably the most frequently found Pleistocene moss,

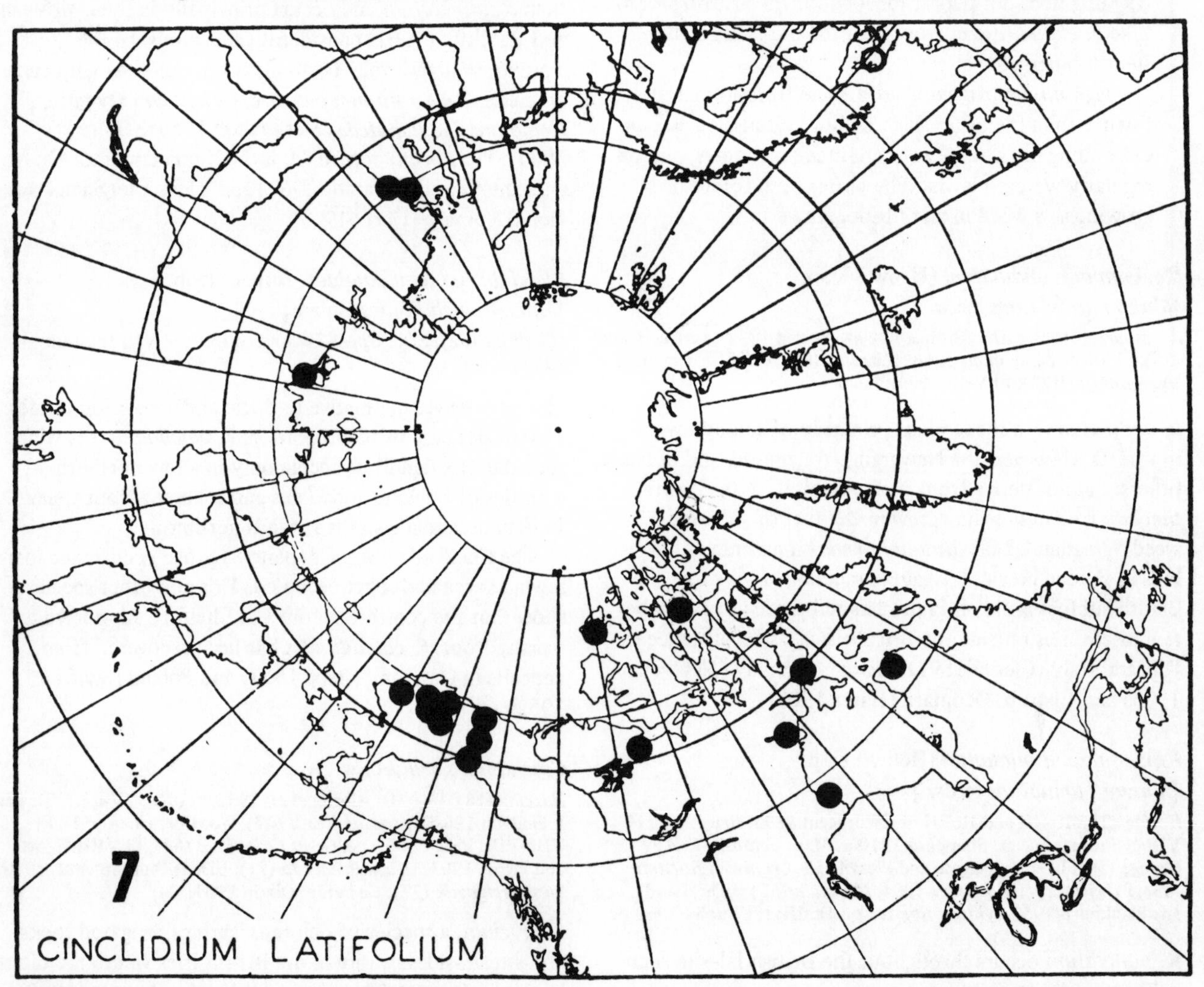

Fig. 43. World range of *Cinclidium latifolium.* This and Fig. 44 are reprinted with the permission of the publisher from 'On the geographical distribution of Arctic bryophytes' by William Campbell Steere, in *Current Biological Research in the Alaskan Arctic*, edited by Ira L. Wiggins (Stanford: Stanford University Press, 1953), pp. 44-5.

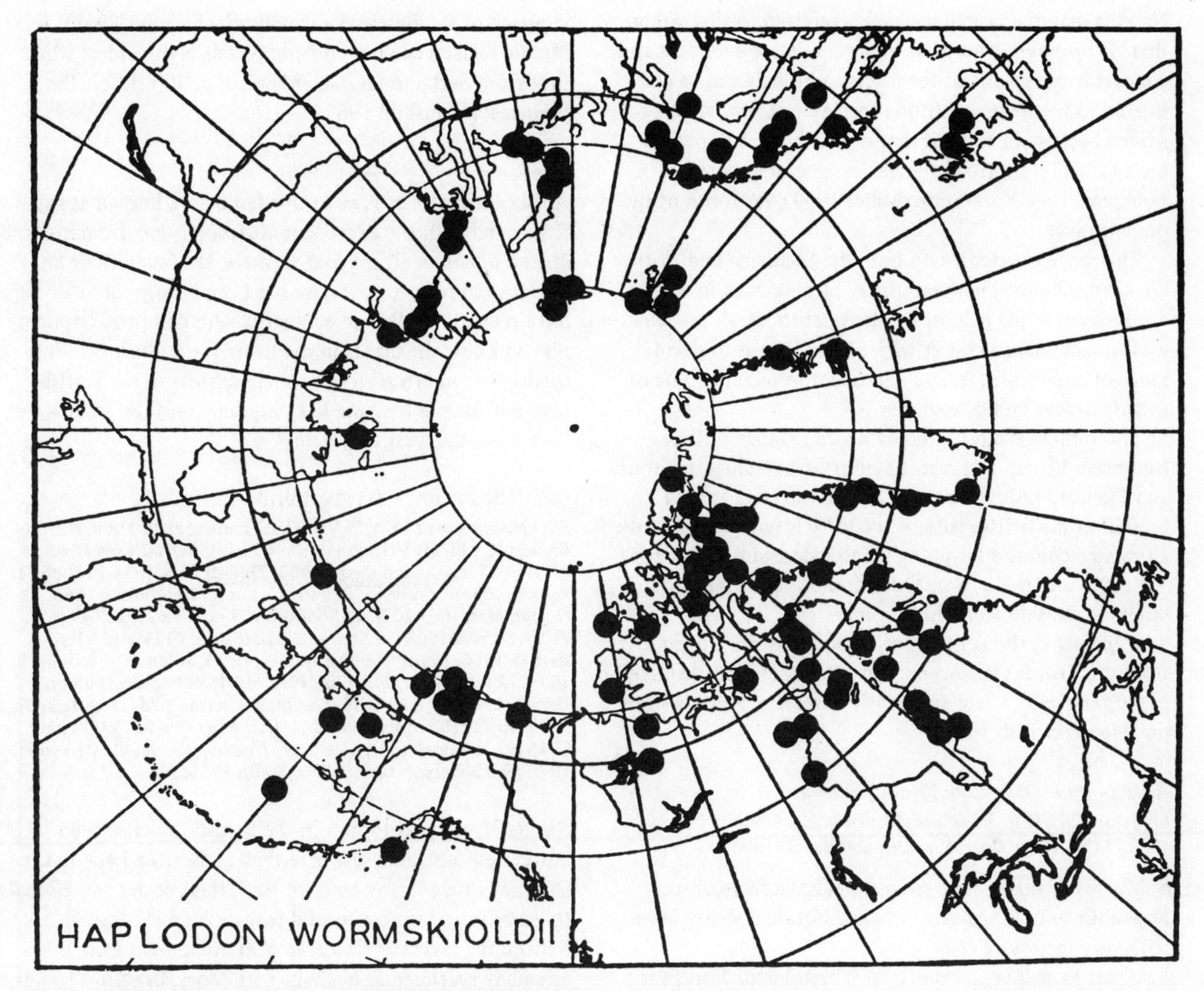

Fig. 44. World range of *Haplodon wormskjoldii.* From Steere 1953.

A. turgidum has a fairly detailed history and *A. androgynum* has no subfossils whatever.

Aulacomnium palustre (Hedw.) Schwaegr.

Hawks Tor (2); LDe II; Conolly *et al.* 1950. !*Stannon* (2); LDe only; t. *Decoy Pool Wood* and *Decoy Pool Drove* (6); Fl VIII; Clapham and Godwin 1948. !*Ashcott Heath* (6); Fl VII*b*. !*Meare Track* (6); VII*b*. !*Honeygore Track* (6); Fl VII*a* and VII*b*. *Weymouth* (9); Fl only; Gepp 1895. !*Amberley Wild Brooks* (13); Fl VIII. !*Nazeing* (21); LDe III; var. *imbricatum*. *Wolvercote* (23); Wo?; Bell 1904. !*Paston* (27); Pa. !*Beeston* (27); Pa. *Woodwalton Fen* (31); Fl VII*b*; Godwin and Clifford 1938; Godwin 1938. *Trundle Mere* (31); Fl VI*a* and VII*b*; Vishnu-Mittre 1959. *Great Billing* (32); MDe; Morgan 1969. !*Fladbury* (39); MDe. *Chartley Moss* (39); Fl VII and VIII; Green 1964 pc. !*Whixall Moss* (40); *Aby Grange* (54); LDe II and III; Suggate and West 1959. !*Chelford* (58); EDe. *Wybunbury Bog* (58); Fl VII and VIII; Green 1963 pc. *Bridgewater Street* (59); Fl VIII*; Roman ar; Roeder 1899. *Chat Moss* (59); Fl IV-V; Birks 1965*a*. *Bag Mere* (59); Fl IV-V; Birks 1965*a*. *Yorkshire Pennines*; Fl VII or VIII*; Burrell 1924. !*Hooks* (61); LDe II. *Malham Tarn Moss* (64); Fl VI; Pigott and Pigott 1963. *Hutton Henry* (66); Ip II, III and IV; Beaumont *et al.* 1969. !*Russland Moss* (69); Fl only; Dickinson 1968 pc. !*Kirkmichael 3a* (71); LDe III; Dickson *et al.* 1970. ! *Garvel Park* (76); LDe/Fl IV*; Robertson 1881; Fl IV*. *West Craigneuk* (77); De? only; Dixon 1907. *Whitrig Bog* (81); LDe II and III; Mitchell 1948; Conolly 1963 pc. *Hailes* or *Corstorphine* (83); LDe only; Bennie 1894*a*. !*Dun Moss* (89); Fl only. ! *Garral Hill* (94); LDe III. *Abernethy Forest* (96); Fl VII*a*; H. H. Birks 1969. *Moine Mohr A and B* (98); Fl VII and VII or VIII; Chesters 1931. *Loch Fada* (104); Fl IV; Birks 1969 pc. *Gort* (H15); Ho III or IV; Jessen *et al.* 1959. *Derrybrennan* (H19); Fl VI?; Synnott 1970 pc. *Timahoe* (H19); Fl VII; Synnott 1970 pc. *Ballybetagh* (H21); LDe II; Jessen 1949. *Canbo* (H25); Fl VIII; Jessen 1949. *Cloonlara* (H26); Fl VIII; Mitchell 1956. *Ardlow Inn* (H30); Fl VII*b* and VIII; Jessen 1949. !*Drumurcher* (H32); LDe III; t. *Cloughmills* (H39); Fl VII*b*; Jessen 1949.

There is little difficulty in recognising this species because of the incrassate cells each bearing large central papillae. The leaf shape is also characteristic though the obtuse-leaved form, var. *imbricatum*, resembles *A. turgidum*. The

Nazeing material, a solitary leaf, is referred to this rather than *A. turgidum* which has even more incrassate cells and concave leaves. None of the glacial subfossils can be considered to be the high latitude species *A. acuminatum*, so named because of the leaf shape. The latter is known in the subfossil state from a single last glacial deposit at Lockport, New York State (Miller 1969), far south of the present range.

The range is extensive in both the Southern and Northern Hemispheres. The species may be abundant in wet tundra even at the highest northern latitudes. *A. palustre* is often abundant in mires where it is tolerant of a wide range of conditions. It may also occur in such habitats as irrigated rocks and lake shores.

The subfossils clearly indicate a wide range in Late Devensian Britain and records from Great Billing, Fladbury and Chelford point to pre-glacial survival (fig. 45). At present in the British Isles, *A. palustre* is unknown in only a few vice-counties including Huntingdonshire where zone VII subfossils are known. This minor disappearance can safely be considered anthropogenic.

A. palustre, though the most frequently recorded moss from the British Pleistocene, is much under-recorded from Flandrian peats. There are many Pleistocene records from outside the British Isles.

Aulacomnium turgidum (Wahl.) Schwaegr.

!*Wretton* (28); EDe. !*Mochras* (48); De? only. !*New Dry Dock* (76); Fl IV*. *Dronachy* (85); LDe only; Bennie 1896.

Subfossils of this species are distinguished from obtuse-leaved forms of *A. palustre* by the obovate concave leaves with very incrassate cells.

Apart from an occurrence on Whernside in Yorkshire (now probably extinct) this species is confined to the Scottish Highlands. Therefore all four localities, especially those from Merioneth and Norfolk, are outside the present British range (fig. 46).

Birks (1968) has described the species in the north-western highlands as 'characteristic of species rich moss heaths occurring on a wide variety of rocks between 1500 and 3500 feet (750-1600 m). It also occurs on the calcareous schists of Perthshire in species-rich *Agrostis-Festuca* grassland.' See p. 106.

In Fennoscandia *A. turgidum* has a calcicolous tendency; in the Torneträsk area it occurs mostly and abundantly in *Dryas* heath rich in mosses (Mårtensson 1956). In the Arctic it is an abundant species. Holmen (1960) considers it to be calcifuge in Peary Land.

There are last glacial subfossils from Michigan (Miller 1969), Poland, Russia and Denmark where the species no longer occurs. Remains of this species with *Drepanocladus flitans* and *Cladonia rangiferia* were found in the stomach of the Beresovka mammoth (Farrand 1961). A Middle Pleistocene record from Russia is the oldest while that from Sermermuit, Greenland, at *c.* 400 B.C. is the youngest (Fredskild 1967).

Meesiaceae

All six European species of this family are known as subfossils and with the exception of *Amblyodon dealbatus* and *Meesia uliginosa* all are peat formers. However, their importance in peat stratigraphy has been poorly understood until recently by British ecologists who had little opportunity to know the plants in the living state. Both *Paludella squarrosa* and *Meesia longiseta* are extinct in the British Isles and *Meesia tristicha* has only one, recently discovered, locality in the west of Ireland.

Paludella squarrosa (Hedw.) Brid.

Buckenham Broad (27); Fl VII-VIII; Lambert *et al.* 1960. *Barnby Broad* (27); Fl VII-VIII; Proctor 1963 pc. !*Malham Tarn Moss* (64); Fl VI; Pigott and Pigott 1963. *Thieves Moss* (64); Fl V or VI; Pigott 1963 pc. *Cronkley Fell* (65); Fl only; Hutchinson 1966. *Dufton Moss* (65); Fl VII*a*; Squires 1969 pc. *Fox Earth Gill* (65); Fl VI; Squires 1968 pc. *Bradford Kaims* (68); Fl IV and VII*a*; Bartley 1966. *Upper Valley Bog* (69); Fl VI; Johnson and Dunham 1963. !*Midgeholme Moss* (70); Fl V or VI; Millington 1965 pc. *Garvel Park* (76); LDe III/IV* only; Robertson 1881. *Renfrew* (76); Fl IV or V?; Thompson 1968 pc. !*Dun Moss* (89); Fl VI; Ingram 1968 pc. !*Straloch* (89); LDe only. *Timahoe* (H19); Fl VI? Synnott 1970 pc. *Fallahogy* (H40); Fl VI; Smith 1958*a*.

Paludella as a peat former in the British Isles has been better demonstrated nowhere than at Burtree Lane in Co. Durham where 'many tons' of both this species and *Homalothecium nitens* were removed from a zone IV deposit. Substantial layers of *Paludella* peat are known from areas as widely scattered as Malham Tarn Moss, Bradford Kaims, Dun Moss and Fallahogy. It is clear that *Paludella* was a common mire moss in northern Britain, especially northern England, during the Flandrian.

When well preserved, *Paludella* is easily identifiable even in the field by its squarrose, acute leaves, broad in proportion to length (plate 18). Possibly it could be confused with *Meesia tristicha* but this has longer, more acuminate leaves. Examined microscopically, *Paludella* is highly distinctive.

The major decline suffered by *Paludella* over the last five thousand years culminated in restriction to three localities. These were Knutsford Moor in Cheshire and Terrington Carr and Skipwith Common, both in Yorkshire. The final extirpation resulted from man's destruction of the habitats. At Skipwith Common the species was first found only fifty-two years ago and perhaps there is some slight chance that it may linger there. At the other two localities, however, it had vanished fifty years earlier. Virtually nothing is known of the British vegetation which supported

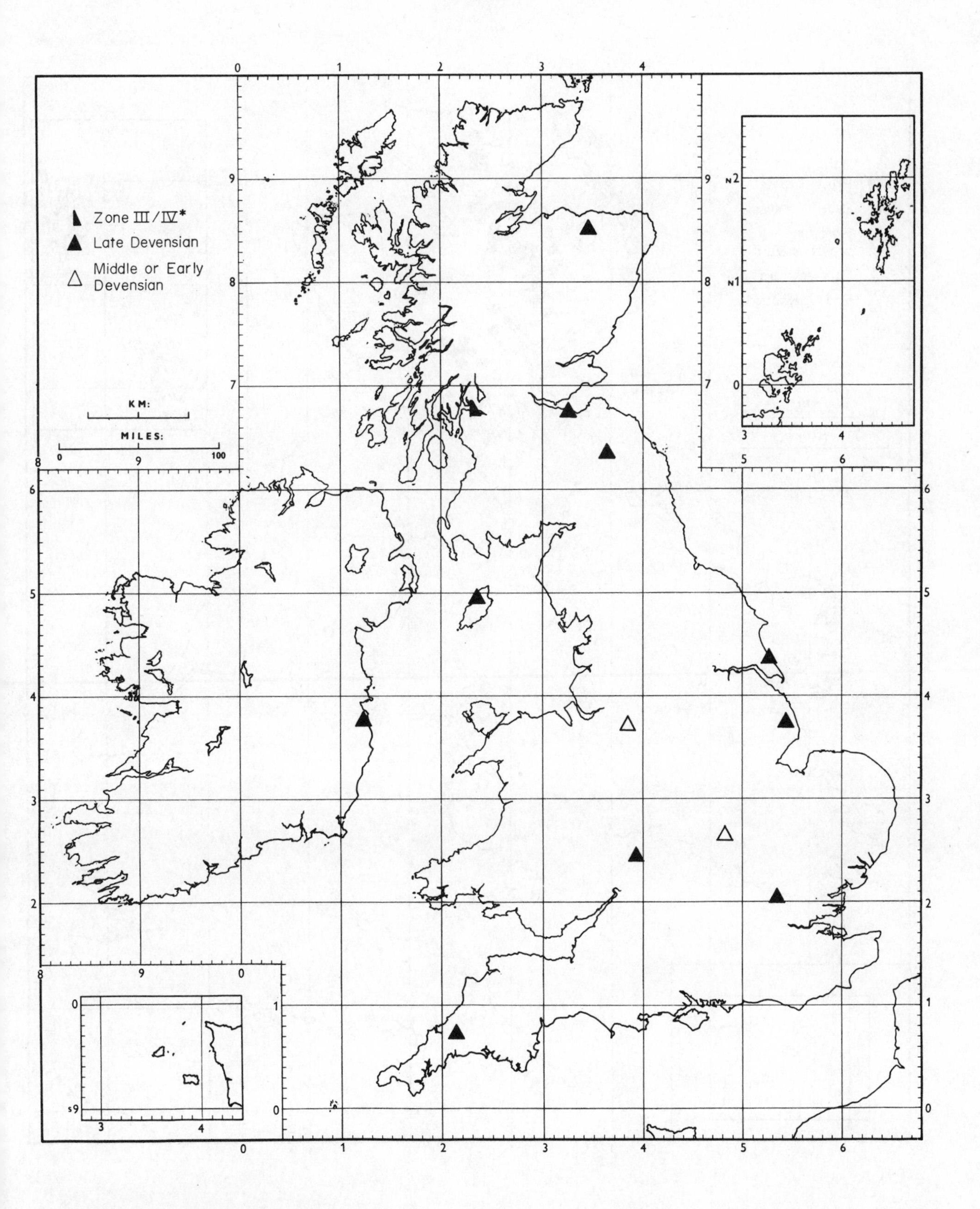

Fig. 45. Devensian subfossils of *Aulacomnium palustre*.

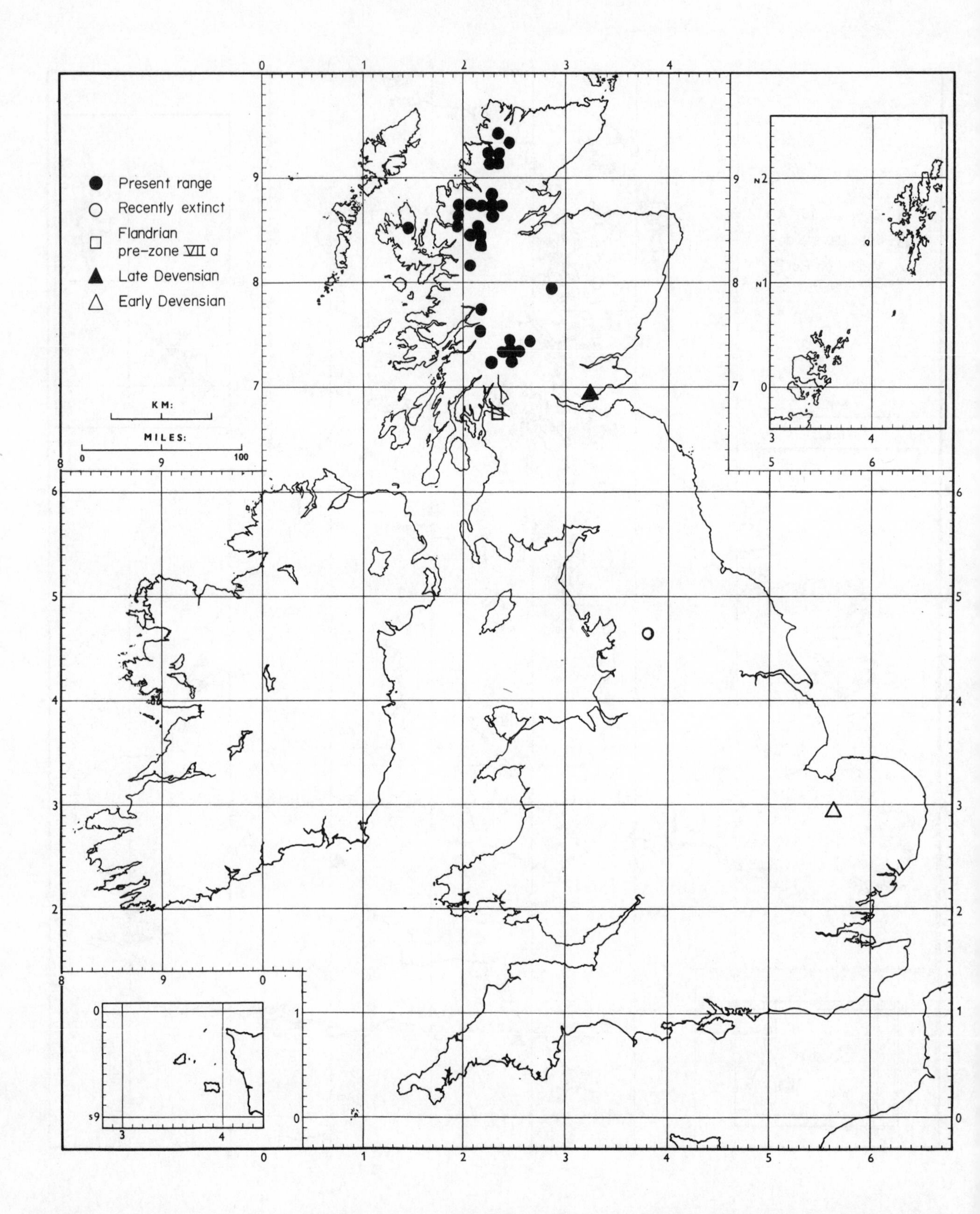

Fig. 46. Late Pleistocene subfossils and present range of *Aulacomnium turgidum*. Partly after H. J. B. Birks, *Trans. Br. bryol. Soc.* 1968.

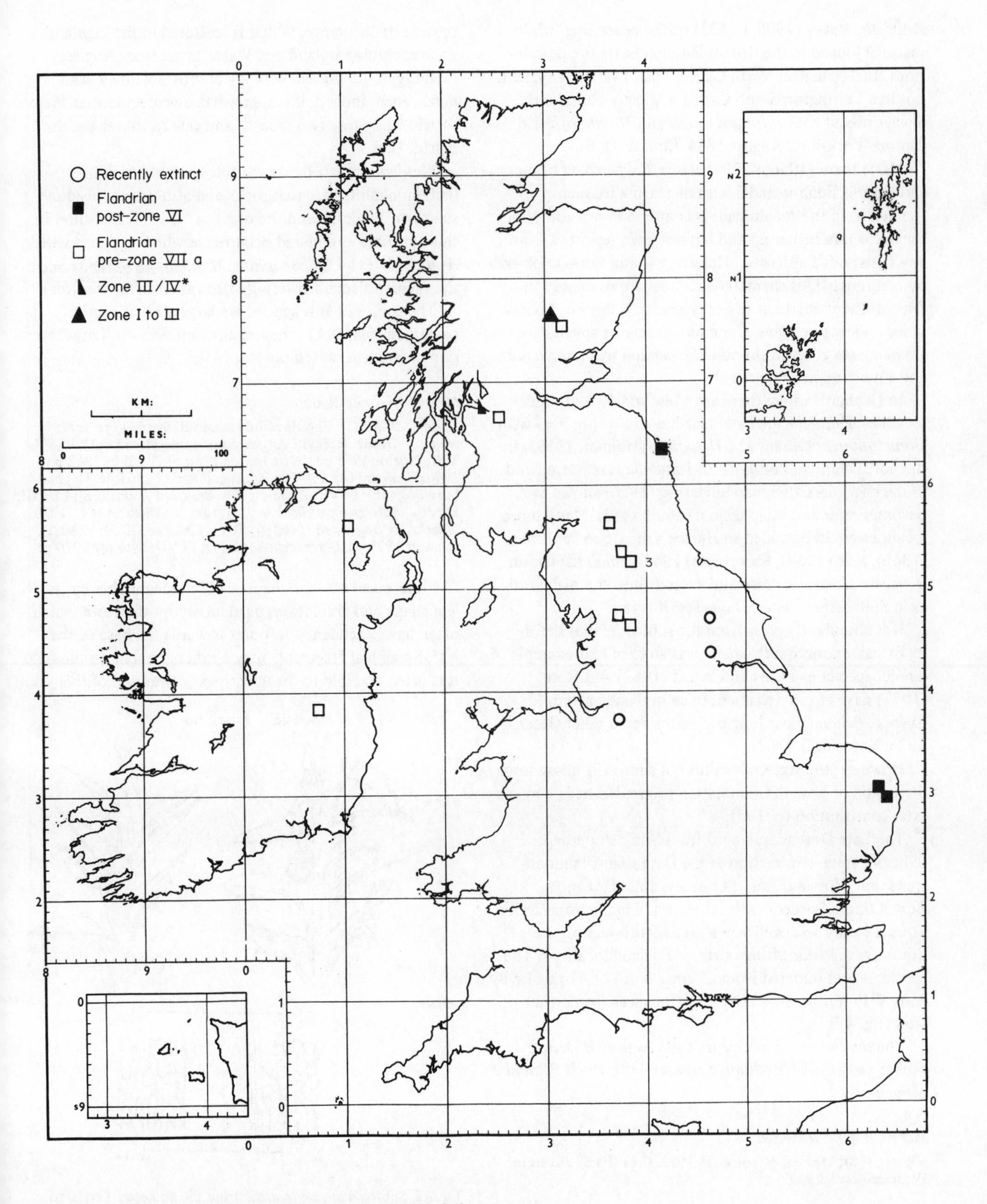

Fig. 47. Late Pleistocene subfossils and present range of *Paludella squarrosa.*

Paludella. Baker (1906, p. 533) states 'peaty bog' while material lodged in the British Museum bears the description: 'In Terrington North Carr, a small swampy hayfield, not like Terrington South Carr, a bog, very abundantly, always mixed with *Hypnum nitens* and *Mnium affine* β. *elatum*. Yorkshire. August 1854. Coll. A. O. Baker.'

Mårtensson (1956, p. 204) states '*P. squarrosa* occurs throughout Fennoscandia with increasing frequency to the north. In the subalpine it is common in not too poor fens. It is rare higher up and has not been reported from any remarkable altitudes.' Referring to the Torneträsk area he continues 'This characteristic moss grows mainly in fens of intermediate or richer types . . . It does not occur in extremely poor fens. Common associated species are *Drepanocladus revolvens* coll., *Sphagnum warnstorfianum* and *Tomenthypnum nitens*.'

In Denmark where there are a few widely scattered localities, *Paludella* grows in marshes and spring bogs with *Homalothecium nitens* and *Helodium* (Holmen 1959). In the north German-Polish plains *Paludella* is scattered and, as the now out-dated map in Herzog (1926) shows, it becomes rarer and more disjunct southwards. Many more localities could be added to Herzog's map. See Pankow (1966), Kuc (1964), Karczmarz (1963). *Paludella* has an extensive Holarctic range and extends into the high arctic as in Spitsbergen, where, however, it is rare.

It is abundantly clear from the subfossils that before its British extinction *Paludella* was an excellent example of a relict species as Pigott and Pigott (1963) and Rose (1957) have already claimed. In central and eastern Europe, *Paludella* has long been considered relict (Herzog 1926; Karczmarz 1963).

However, the sixteen localities of subfossils allow some discussion of how the elimination proceeded before man's fatal contribution (p. 195).

The Late Devensian record from Straloch perhaps indicates some contraction at the Devensian-Flandrian transition. However, the species was abundant in Pre-Boreal times. Subsequently, there was a major contraction. Thus the only Irish and youngest Scottish records are Boreal, as are four of the north of England localities. The species was eliminated from all areas north of Yorkshire by zone VII*b* but the Norfolk extinctions were somewhat later (fig. 47).

The restriction to Yorkshire and Cheshire is strongly similar to that of *Helodium*, a species with which *Paludella* often grows.

Meesia uliginosa Hedw.

Nazeing (18); LDe III; Allison *et al.* 1952. *Gort* (H15); Ho III or IV; Jessen *et al.* 1959.

This solitary Late Devensian record is well south of the present British range, which is scattered in the highland zone excluding Ireland and Wales, apart from Anglesey. Nothing in the present ecology conflicts with such an occurrence. Indeed, it is somewhat unexpected that there should be merely two records and few from outside the British Isles.

M. uliginosa is a freely fruiting species which in Britain inhabits base-rich rocks and also calcareous dune-slacks on Anglesey and the coast of South Lancashire. In this maritime and inland occurrence which it shares with *Amblyodon* and *Catoscopium, M. uliginosa* reminds one of the ranges of such flowering plants as *Armeria maritima.*

The spores of this species are large and distinctively patterned (plate 11). They might well occur in Late Devensian or Boreal rich fen peats.

Meesia tristicha Bruch

!*Huntspill Cut* (6); Fl VII*b*; Godwin and Richards 1946 as *M. triquetra* (Hook. & Tayl.) Angstr. *Buckenham Broad* (17); Fl VII-VIII; Proctor 1963 pc. !*East Walton* (28); Fl IV; Birks 1968 pc. !*Whixall Moss* (40); Fl VII*a*; Pigott and Sinker 1963 pc. !*Wybunbury Bog* (58); Fl VI; Warburg 1958. *Bradford Kaims* (68); LDe III; Bartley 1966. !*Kirkmichael 3a* (71); LDe III; Dickson *et al.* 1970. !*Blackriver Bog B and H* (H19); Fl VI. *Timahoe* (H19); Fl VI?; Synnott 1970 pc. *Derrybrennan* (H19); Fl VI?; Synnott 1970 pc.

Well preserved *M. tristicha* is readily recognisable by its leaf shape and areolation, particularly by the teeth which often have a tendency to point towards the base of the leaf. Small leaf fragments may be recognisable. In the field it is often possible to recognise peat composed of this

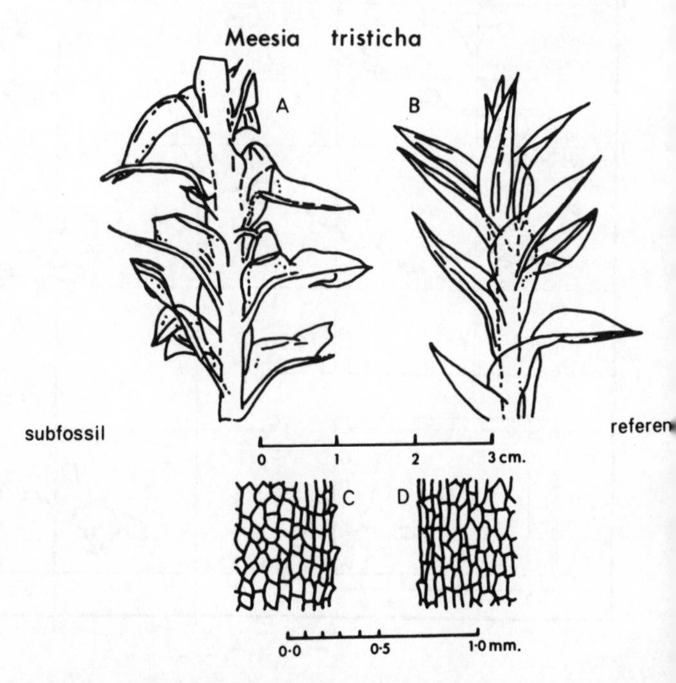

Fig. 48. Subfossil *Meesia tristicha* from the Somerset Levels. After Godwin and Richards 1946.

species. Godwin and Richards (1946) first drew the attention of British bryologists to the presence of *Meesia tristicha* when they recorded subfossil material from Somerset (fig. 48). Some ten years later living plants were discovered in Bellacorick Bog in West Mayo, Ireland (Warburg 1958). This remains the solitary living station in the British Isles but there are ten more localities of subfossils which prove widespread occurrence in England, the Isle of Man and Ireland from Late Devensian times onwards.

King and Scannell (1960) have described the Bellacorick Bog habitat as follows: '*Meesia tristicha* was found in the wettest part of the flush . . . There is some open water and the ground is strongly impregnated with iron. This is the region richest in bryophytes, with *Mnium pseudopunctatum* growing in large pure colonies in a very luxuriant form. The *Meesia* occurred in almost pure patches in five large clumps close together, with some *Sphagnum subsecundum* var. *auriculatum* growing through them.' The species they found in the vicinity are given in table 8.

Table 8 *Species from the* Meesia *locality in Bellacorick Bog*

Flowering plants	
Cardamine pratensis	*Potentilla palustre*
Carex diandra	*Menyanthes trifoliata*
C. lepidocarpa	*Myrica gale*
C. limosa	*Pedicularis palustris*
C. rostrata	*Oxycoccus palustris*
Mosses	
Bryum pseudotriquetrum	*Polytrichum commune*
Calliergonella cuspidata	*Rhizomnium pseudopunctatum*
Campylium stellatum	*Sphagnum flexuosum*
C. polygamum	*S. fuscum*
Campylopus brevipilus	*S. palustre*
C. introflexus	*S. papillosum*
Dicranum bonjeanii	*S. rubellum*
Homalothecium nitens	*S. subsecundum* var. *auriculatum*
Philonotis fontana	*S. teres*
Hepatics	
Odontoschisma sphagni	*Pleurozia purpurea*

The Fennoscandian occurrence is given by Mårtensson (1956, p. 105) as follows. '*M. tristicha* occurs throughout Fennoscandia with increasing frequency to the north. It is rare or unknown from some southern provinces of Sweden and several coastal provinces of Norway. It is not generally common in the Sandes but must be considered as a normal component of subalpine and low-alpine rich fens.' It has been found submerged in northern Sweden (fig. 27).

There were only three localities in Denmark where the species is probably extinct (Holmen 1959). There is one recent locality in Belgium (Delvosalle *et al.* 1969). *Meesia tristicha* is widespread in Eurasia and occurs in the high Arctic as in Spitsbergen, Peary Land and Bathurst Island where subfossils were found in peat samples (Brassard and Steere 1968).

Meesia tristicha is another outstanding example of a relict moss as it has long been considered in its low latitude distribution (Rybnicek 1966; Stefureac 1962; Herzog 1926; Godwin and Richards 1946). Again, like *Paludella*, there is now the possibility of further analysis. The Pre-Boreal and Boreal records from East Walton, Wybunbury Bog, and Ballydermot are reminiscent of *Paludella.* A marked contraction at about the Boreal-Atlantic transition seems indicated. The Somerset and Norfolk occurrences perhaps point to survival until very recent times and then anthropogenic extinction (fig. 49).

Subfossil spores of *M. tristicha* (plate 11) are now known from two English localities, Whixall Moss, Shropshire, and East Walton, Norfolk. In both cases the spores were recovered from peat formed of the species but they are large and distinctive enough in their regular granular pattern to be recognised without such backing.

At East Walton the *Meesia* again was a peat former with *Calliergon trifarium*. The pollen count by Dr H. J. B. Birks indicates a zone IV age. At Whixall Moss, *M. tristicha* was one of the earliest peat-formers; fen peat began to accumulate in zone VII*a*. *M. tristicha* was at least locally abundant. Two bore-holes about 100 m apart revealed the species associated with *Carex* and *Menyanthes. Sphagnum-Calluna-Eriophorum* peat overlies the fen layers.

A pollen count by Miss R. Andrew of *Meesia* peat from a depth of 270 cm was as follows:

	% Total tree pollen		% Total tree pollen
Betula	15.4	Gramineae	3.0
Pinus	2.5	Cyperaceae	74.4
Ulmus	14.1	*Rumex*	1.0
Quercus	37.1	*Scheuchzeria*	1.0
Tilia cordata	6.4	*Equisetum*	8.9
Alnus	23.1	Filicales	3.0
		Sphagnum	3.0
Corylus	25.6	*Meesia tristicha*	76.8

The major interest of these spore discoveries is that in its British station the *Meesia* has not been observed with sporophytes. It is now certain that the species was fertile for a large part of the Flandrian.

Meesia longiseta Hedw.

!*Amberley Wild Brooks* (13); Fl VIII. !*Holme Fen* (31); Fl VII*b*; Dickson and Brown 1966.

M. longiseta has a broadly similar range and ecology to *M. tristicha* but is a rarer species, little known to British bryologists who might, at first thought, be surprised at its former presence. However, the nearest localities are (or were) in Belgium (one locality, extinct, Delvosalle *et al.* 1969) and in France, near Paris (Husnot 1884). Seen in that light the subfossil localities are outposts of the range which thins

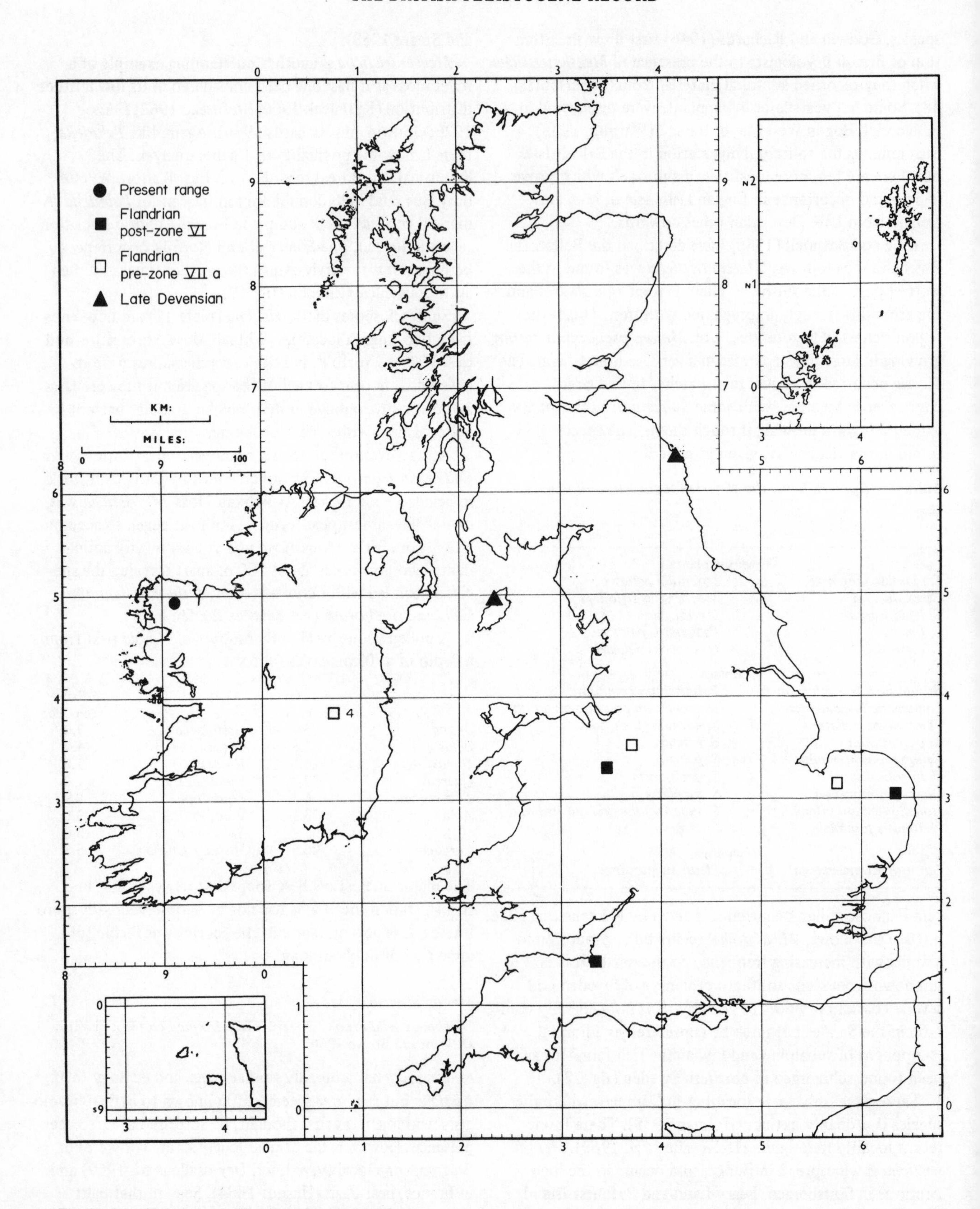

Fig. 49. Late Pleistocene subfossils of *Meesia tristicha.*

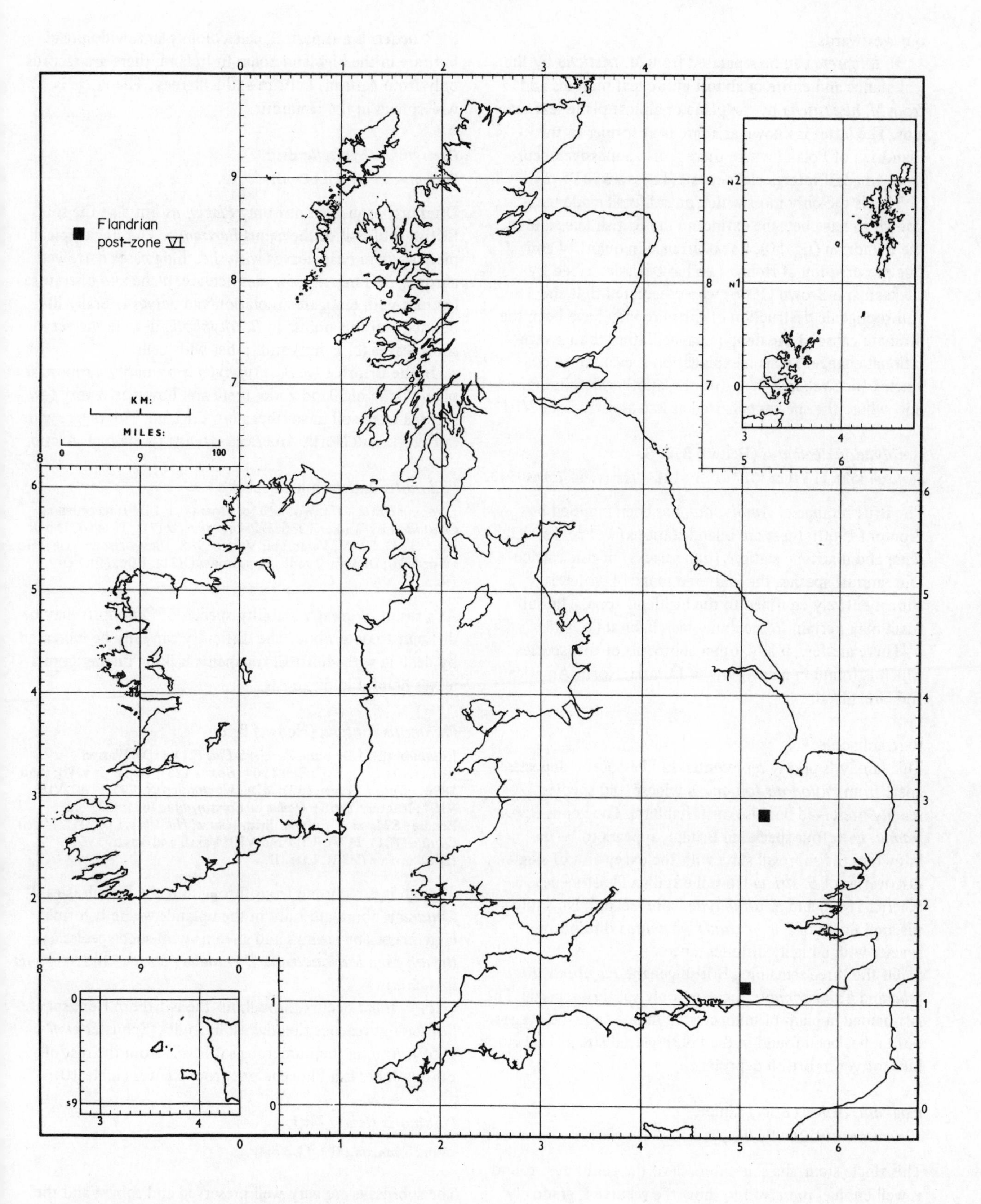

Fig. 50. Flandrian subfossils of *Meesia longiseta*.

out westwards.

M. longiseta can be separated from *M. tristicha* by the leaf stance and entire or almost entire leaf margins and from *M. hexasticha* by the plane or almost plane leaf margins. The latter is known as a rare peat former in the Flandrian of Poland where there is also a massive occurrence in a last interglacial deposit (Jasnowski 1957*b*).

This is the only moss which on subfossil evidence is known to have become extinct in the British Isles late in the Flandrian (fig. 50). Its occurrence in quantity and fine preservation at Holme Fen has been described by Dickson and Brown (1966) who speculated that the anthropogenic destruction of the mire may have been the ultimate cause of the disappearance, rather than a vegetational change. Such an explanation is even more convincing for Amberley Wild Brooks, another derelict raised bog, where the species persisted as late as early zone VIII.

Amblyodon dealbatus (Hedw.) B. & S.

Seaforth (59); Fl VII or VIII*; Travis 1913; Travis and Travis 1913.

The British range of *Amblyodon* has been mapped by Proctor (1960); there are inland stations (wet rocks; rich fens) and maritime stations (dune slacks) of this calcicolous, annual species, the scattered range of which is almost entirely confined to the highland zone. The subfossil may pertain to the dune-slack habitat (p. 188).

There are few, if any, other subfossils of this species which is found in many parts of Eurasia, North America and Greenland.

Bartramiaceae

This family is poorly represented in Pleistocene deposits apart from *Philonotis fontana*, a widespread species readily preserved because of its habitats. The genus *Bartramia*, with four species in Britain, appears to be unknown in the subfossil state with the exception of single discoveries of *B. stricta* from the Italian Pleistocene (Clerici 1892) and *B. ithyphylla* from Ireland. No subfossils are known of *Bartramidula wilsonii*, a diminutive species with a highly disjunct range.

Of the three remaining British genera, *Plagiopus, Breutelia* and *Conostomum*, there are only solitary records. The last-named, a plant often occurring in late snow-bed vegetation, has been found in the Polish glacial site at Ludwinow but not yet in British deposits.

Plagiopus oederi (Brid.) Limpr.

!*Drumurcher* (H32); LDe III.

This single stem, the sole subfossil of the genus ever found, is well enough preserved to show the recurved, gradually tapered leaves with prominent teeth in double rows.

P. oederi is a rupestral, calcicolous plant, widespread but rare in the highland zone. In Ireland, there are records only from Antrim, Leitrim and Killarney. The range is widespread in the Holarctic.

Bartramia ithyphylla Brid.

!*Drumurcher* (H32); LDe III.

Drumurcher yielded not only *Plagiopus* but also the sole British subfossil of the genus *Bartramia.* The single apical piece of stem bears leaves with sheathing bases narrowed abruptly to long, narrow, denticulate, plane and bistratose laminae with elongate areolation and nerves laterally ill-defined. All this points to *B. ithyphylla* despite the very short leaves (*c.* 1 mm) and rather wide cells.

In the British Isles *B. ithyphylla* is primarily a rupestral plant of the highland zone; there are, however, a very few records from soil in southeastern England. The range, wide in Europe and North America, extends to the high Arctic.

Philonotis undetermined species

!*Broome Heath* (27); Wo. !*Abbot Moss* (70); LDe I; recorded as *Ceratodon* by Walker 1966. !*Kirkmichael 3a* (71); LDe III. !*Loch Fada* (104); III-IV; Vasari and Vasari 1968. !*Baggotstown* (H8); Ho IV or EWo; Dickson 1964*a*. *Mapastown* (H31); LDe; Mitchell 1953.

As a result of great variability species of *Philonotis* may be difficult to determine. The difficulty can only be increased by dealing with subfossil fragments lacking the perigonial leaves helpful in diagnosis.

Philonotis fontana (Hedw.) Brid.

!*Stannon* (2); LDe only. *Ponder's End* (21); MDe; Warren 1912. *Wolvercote* (23); Wo?; Bell 1904. *Hoxne* (25); Ho III or EWo; Reid 1896. !*Upton Warren* (37); MDe. *Windmillcroft* (77); Fl VI/VII/VIII*; Mahony 1868*a*. *Hailes* or *Corstorphine* (83); LDe only; Bennie 1894*a* as *P. fontana* Brid. !*Garral Hill* (94); LDe II; t. !*Loch Cuithir* (104); Fl VI, VII*a* and VIII; Vasari and Vasari 1968. !*Drumurcher* (H32); LDe III.

Though it is recorded from throughout the British Isles, *P. fontana* is abundant only in the uplands where it forms large masses by springs and streams with such species as *Bryum pseudotriquetrum, Dicranella palustris* and *Scapania undulata.*

P. fontana occurs throughout the Northern Hemisphere but seldom reaches the highest latitudes (Schuster *et al.* 1959). Among the numerous subfossils from the rest of Europe there is a Pliocene one from Reuver (table 10).

Philonotis seriata Mitt.

!*Nant Ffrancon* (49); LDe only.

The subfossils are very well preserved and robust and the spiral ranks of leaves are clearly shown (plate 19). There

(a)

(b)

Plate 19. (a) *Philonotis seriata.* Well preserved stems, a little over 1 cm long, with evident spiral ranking of the leaves, from the Late Devensian of Nant Ffrancon. (b) Hand-cut section of a leaf of *Polytrichum urnigerum* from Loch Droma. Note the rounded apical cells of the lamellae.

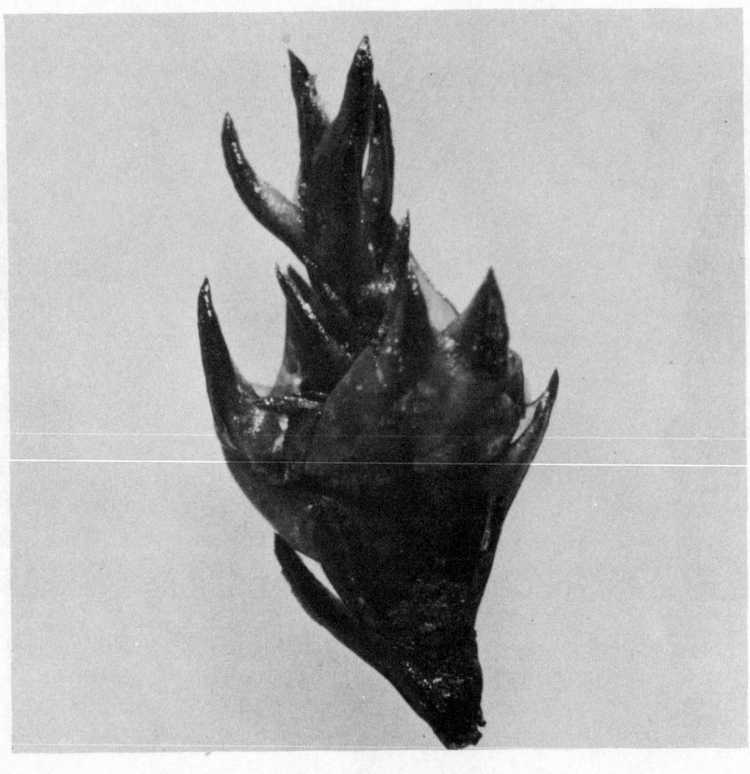

(*c*)

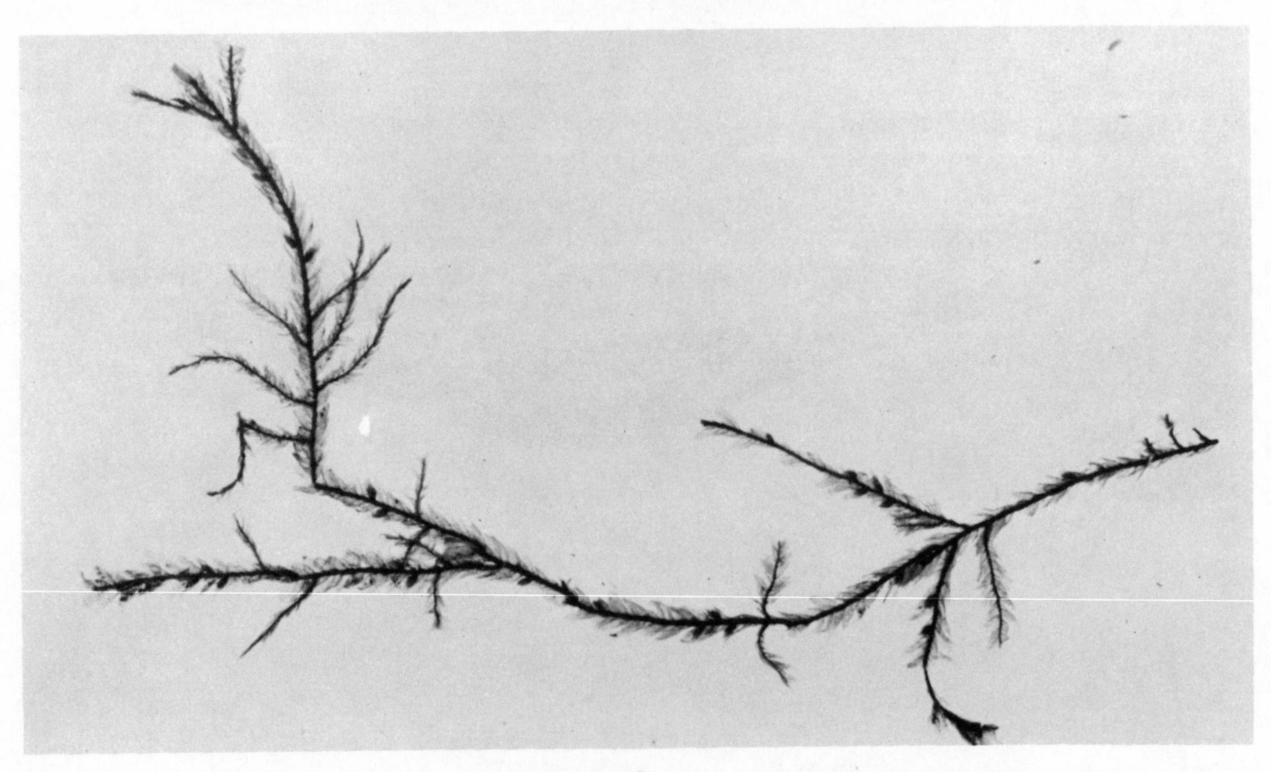

(*d*)

Plate 19. (*c*) *Oligotrichum hercynicum.* Fragment about 5 mm long from the Late Devensian of Nant Ffrancon. (*d*) *Neckera complanata.* The fragment, about 6 cm long, was part of the caulking of North Ferriby boat No. 3. Note the gametangia scattered along the stem.

Plate 20. (*a*) Large snow bed at 3500 ft in granitic corrie, Coire Cais, Cairngorm Mountains, Scotland, early June 1968. *Polytrichum alpinum*, *P. norvegicum* and *Andreaea nivalis* are abundant at the edge of the snow.

Plate 20. (*b*) Tufts of *Pohlia wahlenbergii* var. *glacialis* (right foreground and middle) on stones in running water. High on North Knutshö, Dovrefjell, Norway.

are no Pleistocene records from outside Britain with the exception of a glacial record from Germany (Jovet-Ast 1967).

Ecologically *P. seriata* is similar to *P. fontana* but is now restricted to the Scottish Highlands where it is primarily a species of high altitudes. According to Duncan (1966), it is best developed at *c.* 2800 to 3200 ft in Angus. The Welsh subfossil points to a much wider range in Late Devensian times.

Philonotis calcarea (B. & S.) Schimp.
Dorchester (23); Wo; Duigan 1955.

Proctor found only a single stem but he states (Duigan 1955, p. 234) that the preservation was excellent and that the identification is sure. *P. calcarea* has similar habitats to *P. fontana* but, as the name implies, calcareous conditions are preferred and it is unrecorded from many vice-counties especially in southwestern England. Dixon identified this species tentatively from recent peat in Iceland (Lewis 1911).

Breutelia chrysocoma (Hedw.) Lindb.
!*Seathwaite Tarn* (69); Fl VIII.

This discovery is unique as far as the author is aware. The spreading, plicate leaves and distinctive areolation of this readily recognisable species are easily seen in the well preserved subfossils.

B. chrysocoma, the sole European species of a large genus mainly distributed in the Southern Hemisphere, has a markedly oceanic range (fig. 51). Great abundance is reached on limestone as, for instance, on the Burren pavements, Co. Clare, and on Ben Bulben, Co. Sligo, and also in the extreme west of Norway. *B. chrysocoma* is common too in a variety of more acid habitats in bogs and on rock ledges. In the rest of Europe the species is exceedingly scarce and becoming more so; at the single Belgian locality the last observation was in 1903 (Delvosalle *et al.* 1969). Outside western Europe *B. chrysocoma* is known only from Guatemala.

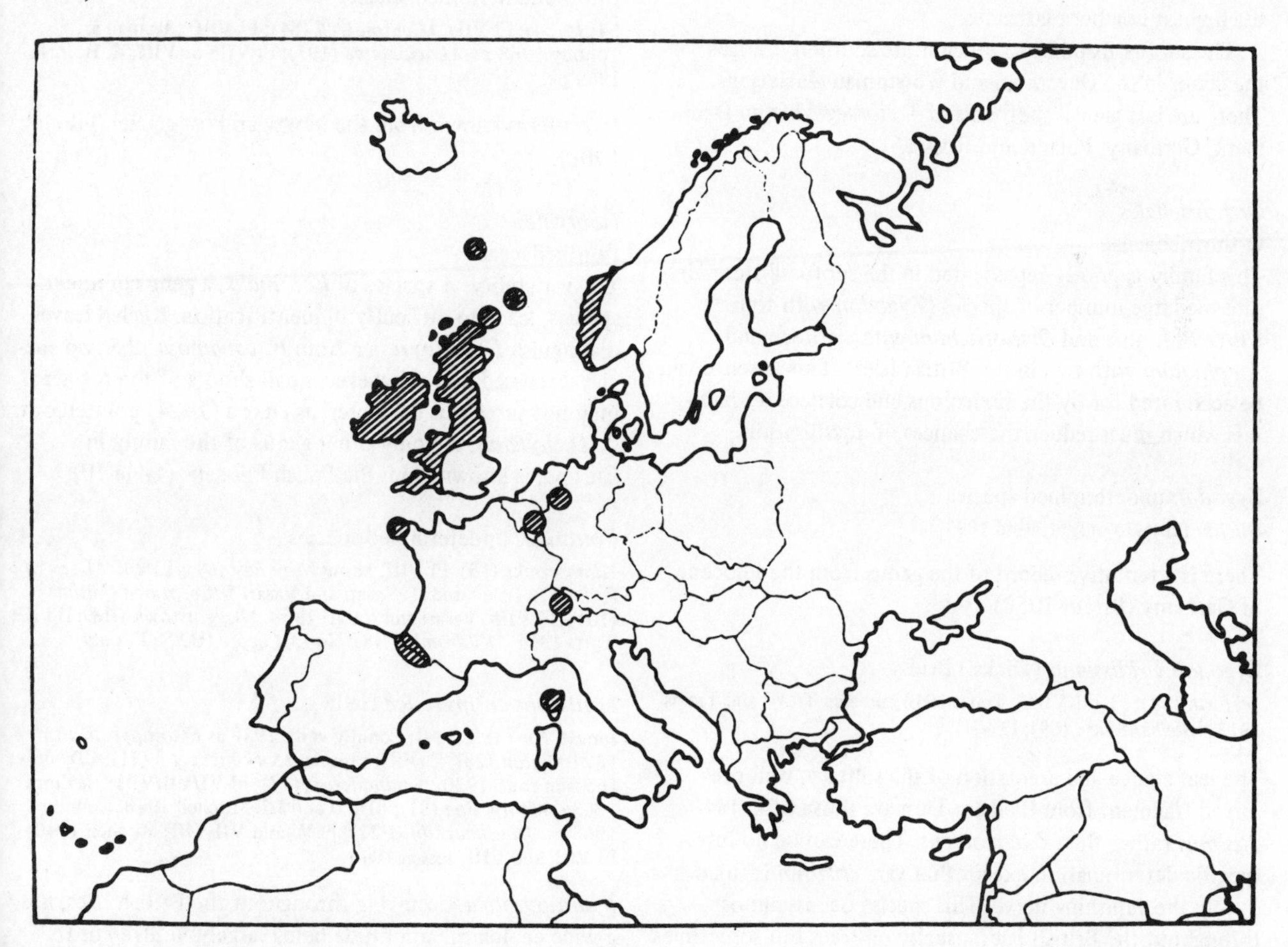

Fig. 51. European range of *Breutelia chrysocoma.* From Störmer 1969.

Timmiales

Timmiaceae

Timmia undetermined species

!*Broome Heath* (27); Wo. !*Whitrig Bog* (81); LDe; Conolly 1963 pc. !*Drumurcher* (H32); LDe III.

T. norvegica and *T. austriaca* are separated on characters of the leaf sheaths and laminas which cannot be seen on the above subfossils, all detached leaves.

T. austriaca has been recovered from the Middle Pleistocene of the Kama region (Abramova and Abramov 1962). The three other European species *T. bavarica, T. megapolitana* and *T. comata* have been recovered from glacial deposits in Denmark and Russia.

Timmia norvegica Zett.

!*Marlow* (24); MDe; Bell 1968 pc. *Ponder's End* (21); MDe; Warren 1912.

Fig. 52 shows the very limited British and Irish range of this calcicolous genus of rock crevices and soil. *T. norvegica* is widespread in Eurasia and North America and reaches the highest northern latitudes.

The subfossils point to much more southerly ranges for the genus in the Devensian and Wolstonian glaciations. There are last glacial subfossils of *T. norvegica* from Denmark, Germany, Poland and Russia.

Orthotrichales

Orthotrichaceae

This family is poorly represented in the subfossil state, despite the large number of species (*Zygodon* with four, *Ulota* with nine and *Orthotrichum* with eighteen and *Amphidium* with two in the British Isles). This absence can be accounted for by the saxicolous and corticolous habitats which must reduce the chances of fossilisation.

Zygodon undetermined species

Hitchin (20); Ho only; t; Reid 1897.

There is a tentative record of the genus from the Pliocene of Germany (Straus 1952).

Zygodon viridissimus (Dicks.) Brid.

Seaforth (59); Fl VII/VIII*; Travis 1913 and also Travis and Travis 1913. !*Blelham Tarn* (69); Fl VII.

The leaf stance and areolation of the solitary, well preserved fragment from Blelham Tarn are those of *Z. viridissimus* rather than *Z. conoideus*. There can be no infraspecific determination except that var. *stirtonii* is ruled out by the vanishing nerve. This species occurs almost throughout the British Isles, usually on trees but sometimes on rock.

Orthotrichum undetermined species

!*Ponder's End* (21); MDe; Dixon unpub. !*Carpow* (88); Fl VIII*; Roman ar.

In the absence of capsules specimens of this and the next genus can be difficult or impossible to determine.

Orthotrichum diaphanum Brid.

!*Ponder's End* (21); MDe; t; Warren 1912.

Distinctive in the hyaline tipped leaves, *O. diaphanum* grows on both wood and rock and is widespread in the British Isles. In Fennoscandia the range is both southern and western (Störmer 1969).

Without explanation Dixon (in Warren 1912) states 'perhaps the identification is open to question . . . ' The two small specimens have deteriorated since Dixon's time. Nevertheless it is doubtful if the hyaline tips or elongate apical cells could ever be seen. The identity cannot be considered fully established.

Ulota undetermined species

!*Aust* (34); Fl VII*a*. !*Denton Well* (54); Fl VIII*; Roman ar; Conolly 1968 pc. !*Loch Maree* (105); Fl VII*b* or VIII; H. H. Birks 1969 pc.

U. crispa is known from the Moravian Post-glacial (Pilous 1968).

Isobryales

Fontinalaceae

The variability of species of *Fontinalis*, a genus of aquatic species, leads to difficulty in identification. Keeled leaves distinguish *F. antipyretica* from *F. squamosa*, the two species discussed here. However, small shoots of the former may not show this character, as Dixon (1924) pointed out.

Dichylema, the only other genus of the family in Europe, is known from the Polish Pliocene (table 10).

Fontinalis undetermined species

!*Hambrooks* (15); Fl VIII. !*Low Wray Bay* (69); LDe II. !*Loch of Park* (92); LDe I and II; Vasari and Vasari 1968. !*Loch Cuithir* (104); Fl VII*b*; Vasari and Vasari 1968. !*Baggotstown* (H8); Ho III; Watts 1964. !*Kildromin* (H8); Ho I. !*Lagore* (H22); Fl only.

Fontinalis antipyretica Hedw.

Hawks Tor (2); LDe II; Conolly *et al.* 1950 as *F. antipyretica* L. !*Whittington* (28); Fl VIII; capsule. !*Kirkmichael 2* (71); LDe only; Dickson *et al.* 1970. *Windmillcroft* (77); Fl VI/VII/VIII*; Mahony 1868*a*. *Whitrig Bog* (81); LDe II and III; Mitchell 1948, Conolly 1963 pc. *Dunshaughlin* (H22); Fl VI and VII-VIII; Mitchell 1940; Fl VII*b* and VIII; Jessen 1949.

F. antipyretica, occurring throughout the British Isles, has a wide ecological amplitude being largely indifferent to water movement and also to trophic state (Spence 1967). Widespread in the Northern Hemisphere, in the Southern

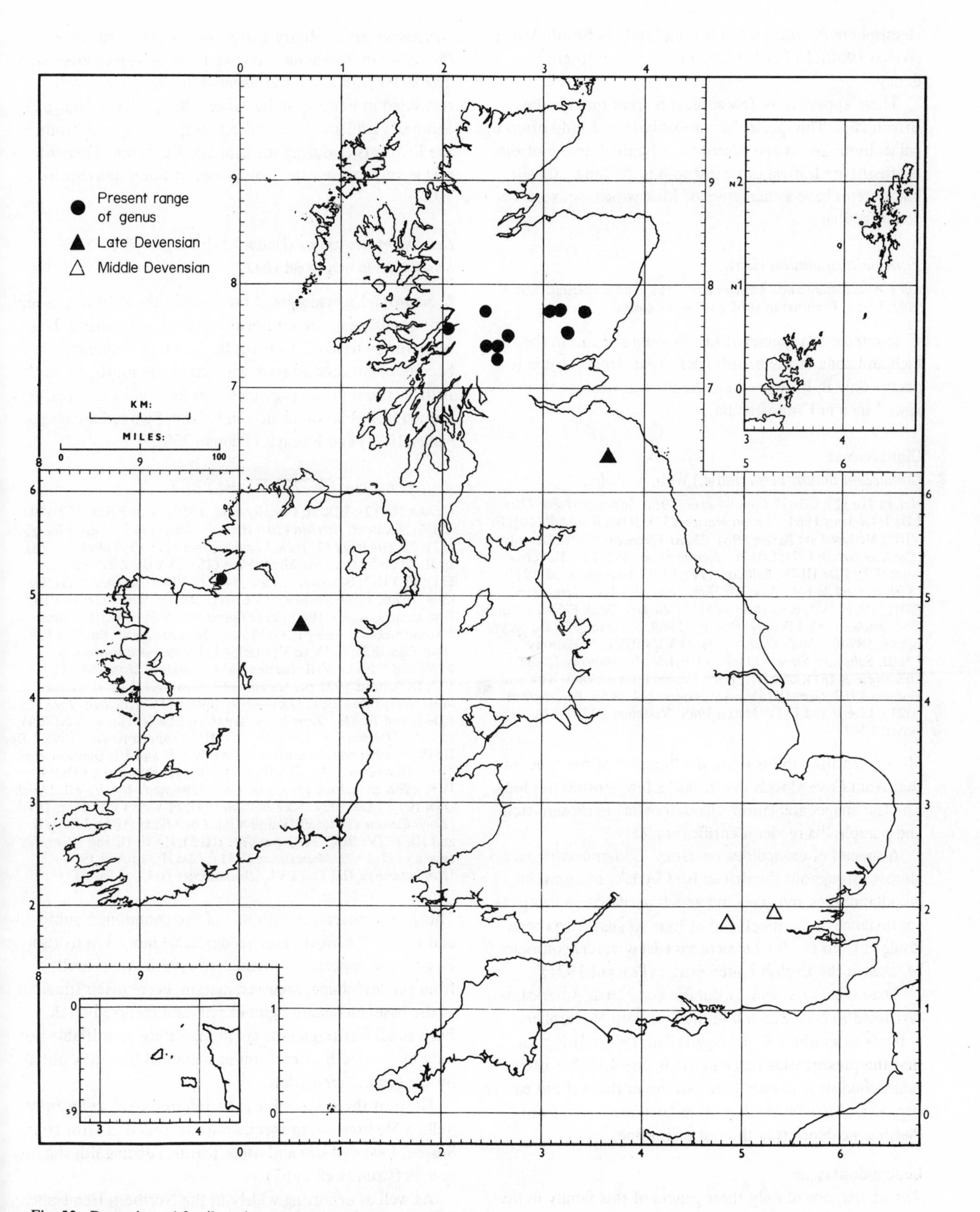

Fig. 52. Devensian subfossils and present range of the genus *Timmia*.

Hemisphere *F. antipyretica* is found only in South Africa (Welch 1960). In Fennoscandia there is a southern tendency.

There appear to be few subfossils from outside the British Isles. The species has probably been found often by palaeolimnologists and passed unrecognised; many of the subfossils are leafless or almost so but the long, straight, black stems have a characteristic look which can soon become familiar.

Fontinalis squamosa Hedw.

Lake Windermere and *Coniston Water* (69) and *Ennerdale Water* (70); LDe I; Pennington 1962 as *F. squamosa* L.

F. squamosa is a species of fast flowing streams in the highland zone of the British Isles. Apart from Algeria it occurs only in Europe where the general western tendency is well seen in Fennoscandia.

Climaciaceae

Climacium dendroides (Hedw.) Web. & Mohr.

Hawks Tor (2); LDe II; Conolly *et al.* 1950. !*Colney Heath* (20); LDe I; Godwin 1964. !*Upton Warren* (37); MDe. *Roushill* (40); Fl VIII*; Medieval ar; Barker 1961. !*Nant Ffrancon* (49); LDe only. !*Burbage Brook* (57); LDe II. !*Kersall Moss* (69); LDe II. !*Cove Moss* (69); LDe III-IV. *Ballaugh* (71); Fl IV; Dickson *et al.* 1970. !*Kirkmichael 3a* (71); LDe III; Dickson *et al.* 1970. *Garvel Park* (76); LDe/Fl IV*; Robertson 1881. !*New Dry Dock* (76); LDe only. *Cowdenglen* (76); Fl? only; Mahony 1869. *Faskine* (77); De only; Bennie 1894*b*. *Windmillcroft* (77); Fl VI/VII/VIII*; Mahony 1868*a*. *Salisbury Street* (77); Fl VI/VII/VIII*; Mahony 1868*a*. !*Whitrig Bog* (81); LDe III. !*Loch Droma* (105); LDe I; Kirk and Godwin 1963. *Gort* (H15); LAn; Jessen *et al.* 1959. *Ballybetagh* (H21); LDe II and Fl IV; Jessen 1949. *Ralaghan* (H30); LDe III; Jessen 1949.

Most often the subfossils are the 'boughs' of the 'tree' but that from Cove Moss is the 'trunk' a few centimetres long. The leaf shape, areolation, characteristic coarse serration and paraphyllia render identification easy.

A species of calcicolous tendency, *C. dendroides* occurs almost throughout the British Isles by lake margins, in woodland, fens and grassland which need not be damp, as for instance in the Breckland of East Anglia. It has been dredged from the *Isoetes lacustris* (deep water) consocies of lakes in the English Lake District (Pearsall 1921).

The range, very wide in Eurasia and North America, but restricted to New Zealand in the Southern Hemisphere, extends to southern Spitsbergen (Kuc 1963). It is clear that the present wide range in the British Isles has been established since at least Late Devensian times if not earlier (fig. 53). *C. dendroides* has often been recovered from Pleistocene deposits in the rest of Europe.

Leucondontaceae

Antitrichia, one of only three genera of this family in the British Isles, has a very extensive Pleistocene history, *Leucodon* has a solitary British subfossil record while *Pterogonium* has none whatever from Britain or elsewhere. The latter with a Mediterranean-Atlantic range is very restricted in Fennoscandia, where the species is thought by Störmer (1969) to have arrived by spore dispersal from the British Isles during the Climatic Optimum. *Forsstroemia* is known from the Pleistocene of Louisiana (Steere 1938).

Leucodon sciuroides (Hedw.) Schwaegr.

Selsey (13); Ip only; Reid 1892.

L. sciuroides is widespread but local in the British Isles and rare or absent in the far north and west of Scotland. It is scattered in Ireland. Towards the north in Fennoscandia it becomes restricted to rock whereas in the south and in the British Isles it grows principally on the trunks of trees. It is known from Flandrian sites in Poland, Switzerland (Neuweiler 1905) and Moravia (Pilous 1968).

Antitrichia curtipendula (Hedw.) Brid.

Hawks Tor (2); LDe II; Conolly *et al.* 1950; !*Winchester* (12); Fl VIII*; Saxon ar. *Hitchin* (20); Ho only; Reid 1897. !*Happisburgh* (17); Pa. !*Beeston* (27); Pa. !*Trimingham* (27); Cr. !*Wretton* (28); Ip II, III and EDe. !*Hockham Mere* (28); Fl VII*a*. *Little Paxton* (31); Fl VIII*; Saxon ar. !*Aust* (34) ; Fl VII*a*. !*Mochras* (48); De? only. !*Nant Ffrancon* (49); LDe only. *Denton Well* (53); Fl VIII*; Roman ar; Conolly 1968 pc. !*Epperstone* (56); Fl VIII*; Roman ar. *Ulrome Lake Dwelling* (61); Fl VII*b*; Bronze Age ar; Smith 1911. !*Star Carr* (62); Fl IV or V; Clark 1954. *Cromwell Bottom* (63); Fl VI and VII*b* or VIII; Bartley 1964. *Russland Moss* (69); Fl V or VI?; Dickinson 1971 pc. !*Seathwaite Tarn* (69); LDe II. !*Cove Moss* (69); LDe III-IV. !*Low Water* (69); Fl IV. !*Blelham Tarn* (69); LDe II and Fl VII. !*Low Wray Bay* (69); LDe II. !*Low Wood* (69); Fl only. !*Burnmoor Tarn* (69); LDe II. !*Keppelcove Tarn* (69); LDe III-IV. !*Kirkmichael 1 and 3a* (71); LDe I, II and III; Dickson *et al.* 1970. !*Burnbrae* (76); Fl VII or VIII. *Garvel Park* (76); LDe/Fl IV*. !*New Dry Dock* (76); LDe only. !*Drymen* (86); Fl VII. !*Loch Oich* (97); LDe? only. *Fort William* (97); Fl V or VI*; Dixon 1910. !*Loch Cuithir* (104); Fl IV and VII*a*. *Loch Fada* (105); LDe I, II and III, Fl IV; Birks 1961 pc. *Gort* (H15); Ho II, III and III or IV; Jessen *et al.* 1959. *Mapastown* (H31); LDe II; Mitchell 1953. !*Rockmarshall* (H13); Fl VI. !*Drumurcher* (H32); LDe III.

This robust species is both one of the commonest subfossils and one of the most easily recognisable mosses in British Pleistocene deposits. A distinctive combination results from the leaf shape, recurved margin, nerve often tripartite at the base, incrassate sigmoid cells and recurved teeth. Even small leaf fragments are immediately identifiable not only by the teeth often forming a grapnel-like apex but also by the general areolation.

None of the subfossils can be referred to *A. californica* Sull., a Mediterranean species once recorded in error from Sussex. Leaf cell size and other features distinguish the species (Crundwell 1957).

As well as occurring widely in the Northern Hemisphere, *A. curtipendula* has been recorded from Patagonia and

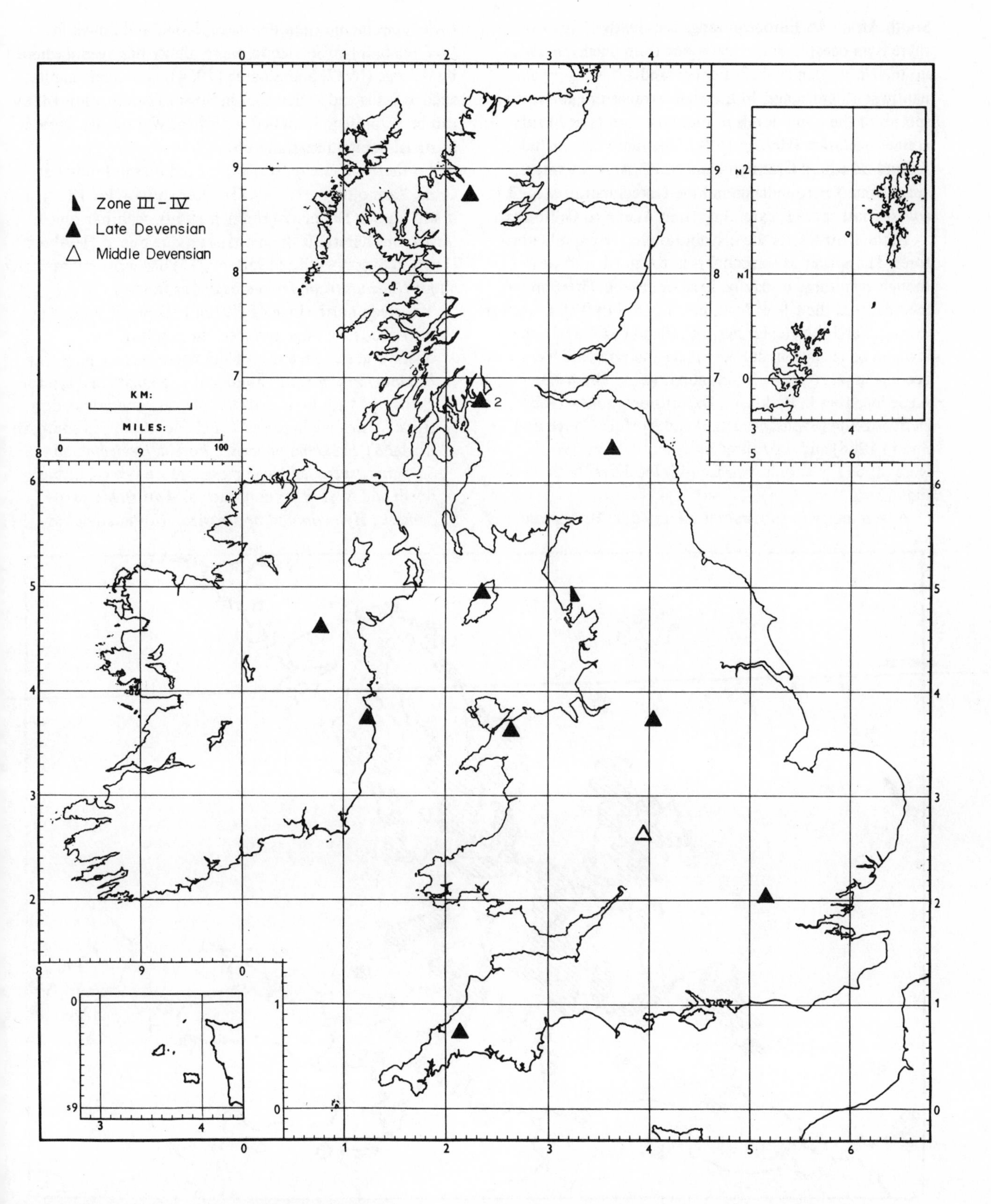

Fig. 53. Devensian subfossils of *Climacium dendroides.*

South Africa. Its European range is extensive (fig. 54); there is an oceanic tendency. Apart from occurrences far up the Norwegian coast in Fennoscandia it has a predominantly southern range. In the Mediterranean region it is a species of the upper levels of montane forests or of sub-alpine vegetation (Herzog 1926). The range also includes Iceland, southern Greenland and North America where, apart from Newfoundland and the Lake Superior area, it is a west coast species, extending from Alaska to California.

In the British Isles the distribution has remarkable features. The species is commonest in the north and west though even there it may be local or absent. There are no records from the Isle of Man, much of South Wales, western Cornwall and the greater part of Ireland (fig. 55). In the lowland zone stations are very local and rare, and have become more so in the last hundred years. There were single localities in Berkshire, Oxfordshire (Jones 1953), Hertfordshire (Swinscow 1959) and Norfolk (Petch and Swann 1968) and only three in Essex (Pettifer 1968). Rose (1951) gives only two Kentish localities, both maritime.

A wide range of substrata is tolerated; in Britain the rock types include granite, schist, basalt and limestone. Preference for rocks rich in micro-niches has been discussed by Allorge (1947) and Sjögren (1964). Man-made habitats such as walls and a tiled roof in Sussex (Braithwaite 1888) can be colonised. Thatched roofs near Warsaw are another artificial habitat (Szafran 1961).

Formerly *A. curtipendula* occurred on sand dunes at Great Yarmouth in Norfolk (Hooker and Taylor 1827), as it still does in Denmark (Holmen 1959). Another type of terricolous habitat is the perhaps unique one at Heyshott Down in Sussex. Tallis (1958, p. 275) has written, primarily about *Rhacomitrium lanuginosum*, as follows:

> The steep north slope of Heyshott Down is broken up into a series of large terraces, the result of former quarrying, and on a few of the upper terrace platforms *R. lanuginosum* forms abundant large mats in a typical chalk turf with luxuriant bryophytes. These include a trio of 'calcicole hepatics' (calcicole, that is, in southern England), *Frullania tamarisci, Porella laevigata* and *Scapania aspera*, and also mosses more typical of the north and west of Britain, such as *Antitrichia curtipendula, Hylocomium brevirostre, Rhytidiadelphus*

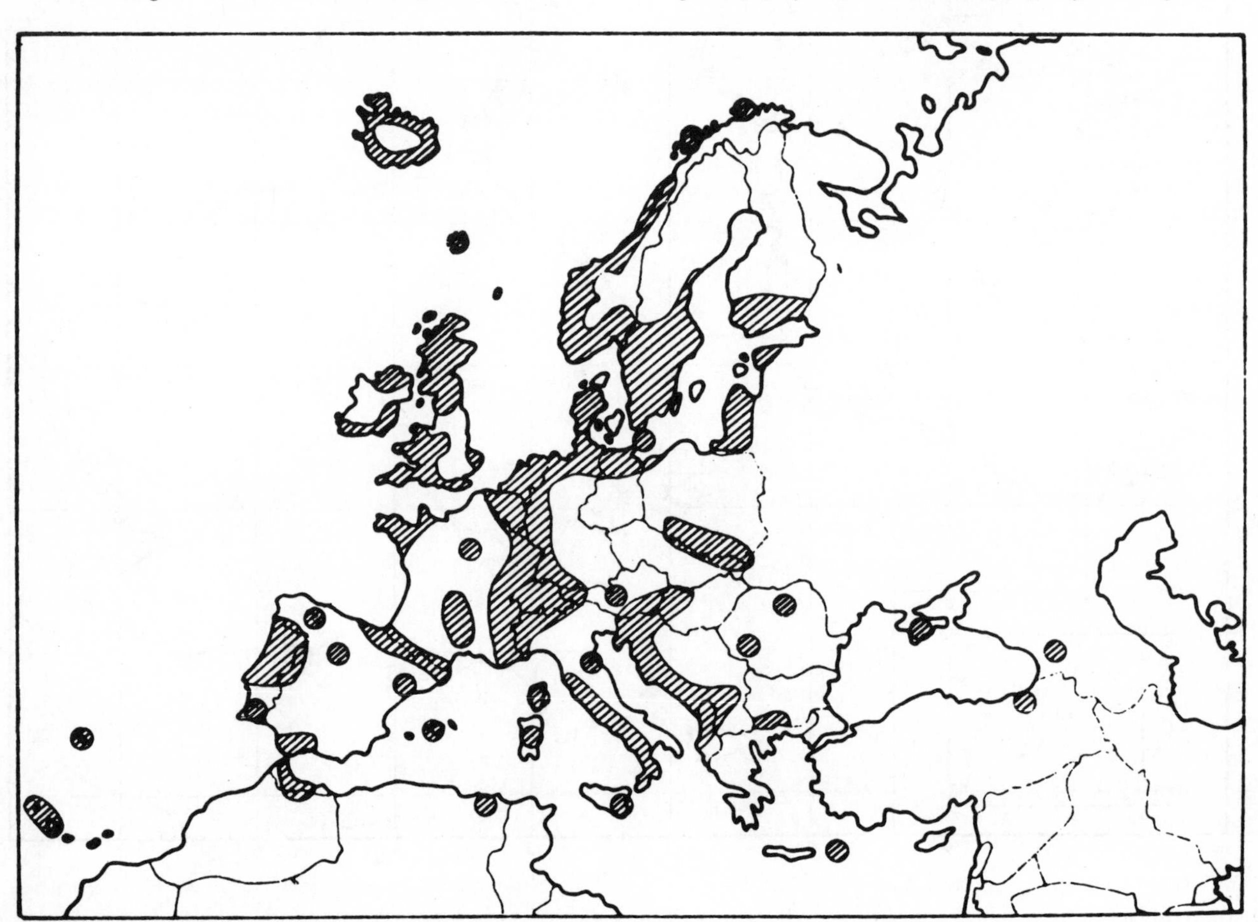

Fig. 54. European range of *Antitrichia curtipendula*. From Störmer 1969.

Fig. 55. Present range of *Antitrichia curtipendula* in the British Isles. Black shows the vice-counties where *Antitrichia* is present, hatching where recently extinct.

loreus, Neckera crispa and *Tortella tortuosa.* Together they constitute an assemblage of bryophytes hardly paralleled elsewhere on the chalk in the south of England. *R. lanuginosum* presumably became established after the quarries ceased activity, which appears to have been around 1900.

At Dungeness and Hythe in Kent (Rose 1951). *A. curtipendula* occurs on shingle. Perhaps paralleling this is the occurrence in southern Greenland where Lee (1944) found the species on 'moist, somewhat protected places where there is considerable humus in the gravel' and on 'a wet humus area along the edge of a small pond on the moraine'.

In Britain the range of trees inhabited includes *Alnus, Betula, Corylus, Fagus, Fraxinus, Pinus, Quercus, Prunus spinosa, Sorbus aucuparia* and, no doubt, others. Similarly in Norway a wide variety of trees are colonised (Störmer 1969). It seems *A. curtipendula* is largely indifferent to bark type. Trunks and upper branches may be colonised. The occurrences at Dungeness is worth quoting. Rose (1949, p. 210) states 'The strange copses of stunted blackthorn and holly, in spite of their exposure, have also many epiphytes. The most remarkable of these is the moss *Antitrichia curtipendula*, here growing very finely in great tassels, almost at sea level.'

The luxuriant abundance sometimes attained by *A. curtipendula* has impressed many bryologists before Rose. Dixon (1924, p. 408) wrote 'This plant grows nowhere, perhaps, more finely in our islands than in Wistman's Wood, Dartmoor, where it clothes the limbs of old and stunted oaks with large masses, hanging down to a length of a foot or more, and producing fruit in abundance.' As Dixon's graphic passage indicates it is rocky woodland in the mountains that is the optimal habitat. Typical is a birch woodland on hummocky moraine at Coille 'a' Choire (NN 464898) by Loch Laggan, Inverness-shire. Here the rocks and trunks of *Betula* are covered abundantly; the trunks of *Alnus* and *Sorbus aucuparia* are less favoured.

In southern and eastern England most occurrences are or were corticolous, with noteworthy exceptions pointed out above. In the southwest trees are preferred, as shown by Wistman's Wood where Courtney and Hardy (1969) found 271 distinct patches on 75 trees (oaks) and only two patches on rock (granite). By contrast, in the oak woodland on a steep hillside at Keskadale in Cumberland the occurrence is solely rupestral. *A. curtipendula* is abundant on one small outcrop but is absent from the trees. In North Wales the occurrences are largely, if not solely, rupestral, as is true of northeastern Ireland (Taylor 1951).

In many areas of both western Britain and Norway, *A. curtipendula* occurs in open, quite treeless situations, often on cliffs, as for instance on Eigg and Rhum (Lodge 1963), Skye, Ailsa Craig (Vevers 1936), and in the Lake District and North Wales.

This change of substrate preference with geography, which occurs in other areas apart from the British Isles, has been commented on by many authors. The causes are far from obvious. In the British Isles one can say that towards the northwest there is a tendency to inhabit rock rather than trees.

In view of the wide tolerance of rupestral, terricolous and corticolous habitats the localness of *A. curtipendula* is puzzling, especially when the occurrence is profuse over a small area and seemingly suitable or identical habitats all around are totally unoccupied. The most striking example of this known to the author is in Briksdalen close to a lobe of the Jostedal glacier in Sogne and Fjordane county of Norway. Here in well developed *Populus tremula-Alnus incana* forest *A. curtipendula* occurs in vast profusion on boulders but seemingly nowhere else in the vicinity though there is much *Betula pubescens* and *Alnus incana* woodland.

There are five interglacial occurrences in the British Isles. Three different ages are represented: the Pastonian, Hoxnian and Ipswichian. There are no glacial occurrences before the last glaciation and then, with the possible exception of Wretton, all are of Late Devensian age. The Wretton subfossil may well be derived from the underlying Ipswichian deposits.

It is clear that *Antitrichia* was widespread in Late Devensian Britain; there are records from Cornwall, North Wales, the Isle of Man, Renfrewshire and Skye as well as Co. Louth and Co. Monaghan. The location of these deposits along the western seaboard of Britain is striking, as is the absence from all glacial sites elsewhere in the country (fig. 57).

Moreover, it seems safe to state that in Late Devensian times *A. curtipendula* was abundant over much of its British range. The Nant Ffrancon subfossils were so numerous, many being large and well preserved, that the conclusion seems inescapable that *A. curtipendula* was a dominant of some facies of the bryophytic vegetation. Similarly in the Lake District *A. curtipendula* is represented, often in quantity in most Late Devensian and early Flandrian deposits; particular interest attaches to the occurrence at Keppelcove because growth at almost 2000 ft, or possibly higher in zone III to IV, is established. Present ecology renders it unlikely that *A. curtipendula* suffered a decrease, as did many other species frequent in Late Devensian deposits when the forests spread in response to climatic amelioration. The reverse may well have been the case.

If the Late Devensian restriction to western areas as shown by the subfossils is true, *A. curtipendula* must have spread rapidly to new areas in the earliest part of the Flandrian; there is a zone IV or V record in eastern Yorkshire. There are interglacial records from mainland Europe as well as Flandrian such as in Poland (Szafran 1964 pc) and Czechoslovakia (Pilous 1968). However, despite the investigation of glacial moss floras in Sweden, Denmark, Belgium, Germany, Czechoslovakia, Poland and European Russia, *A. curtipendula* has never been found. Vanden Berghen (1951) found *A. curtipendula* in an assemblage, consisting almost exclusively of rich fen species, from a site ten miles northwest of Brussels. He believed the age to be Weichselian. However, according to Dr A. V. Munaut (personal communication) the stratum in question may be of last interglacial or earlier age.

It may be that *A. curtipendula* can be envisaged as occurring not just in western Britain but along the western margins of Europe as a whole in Late Devensian times and then there was a spreading eastwards in the early Flandrian. However, I know of no glacial moss floras from Norway, western France or Iberia.

One can do little more than speculate as to what the Pleistocene habitats were. In the case of interglacial deposits woodland provenance seems more than likely. However, for Late Devensian occurrences there can be no such conviction. There would have been abundant rupestral situations available and also birch woodland, if only locally, and possibly *Populus tremula* woodland. It is tempting to think of the Coille 'a' Choire and Briksdal habitats mentioned above as offering parallels. A striking feature of the Late Devensian assemblages from the Lake District, as well as Nant Ffrancon, Loch Oich and also Mochras, is the preponderance not just of *Antitrichia* but also of *Hylocomium spendens* and *Rhacomitrium*, presumably *R. lanuginosum*. If one imagines all three growing together, this leads to a rupestral rather than corticolous and open rather than shaded habitat.

Though the present range is by no means confined to the most oceanic areas of Europe, the abundance in western Britain and the western fringe of southern Norway is marked. Perhaps, therefore, the Late Devensian abundance in western Britain speaks for an oceanicity of climate. It is clear that subfossil *A. curtipendula* points to temperate and certainly not arctic or subarctic conditions.

Bryophyte recording in the last 200 years has revealed a marked decrease or extinction of *A. curtipendula* in southern and eastern England and led to the speculation that atmospheric pollution is the agent. At Wistman's Wood the species no longer has the luxuriance it once had, for what reason it is hard to say, possibly pollution or over-collecting (Courtney and Hardy 1969).

No less than nine of the Flandrian records prove occurrence in areas no longer occupied. Those from Ulrome Lake Dwelling, southeast Yorkshire, Epperstone, Nottinghamshire, Denton, South Lincoln, and Little Paxton, Huntingdonshire, are particularly important in this respect. In common with that from Winchester these four are all from archaeological deposits (one Bronze Age, two Roman and two Saxon). Therefore they indicate that the decrease revealed by recent field study has been more extensive and perhaps of greater duration than has been realised hitherto.

Neckeraceae

This family is well represented as Pleistocene subfossils. It is striking and certainly significant that the five species (out of eight recorded from the British Isles) are lacking in glacial assemblages but occur freely in interglacial, especially Flandrian, deposits.

The three absentees from the fossil record are all species of particular bryogeographical importance. *Leptodon smithii* is a Mediterranean species of markedly southern British range (fig. 9), *Neckera pennata* was known in Britain only from Forfarshire but is now extinct and *Thamnobryum angustifolium*, known in the British Isles with certainty only from Derbyshire, also occurs on Madeira.

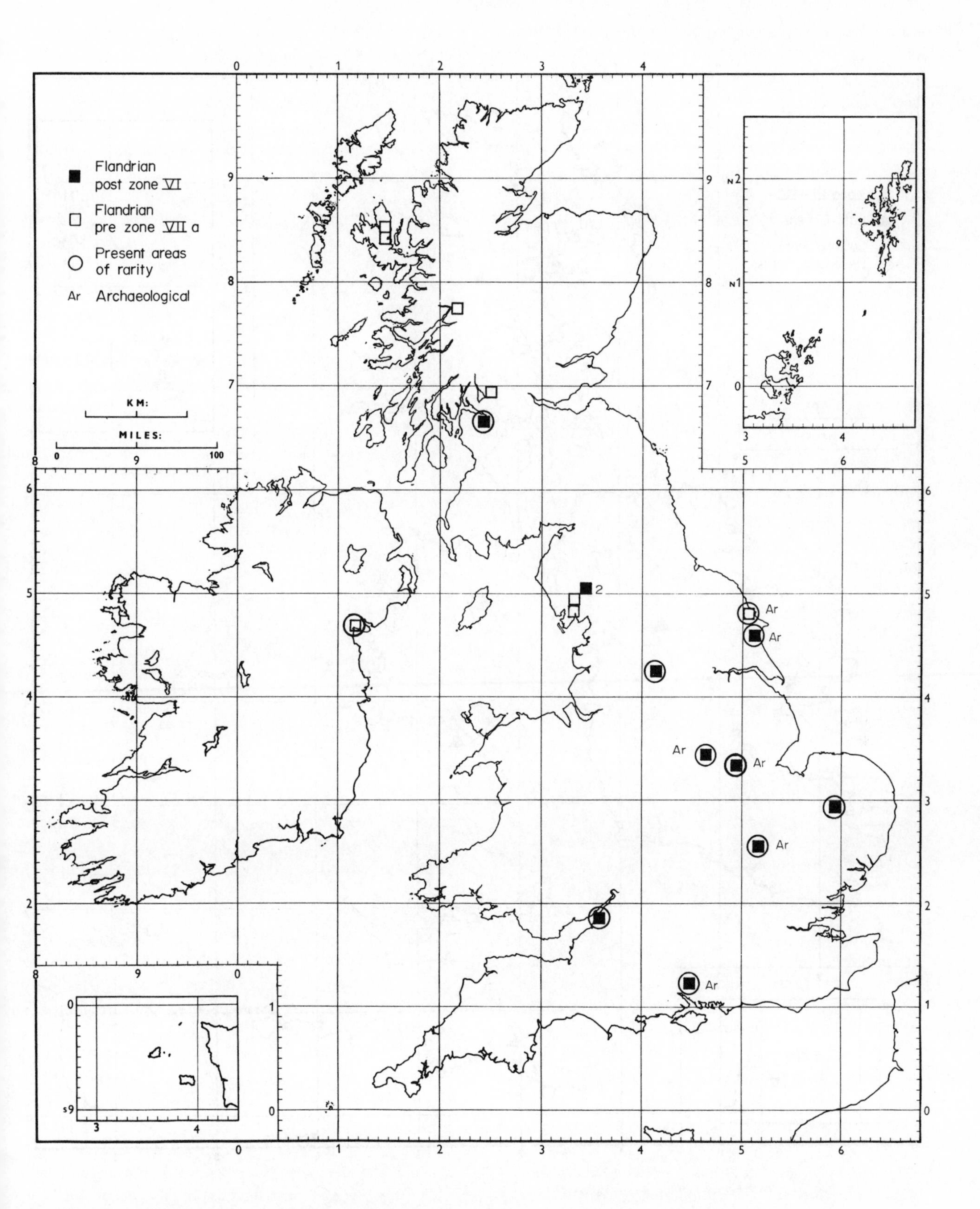

Fig. 56. Flandrian subfossils of *Antitrichia curtipendula.*

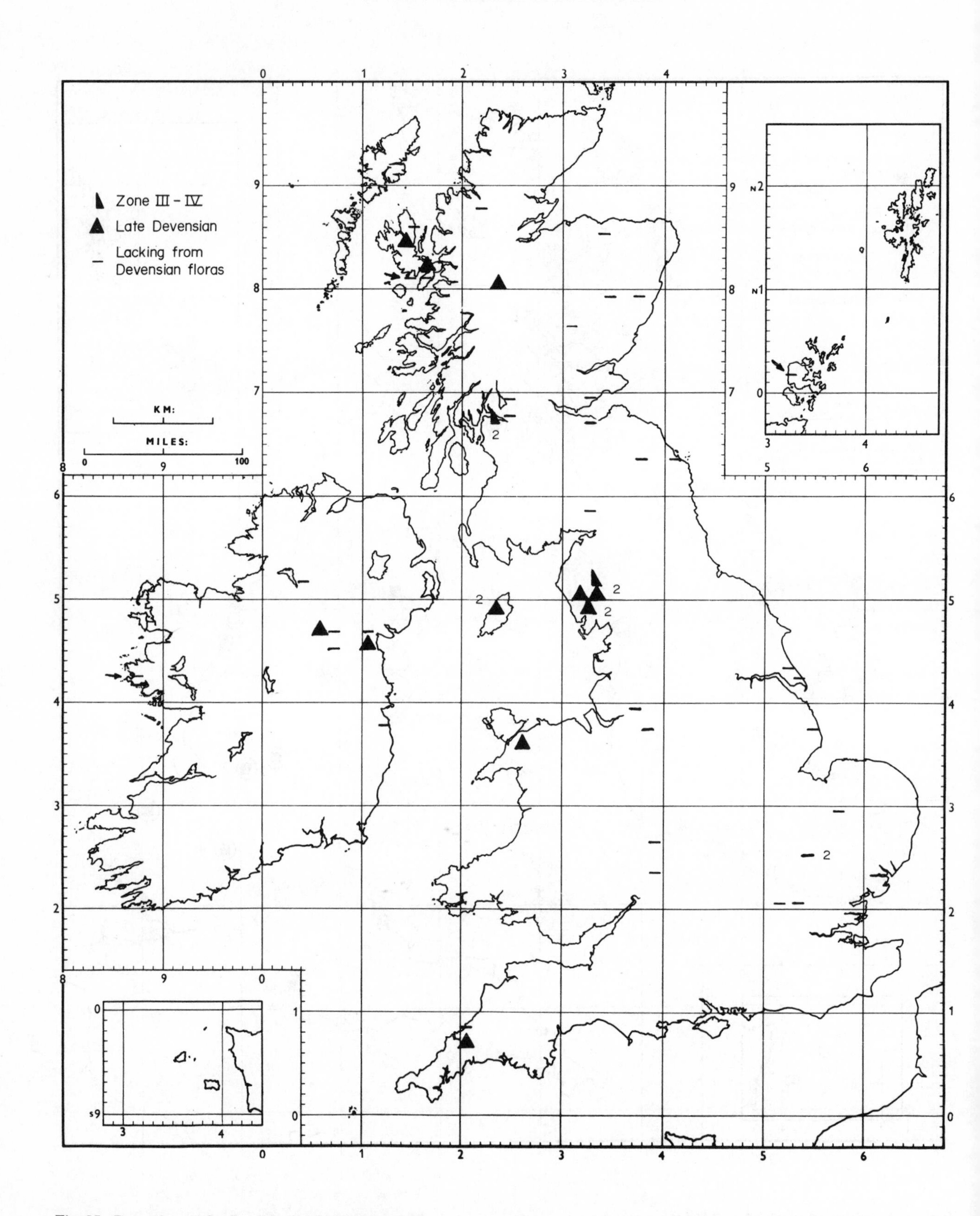

Fig. 57. Devensian subfossils of *Antitrichia curtipendula*.

Neckera crispa Hedw.

!*Bryn-y-Mor* (48); Fl VII*a*. *Windmillcroft* (77); Fl VI/VII/VIII*; Mahony 1868*a*. *Gort* (H15); Ho III or IV; Jessen *et al.* 1959. !*Winetavern Street* (H21); Fl VIII*; Medieval ar.

Easily recognisable by its robustness and undulate leaves *N. crispa* has been recorded throughout the British Isles with the principal exceptions of East Anglia and much of eastern and southern Ireland. The range, extensive in Europe, is southern in Fennoscandia. Calcareous rocks are the usual habitats though sometimes tree trunks are colonised.

N. crispa has been recovered no less than nine times from Flandrian deposits in Switzerland (five Neolithic, four Bronze Age) as well as from two Bronze Age sites in Italy and one Neolithic site in Austria (Neuweiler 1905). There are Pliocene records from Poland and the Caucasus (table 10).

Neckera pumila Hedw.

Windmillcroft (77); Fl VI/VII/VII*; Mahony 1868*a*. *Fort William* (97); Fl V or VI*; Dixon 1910.

N. pumila is readily separated from other members of the genus by the recurved leaf margins. Like the other two species discussed here, *N. pumila* can be either corticolous or saxicolous and is widespread in the British Isles and southern in Fennoscandia.

Neckera complanata (Hedw.) Hüben

!*Bell Track* (6); Fl VII*b*; *Wilsford Shaft* (8); Fl VII*b**; Bronze Age ar; Dimbleby 1967 pc. *Brook* (10); VII*a*; Clifford 1936. !*Winchester* (12); Fl VIII*; Saxon ar. *West Wittering* (13); Ip only; Reid 1892. !*Tilbury Dock* (18); Fl VII*a*. *Royal Albert Dock* (18); Fl VII*; Dixon 1914. !*Marks Tey* (19); Ho II and EWo. *Hitchin* (20); Ho only; Reid 1897. !*Wretton* (18); Ip II. !*Shippea Hill* (29); Fl VII*b*; also recorded by Clark *et al.* 1934. !*Soham Lode* (29); Fl VII*b*. !*Old Decoy* (29); Fl only. !*Little Paxton* (31); Fl VIII*; Saxon ar. !*Gloucester* (33); Fl VIII*; Medieval or Roman ar. !*Barnsley Park* (33); Fl VIII*; Roman ar. *Bridgewater Street* (59); Fl VIII*; Roman ar; Roeder 1899. !*North Ferriby Boats 2 and 3* (61); Fl VII*b**; Bronze Age ar; with sporophytes; Wright and Wright 1947; Wright and Churchill 1965. *Cromwell Bottom* (64); Fl VII*b* or VIII; Bartley 1964. *Castle Eden* (66); EP; Reid 1920. !*Burnbrae* (76); Fl VII or VIII. *Windmillcroft* (77); Fl VI/VII/VIII*; t; Mahony 1868*a*. *Salisbury Street* (77); Fl VI/VII/VIII*; Mahony 1868*a*. *Fort William* (97); Fl V or VI*; Dixon 1910. *Gort* (H15); Ho III; Jessen *et al.* 1959. !*Winetavern Street* (H21); Medieval ar. !*High Street* (H21); Medieval ar. *Cushendun* (H40); Fl VI; Jessen 1949.

Subfossils of *N. complanata* are readily separated from other members of the genus by size and leaf shape. Many are large and particularly fine; those from North Ferriby Boats 2 and 3 are several centimetres long and one of the Marks Tey subfossils is tiny, about 0.5 mm with twelve leaves; it is the apex of a filiform shoot often produced by this species.

N. complanata occurs throughout the British Isles in such habitats as shaded, often calcareous, rocks and trunks of trees in woodland. Sometimes it occurs in chalk grassland or on shaded calcareous cliffs at low or moderate altitudes. In Fennoscandia it is a lowland plant of southern distribution.

The subfossils are surprisingly numerous. There are no less than twenty Flandrian records from widely scattered areas (fig. 58). Wood peats often yield this species as at Bell Track, Tilbury and Royal Albert Docks, Shippea Hill, Soham Lode and Old Decoy. In these cases the habitat must have been corticolous. Of the six archaeological localities the North Ferriby occurrences are the most remarkable. Here were found boats built in the late Bronze Age by workmen who used *N. complanata* as caulking material (p. 192).

N. complanata has a very detailed history beginning in the Miocene of France at Arjuzanx, of Poland at Stare Gliwice and of Russia at Pshekha (table 9), and extending through the Pliocene, at Kroscienko and Czorsztyn in Poland (table 10), to the Pleistocene when the records become frequent in interglacial deposits. In accord with the present ecology and range of the species there are no glacial records, apart from the Early Wolstonian one from Marks Tey which can reasonably be considered as derived from the underlying Hoxnian strata. If one divided the Pleistocene into moss periods the Flandrian would be the *Neckera complanata* period.

Homalia trichomanoides (Hedw.) B., S. & G.

Wolvercote (23); Ho?; Bell 1904. !*Happisburgh* (27); Ba or earlier. !*Wretton* (28); Ip II. *Castle Eden* (66); EP; Reid 1920. *Windmillcroft* (77); Fl VI/VII/VIII*; Mahony 1868*a*.

H. trichomanoides, widespread in England, Wales, Ireland and much of Scotland, becomes rare or absent in far north and west of Scotland. It is a woodland species of tree bases, soil and rocks. Like the species of *Neckera* discussed above, it is southern in Fennoscandia and absent from glacial deposits.

Again like the *Neckera* species it is well known in Tertiary deposits. There are Pliocene records from Bashkir, Russia, Kroscienko and Czorsztyn, Poland (table 10). From the Miocene deposits at Arjuzanx, France, this species and also *H. lusitanica* were recorded (table 9). It has recently been recorded from Post-glacial deposits in Moravia (Pilous 1968).

Thamnobryum alopecurum (Hedw.) Nieuwl.
[*Thamnium alopecurum* (Hedw.) B., S. & G.]

!*Bell Trackway* (6); Fl VII*b*. !*Lowham* (6); Fl VIII*; Roman ar. *Wilsford Shaft* (8); Fl VII*b**; Bronze Age ar; Dimbleby 1967 pc. *Royal Albert Docks* (18); Fl VII*; Dixon British Museum collection. !*Shippea Hill* (29); Fl VII*b*; recorded in error as *Aulacomnium androgynum* by Godwin and Clifford 1938. !*Barnsley*

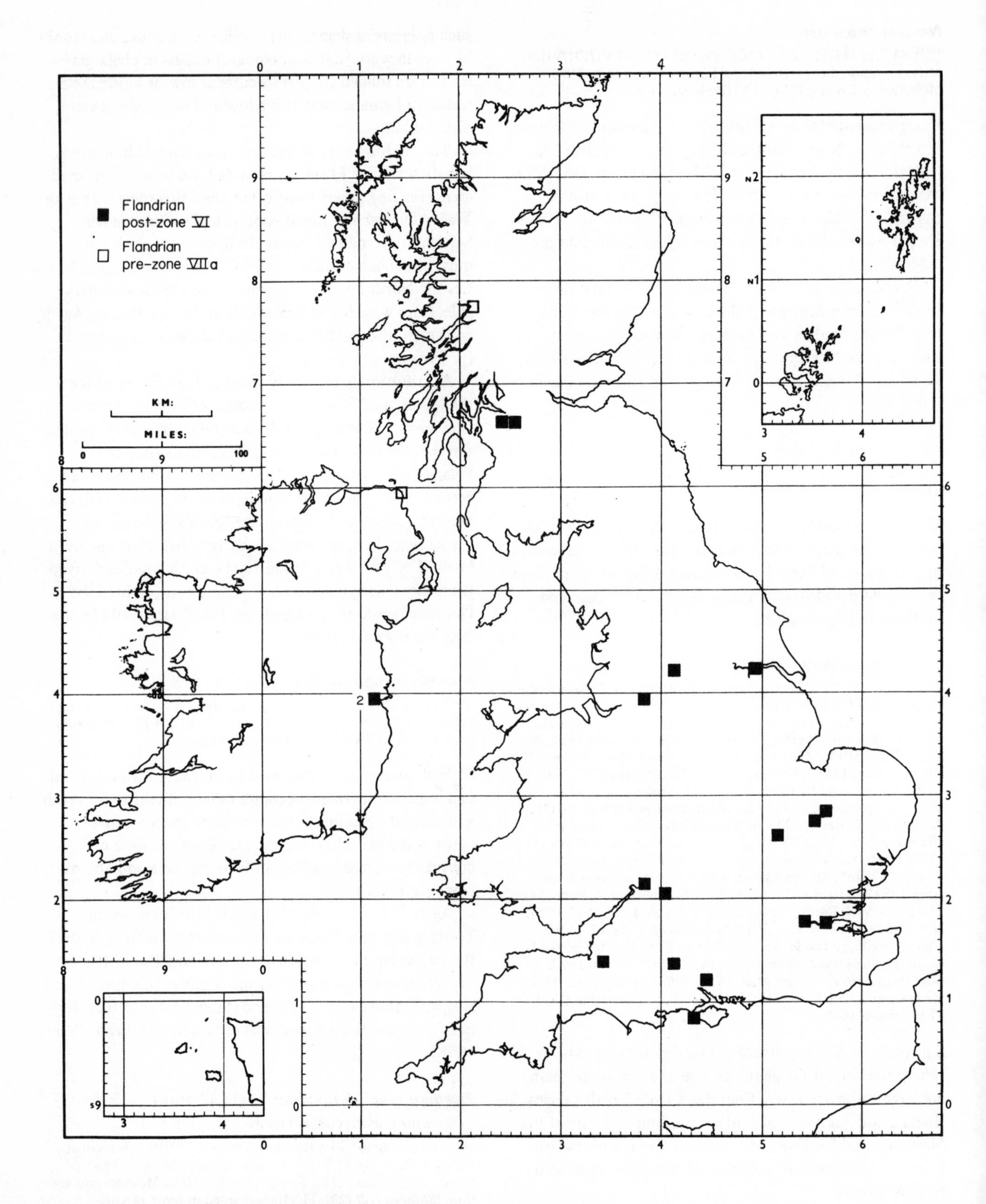

Fig. 58. Flandrian subfossils of *Neckera complanata*.

Park (33); Fl VIII*; Roman ar. *Low Wood* (69); Fl only; Pennington 1943 as *Porotrichum alopecorum* Mitt. !*Seathwaite Tarn* (69); Fl VIII. *Fugla Ness* (112); Ho II; Birks and Ransom 1969. !*Winetavern Street* (H21); Fl VIII*; Medieval ar. *Cushendun* (H40); Fl VI; Jessen 1949.

The leaves of this species are highly characteristic in shape and areolation; well preserved subfossils are immediately identifiable. Particularly fine preservation is shown by the Barnsley Park material which is large enough to show the dendroid habit (plate 9).

There is no surprise in finding this markedly shade-tolerant species among the plant debris from two Roman wells. It makes up the bulk of the assemblage from Barnsley Park where very probably it grew on the inside walls.

T. alopecurum occurs throughout the British Isles in such habitats as the ground in woodlands and on rocks by streams. It has an extensive European range with a southern range in Fennoscandia (fig. 59). There are few Pleistocene records from the rest of Europe but two Neogene records are worthy of mention; they are from the Pliocene of Frankfurt (table 10), and the Miocene of Zatoka Gdowska, Poland (table 9).

Thamnites marginatus Jovet-Ast & Huard, a new genus and species of the Neckeraceae, was recovered from the Miocene deposit at Arjuzanx, France.

Lembophyllaceae

Isothecium undetermined species

Hitchin (20); Ho only; t; Reid 1897.

Isothecium myurum Brid.

!*Winchester* (12); Fl VIII*; Saxon ar. !*Wretton* (28); Ip II. *Fort William* (97); Fl V or VI*; Dixon 1910 as *E. myurum* Dixon. *Gort* (H15); Ho III or IV; Jessen *et al.* 1959.

Isothecium myosuroides Brid.

Star Carr (61); Fl IV or V; Clark 1954. *Windmillcroft* (77); Fl VI/VII/VIII*; Mahony 1868*a*. *Fort William* (97); Fl V or VI*; Dixon 1910 as *E. myosuroides* Schp. *Gort* (H15); Ho III; Jessen *et al.* 1959.

Both these species are recorded throughout the British Isles but are much more abundant in the north and west than in the south and east. Both can grow on rock or trees.

I. myurum outside Europe is known in Algeria, the Canaries and the Caucasus whereas *I. myosuroides* occurs

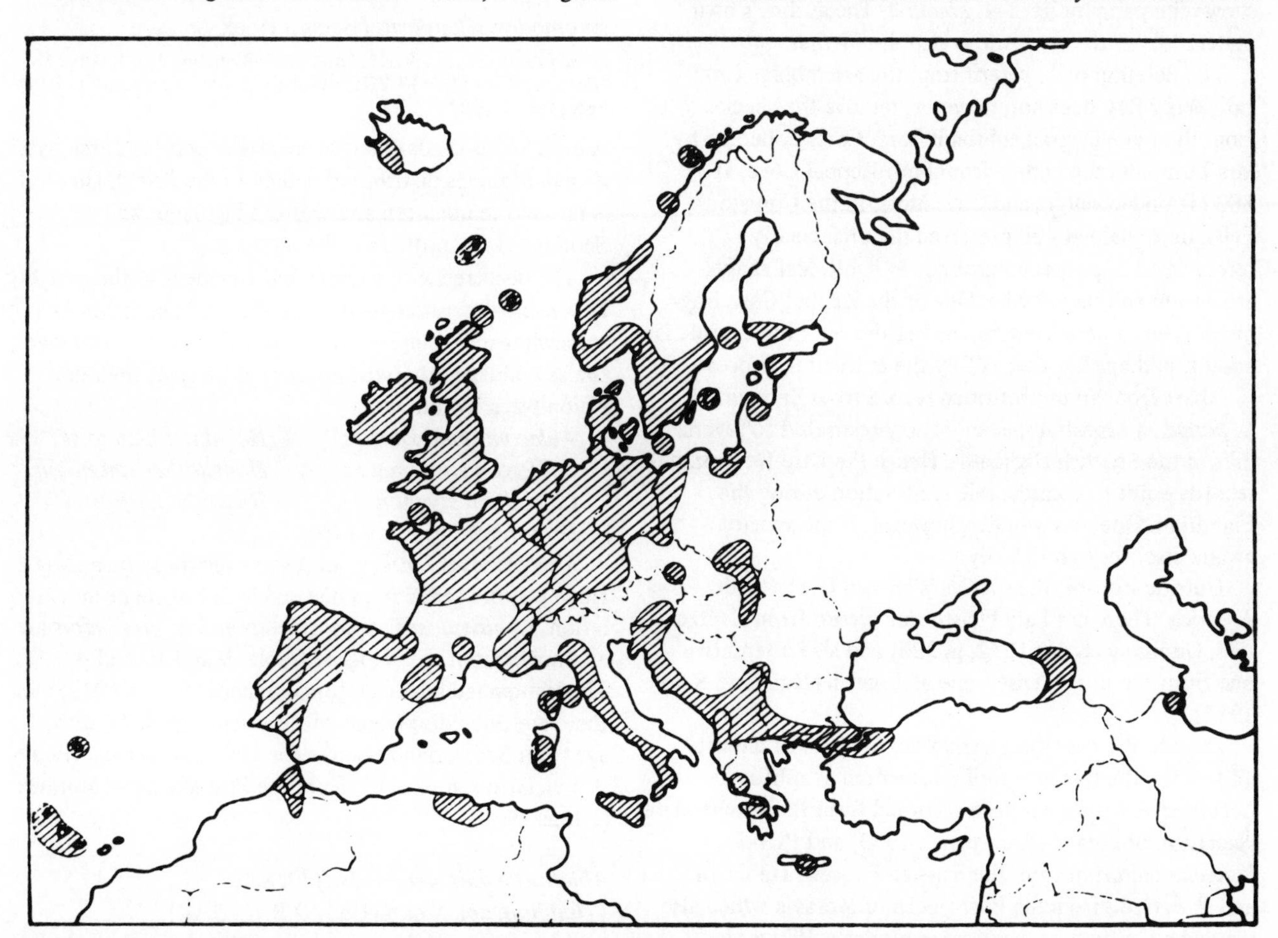

Fig. 59. European range of *Thamnobryum alopecurum*. From Störmer 1969.

also in North America and Formosa. In Fennoscandia both show a southern tendency. As in the British Isles the few subfossils from the rest of Europe are Flandrian or interglacial.

Hypnobryales
Leskeaceae
Lescuraea patens (Lindb.) Arn. & C. Jens
[*Pseudoleskea patens* (Lindb.) Limpr.]
!*Cove Moss* (69); LDe III-IV. !*Keppel Cove* (69); LDe III-IV.

That there has been confusion between this species and *Abietinella abietina* may seem surprising because the species are very different in habit. However, many of the subfossils are small fragments which in the case of *Abietinella* do not show the pinnate habit well, if at all.

However similar the leaf shape and structure of the two species may be, the paraphyllia are completely distinct; in *L. patens* often lanceolate and up to four cells broad, in *A. abietina* narrow, filamentous, branched and papillose. Mrs Tutin recorded *L. patens* from zone II of Low Wray Bay (Pennington 1962). However, her illustration clearly shows the paraphyllia of *A. abietina.* The author's own material from the same site is also *A. abietina.*

The deletion of *L. patens* from the assemblage from Low Wray Bay does not, however, remove the species from the Lake District subfossil flora. Cores collected by Mrs Tutin and the author from both Keppel Cove, at 1800 ft on Helvellyn, and Cove Moss, in the Coniston Fells, have yielded well preserved material clearly referable to *L. patens* on grounds of habit, leaf shape, areolation and paraphyllia. One of the Keppel Cove fragments bears a perichaetium and mature seta; the capsule is absent, perhaps knocked off by the extraction process.

Apart from an unconfirmed record from Snowdonia, *L. patens*, a rupestral species, is now restricted to several hills in the Scottish Highlands. Hence the Late Devensian records point to considerable contraction during the Flandrian. One may wonder, however, if the species awaits discovery on Helvellyn.

Outside Europe, the species is known from North America. There is a Late Pleistocene record from Krutzelried, Germany (Gams 1932, p. 320) and also a tentative one from the Early Pleistocene of Tegelen (Reid and Reid 1915).

Despite the rupestral, terricolous habitats of almost all species, the family is well represented as subfossils; *Lecuraea incurvata* has been recorded from the penultimate glaciation of Poland (Szczepanek 1960) and *Pseudoleskeella tectorum* from the Russian Pliocene (table 10), and *P. nervosa* from the Post-glacial of Moravia which also yielded an extinct species *Leskea moravica* Pilous (p. 213). However, one might have anticipated more than two subfossils of *Leskea polycarpa* (Andersson 1896; Pilous 1968) in view of the stream-side habitats.

Thuidiaceae
Thuidium tamariscimum and *Helodium blandowii* have often been recovered from Late Pleistocene deposits in Britain. They are the commonest subfossils of this family which, with several species of more than usual bryogeographical interest, is well represented not only in Pleistocene but also in Neogene deposits. Particular significance attaches to the well authenticated European Neogene occurrences of *Claopodium* (p. 172).

Anomodon is well represented in Neogene and Pleistocene deposits. *Heterocladium*, unknown from the British Pleistocene, has a Pliocene record from Germany (Mai *et al.* 1963) and a Miocene one from Poland (table 9) as well as a Pleistocene one from that country and also Moravia (Pilous 1968).

Anomodon unidentified species
!*Trimingham* (27); Cr.

Anomodon viticulosus (Hedw.) Hook. & Tayl.
!*Winchester* (12); Fl VIII*; Saxon ar. !*Wretton* (28); Ip zone II. !*Barnsley Park* (33); Fl VIII*; Roman ar. *West Craigneuk* (77); De? only; Dixon 1907.

A calcicolous shade-tolerant species of rock and tree bases, *A. viticulosus* is distributed widely in the British Isles but is rare in the northern and western highlands and islands of Scotland. It is southern in Fennoscandia.

The occurrence of a single leaf fragment in the possibly Devensian deposit at West Craigneuk in Lanarkshire is somewhat surprising, as Dixon himself realised. Moreover, the assemblage otherwise consists entirely of the mire bryophytes, as follows:

Aulacomnium palustre
Calliergon stramineum
Cinclidium stygium
Drepanocladus exannulatus
Helodium blandowii
Homalothecium nitens
Riccardia latifrons (t)

This makes the *Anomodon* leaf even more surprising. However, the genus is unmistakable in leaf shape and areolation. *A. viticulosus, A. longifolius* and *A. attenuatus* are all known from the Neogene (tables 9 and 10 and p. 172). The last-named has an Aftonian record (Steere 1942) and there are three discoveries of *A. viticulosus* of Neolithic age from Switzerland (Neuweiler 1905). *A. attenuatus* and *A. viticulosus* are known from the Post-glacial of Moravia (Pilous 1968).

Abietinella abietina (Hedw.) Fleisch.
[*Thuidium abietinum* (Hedw.) B., S. & G.]
!*Colney Heath* (20); LDe I; Godwin 1964. !*Marlow* (24); MDe; Bell 1968 pc. !*Barnwell Station* (29); LDe pre-I. ! *Upton Warren*

(37); MDe. !*Low Wray Bay* (69); LDe II. !*Kirkmichael 3a* (71); LDe I. !*Drumurcher* (H32); LDe III.

None of the subfossils approach *Abietinella histricosa* All seem to be derived from small slender plants with leaves and areolation too short for *A. histricosa* which in any case is given only subspecific rank by Dixon (1924), a view upheld by Pospisil (1967). See *Lescuraea patens* (p. 130) for further discussion of the identifications. In Britain *A. abietina* has a scattered mainly southeastern range but extends north to Easter Ross. There are records from only a few Irish vice-counties, mainly in the northwest.

The range is very wide in the Northern Hemisphere and extends north to Spitsbergen and Ellesmere Island (Kuc 1969). In the Torneträsk area *A. abietina* grows mainly in *Dryas* heath on slopes of southern aspect. The map of the range in Denmark shows a scattered occurrence in the north and west (Holmen 1959); there the habitats are 'dry open grass fields in dunes and slopes'. In the British Isles as elsewhere *A. abietina* is often associated with such species as *Rhytidium rugosum* and *Homalothecium lutescens* in calcicolous grassland.

The continental bias of the range has been stressed by various authors such as Herzog (1926) and Boros (1962). Herzog stated (p. 261) 'Im südlichen sowohl wie (seltener) im nördlichen Sibirien ist auf trockenen Flächen die associationsfeste Gesellschaft *Rhythidium, Entodon orthocarpus* und *Thuidium abietinum* häufig, genau unter den gleichen Verhältnissen wie in der pontisch-pannonischen Formation Europas, wo sich ihnen noch *Camptothecium lutescens*, ein xerothermes Element von mediterran-atlantischem charakter, beigeselt.'

Subfossils are well known from the last glaciation of Poland and North America (Saskatchewan; Wisconsin; New York). There are two Pliocene records from Russia (table 10).

Thuidium undetermined species

Ponder's End (21); MDe; Warren 1912. *Gort* (H15); Ho III or IV; Jessen *et al.* 1959.

Possibly the Ponder's End subfossil is *Abietinella*; this would agree well with the other three Middle Devensian records.

Thuidium tamariscinum (Hedw.) B., S. & G.

Meare Lake Village (6); Fl VII*b*; t; Godwin 1941. !*Stanford* (15); Fl VIII. *Hoxne* (25); Ho III; Reid 1896. *Woodwalton Fen* and *Trundle Mere* (31); Fl VII*b*; Godwin and Clifford 1938. !*Barnsley Park* (33); Fl VIII* Roman ar. *Brigg* (54); Atkinson 1887; Sheppard 1910 as *T. tamariscifolium* (Neck.) Lindb. !*Bunny* (56); Fl VIII*; Roman ar. *Bridgewater Street* (59); Fl VIII*; Roman ar; Roeder 1899. *Lady Bridge Stack* (62); Fl VII*b*; Simmons 1968 pc. *Cromwell Bottom* (63); Fl VII*b*; Bartley 1964. *Castle Eden* (66); EP; Reid 1920. !*Russland Moss* (69); Fl only; Dickinson 1968 pc. *Cowdenglen* (76); Fl? only; Mahony 1869. *Garvel Park* (76); LDe/Fl IV*; Robertson 1881. *Windmillcroft* (77); Fl VI/VII/VIII*; Mahony 1868*a* as *Hypnum tamariscinum*. !*Carpow* (88); Fl VIII*; Roman ar. *Fort William* (97); Fl V or VI*; Dixon 1910. !*Loch Maree* (105); Fl VII*a*; H. H. Birks 1969 pc. !*Fugla Ness B* (112); Ho IV. *Gort* (H15); Ho I, II, III, and III or IV; Jessen *et al.* 1959. !*Barnaran Bog* (H19); Fl. !*Winetavern Street* (H22); Fl VIII*; Medieval ar. *Raheelin* (H29); Fl VII*b*; Jessen 1949. !*Moville* (H34); Fl only; McMillan 1957. *Cushendun* (H40); Fl VI; Jessen 1949 as *T. tamariscifolium.*

T. tamariscinum is easily separated from the closely related species by the apical cells of the branch leaves. It occurs throughout the British Isles in a variety of habitats but is best developed in woodland. It ascends to over 3000 ft in the hollows of block scree on Ben Lawers (Watson 1925). Outside Europe it occurs in Japan, Macaronesia and Jamaica. In Fennoscandia the range is southern.

The detailed British history is exclusively interglacial and Flandrian, with the possible exception of the record from Garvel Park. There is a Miocene record from Arjuzanx (table 9), a Pliocene record from Kroscienko (table 10) and an Eemian one from Belgium (Paepe and Vanhoorne 1967). An interesting record is that belonging at Bialka Tatranska in the Polish Carpathians (Sobolewska and Srodon 1961).

Thuidium delicatulum (Hedw.) Mitt

Bridgewater Street (59); Fl VIII*; Roman ar; Roeder 1898. !*Carpow* (88); Fl VIII; Roman ar. *Fort William* (97); Fl V or VI; Dixon 1910. !*Loch Cuithir* (104); Fl V; Vasari and Vasari 1968. !*Loch Fada* (104); Fl VIII; Vasari and Vasari 1968. *Loch Mealt* (104); LDe II and III; t; Birks 1969 pc. !*Loch Maree* (105); Fl VII*b* or VIII; H. H. Birks 1969 pc.

Careful examination of the stem leaves and paraphyllia is necessary before this species can be separated from *T. philibertii* and *T. recognitum* (Tallis 1961).

The British distribution is markedly northern and western (Tallis 1961). The characteristic habitat is wet, shaded, siliceous rock in woods or on mountains. Confirmation of the Bridgewater Street record would be needed before much could be made of its lying outside the present range. Outside the Northern Hemisphere the species extends into South America.

There is a Miocene record from Arjuzanx (table 9), a Pliocene one from Poland (table 10) and Weichselian occurrences in that country and Denmark, as well as an Aftonian record from Iowa (Steere 1942) and Post-glacial records from Moravia (Pilous 1968) and the Vosges (Firbas *et al.* 1948).

Thuidium philibertii Limpr.

Fort William (97); Fl V or VI*; Dixon 1910.

T. philibertii, scattered in the British Isles, but most

abundant in the south, occurs on basic soil, rock and dunes (Tallis 1961). There are a few Pleistocene records but two from the Pliocene of Russia (table 10).

Thuidium recognitum (Hedw.) Lindb.

Wolvercote (23); Ho?; Bell 1904. *Gort* (H15); Ho II or III; Jessen *et al.* 1959.

T. recognitum is the rarest of the British species, being confined to widely scattered areas of the north and west (Tallis 1961). Like *T. philibertii* it is a calcicole of rock and sand. There is a Pliocene record from Russia (table 10) but few, if any, Pleistocene records.

Helodium blandowii (Web. & Mohr). Warnst.

!*Thrapston* (32); LDe only; Bell 1968 pc. *Aby Grange* (54); LDe II; Suggate and West 1959. !*Chelford* (58); EDe. *Hooks* (61); LDe II. !*Burtree Lane* (66); Fl IV; Bellamy *et al.* 1966. !*Foolmire Syke* (69); Fl VII*a*. !*Scaleby Moss* (70); Fl IV or V. !*Kirkmichael 3a* (71); LDe III; Dickson *et al.* 1970. *West Craigneuk* (77); De? only; Dixon 1907. !*Dun Moss* (89); Fl VI. !*Burreldale Moss* (93); Fl VI; Durno 1964 pc.

This peat-forming species is often found well enough preserved to show the habit clearly (plate 18). Even detached leaves are readily recognisable by the paraphyllia arising from the leaf bases.

H. blandowii became extinct in the British Isles in the last 100 years. By the last century it was reduced to four localities, Malham Tarn, Terrington North Carr and Halnaby Carr, all in Yorkshire, and Knutsford Moor in Cheshire (fig. 60). It shared the last-named and Terrington North Carr with *Paludella squarrosa*. Little precise is known of the habitats. Baker (1906, p. 530), referring to Halnaby Carr and Terrington North Carr, states 'wet moorland bogs'. In Denmark where the species is scattered the ecology is described by Holmen (1959) as follows. 'The habitat of the species is rather poor neutral spring-bogs with vegetation of *Carex diandra, C. rostrata, Tomenthypnum nitens* and *Sphagnum teres*.' In the Torneträsk area the species is characterised by Mårtensson (1956) as a moss of well nutrified wet birch forest. He states (p. 242), 'There it often grows in abundance, associated with *Mnium punctatum* coll., *Paludella squarrosa, Sphagnum teres, S. warnstorfianum* and *Tomenthypnum nitens.* When the moss is found in the lower parts of the low-alpine belt, the habitat is very similar, but instead of birch forest there is tall willow scrub. The moss is rare in fens without tree or scrub layer.'

Throughout most if not all of the Devensian and until the Boreal *H. blandowii* had a very different range in the British Isles from that of the last century. The contraction to Yorkshire and Cheshire closely parallels that of *Paludella squarrosa*, although the details of the histories are different. Most records of *H. blandowii* are Devensian, most of *Paludella* are Flandrian. In essential the histories are the same; they indicate a contraction at the Boreal-Atlantic transition. Particularly noteworthy are the records from Aberdeenshire and the Isle of Man. The former proves occurrence far to the north of Scotland in the Boreal while the latter points to past occupation of Ireland.

Before rendered extinct by man, *H. blandowii* was a relict species greatly reduced from former periods of much greater abundance and range. Unlike many relict species it produced sporophytes fairly freely until its demise. Sporophytes are known from Halnaby, Terrington and Knutsford. At the last-named at least sporophytes were formed in some quantity.

Herzog (1926) considered *H. blandowii* a subarctic glacial relict and gave a map of its scattered range in Middle Europe. Later Gams (1932) added Pleistocene fossils, thus strengthening Herzog's case and Stefureac (1956) added present localities in Romania. Jasnowski (1957*b*) has discussed the Flandrian history in Poland. There are solitary living stations in the Alps, and in the Jugoslavian mountains (Pavletic 1955) while the southernmost subfossils were recently discovered in zone II and III deposits by Lake Garda in the southern Alps in Italy (Grüger 1968*b*).

The range is extensive in Eurasia and North America but does not include the high Arctic.

Amblystegiaceae

The abundance of this family in the Pleistocene is well illustrated by the genera *Calliergon* and *Drepanocladus*, both of which have all the British species known as subfossils. Indeed in addition there are records of *C. richardsonii* and *D. capillifolius*, both no longer British taxa. Species of *Amblystegium, Campylium, Cratoneuron* and *Scorpidium* are very frequently encountered.

Cratoneuron undetermined species

Gort (H15); Ho III; Jessen *et al.* 1959.

Cratoneuron filicinum (Hedw.) Spruce

Hawks Tor (2); LDe II; t; Conolly *et al.* 1950 as *Amblystegium filicinum* De Not. !*Lexden* (18); Ho or Ip; Shotton *et al.* 1962. !*Colney Heath* (20); LDe I; Godwin 1964. *Ponder's End* (21); MDe; Warren 1912 as *Amblystegium filicinum*, also var. *vallisclausae. Dorchester* (23); Wo; Duigan 1955. !*Hoxne* (25); LAn; Turner 1968. !*Broome Heath* (27); Wo; var. *curvicaule* (Jur.) Monk. *Wretton* (28); Ip III and EDe. !*Soham Lode* (29); Fl VII*b*. !*Shippea Hill* (29); Fl VII*b*. !*Barnwell Station* (29); LDe pre-I. *Histon Road* (29); Ip IV; Proctor 1959*a*. !*Barnsley Park* (33); Roman ar; Fl VIII*. !*Upton Warren* (37); MDe. *Wallasey* (58); Fl VII/VIII*; Travis 1922 as *Amblystegium filicinum* De Not. *Malham Tarn Moss* (64); Fl VI; Proctor and Birks 1963 pc. !*Bradford Kaims* (68); Fl IV; Bartley 1966. !*Low Wray Bay* (69); LDe II. !*Kirkmichael 3a* (71); LDe III; Dickson *et al.* 1970. *Windmillcroft* (72); Fl VI/VII/VIII*; Mahony 1868 as *Hypnum filicinum. Garvel Park* (76); LDe/Fl IV*; Robertson 1881 as *Hypnum filicinum.* !*New Dry*

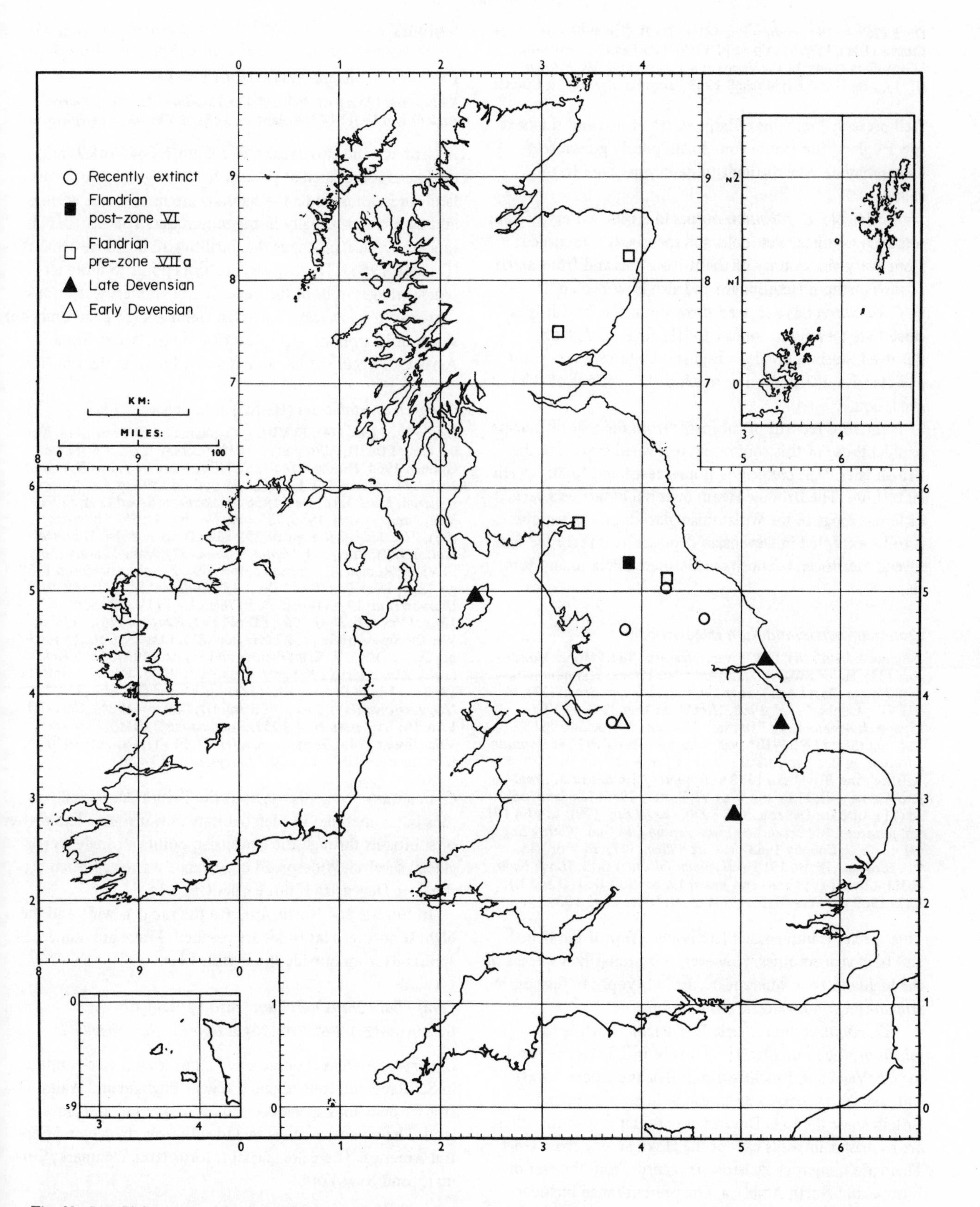

Fig. 60. Late Pleistocene subfossils of *Helodium blandowii.*

Dock (76); Fl IV*. *Whitrig Bog* (81); LDe II; Conolly 63 pc. !*Loch Cuithir* (104); LDe III-IV and Fl VII*b*; Vasari and Vasari 1968. !*Loch Fada* (104); Fl IV; Vasari and Vasari 1968. *Mapastown* (H31); LDe II and III; Mitchell 1953. !*Drumurcher* (H32); LDe III.

Well preserved remains of large states of this and the next species give little trouble but small, poorly grown fragments may be very difficult to determine even to the genus.

A calcicole, *C. filicinum* occurs in a wide variety of habitats such as mires, wet rocks and springs. It is recorded from every vice-county of the British Isles and from much of the Northern Hemisphere and also New Zealand.

C. filicinum has occurred throughout the British Isles since Late Devensian zones I to III; the records from Barnwell Station, Ponder's End and Upton Warren point to extensive distribution at much earlier periods of the last glaciation.

Numerous records of subfossils from the rest of Europe parallel those of this commonly recovered species in the British Isles. Var. *curvicaule* is now restricted to Mid-Perth and Rhum. The Broome Heath material indicates a very different range in the Wolstonian glaciation; the variety is to be expected in Devensian deposits. It is recorded from several Pleistocene deposits in Germany, Poland and Russia.

Cratoneuron commutatum (Hedw.) Roth.

!*Shapwick Heath* (6); Fl VII; var. *falcatum* (Brid.) Mönk. *Wolvercote* (23); Ho or EWo; var. *falcatum*; Bell 1904 as *Hypnum falcatum*. *Hoxne* (25); LAn; Turner 1968. *Buckenham Broad* (27); Fl VII-VIII; Lambert *et al.* 1960. !*Hockham Mere* (28); Fl VII*a*. !*Sidgwick Avenue* (29); EDe; var. *falcatum*; t; Dickson 1964. *Wallasey* (58); Fl VII/VIII*; var. *falcatum*; Travis 1922 as *Hypnum falcatum*. Brid. *Linton Mires* (63); Fl IV or V var. *falcatum*; Raistrick and Blackburn 1938 as *Hypnum falcatum* Brid. !*Bradford Kaims* (68); Fl IV and V or VI; Bartley 1966. !*Kirkmichael 3a* (71); LDe III; Dickson *et al.* 1970. *Garvel Park* (76); LDe/Fl IV*; var. *falcatum*; Robertson 1881 as *Hypnum falcatum*. *Whitrig Bog* (81); LDe II; Conolly 1963 pc. *Fort William* (97); Fl V or VI*; var. *falcatum*; Dixon 1910 as *Hypnum falcatum* Brid. !*Loch Fada* (104); LDe III-IV; Vasari and Vasari 1968; also LDe I, II and III; Birks 1969 pc. *Gort* (H15); Ho II or III; Jessen *et al.* 1959.

This species is unrecorded from only a few of the British and Irish vice-counties. However, it is primarily a species of the highland zone where it occurs in bryophyte flushes, by stream-sides and waterfalls and in fens.

Var. *commutatum* is strictly calcicolous while var. *falcatum*, the more often found subfossil, is less demanding (McVean and Ratcliffe 1962; Bell and Lodge 1963). The species has been widespread at least in mainland Britain since the Late Devensian period if not earlier; there are no Irish subfossils except the Hoxnian one from Gort. There are numerous Pleistocene records from the rest of Europe and North America. The present range includes much of the Northern Hemisphere, but not the highest latitudes.

Cratoneuron decipiens (De Not.) Loeske

Wolvercote (23); Wo?; Bell 1904 as *Thuidium decipiens*. *Garvel Park* (76); LDe/Fl IV*; Robertson 1881 as *Thuidium decipiens*.

Apart from the two eastern vice-counties of Yorkshire, *C. decipiens* is confined at present to the Highlands of Scotland. In Fennoscandia the habitats are in base-rich springs and by streams, usually in the mountains. The British habitats are similar; in Angus the localities all lie above 1800 ft (Duncan 1966). The British subfossils point to more extensive ranges in both the penultimate and last glaciations. Similarly the species occurred in Denmark (at Skaerumhede), south of its present Fennoscandian range. Apart from Alaska, *C. decipiens* has an extensive Eurasian distribution.

Campylium stellatum (Hedw.) J. Lange & C. Jens.

Decoy Pool Wood (6); Fl VIII; t; Clapham and Godwin 1948. *Nazeing* (18); LDe III; Allison *et al.* 1952. !*Colney Heath* (20); LDe I; Godwin 1964. *Ponder's End* (21); MDe; Warren 1912 as *Hypnum stellatum*. *Hoxne* (25); Ho III or EWo; Reid 1896 as *Amblystegium stellatum* Mitt.; LAn; West 1956. *Buckenham Broad* (27); Fl VII-VIII; Lambert *et al.* 1960. !*Broome Heath* (27); Wo. !*Wretton* (28); EDe. *Sidgwick Avenue* (29); EDe; Dickson 1964. !*Barnwell Station* (29); LDe pre-I. !*Upton Warren* (37); MDe. !*Bunny* (56); Fl VIII*; Roman ar. !*Russland Moss* (69); Fl only; Dickinson 1968 pc. !*Low Wray Bay* (69); LDe II. !*Kirkmichael 3a* (71); LDe III; Dickson *et al.* 1970. *Garvel Park* (76); LDe/Fl IV*; Robertson 1881. !*New Dry Dock* (76); LDe/Fl IV*. *Renfrew* (76); Fl IV or V*; Thompson 1968 pc. *Whitrig Bog* (81); LDe III; Conolly 1963 pc. *Canna* (104); Fl VII*b*; Flenley and Pearson 1967. *Loch Fada* (104); LDe I; Birks 1969 pc. *Loch Mealt* (104); LDe I and II; Birks 1969 pc. !*Loch Droma* (105); LDe I; Kirk and Godwin 1963. !*Baggotstown* (H8); LAn, Ho II and III; Dickson 1964*a*. *Gort* (H15); LAn, Ho III; Jessen *et al.* 1959. *Roundstone 2* (H16); Fl V and VII*a*; Jessen 1949. *Derrybrennan* (H19); Fl VI?; Synnott 1970 pc. !*Drumurcher* (H32); LDe III. !*Derryvree* (H33); MDe.

C. stellatum occurs throughout the British Isles usually in fens but sometimes in such habitats as wet rocks. The scatter of subfossils through the Devensian points strongly to perglacial survival. Widespread occurrence was established by the Late Devensian if not earlier (fig. 61).

In Eurasia and North America the range is wide and the highest northern latitudes are reached. There are numerous subfossils from outside the British Isles.

Campylium chrysophyllum (Brid.) J. Lange

Wolvercote (23); Wo?; Bell 1904 as *Hypnum chrysophyllum*.

C. chrysophyllum is unrecorded from several vice-counties in Scotland and Ireland and a few in England and Wales. It is a calcicolous species found on rock, sand and grassland. The extensive range includes Eurasia, North and Central America. There are glacial records from Denmark, Germany and New York.

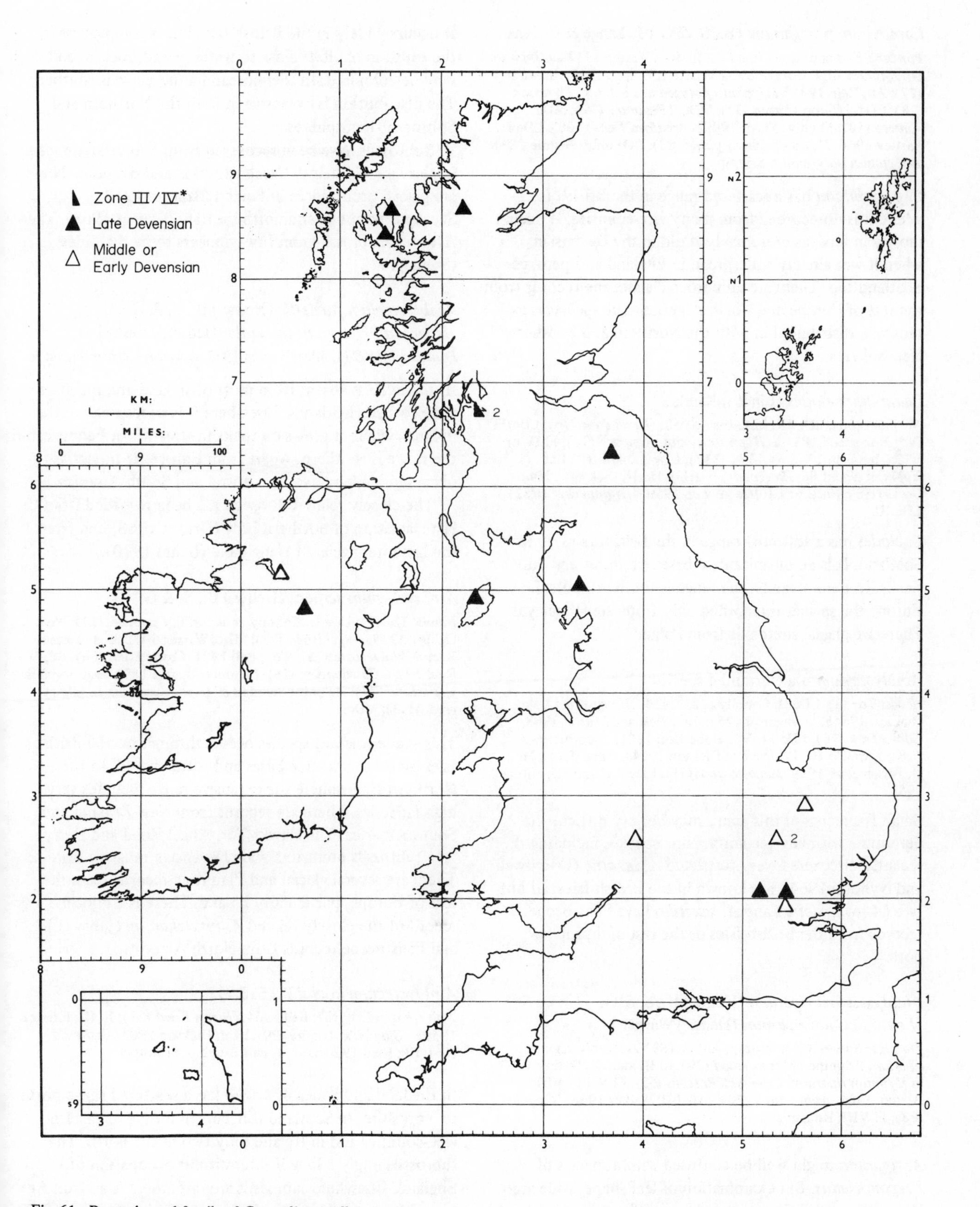

Fig. 61. Devensian subfossils of *Campylium stellatum.*

Campylium polygamum (B., S. & G.) J. Lange & C. Jens.

Ponder's End and *Angel Road* (21); MDe; Warren 1912 as *Hypnum polygamum* Brid. *Dorchester* (23); Wo; t; Duigan 1955. *Mundesley* (27); An; Reid 1904 as *Hypnum polygamum* Schimp. !*Wretton* (28); EDe. !*Upton Warren* (37); MDe. !*Fladbury* (37); MDe. *Aintree* (59); Fl only; Travis 1909. !*Bradford Kaims* (68); LDe II; Bartley 1966. *Hailes* or *Corstorphine* (83); LDe only; Bennie 1894*a* as *Hypnum polygamum* Schimp.

C. polygamum has a scattered range in the British Isles where it is unrecorded from many vice-counties. It occurs mostly in fens, as presumably it did in the Devensian when it was already widespread in England and perhaps Scotland too. There are numerous Pleistocene records from the rest of Europe and North America. The species is extensively distributed in both the Northern and Southern Hemispheres.

Campylium elodes (Lindb.) Kindb.

Neasham (66); LDe III; Blackburn 1952. *Garvel Park* (76); LDe/Fl IV*; Robertson 1881 as *Hypnum elodes*. *Renfrew* (76); Fl IV or V*?; Thomson 1968 pc. *Walls* (112); LDe?; Dixon 1911 as *Hypnum elodes* Spr. *Baggotstown* (H8); Ho II; Dickson 1964*a*. *Gort* (H15); Ho II or III; Jessen *et al.* 1959. !*Drumurcher* (H32); LDe III.

C. elodes has a scattered range in the British Isles, north to Shetland. It is an inhabitant of base-rich mires and dune-slacks. In Fennoscandia the range is southern; outside Europe the species is recorded only from the Himalayas. There are glacial subfossils from Poland.

Amblystegium undetermined species

Hawks Tor (2); LDe II; Conolly *et al.* 1950. !*Ilford* (18); LWo; Dickson 1964*b*. !*Lowestoft* (25); An?; West and Wilson 1968. *Garvel Park* (76); LDe/Fl IV*; Robertson 1881. !*Baggotstown* (H8); LAn, Ho III, IV or EWo; Dickson 1964*a*. *Gort* (H15); Ho II; Jessen *et al.* 1959. *Mapastown* (H31); LDe I, II and III; Mitchell 1953.

Small fragments of this genus may be very difficult to determine with any certainty. Nine species, including the recently discovered, very restricted *A. saxatile* (Crundwell and Nyholm 1964), are known in the British Isles; all but two (*A. compactum* and *A. saxatile*) have Pleistocene records from the British Isles or the rest of Europe or both.

Amblystegium riparium (Hedw.) B., S. & G.
[*Leptodictyum riparium* (Hedw.) Warnst.]

Trafalgar Square (21); Ip only; Abbot 1892 as *Amblystegium riparium* Schimper. *Histon Road* (29); Ip III and IV; Walker 1953 as *Hypnum riparium*. *Cromwell Bottom* (63); Fl VII or VIII; Bartley 1964. *Abbot Moss* (70); LDe I-II; Walker 1966. !*Carpow* (88); Fl VIII; Roman ar.

A. riparium might well be confused with a species of *Drepanocladus*. But examination of leaf shape, wide areolation, entire margin and broad nerve allows separation. It occurs widely in the British Isles but is commonest in the south in habitats close to water; wood, rock or soil may be colonised. In Fennoscandia the range is southern. The distribution is extensive in both the Northern and Southern Hemispheres.

Subfossils have been recovered from Flandrian deposits in Norway, Germany, Czechoslovakia and Belgium. There are glacial occurrences in Poland (Szafran 1952; Szczepanek 1960) and northern Italy (Grüger 1968). The Aftonian subfossil from Iowa appears to be the oldest (Steere 1942).

Amblystegium fluviatile (Hedw.) B., S. & G.
[*Hygroamblystegium fluviatile* (Hedw.) Loeske]

Ponder's End (21); MDe; Warren 1912 as *Amblystegium fluviatile*.

A. fluviatile is absent from most of Ireland and much of the Scottish Highlands. Elsewhere it is widespread in the uplands where it grows on rocks in streams. In Fennoscandia the range is southern. Apart from Europe *A. fluviatile* occurs in Madeira, North, Central and South America.

The closely related *A. tenax* has been recorded from the last glaciation of northern Italy (Grüger 1968) and from the last interglacial of Heligoland (Behre 1970).

Amblystegium serpens (Hedw.) B., S. & G.

Hawks Tor (2); LDe II; Conolly *et al.* 1950. *Farnham* (17); Wo?; Oakley 1939. *Angel Road* (21); MDe; Warren 1912 as *A. serpens* B. & S. *Wolvercote* (23); Wo?; Bell 1904. *Castle Eden* (66); EP; Reid 1920. *Cowdenglen* (76); Fl? only; Bennie 1891 as *A. serpens* L. *Faskine* (77); De only; Bennie 1894*b* as *A. serpens* L. !*Derryvree* (H33); MDe.

This shade-tolerant species occurs throughout the British Isles on soil, rock, tree bases and rotten wood. In the Northern Hemisphere the extensive range includes very high latitudes. There are records from New Zealand and South America. The Hawks Tor, Angel Road and Derryvree subfossils point to a wide Devensian range in Britain. There are several glacial and Flandrian records from the rest of Europe, particularly Poland. There is a Pliocene record of the closely related *A. juratzkanum* (table 10) and Pleistocene records from North America.

Amblystegium kochii B., S. & G.

Wolvercote (23); Ho?; Bell 1904. *Histon Road* (29); Ip III; Proctor 1959*a*. !*Sidgwick Avenue* (29); EDe; Dickson 1963. *Castle Eden* (66); EP; Reid 1920 as *A. kochii* Bruch & Schimper.

In England and Wales *A. kochii* has a scattered range north to Yorkshire. In Scotland it occurs in only two eastern vice-counties and in Ireland only two in the north. The subfossils imply a long if intermittent occupation of England. Elsewhere subfossils are rare; there is an Iron Age record from northwestern Germany (Körber-Grohne 1967) and a Late Wisconsin one from Saskatchewan (De Vries and Bird 1965).

Amblystegium varium (Hedw.) Lindb.
Mundesley (27); An; Reid 1904. !*Wretton* (28); Ip II; t.

A. varium, in habitats broadly similar to those of *A. serpens*, is widespread in England but absent from much of Ireland, Scotland and Wales. Reaching the highest latitudes it occurs widely in Eurasia and North America. There are last glacial records from Poland (Szafran 1952, Szafer 1954; Sobolewska and Srodon 1961; Ralska-Jasiewiczowa 1968).

Drepanocladus undetermined species
!*Ilford* (18); LWo, Ip II; Dickson 1964*b*. !*Marks Tey* (19); EWo. !*Lowestoft* (25); An?; West and Wilson 1968. *Hoxne* (25); LAn; West 1956. !*Mundesley* (27); Ba, Pa, LPa or EBe, EAn. !*Paston* (27); Pa, EAn. !*Beeston* (27); Pa. !*Hockham Mere* (28); Fl VI and VII*a*. *Great Billing* (32); MDe; Morgan 1969.

Drepanocladus species, all of which show great phenotypic plasticity are among the most frequently recovered mosses from Pleistocene deposits. Considerable care is needed in identifying the species; Nyholm's account (1954-69), which is the most up-to-date treatment, covers all the British species and in addition foreign species such as *D. capillifolius*, well known from the British Pleistocene and also the following, any or all of which may turn up in the British Pleistocene:

D. badius; Pleistocene of Poland (Szafran 1952) and Russia (Savicz-Ljubitzkaja and Abramova 1954).
D. pseudostramineus; Pleistocene of Sweden (Szafran 1952).
D. purpurascens; Pleistocene of Sweden (Lundquist 1964).
D. tundrae; Pleistocene of Poland (Szafran 1952; Koperawa 1958), Sweden (Lundquist 1964) and Canada (Persson and Sjörs 1960).

Drepanocladus aduncus (Hedw.) Warnst.
West Wittering (13); Ip only; Reid 1892 as *Hypnum aduncum. Nazeing* (18); LDe III; Allison *et al.* 1952. *Ponder's End* (21); MDe; Warren 1912 as *Hypnum aduncum* and *H. aduncum* (group *pseudofluitans*). *Wolvercote* (23); Wo or Ho?; Bell 1904 as *H. aduncum* and var. *pseudofluitans*. *Hockham Mere* (28); t; Fl VIII; Godwin and Tallantire 1951. *Histon Road* (29); Ip III; t; Proctor 1959*a*. *Wallasey* (58); Fl VII/VIII*; Travis 1922 as *Hypnum aduncum* Hedw. *Aintree* (59); Fl; Travis 1909 as *H. aduncum*. *Helton Tarn* (69); Fl VII*b*; Smith 1958*b*. !*Kirkmichael 3a* (71); LDe I; Dickson *et al.* 1970. *Garvel Park* (76); LDe/Fl IV*; Robertson 1881 as *D. aduncum* and *D. kneiffii*. *Timahoe* (H19); Fl VI?; Synnott 1970 pc. *Mapastown* (H31); LDe III; Mitchell 1953 as *D. aduncus* (Hedw.) Mönk.

D. aduncus is unrecorded from several vice-counties in Ireland, northern Scotland and Wales. It is a plant of low altitudes where it inhabits eutrophic waters of streams, fens and dune-slacks. Presence in the last glaciation is well established by three Late Devensian and one Middle Devensian record.

Apart from many Pleistocene records, there is a Pliocene record from Russia (table 10) of this species which is widely spread in the Northern Hemisphere and also occurs in New Zealand.

Drepanocladus sendtneri (Schimp.) Warnst.
Ponder's End (21); MDe; Warren 1912 as *Hypnum sendtneri. Dorchester* (23); Wo; Duigan 1955. *Wolvercote* (23); Wo or Ho?; t; Bell 1904 as *Hypnum sendtneri. Hoxne* (25); LAn; t; West 1956. *Mundesley* (27); An; also var. *wilsoni*; Reid 1904 as *Hypnum sendtneri* Schimp. and *H. wilsoni* Schimp. !*Broome Heath* (27); Wo; t. *Hockham Mere* (28); LDe III and Fl IV; t; Godwin and Tallantire 1951. !*Upton Warren* (37); MDe. *Wallasey* (58); Fl VII/VIII*; also var. *wilsoni*. Travis 1922 as *Hypnum sendtneri* Schimp. and *H. wilsoni* Schimp. *Aintree* (59); Fl VII or VIII*; also var. *wilsoni*; Travis 1909 as *H. sendtneri* and *H. wilsoni*.

D. sendtneri is a local plant of scattered distribution in the British Isles where habitats include dune-slacks and fens. There are many subfossils from the rest of Europe including a Pliocene one from Poland (table 10). The species has a wide range in Eurasia and North America.

Drepanocladus lycopodioides (Brid.) Warnst.
Decoy Pool Wood (6); Fl VIII; Clapham and Godwin 1948. *Wolvercote* (23); Wo or Ho; t; Bell 1904 as *Hypnum lycopodioides*. !*Broome Heath* (27); Wo; t. *Trundle Mere* (31); Fl VIII; t; Vishnu-Mittre 1959. *Wallasey* (58); Fl VII or VIII*; Travis 1922 as *Hypnum lycopodioides* Schwaegr.

This species has a scattered range in the British Isles; it is rare in both Scotland and Ireland. Dune-slacks are the principal habitat in these islands; inland, shallow rich fens sometimes support this species, easy of determination in its robustness and areolation.

With the exception of the tentative one from Wolvercote, the subfossils belong to the latter part of the Flandrian. The Wallasey specimens derive from a dune-slack, the rest from fens. There are interglacial records from Russia and Poland and glacial ones from the latter and Germany.

Drepanocladus fluitans (Hedw.) Warnst.
!*Stannon* (2); LDe only. *West Wittering* (13); Ip only; Reid 1892 as *Hypnum fluitans* Hedw. *Ponder's End* (21); MDe; also var. *falcatus*; Warren 1912 as *H. fluitans* and *H. fluitans* (group *falcatum*). *Wolvercote* (23); Wo or Ho?; Bell 1904 as *H. fluitans. Hoxne* (25); Ho III or EWo; Reid 1896 as *Amblystegium fluitans* Mitt.; West 1956. *Pakefield* (25); Cr?; Blake 1890 as *A. fluitans* Mitt. *Saint Cross* (25); Ho?; Candler 1889 as *Amblystegium fluitans*. *Mundesley* (27); An; Reid 1904 as *H. fluitans* L. *Histon Road* (29); Ip III; t; Proctor 1959. !*Upton Warren* (37); MDe. *Bag Mere* (59); Fl IV-V; Birks 1965*a*. *Linton Mires* (63); Fl IV or V; Raistrick and Blackburn 1932 as *Hypnum fluitans* L. *Fox Earth Gill* (65); Fl VI; Squires 1968 pc. !*Bigholm Burn* (72); LDe III; Moar 1969*a*. *Garvel Park* (76); LDe/Fl IV*; Robertson 1881 as *H. fluitans. Whitrig Bog* (81); LDe III; Mitchell 1948 as *H. fluitans*; LDe III;

t; Conolly 1963 pc. *Hailes* or *Corstorphine* (83); LDe only; Bennie 1894*a* as *H. fluitans* L. *Dronachy* (85); LDe only; Bennie 1896 as *A. fluitans. Garral Hill* (94); LDe III; Donner 1957. *Fort William* (97); Fl V or VI*; Dixon 1910 as *Hypnum fluitans* Brid. *Derrycassan* (H24); Fl VII*a*; Jessen 1949. !*Derryvree* (H33); MDe. *Fallahogy* (H40); Fl VI; t; Smith 1958.

D. fluitans is extensively distributed in both the Northern and Southern Hemispheres. In the British Isles it is unrecorded only from a few vice-counties in Britain and several in Ireland. Primarily it is a species of base-poor pools. The widespread occurrence relates to the Late Devensian period if not earlier. There are many subfossils from outside the British Isles.

Drepanocladus exannulatus (B., S. & G.) Warnst.

!*Stannon* (2); LDe only. !*Ilford* (18); LWo; t; Dickson 1964*b*. !*Colney Heath* (20); LDe I; var. *rotae*; Godwin 1964. *Ponder's End* and *Angel Road* (21); MDe; Warren 1912 as *Hypnum exannulatum*, also group *rotae* and var. *orthophyllum* (Milde). !*Broome Heath* (27); Wo; var. *rotae*. !*Sidgwick Avenue* (29); EDe; var. *rotae*; Dickson 1963. *Histon Road* (29); Ip III; t; Proctor 1959*a*. *Bridgewater Street* (59); Fl VIII*; Roman ar; Roeder 1899 as *Hypnum exannulatum. Skelsmergh Tarn* (70); Fl IV; Walker 1955. *West Craigneuk* (77); De? only; Dixon 1907 as *Hypnum exannulatum* var. *brachydictyon Ren.* !*Loch of Park* (92); Fl VII*a*; Vasari and Vasari 1968. !*Garral Hill* (94); LDe II; var. *rotae*. *Loch Fada* (105); LDe I. !*Loch Droma* (105); LDe I; Kirk and Godwin 1964. *Roundstone 1* (H16); LDe III; Jessen 1949.

D. exannulatus is unrecorded from several vice-counties of Britain and of central and western Ireland. Records from the Early, Middle and Late Devensian point to per-glacial survival. The widespread distribution was established at least by the Late Devensian. It is more characteristic of upland areas than *D. fluitans* and occurs in pools on heaths and in springs and flushes. There are many subfossils of this bipolar species which is extensively spread in the Holarctic.

Drepanocladus revolvens (Sw.) Warnst.

Dogger Bank; Fl IV or V; var. *intermedius*; Whitehead and Goodchild 1909 as *Hypnum intermedium. Hawks Tor* (2); LDe II; Conolly *et al.* 1950 as *Hypnum revolvens* Swartz. !*Colney Heath* (20); LDe I; Godwin 1964. *Ponder's End* (21); MDe; Warren 1912 as *Hypnum revolvens. Dorchester* (23); Wo; Duigan 1955. *Wolvercote* (23); Wo or Ho?; Bell 1904 as *Hypnum revolvens. Buckenham Broad* (27); Fl VII-VIII; var. *intermedius*; Lambert *et al.* 1960. *Mundesley* (27); as vars. *revolvens* and *intermedius*; Reid 1904 as *Hypnum revolvens* Swartz and *H. intermedius* Lind. !*Barnwell Station* (29); LDe pre-I. !*Sidgwick Avenue* (29); EDe; Dickson 1963. !*Holme Fen* (31); Fl VII*b*. *Upton Warren* (37); MDe. *Aby Grange* (54); LDe III; Suggate and West 1959. *Bag Mere* (58); Fl IV-V; Birks 1965. *Wallasey* (58); Fl VII/VIII*; var. *intermedius*; Travis 1922 as *Hypnum intermedium* Lindb. *Chat Moss* (59); Fl IV-V; Birks 1965. *Malham Tarn Moss* (64); Fl VI; t; Proctor and Birks 1963 pc. !*Kirkmichael 3a* (71); LDe III; Dickson *et al.* 1970. *Garvel Park* (76); LDe/Fl IV*; Robertson 1881 as *Hypnum revolvens. Whitrig Bog* (81); LDe II and III; t; Conolly 1963 pc. *Hailes* or *Corstorphine* (83); LDe only; Bennie 1894*a* as *Hypnum revolvens* Swartz. !*Loch of Park* (92); LDe I; Vasari and Vasari 1968. *Canna* (104); Fl VII*b*; t; Flenley and Pearson 1967. !*Loch Cuithir* (104); LDe III, and Fl VII*b* and VIII; Vasari and Vasari 1968. *Loch Mealt* (104); LDe II; Birks 1969 pc. !*Loch Droma* (105); LDe I; Kirk and Godwin 1964. !*Baggotstown* (H8); Ho III; Dickson 1964*a*. !*Kildromin* (H8); Ho I. *Derrybrennan* (H19); Fl VI?; Synnott 1970 pc. *Gort* (H15); LAn and Ho III or IV; Jessen *et al.* 1959. *Timahoe* (H19); Fl VI?; Synnott 1970 pc. *Mapastown* (H31); LDe I, II and III; Mitchell 1953 as *D. revolvens* (Sw.) Warnst.

D. revolvens is unrecorded from several vice-counties chiefly in the southeast of England. It is most abundant in the highland zone where it is often found in mesotrophic and eutrophic mires and flushes. In the lowlands it inhabits eutrophic fens, as in Norfolk. The widespread Holarctic range includes the highest latitudes. This species is also recorded from South America. There are many Pleistocene records from mainland Europe, just as there are in Britain, where the scatter of records through the Devensian implies per-glacial survival and a widespread range at least since the Late Devensian (fig. 62).

Drepanocladus vernicosus (Mitt.) Warnst.

!*Colney Heath* (20); LDe I; Godwin 1964.

This rare and local species has a scattered distribution in the British Isles. The great bulk of localities are in the highland zone where the habitats include 'sub-alpine' and 'mountain bogs' (Dixon 1924, p. 321). Bellamy and Rose (1960) referring to Thelnetham Old Fen, Suffolk, state 'The bryophyte flora of this fen system is rich in calcicole species characteristic of *Schoenus-Cladium* fens. The most noteworthy species present are *Preissia quadrata* (a long overlooked East-Anglian species, now known to be also in at least 3 places in Norfolk valley fens); *Moerckia flotoviana*; *Philonotis calcarea* and *Drepanocladus vernicosus.* These bryophytes may well be relics here, of comparatively early post-glacial times, as they are characteristic of open fens at fairly high altitudes and latitudes.'

There are no subfossils relevant to the three first-named species but the Colney Heath material of *D. vernicosus* points to occupation of southeastern England since Late Devensian times. Hence the Suffolk locality may be relict.

The wide Holarctic range of *D. vernicosus* excludes the highest latitudes. There are numerous subfossils of interglacial, glacial and post-glacial age from the rest of Europe and there is also a Pliocene record from Kroscienko, Poland (table 10).

Drepanocladus uncinatus (Hedw.) Warnst.

!*Nant Ffrancon* (49); LDe only. !*Chelford* (58); EDe. !*Low Wray Bay* (69); LDe II. *Cowdenglen* (76); Fl? only; Mahony 1869 as *Hypnum uncinatum. Fort William* (76); Fl V or VI*; Dixon 1910 as *Hypnum uncinatum* Hedw. *Faskine* (77); De? only; Bennie 1894*b* as *Hypnum uncinatum* Hedw. !*Drumurcher* (H32); LDe III.

D. uncinatus is readily separated from the other species considered here by the strongly falcato-secund, dentate

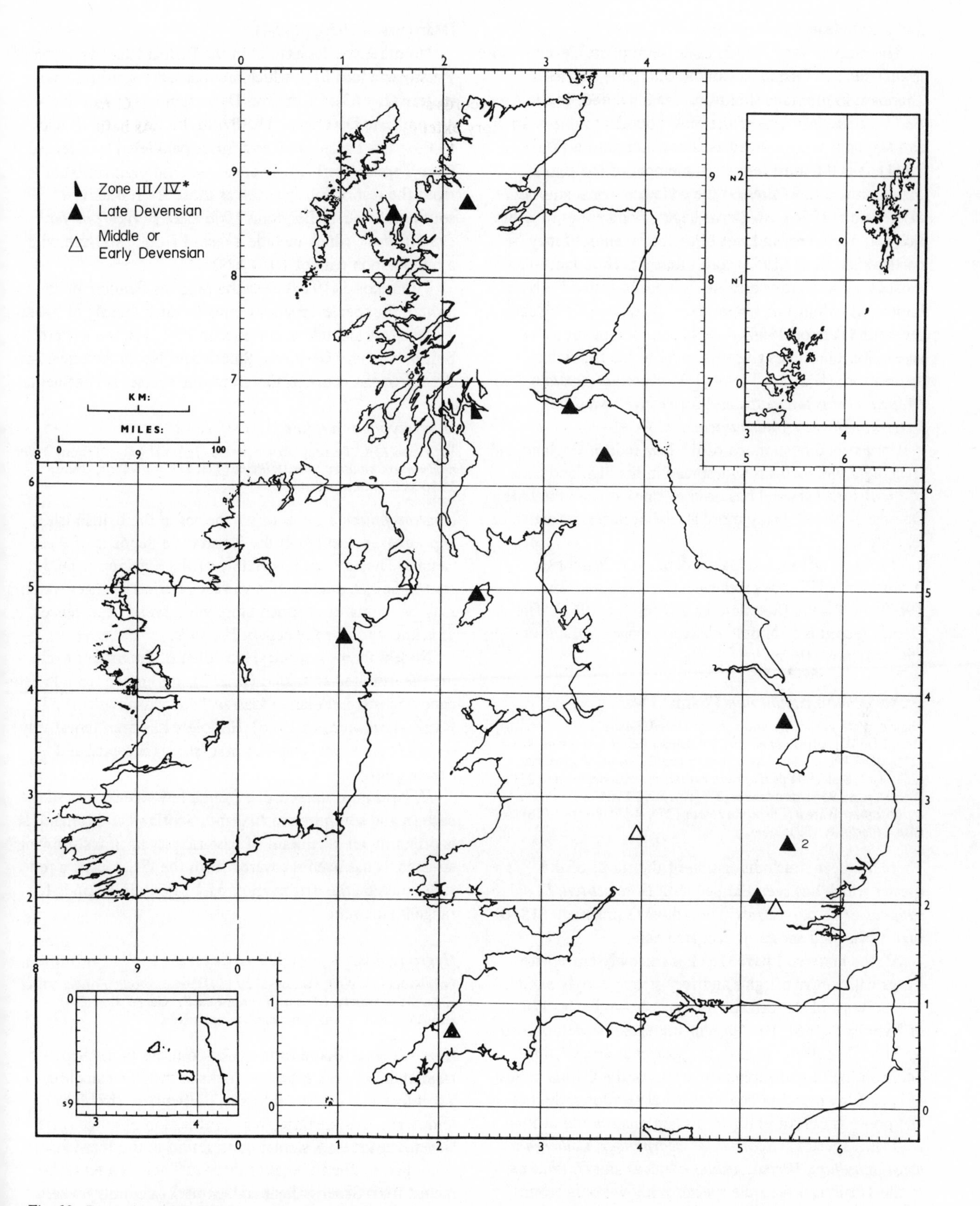

Fig. 62. Devensian subfossils of *Drepanocladus revolvens.*

and plicate leaves.

The range is very wide in both Hemispheres; even remote islands such as Tristan da Cunha, where the species is abundant in montane situations, are inhabited. In the Arctic *D. uncinatus* is one of the most abundant mosses. In Jan Mayen it is considered as almost ubiquitous (Watson 1964). It is the most terrestrial member of the genus. Many authors have stressed the wide ecological amplitude. Mårtensson (1956) lists *Dryas* heath, flood zones of rivers, rich and poor fens, and wet birch forest where it may be corticolous. Rune (1953) states that it is abundant in almost every serpentine area he studied. In Iceland it is an important colonist of ground recently exposed by glacier recession (Persson 1964). It is so common in snow-bed vegetation that it has given its name to the following associations (Gjaerevoll 1956); *Deschampsia flexuosa-Drepanocladus uncinatus* and *Carex bigelowii-Carex lachenalii-Drepanocladus uncinatus.*

Unrecorded from much of the lowlands of England and Ireland, *D. uncinatus* is widespread in the highland zone of the British Isles and has been so since at least the Late Devensian period. Many more glacial occurrences are to be expected.

It is known from last glacial deposits in Wisconsin (Culberson 1955), Saskatchewan (De Vries and Bird 1965) and Poland (Sobolewska and Srodon 1961). The earliest record is of Middle Pleistocene age (Abramova and Abramov 1962).

Drepanocladus capillifolius (Warnst.) Warnst.

Thames Valley; locality and age not stated; Dixon 1924. *Nazeing* (18); LDe III; Allison *et al.* 1952. !*Ponder's End* and *Angel Road* (21); MDe; Warren 1912 as *Hypnum capillifolium. Wolvercote* (23); Wo?; Bell 1904 as *Hypnum capillifolium. Dorchester* (23); Wo; Duigan 1955. *Mundesley* (27); An; Reid 1904 as *Hypnum capillifolium* Warnst. !*Broome Heath* (27); Wo. !*Wretton* (28); EDe. !*Fladbury* (37); MDe.

There has been much discussion of the status of this species which has been lumped with *D. sendtneri, D. aduncus* or *D. polycarpus.* Tuomikoski's judgment (1940) that it is a good species is accepted here.

All the material I have seen has narrowly lanceolate leaves with entire margins and very strong, longly excurrent nerves and alar cells of the type shown in fig. 278 of Nyholm (1965). The Nazeing material was determined by Tuomikoski who regards the species as an exacting one, confined to eutrophic situations in the littoral zones of lakes (down to 2 m depth), littoral meadows, ditches and pools. It almost never forms large stands and occurs with such species as *Scorpidium scorpioides, Calliergon megalophyllum, Drepanocladus aduncus* and *D. tundrae.* In the Torneträsk area the species is known only from permanently submerged positions down to *c.* 5 m (Mårtensson 1956, p. 254).

No other species extinct in the British Isles has so many Pleistocene localities, which are markedly southern and eastern (fig. 63). Of the five Devensian records all but one are pre-Late Devensian. The British history extends back to the early Pleistocene, and this is paralleled by a record from Tegelen (p. 175). There are several younger records from the rest of Europe such as those from Blekinge, southern Sweden (Berglund 1966 as *D. polycarpus* var. *capillifolius*), which include a record from a stratum with a radiocarbon date of 1010 ± 60 B.C.

Tuomikoski (1940) gives the range as Fennoscandia (where the species reaches Finmark but is mainly of southern and low altitude occurrence in Finland), the eastern Baltic, northern Germany, Siberia and North America. In the European context, therefore, the species is continental.

Hygrohypnum luridum (Hedw.) Jenn.

!*Stannon* (2); LDe only. *Cowdenglen* (76); Fl? only; Mahony 1869 as *Hypnum palustre. Fort William* (97); Fl V or VI*; t; Dixon 1910 as *Hypnum palustre* L.

Hygrohypnum, a genus of six species in the British Isles, is poorly represented in the Pleistocene deposits of these islands. They are species mainly of the highland zone. *H. smithii* and *H. dilatatum* have very restricted ranges which may be remnants of much more widespread occurrences, as is known to be the case in *H. molle.*

No less than six species, including *H. polare* but excluding the disputed *H. szaferi* (p. 212), are known from Pleistocene deposits in Poland (Szafran 1952; Jasnowski 1957; Ralska-Jasiewiczowa 1966). *H. polare* has been tentatively recorded from peats on Bathurst Island (Brassard and Steere 1968).

H. luridum is unrecorded from a few vice-counties in eastern and southeastern England, Scotland and Ireland. It is primarily an inhabitant of base-rich rocks in fast-flowing streams. It has been recovered from the Pleistocene a few times outside the British Isles, particularly in Poland. The range is Holarctic.

Hygrohypnum molle (Hedw.) Loeske

Farnham (17); Wo?; Oakley 1939 as *Hypnum molle. Ponder's End* (21); MDe; Warren 1912 as *Hypnum molle. Garvel Park* (76); LDe/IV*; Robertson 1881 as *Hypnum molle.*

H. molle is restricted to four vice-counties in the Scottish Highlands where it grows on rocks in montane streams. Therefore two of the subfossils, particularly that from Middlesex, demonstrate great contraction of range since Devensian times. A similar contraction in the Ipswichian is implied by the Farnham subfossil. There is a tentative record from Skaerumhede in Denmark, a country where the species no longer grows. The range is extensive in

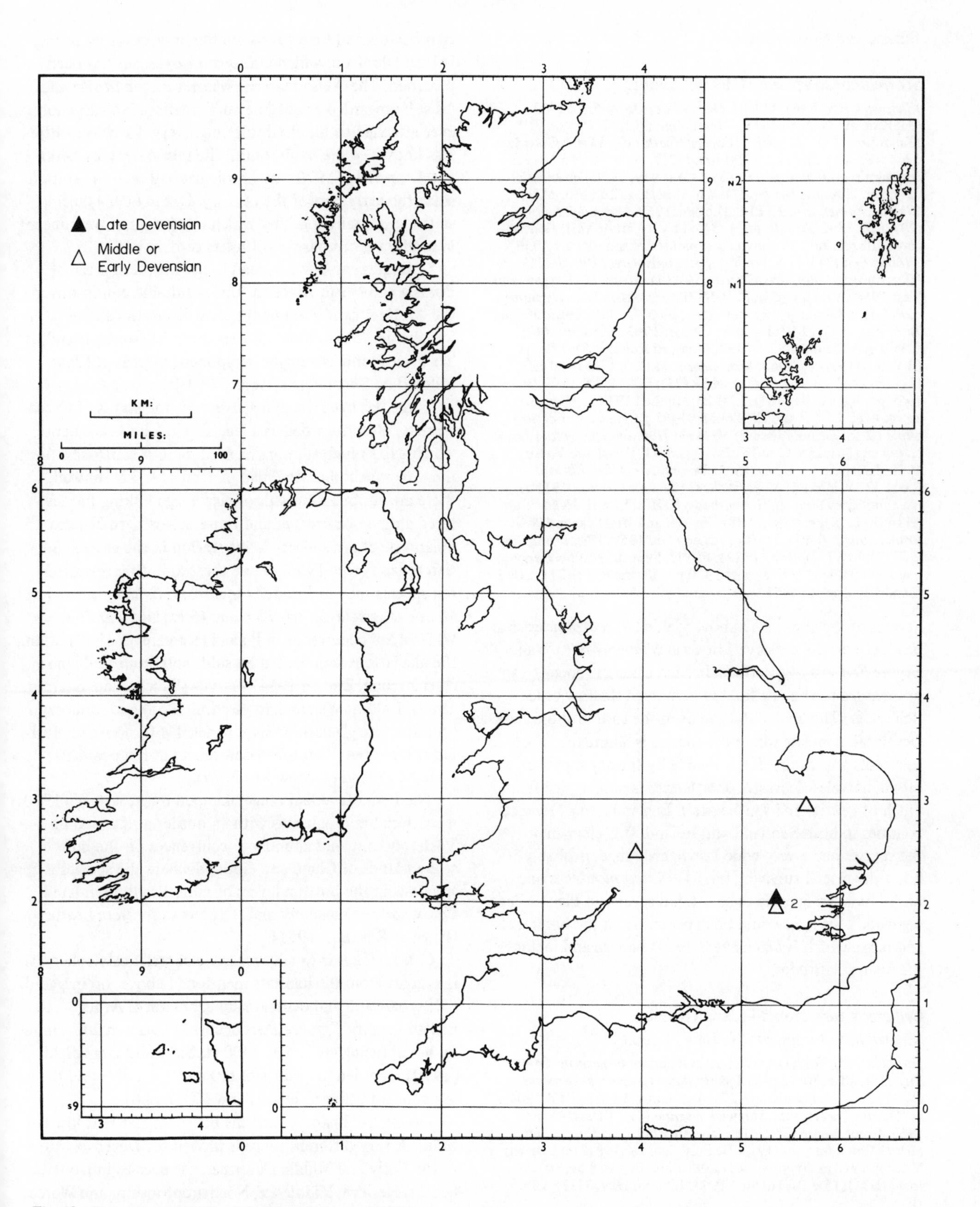

Fig. 63. Devensian subfossils of *Drepanocladus capillifolius*.

Eurasia and North America.

Scorpidium scorpioides (Hedw.) Limpr.

!*Shapwick Heath* (6); Fl VIII. *Decoy Pool Drove* (6); Fl VIII; t; Clapham and Godwin 1948. !*Meare Track* (6); Fl VII*b* or VIII. !*Sandhurst* (17); LDe only. !*Colney Heath* (20); LDe I; Godwin 1964. *Ponder's End* and *Angel Road* (21); MDe; Warren 1912 as *Hypnum scorpioides. Hoxne* (25); LAn; West 1956. !*Mundesley* (27); An. !*Broome Heath* (27); Wo. !*Wretton* (28); E or MDe. !*Sidgwick Avenue* (29); EDe; Dickson 1963. !*Barnwell Station* (29); LDe pre-I. *Trundle Mere* (31); Fl VII-VIII or VIII; Godwin and Clifford 1968 as *Hypnum scorpioides*; Vishnu-Mittre 1959. !*Holme Fen* (31); Fl VII or VIII; perigonia. *Great Billing* (31); MDe; Morgan 1969. *Aby Grange* (54); LDe II and III; Suggate and West 1958. *Wallasey* (58); Fl VII/VIII*; Travis 1922 as *Hypnum scorpioides* Schwaegr. *Bag Mere* (58); Fl IV-V; Birks 1965. *Malham Tarn Moss* (64); Fl VI; Pigott and Pigott 1963. *Neasham* (66); LDe II and III; Blackburn 1952. *Bradford Raims* (68); LDe III, Fl V or VI; Bartley 1966. *Kirkmichael 3a* (71); LDe I and III; Dickson *et al.* 1970. *Loch Dungeon* (73); Fl V or VI; H. H. Birks 1969 pc. *Whitrig Bog* (81); LDe III; Mitchell 1948 as *Hypnum scorpioides*; LDe II and III; Conolly 1963 pc. *Hailes* or *Corstorphine* (83); LDe only; Bennie 1894*a* as *Hypnum scorpioides* Linn. !*Drymen* (86); LDe II and Fl IV, V and VI; Vasari and Vasari 1968. !*Garsgadden Mains* (99); LDe only. *Loch Cill Chriosd* (104); Fl IV; Birks 1969 pc. *Walls* (112); LDe?; Dixon 1911 as *Hypnum scorpioides* L. !*Dromshallagh* (H8); LDe II. !*Kildromin* (H8); Ho I. !*Baggotstown* (H8); Ho I, II and III; Dickson 1964*a*. *Roundstone 2* (H16); Fl VII*a*; Jessen *et al.* 1959. !*Blackriver Bog B* (H19); Fl VI. *Timahoe* (H19); Fl VI?; Synott 1970 pc. *Derrybrennan* (H19); Fl VI?; Synott 1970 pc. *Mapastown* (H31); LDe I, II and III; Mitchell 1953.

Several of the sites listed above yielded *S. scorpioides* as a peat former; the preservation, as at *Shapwick Heath* and *Holme Fen*, may be perfect. In such cases the species can be envisaged as having floated in mats in shallow base-rich water. This may also have been the case in those Late Devensian deposits where the species is abundant. Most subfossils are readily determinable by the asymmetric, broad, nerveless leaves. *S. scorpioides* is unrecorded from much of central and southeastern England and a few vice-counties in both Scotland and Ireland. It is clear that the species had a very wide last glacial range; probably it was a per-glacial survivor (fig. 64). *S. scorpioides* is one of the most frequently encountered species in Pleistocene deposits. This is as would be expected from the habitat and range which is extensive in the Holarctic and includes the highest latitudes.

Calliergon turgescens (T. Jens.) Kindb.
[*Scorpidium turgescens* (T. Jens.) Loeske]

Farnham (17); Wo?; Oakley 1939 as *Hypnum turgescens. Ponder's End* (21); MDe; Warren 1912 as *Hypnum turgescens.* !*Broome Heath* (27); Wo. *Mundesley* (27); An; Nathorst 1873; Reid 1904 as *Hypnum turgescens.* !*Sidgwick Avenue* (29); EDe; Dickson 1963. *Great Billing* (32); MDe; Morgan 1969. !*Fladbury* (37); MDe. !*Nant Ffrancon* (49); LDe only. *Whitrig Bog* (81); LDe III; Mitchell 1948 as *Hypnum turgescens*; Conolly 1968 pc. *Mapastown* (H31); LDe III; Mitchell 1953. !*Drumurcher* (H32); LDe III.

Ben Lawers and Cader Idris are the only localities in the British Isles from which *Calliergon turgescens* has been recorded. The Welsh material was never seen *in situ* and must be regarded as needing confirmation. No suspicion ever attached to the Perthshire discovery. However, little was known of the habitat until the rediscovery by Birks and Dransfield (1970); no bryologist had seen the station since the early part of the century. Dixon (1924) merely states 'mountain bogs', by which one imagines were meant in modern terms base-rich flushes or rich fens.

C. turgescens is widespread in the northern areas of Eurasia and North America. Under suitable conditions in the Arctic it can be abundant, as at Kongsfjorden in Spitsbergen, Peary Land and northern Ellesmere Island. It has been found as a major component of peat at Klaas Billen Bay, Spitsbergen (Acock 1940).

There has been much discussion of the status of *C. turgescens*, often regarded as a glacial relict, an assessment which gains much support from numerous subfossils. As the now much outdated map in Gams (1932) showed, there are various subfossils of glacial age linking the scattered alpine occurrences and those in northern Europe. In Poland there is a solitary living station in the eastern Silesian upland (Kuc 1956) where there are other relict rich fen species such as *Paludella squarrosa, Calliergon trifarium, Meesia longiseta, M. tristicha* and *Helodium blandowii.* Widespread occurrence in Poland (Karczmarz 1969) during the glaciations is indicated by subfossils from Ludwinow (Weichselian; Zmuda 1914), Barycz (Weichselian; Szafran 1952), Delora (Weichselian; Srodon 1968) and Mokoszyn (penultimate glaciation; Szczepanek 1960). More interesting still are two Flandrian occurrences in fen peats in lowland Poland (Jasnowski 1957*b*).

The Fennoscandian range, mapped by Albertson (1940*a*) is scattered in mountains with an outlier at Kinnekulle, Västergötland, and abundant occurrences on the limestone islands of Öland and Gotland where the species is conspicuous in the bottom layers of periodically inundated *Carex panicea* meadows and *Agrostis stolonifera* heaths (Horn af Rantzien 1951).

C. turgescens is usually regarded as strict calcicole growing, apart from the habitats mentioned above, on irrigated rocks, bare soil and occasionally submerged. At its British locality *C. turgescens* occurs in *Carex saxatilis* mires on north-facing ground at 3300 ft. Birks and Dransfield (1970) recorded pH values of 6.0 and 6.1 and found the Ca ion concentration to be very low (0.80 mg/l).

The British history confirms the relict status at low latitudes. A very different range is proved for Devensian times. In the Early and Middle Devensian the species inhabited Cambridgeshire, Middlesex, Northamptonshire and Worcestershire, in Late Devensian times Caernarvonshire, Co.

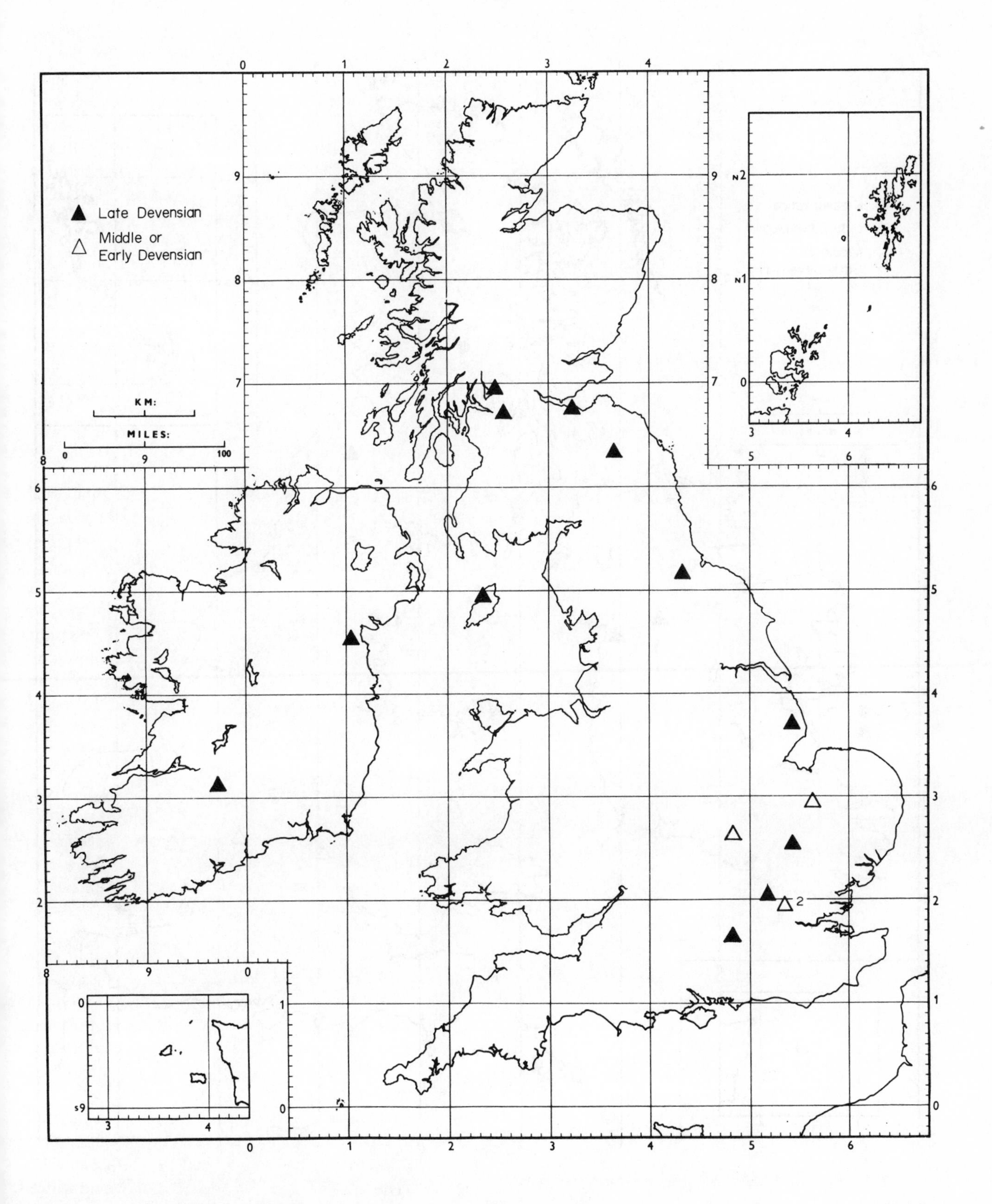

Fig. 64. Devensian subfossils of *Scorpidium scorpioides*.

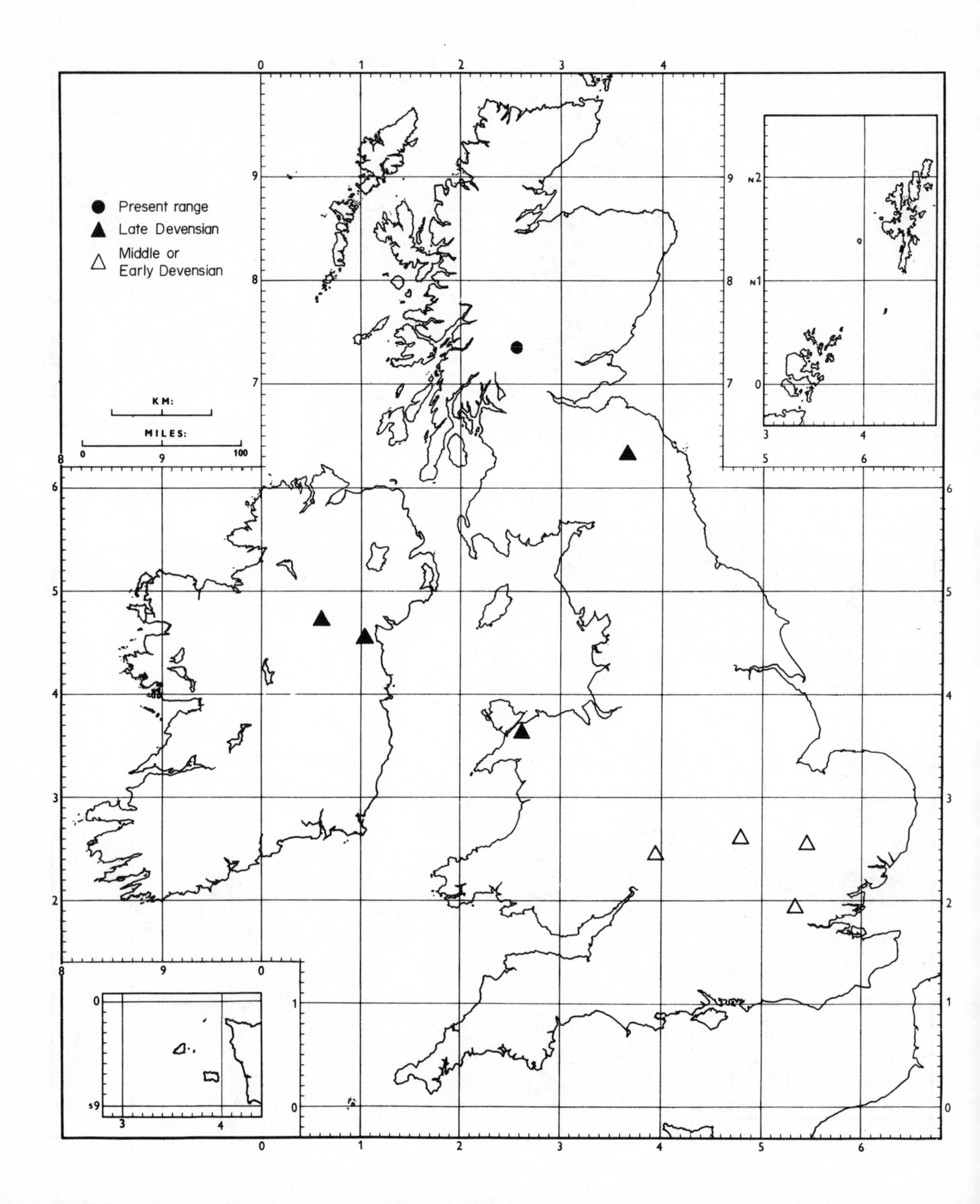

Fig. 65. Devensian subfossils and present range of *Calliergon turgescens*.

Louth, Co. Monaghan and Berwickshire. Subfossils from Norfolk and Surrey demonstrate southern occurrences in earlier glaciations.

C. turgescens is an excellent example of a relict species. As befits its present European status the reproductive capacity is very low. Sporophytes are unknown over much of the range as, for instance, in Britain, Poland and the Torneträsk region. Some authors, however, refer to apical buds as vegetative propagules.

Calliergon stramineum (Brid.) Kindb.
[*Acrocladium stramineum* (Brid.) Rich. & Wall.]

!*Stannon* (2); LDe only. !*Honeygore Track* (6); Fl VII*b*. *Trafalgar Square* (21); Ip only; Abbot 1892. *Ponder's End* (21); MDe; t; Warren 1912 as *Hypnum stramineum. Dorchester* (23); Wo 3; Duigan 1955. *Wolvercote* (23); Ho or Wo?; Bell 1904 as *Hypnum stramineum. Woodwalton* (31); Fl VII*b*; Godwin and Clifford 1938. *Chartley Moss* (39); Fl VII and VIII; Green 1964 pc. *Wybunbury Bog* (58); Fl VII and VIII; Green 1963 pc. !*Chelford* (58); EDe. *Linton Mires* (63); Fl IV or V, VI and VII*a*; Raistrick and Blackburn 1932 as *Hypnum stramineum* Dicks. *Swang Moss* (62); Fl VII*b*; Simmons 1968 pc. *Neasham* (66); LDe II; Blackburn 1952. *West Craigneuk* (77); LDe only; Dixon 1907 as *Hypnum stramineum.* !*Dun Moss* (89); Fl VI.

C. stramineum, primarily a species of oligotrophic mires, is known from almost all parts of Britain but is unrecorded from much of Ireland. It has been widespread in Britain since the Late Devensian period. The extensive Holarctic range reaches very far to the north (Franz Joseph Land, Störmer 1940); in Spitsbergen habitats include moss fields below bird cliffs; rock crevices and morainic ground. From outside the British Isles there are numerous subfossils including a Pliocene one from Poland (table 10).

Calliergon trifarium (Web. & Mohr) Kindb.
[*Acrocladium trifarium* (Web. & Mohr) Rich. & Wall]

!*East Walton* (28); Fl IV. !*Kirkmichael 3a* (71); LDe III; Dickson *et al.* 1970. *Hailes* or *Corstorphine* (83); LDe only; Bennie 1894*a* as *Hypnum trifarium* Web. & Mohr. !*Drymen* (86); Fl IV and V; Vasari and Vasari 1968. !*Blackriver Bog B* (H19); Fl VI. *Timahoe* (H19); Fl VI?; Synnott 1970 pc.

I have been unable to trace the source of the record of this species given by Godwin (1956, p. 331) as zone VII*b* of Woodwalton Fen, Hunts.

The terete, sparsely branched stems and very concave and obtuse leaves make *C. trifarium* easy to recognise, even in the subfossil state. Apart from the few Irish stations in the Burren, Co. Clare, the species is now restricted to the Scottish Highlands (fig. 66). Last century it was known at Whittlesey Mere, Hunts. Proctor (1959*b*) has discussed the British and Irish habitats in some detail. *C. trifarium*, a species of rich fens, has a wide range in the Northern Hemisphere, and extends to the highest latitudes.

At East Walton the species grew with *Meesia tristicha*; stems are scattered through the peat composed mostly of the *Meesia*. At Blackriver Bog *M. tristicha* and *Scorpidium scorpioides* are both recorded. At Drymen *S. scorpioides* was the dominant, while at Kirkmichael the range of rich fen species includes *Meesia tristicha, Helodium blandowii, Homalothecium nitens* and *Calliergon giganteum.*

The subfossils give evidence of a contraction of range since zone VI, if not earlier. There can be no doubt that the present range is a relict one, as Proctor (1959) pointed out in the then almost complete absence of subfossils from the British Isles.

Calliergon cordifolium (Hedw.) Kindb.
[*Acrocladium cordifolium* (Hedw.) Rich & Wall.]

Dorchester (23); Wo; Duigan 1955. *Wolvercote* (23); Ho or Wo; Bell 1904 as *Hypnum cordifolium.* !*Hoxne* (25); LAn; Turner 1968. *Histon Road* (29); Ip III; Proctor 1959*a*. !*Chelford* (58); EDe. *Cromwell Bottom* (63); Fl VII*b*; Bartley 1964. !*Low Wray Bay* (69); LDe II. !*Kirkmichael 3a* (71); LDe III; Dickson *et al.* 1970. !*Bigholm Burn* (72); LDe II; Moar 1969*a*. *Cowdenglen* (76); Fl? only; Mahony 1869; Bennie 1891 as *Hypnum cordifolium. Winterhope Burn* (79); Fl only; Lewis 1905 as *Hypnum cordifolium* Hedw. !*Loch Kinord* (92); Fl VII*a* and VII*b*; Vasari and Vasari 1968.

C. cordifolium is unrecorded from a few vice-counties in Britain and several in central Ireland. The principal habitat is wet woodland such as *Alnus* and *Salix* carr, the kind of vegetation which might well relate to the records from Chelford and Cromwell Bottom and probably others. The distribution has been extensive since Late Devensian times if not earlier. There are many extra-British subfossils of this bipolar species, which occurs in New Zealand as well as widely in the Holarctic.

Calliergon giganteum (Schimp.) Kindb.
[*Acrocladium giganteum* (Schimp.) Rich. & Wall.]

Dogger Bank; Fl IV or V; Whitehead 1921 as *Hypnum giganteum.* !*Chilton* (6); Fl VII*b*. !*Honeygore Track* (6); Fl VII*a* and VII*b*. !*Stanford* (15); Fl VIII. !*Lexden* (18); Wo or Ip; Shotton *et al.* 1962. *Nazeing* (18); LDe III; Allison *et al.* 1952 as *Calliergon giganteum* (Schimp.) Kindb. !*Marks Tey* (19); EWo. !*Colney Heath* (20); LDe I; Godwin 1964. *Ponder's End* and *Angel Road* (21); MDe; Warren 1912 as *Hypnum giganteum. Cothill Fen* (22); Fl IV-V and VI; Clapham 1939. *Wolvercote* (23); Ho or Wo; Bell 1904 as *Hypnum giganteum.* !*Hoxne* (25); LAn; Turner 1968. *Buckenham Broad* (27); Fl VII-VIII. Lambert *et al.* 1960. !*Broome Heath* (27); Wo. !*Mundesley* (27); EAn. !*Wretton* (28); EDe. !*Sidgwick Avenue* (29); EDe; Dickson 1963. !*Barnwell Station* (29); LDe pre-I. *Woodwalton Fen* (31); Fl VII*b*; Godwin and Clifford 1938. !*Holme Fen* (31); Fl VII*b*. !*Upton Warren* (37); MDe. !*Fladbury* (37); MDe. *Aby Grange* (54); LDe II and III; Suggate and West 1959. *Wallasey* (58); Fl VII/VIII*; Travis 1922 as *Hypnum giganteum* Schimp. !*Whixall Moss* (58); VII*a*. !*Hooks* (61); LDe II. *Swang Moss* (62); VII*b*; Simmons 1968 pc. !*Seamer Carr* (62); Fl only; with perigonia. *Linton Mires* (63); Fl V or VI; Raistrick and Blackburn 1932 as *Hypnum giganteum* Schimp. *Burtree Lane* (66); Fl IV; Bellamy *et al.* 1966. !*Kersall Moss* (69); LDe II and III. !*Russland Moss* (69); Fl only; Dickinson 1968 pc. !*Kirkmichael 3a* (71); LDe III; Dickson *et al.* 1970.

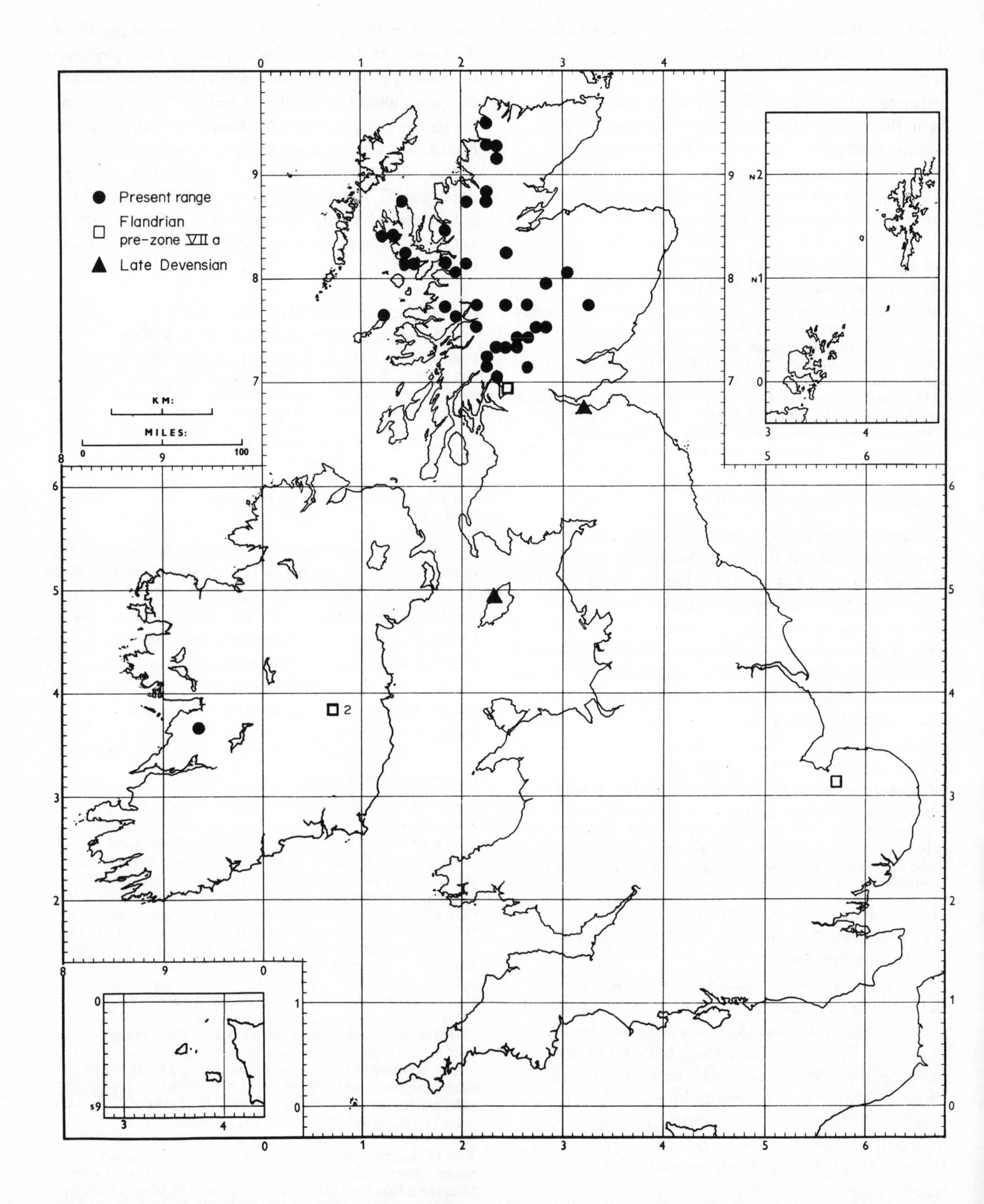

Fig. 66. Late Pleistocene subfossils and present range of *Calliergon trifarium*. Partly after Proctor, *Trans. Br. bryol. Soc.* 1963.

!*Loch Dungeon* (73); Fl V or VI; H. H. Birks 1969 pc. *Garvel Park* (76); LDe Fl IV*; Robertson 1881 as *Hypnum giganteum. Whitrig Bog* (81); LDe III; Mitchell 1948 as *Hypnum giganteum.* !*Drymen* (86); LDe II, III-IV and Fl IV, V and VI; Vasari and Vasari 1968. !*Loch Kinord* (91); LDe III; Vasari and Vasari 1968. *Canna* (104); Fl VII*b*; t; Flenley and Pearson 1967. *Loch Mealt* (104); LDe II; Birks 1969 pc. *Loch Cill Chriosd* (104); LDe III; Fl IV. Birks 1969 pc. !*Baggotstown* (H8); Ho I; Dickson 1964*a*. !*Kildromin* (H8); Ho I. !*Blackriver Bog A* (H19); Fl VI?. !*Lullymore Bog* (H19); Fl VII*b*. !*Derrybrennan* (H19); Fl VI? also Synnott 1970 pc. *Timahoe* (H19); Fl VI?; Synnott 1970 pc. !*Drumurcher* (H32); Fl Iv.

All the subfossils have stout nerves and, in contrast to *C. cordifolium*, clear-cut auricular cells. Most of the subfossils are only fragments. However, some of the material from Holme Fen and Drymen shows the densely pinnate habit.

This local species is absent from about twenty vice-counties scattered in Britain and seventeen in Ireland, mainly in the south. Hence the forty-seven localities of subfossils, making it one of the commonest Pleistocene species, point to greater abundance in the past. Of course, preservation is favoured by the habitats, primarily rich fens.

The Early, Middle and many Late Devensian localities indicate survival through the last glaciation when the range was widespread (fig. 68). Disappearance from East Kent is very recent (zone VIII) and may be anthropogenic. The same may be said for Huntingdonshire but not in the cases of the Isle of Man, Hertfordshire and South Essex where the extinctions may be much earlier.

There are many subfossils from outside the British Isles (e.g. Schmitz 1967) and the present distribution, extensive in the Holarctic, includes the highest latitudes.

T *Calliergon richardsonii* (Mitt.) Kindb.

Dogger Bank; Fl IV or V; Whitehead and Goodchild 1909 as *Hypnum richardsonii. Thames Valley*; age and locality unclear; t; Dixon 1924. *Mundesley* (27); An; t; Reid 1904; Dixon 1924 as *Hypnum richardsonii* (Mitt.) Lesq. & James.

In the absence of re-examination some doubt attaches to the above records of a species unknown in the living state in the British Isles. These few records contrast markedly with the abundance of *C. giganteum.* However, I have seen nothing to indicate *C. richardsonii* with any certainty on the basis of characters of the nerve (weak, short and often branched) referred to in modern treatments (Nyholm 1965; Karczmarz and Kuc 1966). Nor have I seen any subfossils which might be referred to *C. megalophyllum*, a species of northwestern Europe, North Asia and Alaska.

The range is widespread in Eurasia and North America. In Fennoscandia there is a northern tendency, according to Nyholm (1965) who states that the habitats of this calcicole include fens, lake-sides and pools into the subalpine region. Subfossils are known from glacial deposits in Europe, particularly Poland. Steere (1942) recovered the species from the Aftonian deposits of Iowa.

There is nothing improbable about the past occurrence of *C. richardsonii* in the British Isles, where it may turn up in the living state.

Calliergon sarmentosum (Wahlenb.) Kindb.
[*Acrocladium sarmentosum* (Wahlenb.) Rich. & Wall.]

Bovey Tracey (3); LDe? only; Nathorst 1873. *Ponder's End* (21); MDe; t; Warren 1912 as *Hypnum sarmentosum. Hoxne* (25); Ho III or EWo; Reid 1896 as *Acroceratium sarmentosum* Mitt. *Hartford* (31); LDe; Godwin 1959. *Malham Tarn Moss* (64); Fl VI; Pigott and Pigott 1963. *Cross Fell* (70); Fl V or VI; Lewis 1905; Godwin and Clapham 1951. *Garvel Park* (76); LDe/Fl IV*; Robertson 1881 as *Hypnum sarmentosum. Hailes* or *Corstorphine* (83); LDe only; Bennie 1894*a* as *Hypnum sarmentosum* Wahl. *Loch Fada* (104); LDe I; Birks 1969 pc. *Lochain Coir 'a' Ghobhain* (104); LDe II; Birks 1969 pc. *Timahoe* (H19); Fl VI?; Synnott 1970 pc. *Derrybrennan* (H19); Fl VI; Synnott 1970 pc. *Ralaghan* (H30); LDe III; Jessen 1949.

C. sarmentosum is a bipolar species occurring in Antarctica, New Zealand and South America as well as widely in the Holarctic. In the British Isles it is exclusive to the highland zone where it often grows in mires, flushes and on rock where there is trickling base-poor water. It is rare in Ireland. The lowland zone as well as southeastern Scotland and central Ireland have been vacated since the last glaciation. There are several subfossils from the rest of Europe including interglacial and Flandrian ones apart from those from glacial deposits.

Calliergonella cuspidata (Hedw.) Loeske
[*Acrocladium cuspidatum* (Hedw.) Lindb.]

Wilsford Shaft (8); Fl VII*b*; Bronze Age ar; Dimbleby 1967 pc. !*Winchester* (12); Saxon ar. !*Colney Heath* (20); LDe I; Godwin 1964. *Ponder's End* (21); MDe; Warren 1912 as *Hypnum cuspidatum. Wolvercote* (23); Wo; Bell 1904 as *Hypnum cuspidatum. Hoxne* (25); Ho III and Ho III or EWo; Reid 1896; also !; LAn; Turner 1968. !*Broome Heath* (27); Wo. !*Wretton* (28); Ip II. !*Soham Lode* (29); Fl VII*b*. !*Godmanchester* (31); Fl VIII*; Roman ar. !*Bunny* (56); Fl VIII*; Roman ar. *Bag Mere* (58); Fl IV-V; Birks 1965. *Bridgewater Street* (59); Fl VIII*; Roman ar; Roeder 1899 as *Hypnum cuspidatum. Wallasey* (58); Fl VII/VIII*; Travis 1922 as *Hypnum cuspidatum* L. *Northallerton* (62); interglacial? only; Hawkesworth 1912. *Malham Tarn Moss* (64); Fl VI; Proctor and Birks 1963 pc. *Burtree Lane* (66); Fl IV; Bellamy *et al.* 1966. *Halton* (68); Fl VIII*; Roman ar; Simpson and Richmond 1937 as *Hypnum cuspidatum* L. *Upper Valley Bog* (69); Fl VI; Johnson 1960 pc. !*Foolmire Sike* (69); Fl VII*b*. !*Seathwaite Tarn* (69); Fl VIII. !*Skelsmergh Tarn* (69); LDe III. *Moorthwaite Moss* (70); Fl V; t; Walker 1966. *Hailes* or *Corstorphine* (83); LDe only; Bennie 1894*a* as *Hypnum cuspidatum* L. *Fort William* (97); Fl V or VI*; Dixon 1910 as *Hypnum cuspidatum* L. !*Loch Cuithir* (104); Fl VII*b* and VIII; Vasari and Vasari 1968. !*Loch Fada* (104); Fl VIII; Vasari and Vasari 1968; LDe I and II; Birks 1969 pc. !*High Street* and *Winetavern Street* (H21); Medieval ar. *Mapastown* (H31); LDe I and II; Mitchell 1953.

C. cuspidata is a bipolar species, occurring in New Zealand as well as very extensively in the Northern Hemisphere. It

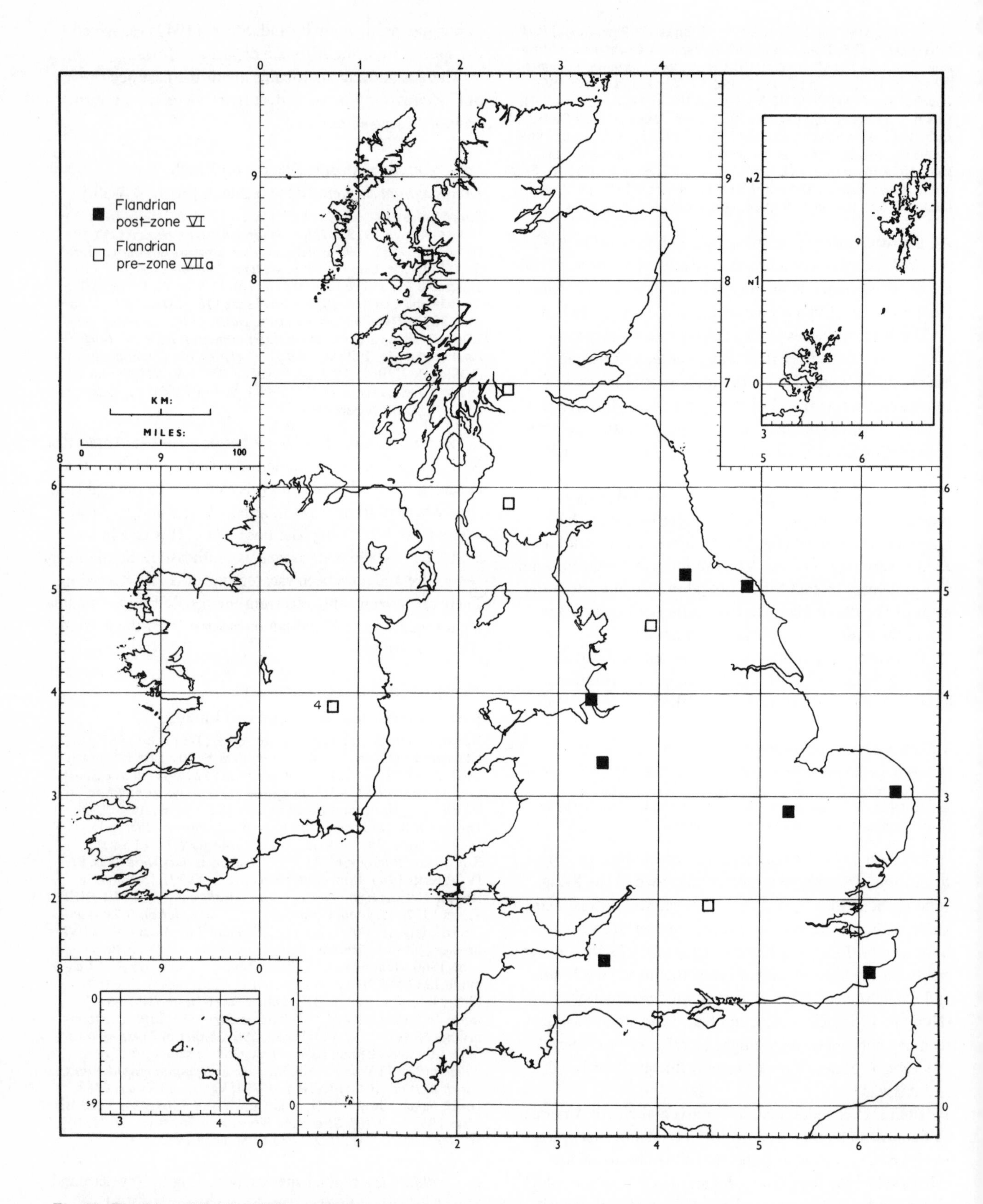

Fig. 67. Flandrian subfossils of *Calliergon giganteum*.

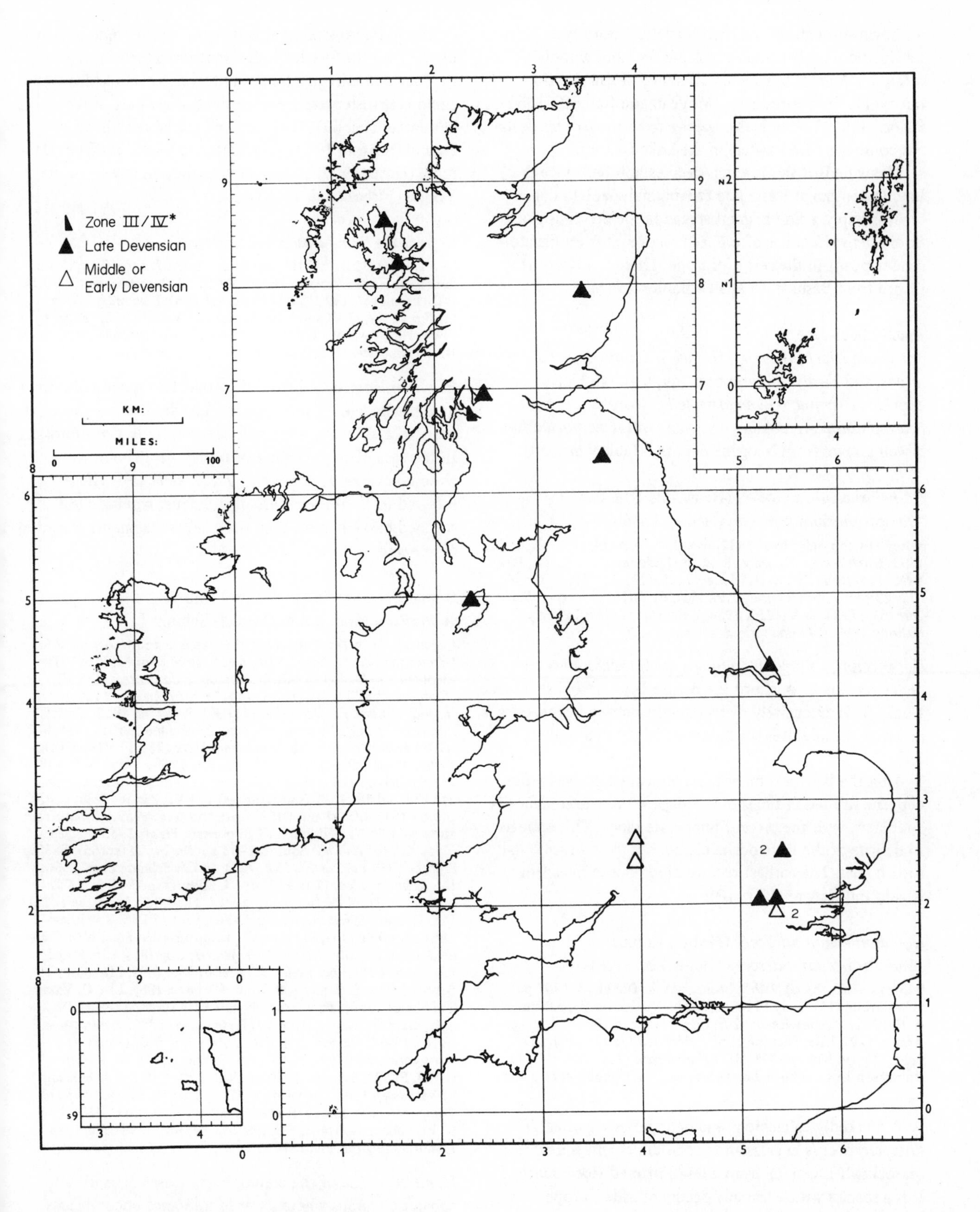

Fig. 68. Devensian subfossils of *Calliergon giganteum.*

is known throughout the British Isles in a variety of habitats including calcicolous grassland, fens and woodland, often of *Alnus*. It has a wide tolerance of soil reaction but avoids the poorest soils (McVean and Ratcliffe 1962). Pearsall (1921) records this species from the *Isoetes lacustris* consocies (deep water) in the Lake District.

The wide British range has been established since the Late Devensian at least. The two possibly Wolstonian records imply a similar great spread in the previous glaciation. *C. cuspidata* has often been recovered from Pleistocene deposits in the rest of Europe. There is a Pliocene record from Bashkir, Russia (Abramov 1965).

Brachytheciaceae

Species of *Brachythecium*, *Homalothecium*, especially *H. nitens*, and *Eurhynchium* occur frequently in Pleistocene deposits. *Brynhia*, *Rhynchostegiella*, *Scorpiurium* and *Scleropodium* are unknown as subfossils. *Palamocladium* is well known from Neogene deposits (table 9 and 10).

Homalothecium sericeum (Hedw.) B., S. & G.
[*Camptothecium sericeum* (Hedw.) Kindb.]

Selsey (14); Ip only; Reid 1892. *West Wittering* (14); Ip only; Reid 1892. *Hitchin* (20); Ho only; Reid 1897. *Hoxne* (25); Ho III; Reid 1896. !*Wretton* (28); Ip II. !*Soham Lode* (29); Fl VII*b*. !*Shippea Hill* (29); Fl VII*b*. !*Barnsley Park* (33); Fl VIII*; Roman ar. *Star Carr* (61); Fl IV or V; Clark 1954. *Cowdenglen* (76); Fl? only; Mahony 1869 as *Leskia sericea*.

H. sericeum, a shade-tolerant species of calcicolous tendency, occurs throughout the British Isles on rocks and tree trunks. It is extensively distributed in Eurasia but reaches no great latitude; it is southern in Finland (Tuomikoski 1939).

That the British subfossils are almost all derived from deposits formed in times of deciduous forest dominance is consistent with the present temperate range. This appears to be true of the Pleistocene discoveries from outside the British Isles. The earliest records are Pliocene ones from Russia and Reuver (table 10).

Homalothecium lutescens (Hedw.) Robins.
[*Camptothecium lutescens* (Hedw.) B., S. & G.]

Wilsford Shaft (8); Fl VII*b**; Bronze age ar; Dimbleby 1967 pc. !*Folkestone* (15); early Fl only. *Nazeing* (18); LDe III; Allison *et al.* 1952. !*Colney Heath* (20); LDe I; Godwin 1964. !*Sidgwick Avenue* (29); EDe; Dickson 1963. !*Barnwell Station* (29); LDe pre-I. ! *Upton Warren* (37); MDe. *Garvel Park* (76); LDe Fl IV*; Robertson 1881. *Whitrig Bog* (81); LDe II; t; Conolly *et al.* 1963 pc.

A markedly calcicolous and heliophilous species, *H. lutescens* occurs in grasslands, sand-dunes and sometimes on rocks. It is lacking from a few scattered vice-counties. It is a species which scarcely occurs outside Europe. Nyholm (1965) gives the Canary Islands, the Caucasus and Persia. In Fennoscandia it is strongly southern, as is shown by the map for Sweden in Krusentsjerna (1945).

The subfossils go a long way towards establishing a continuous British presence since the earliest part of the Devensian (fig. 69). There are last glacial records from Poland (Szafran 1952), a last interglacial record from Denmark (Hartz 1909) and an early Flandrian one from the Vosges (Firbas *et al.* 1948).

Homalothecium lutescens or *sericeum*

!*Winchester* (12); Fl VIII*; Saxon ar. !*Little Paxton* (31); Fl VIII*; Saxon ar. !*Low Wray Bay* (69); LDe II. !*Kirkmichael 3a* (71); LDe I; Dickson *et al.* 1970. !*New Dry Dock* (76); LDe only. !*Loch Cuithir* (104); Fl VI and VII*a*; Vasari and Vasari 1968. !*Baggotstown* (H8); LAn; Ho III; Dickson 1964*a*. !*Newtonbabe* (H31); LDe II. !*Drumurcher* (H32); LDe III.

H. nitens is separated from the other two species discussed here by the less distinct angular cells, by the tomentum, if present, and by the habit when shown by *in situ* subfossils. In the case of small fragments it may be difficult to choose between *H. lutescens* and *H. sericeum*. Those referred to *H. lutescens* are more robust and have less toothed basal margins than the smaller fragments placed in *H. sericeum*.

Homalothecium nitens (Hedw.) Robins.
[*Camptothecium nitens* (Hedw.) Schimp.]

Blakeway Farm Track (6); Fl VII*b*; Clapham and Godwin 1948. !*Honeygore Track* (6); Fl VII*b*. *Decoy Pool Drove* (6); Fl VIII; Clapham and Godwin 1948. !*Amberley Wild Brooks* (13); Fl VIII. !*Stanford* (15); Fl VIII. *Farnham* (17); Wo?; Oakley 1939 as *Hypnum nitens*. *Ponder's End* (21); MDe; Warren 1912. *Cothill Fen* (22); Fl only; Warburg 1963 pc. *Buckenham Broad* (27); Fl VII-VIII; Lambert *et al.* 1960. !*Hockham Mere* (28); Fl VI and VII*a*. *Histon Road* (29); Ip III and IV; Proctor 1959*a*. *Ramsay Fen* (31); Fl only; Dixon 1895. *Woodwalton Fen* (31); Fl VII*b*; t; Godwin and Clifford 1939. *Whixall Moss* (40); Fl VI; Pigott 1963 pc. *Aby Grange* (54); LDe II and III; Suggate and West 1959. *Malham Tarn Moss* (64); Fl VI; VII and VIII to present; Pigott 1963 pc. *Fox Earth Gill* (65); Fl VI; Squires 1968 pc. *Dufton Moss* (65); Fl VII*a*; Squires 1969 pc. *Burtree Lane* (66); Fl IV; Bellamy *et al.* 1966. !*Midgeholme Moss* (70); Fl V or VI; Millington 1965 pc. !*Kirkmichael 3* (71); LDe III; Dickson *et al.* 1970. !*Bigholm Burn* (72); LDe II; Moar 1969*a*. *Garvel Park* (76); LDe/Fl IV*; Robertson 1881. *Renfrew* (76); Fl IV or V?; Thompson 1968 pc. *West Craigneuk* (77); De only; Dixon 1907. *Whitrig Bog* (81); LDe II and III; Conolly 1963 pc. *Hailes* or *Corstorphine* (83); LDe only; Bennie 1894*a* as *C. nitens* Schimp. !*Drymen* (86); LDe II; Vasari and Vasari 1968. !*Dun Moss* (89); Fl VI. !*Straloch* (89); LDe only. *Gort* (H15); Ho III or IV; Jessen *et al.* 1959. *Roundstone 2* (H16); LDe II; Jessen 1949. !*Blackriver Bog B* (H19); Fl VI. !*Derrybrennan* (H19); Fl VI?; also Synnott 1970 pc. *Timahoe* (H19); Fl VI?; Synnott 1970 pc. *Ballybetagh* (H21); LDe II and Fl VI; Jessen 1949. *Mapastown* (H31); LDe II; Mitchell 1953 as *Tomenthypnum nitens* (Hedw.) Loeske. *Sharvogues* (H39); Fl V or VI. *Fallahogy* (H40); Fl VI; Smith 1958*a*. Three other Irish Flandrian records, Pigott 1963 pc.

H. nitens is one of the commonest mosses of arctic and subarctic regions where its wide tolerance encompasses

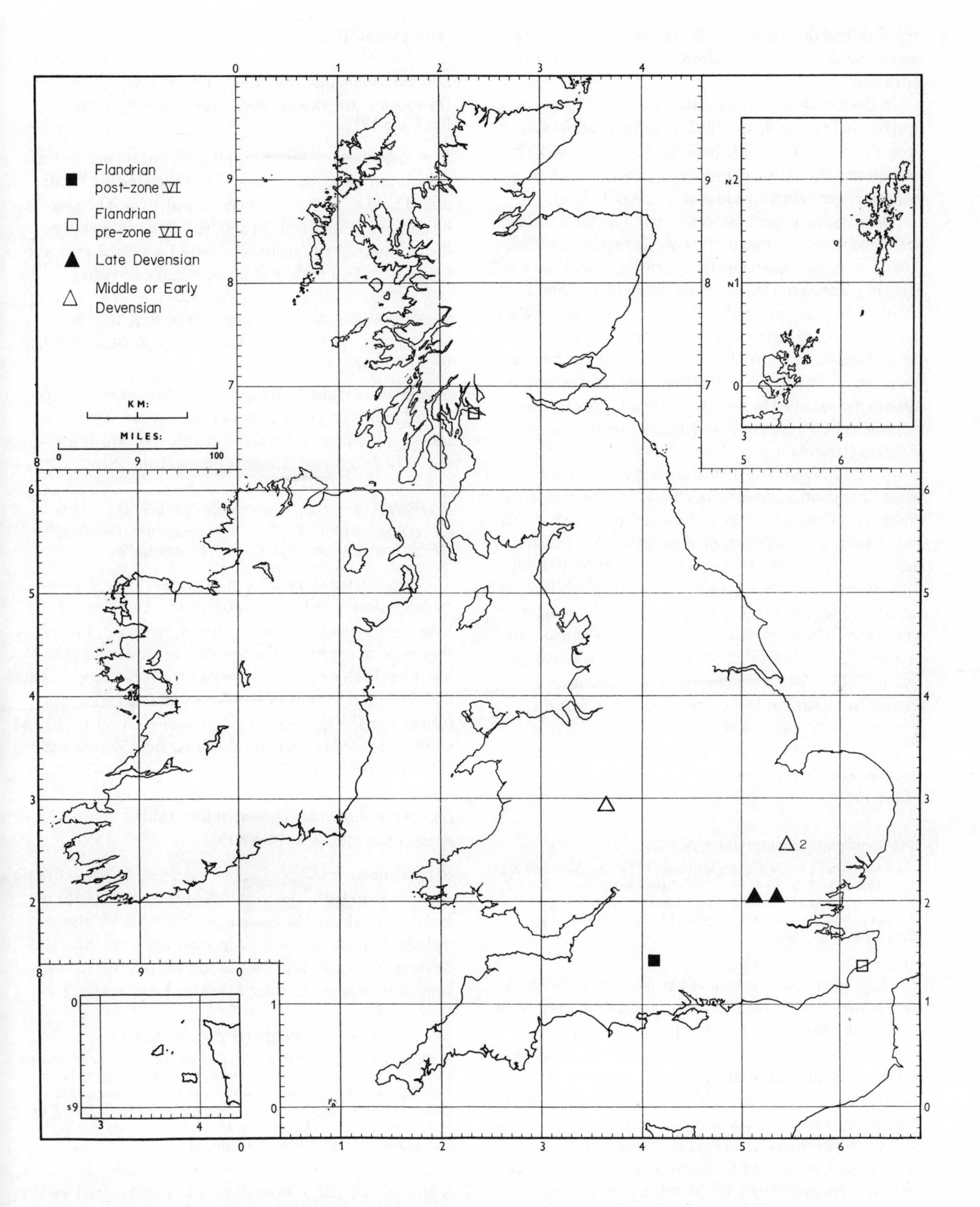

Fig. 69. Late Pleistocene subfossils of *Homalothecium lutescens*.

rich fens and dry habitats such as *Dryas* heath. Its range is very extensive in the Holarctic and includes the highest latitudes.

In the British Isles the species is restricted to wet mesotrophic and eutrophic sites such as springs and flushes in Angus (Duncan 1966) and base-rich fens in Norfolk (Petch and Swann 1968). It has a scattered, mainly northern and eastern range in Britain and is very rare in Ireland.

The detailed Late Pleistocene history demonstrates a less discontinuous range in the Late Devensian and Flandrian and reveals an elimination northwards and eastwards (fig. 70). The extinction from the south of England can be ascribed, in part, perhaps in major part, to man; there are zone VIII records from Somerset, Sussex and Kent, all from mires now grossly altered. Moreover, there were six localities in the Cheshire-Shropshire area from which the species has vanished in the last 100 years. This anthropogenic contraction can be paralleled elsewhere as in Belgium (Delvosalle *et al.* 1969).

In the highland zone, contraction, possibly a widespread contraction, took place about the Boreal-Atlantic transition. Thus at Whixall Moss in Shropshire, Dun Moss in Perthshire and Ballybetagh, Sharvogues and Fallahogy, all in Ireland, remains are known from zone VI but not later zones. At Malham Tarn Moss the Boreal occurrence might well have been more abundant than the present few stands. The eastwards contraction relates mostly to this period. It may be that some areas such as Lincolnshire and the Isle of Man were vacated at the Devensian-Flandrian transition. There can be no doubt that the present British range is a relict one.

H. nitens has often been recovered from Pleistocene deposits elsewhere. There is a Pliocene record from Russia (table 10).

Brachythecium undetermined species

!*Broome Heath* (27); Wo?. !*Happisburgh* (27); Pa. !*Barnwell Station* (29); MDe. ! *Upton Warren* (36); LDe. *Castle Eden* (66); EP; Reid 1920. !*New Dry Dock* (76); LDe. *Loch Fada* (104); Fl IV; Birks 1969 pc. *Loch Mealt* (104); LDe II; Birks 1969 pc. !*Derryvree* (H33); MDe.

The large genus *Brachythecium* has thirteen species in the British Isles. No less than eleven of these are known from the British Pleistocene, though there are only a few records or merely one record of each. Because many taxonomic problems exist in the genus, one must be unusually careful in determining the subfossils. Nevertheless, there is no strong reason for discarding any of the records; a few of the records from Garvel Park and Cowdenglen, especially *B. salebrosum* and *B. reflexum*, perhaps should be treated cautiously. There is a Miocene record of the genus from Poland and there are Pliocene records from Russia (tables 9 and 10).

Brachythecium albicans (Hedw.) B., S. & G.

!*Ilford* (18); LWo; Dickson 1964*b*. *Garvel Park* (76); LDe/Fl IV*; Robertson 1881.

B. albicans occurs throughout Britain but is unrecorded from much of central Ireland. The characteristic habitat is acid sand of both coastal and inland areas. The range is bipolar including Australia and New Zealand as well as the Northern Hemisphere. There are Late Pleistocene subfossils from Denmark, Poland and Czechoslovakia.

Brachythecium glareosum (Spruce) B., S. & G.

Wilsford Shaft (8); Fl VII*b**; t; Bronze Age ar; Dimbleby 1967 pc. ! *Wolvercote* (23); Ho?; Bell 1904.

B. glareosum is absent from much of Ireland but is widespread in Britain on calcareous rock and soil. There are few, if any, other subfossils of this species, which is extensively distributed in the Northern Hemisphere.

Brachythecium salebrosum (Web. & Mohr) B., S. & G.

Garvel Park (76); LDe/Fl IV*; t; Robertson 1881. *Cowdenglen* (76); Fl? only; Bennie 1891 as *B. salebrosum* Hoffm.

The record from Garvel Park was given only tentatively by Fergusson. One may wonder if the material could have been the next species or even *B. turgidum*, the robust species common in the Fennoscandian mountains. The latter has been recovered from Polish glacial deposits. There are last glacial records of *B. salebrosum* from Wisconsin (Culberson 1955), Poland (Ralska-Jasiewiczowa 1966) and a Miocene record of section *Salebrosa* from Zatoka Gdowska, Poland (table 9).

Brachythecium mildeanum (Schimp.) Milde

Ponder's End (21); MDe; Warren 1912.

B. mildeanum is widespread but mainly southern in Britain and rare in Ireland. The map in Krusentsjerna (1945) reveals the southern distribution in Fennoscandia; the range includes Eurasia and North America. There are three Devensian records from Poland and an Iron Age record from northwestern Germany (Körber-Grohne 1967).

Brachythecium rutabulum (Hedw.) B., S. & G.

Wolvercote (23); Ho?; Bell 1904. *Hoxne* (25); Ho III or EWo; Reid 1896. *Histon Road* (29); Ip IV; t; Proctor 1959*a*. *Brigg* (54); Fl VII*b**; Bronze Age ar; Sheppard 1910 as *Hypnum rutabulum* L. *Bridgewater Street* (59); Fl VIII*; Roman ar; Roeder 1899. *Fort William* (97); Fl V or VI*; Dixon 1910 as *B. rutabulum* B. & S. !*Newgrange* (H22); Fl VII*b*; Neolithic ar.

A protean species, *B. rutabulum* is found in a great variety of habitats but reaches its greatest luxuriance in moist

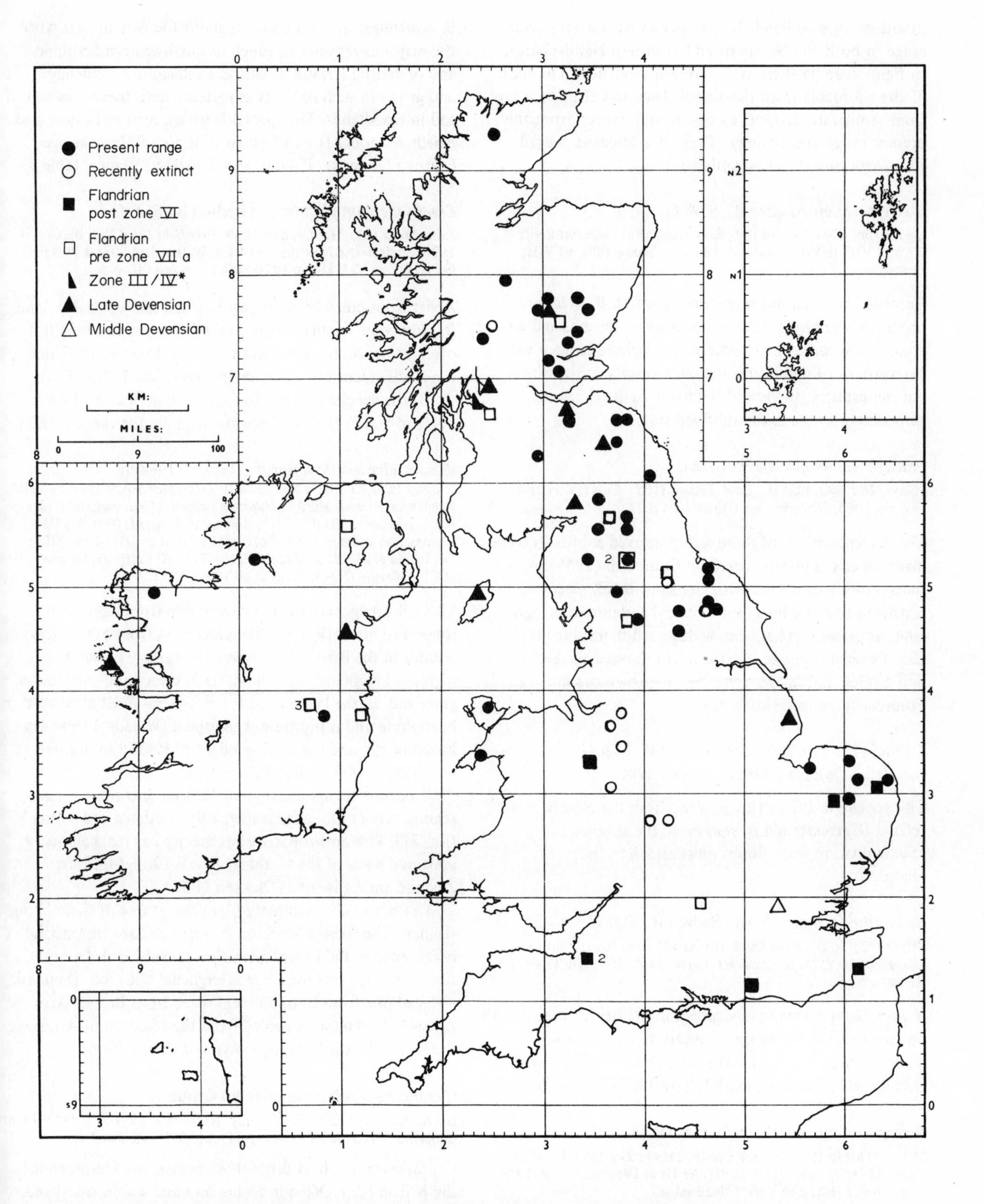

Fig. 70. Late Pleistocene subfossils and present range of *Homalothecium nitens.* Partly after Pigott, *Trans. Br. bryol. Soc.* 1963.

situations in woodland. It is a species with a very wide range in both the Northern and Southern Hemispheres. In Fennoscandia there is a southern tendency. The bulk of the subfossils from the British Isles and elsewhere are from temperate deposits as one would expect from the present range and ecology. There is a Miocene record from Arjuzanx, France (table 9).

Brachythecium rivulare B., S. & G.

Garvel Park (76); LDe Fl IV*; Robertson 1881. *Windmillcroft* (76); Fl VI/VII/VIII*; Mahony 1868*a*. !*Carpow* (88); Fl VIII; Roman ar.

Recorded from all but seven vice-counties, *B. rivulare* occurs in bryophyte flushes, on rocks by streams and wet ground in woods. The Northern Hemisphere range is very extensive but Kerguelen is the only locality in the Southern Hemisphere. In view of the habitats the species is surprisingly scarce in the subfossil state.

Brachythecium glaciale B., S. & G.

!*Garral Hill* (94); LDe II. !*Loch Droma* (105); LDe I; Kirk and Godwin 1963. !*Drumurcher* (H32); LDe III; t.

The determinations of these well preserved subfossils were made on characters discussed by Crundwell (1959). A species often associated with late snow lie, *B. glaciale* is restricted to a few hills in the central and northern highlands at present. It has been widespread there since the Late Devensian period. To the north the species reaches Spitsbergen and Jan Mayen. The range includes much of Eurasia and eastern North America.

T *Brachythecium reflexum* (Starke) B., S. & G.

Garvel Park (76); LDe/Fl IV*; Robertson 1881.

This species is difficult to separate from the closely related *B. glaciale* and *B. starkei*. In the absence of re-examination some doubt must attach to the record.

Brachythecium populeum (Hedw.) B., S. & G.

Selsey (13); Ip II?; Reid 1892. *Hoxne* (25); Ho III; t; Reid 1896. !*Happisburgh* (27); Ba or earlier. *Garvel Park* (76); LDe/Fl IV*; Robertson 1881.

B. populeum is extensively spread in the British Isles usually on moist rocks. The range, southern in Fennoscandia, includes Eurasia and North America. There is a last interglacial record from Denmark (Hartz 1909).

Brachythecium velutinum (Hedw.) B., S. & G.

!*Ilford* (18); Ip II; t; Dickson 1964*b*. !*Marks Tey* (19); Ho III; t. *Gort* (H15); Ho LAn, Ho I, II, III and III or IV; Jessen *et al.* 1959. !*High Street* (H21); Fl VIII*; Medieval ar.

B. velutinum is recorded throughout the British Isles with the major exceptions of much of northwestern Scotland and of central Ireland. It shows a calcicolous tendency and grows in such habitats as rotten wood, tree bases and soil in woodlands. The species is widespread in Eurasia and North America. The earliest records are of Miocene age (Zatoka Gdowska, Poland, and Arjuzanx, France; table 9).

Brachythecium plumosum (Hedw.) B., S. & G.

Saint Cross (25); Ho; Candler 1889. *Pakefield* (25); Cr?; Blake 1890. *Skelsmergh Tarn* (69); Fl VI; t; Walker 1955. *Fort William* (97); Fl V or VI*; Dixon 1910 as *B. plumosum* B. & S.

Apart from much of southeast England and central Ireland, *B. plumosum* occurs widely in the British Isles where it often grows on boulders and tree roots by streams. There is a glacial record from Denmark (Hesselbo 1910). Southwards the species extends to Central America and New Zealand and in the Northern Hemisphere the range is wide.

Pseudoscleropodium purum (Limpr.) Fleisch.

!*Bunny* (56); Fl VIII*; Roman ar. *Bridgewater Street* (59); Fl VIII*; Roman ar; Roeder 1899 as *Hypnum purum*. !*Grisedale* (69); LDe II-III. !*Carpow* (88); Fl VIII*; Roman ar. *Canna* (104); Fl VII*a*; Flenley and Pearson 1967. *Gort* (H15); Ho II or III, III and III or IV; Jessen *et al.* 1959. !*Lissue* (H30); Fl VIII; Celtic ar; Mitchell 1951. !*Moville* (H34); Fl only; McMillan 1957.

This robust species, easily recognisable from habit, leaf shape and areolation, has been recorded from every vice-county in the British Isles where the variety of habitats includes heathland, woodland, fixed dunes and calcicolous grassland. In the last named it is often the most abundant bryophyte and is indifferent to aspect (Watson 1960). On broad north- and east-facing ledges of Helvellyn it grows at 2600 to 3000 ft (Ratcliffe 1960).

P. purum occurs throughout Europe, including Fennoscandia where the range is markedly southern and western (fig. 71). This anthropochorous species has reached many scattered parts of the world such as Washington, New Zealand and St Helena (Dickson 1967).

The subfossils are mostly Flandrian, three of them being Roman. The Grisedale record establishes Late Devensian occurrence in the British Isles. Several subfossils from the rest of Europe include a last interglacial one from Denmark, a glacial one from Germany and three from Bronze Age graves in Denmark. A record from the Miocene of Arjuzanx, France, is the earliest (Jovet-Ast and Huard 1966).

Cirriphyllum piliferum (Hedw.) Grout

Garvel Park (76); LDe/Fl IV* only; Robertson 1881 as *Eurhynchium piliferum*. !*Winetavern Street* (H21); Fl VIII*; Medieval ar.

C. piliferum, a robust distinctive species, is widespread in the British Isles. Often it occurs on basic soil in woodland. The distribution, wide in the Northern Hemisphere, in-

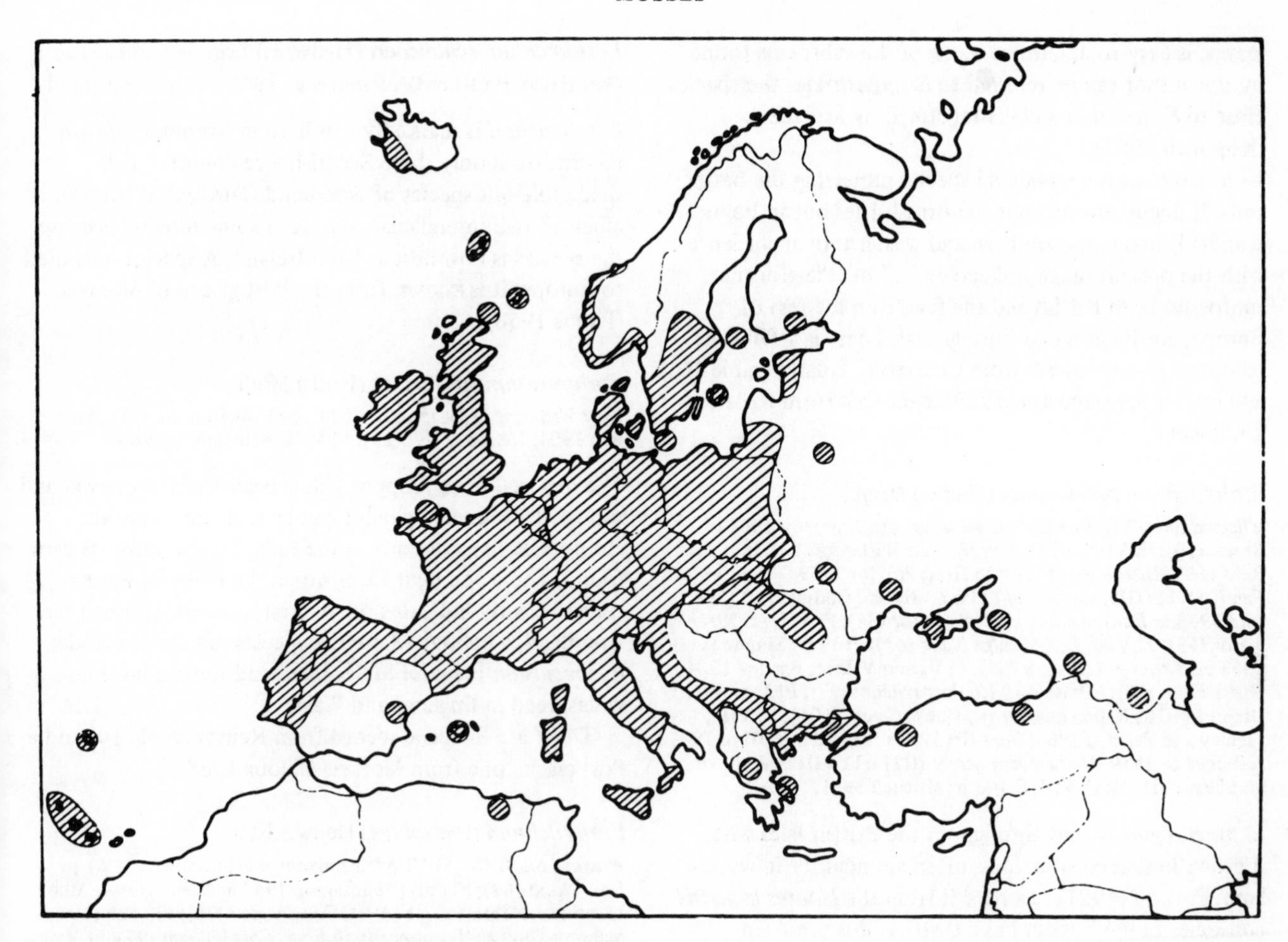

Fig. 71. European range of *Pseudoscleropodium purum*. From Störmer 1969.

cludes Fennoscandia where the range is southern. Both these records are unusual. *C. piliferum* seems a rather unlikely species to be found in a glacial deposit. One may wonder if Fergusson considered the possibility of *C. cirrosum* now known in Britain only high on Ben Lawers. The Irish material was extracted from a large mass of mosses thrown into a latrine in Medieval Dublin.

There are Miocene records of *C. piliferum* from Arjuzanx, France (Jovet-Ast and Huard 1966), and Poland. *C. cirrosum* is recorded from glacial deposits in Poland, Russia and Denmark.

Eurhynchium undetermined species

Hoxne (25); Ho III; Reid 1896. *Histon Road* (29); Ip IV; t; Proctor 1959*a*. *Woodwalton Fen* (31); Godwin and Clifford 1938. !*Blelham Tarn* (69); Fl VI-VII*a*. *Ehenside Tarn* (70); Fl VII*b* or VIII; Walker 1966. *Loch Fada* (104); LDe II; Birks 1969 pc. *Moville* (H34); Fl only; McMillan 1957. *Cushendun* (H39); Fl VI; Jessen 1949.

The Cushendun material is drawn and described in detail by Hesselbo (Jessen 1949, p. 138) who states that the specimens do not match any European species of *Eurhynchium*. Gams (1952, p. 153) suggests that *Brynhia novae-angliae* may be the species in question. This would be a most interesting discovery if correct, because *B. novae-angliae* occurs only in southern Norway and Sweden, and eastern North America, apart from eastern Asia (Störmer 1939). However, Gams did not examine the material which is now lost.

The very rare British *E. pulchellum*, which might well be relict from the Devensian, has an extensive Neogene history (tables 9 and 10). There is also a Polish last glacial record (Ralska-Jasiewiczowa 1966) and a Pleistocene one from Louisiana (Steere 1938).

Eurhynchiaum striatum (Hedw.) Schimp.

!*Winchester* (12); Fl VIII*; Saxon ar. !*Folkestone* (15); early Fl only. *Hoxne* (25); Ho III; Reid 1896. !*Wretton* (28); Ip II. !*Godmanchester* (31); Fl VIII*; Roman ar. *Brigg* (54); Fl VII*b*; Bronze Age ar; Sheppard 1910 as *Hypnum striatum* Schreb. !*Bunny* (56); Fl VIII*; Roman ar. *North Ferriby Boat 1* (61); Fl VII*b**; Bronze Age ar; Wright and Wright 1947. !*North Ferriby Boat 3* (61); Fl VII*b**; Bronze Age ar. *Fort William* (97); Fl V or VI*; Dixon 1910 as *E. striatum* B. & S. !*Winetavern Street* (H21); Fl VIII*; Medieval ar.

This robust species with acuminate, somewhat plicate

leaves, is easy to determine; none of the subfossils found by the author can be referred to *E. angustirete*, the species close to *E. striatum* and distinguished by leaf shape (Koponen 1967).

E. striatum is a woodland species preferring the better soils. It occurs throughout the British Isles but in Fennoscandia is markedly southern and western. In accordance with the present range and ecology all the Pleistocene subfossils, both British and the few from the rest of Europe, are Flandrian or interglacial. There is a Pliocene record of *E. angustirete* from Czorsztyn, Poland (table 10), and one of the same age of *E. meridionale* from Kroscienko, Poland.

Eurhynchium praelongum (Hedw.) Hobk.

Wilsford Shaft (8); Fl VII*b**; Bronze Age ar; Dimbleby 1967 pc. *Brook* (10); Fl VII*a*; Clifford 1936. *West Wittering* (14); Ip only; Reid 1892. *Histon Road* (29); Ip III; t; Proctor 1959*a*. *Woodwalton Fen* (31); Fl VII*b*; var. *stokesii* (Turn.) Hobk.; Godwin and Clifford 1938 as *Eurhynchium stokesii*. !*Aust* (34); Fl VII*a*; t. !*Bryn-y-Mor* (49); Fl VII*a*. *Lady Bridge Slack* (62); Fl VII*b*; Simmons 1968 pc. *Cromwell Bottom* (63); Fl VI and VII*b*; t; Bartley 1964. *Castle Eden* (66); EP; Reid 1920. *Fort William* (97); Fl V or VI*; Dixon 1910 as *E. praelongum* (L) Hobk. *Canna* (104); Fl VII*a*; Flenley and Pearson 1967. *Gort* (H15); Ho II or III and III or IV; Jessen *et al.* 1959. !*Winetavern Street* (H21); Fl VIII*; Medieval ar. !*Lissue* (H39); Fl VIII; Celtic ar; Mitchell 1951.

E. praelongum occurs throughout the British Isles where it grows in shaded situations, often abundantly in woodland. Pearsall (1921) recorded it from the *Isoetes lacustris* consocies in the English Lake District; this is a deep water community into which the moss must have been washed. It has been discovered in the hollows of block screes above 3000 ft on Ben Lawers.

All fourteen localities of subfossils are interglacial or Flandrian; this accords with the present ecology and range which, southern in Fennoscandia, includes Eurasia, North Africa and western North America. *E. praelongum* may well have benefited from the spread of deciduous forest in Boreal times.

Eurhynchium swartzii (Turn.) Carn.

Wilsford Shaft (8); Fl VII*b**; Bronze Age ar; Dimbleby 1967 pc. *Wolvercote* (23); Ho?; Bell 1904. *Pakefield* (27); Cr?; Blake 1890 as *E. swartzii* Turn. !*Wretton* (28); Ip II. *Shippea Hill* (29); Fl VII*b*; Clark *et al.* 1934; Godwin and Clifford 1938. !*Barnsley Park* (33); Fl VIII*; Roman ar. *Castle Eden* (66); EP; Reid 1920 as *E. swartzii* (Turn.) Hobk.

E. swartzii has a calcicolous tendency growing in such habitats as chalk grassland and deciduous woodland. The Barnsley Park material is large and finely preserved; it probably grew on the inside of the stonework of the well. It is known from almost throughout the British Isles.

Eurhynchium schleicheri (Hedw. f.) Lor.

Gort (H15); Ho III or IV; Jessen *et al.* 1959.

E. schleicheri is markedly southern in Britain; there are records from only three Scottish vice-counties. It is a shade-tolerant species of woodland. The record from Gort, albeit of two interglacials ago, is of some interest because the species is now absent from Ireland. A species restricted to Europe, it is known from the Post-glacial of Moravia (Pilous 1968).

Eurhynchium speciosum (Brid.) Milde

West Wittering (13); Ip only; Reid 1892. *Wolvercote* (23); Wo?; Bell 1904. !*Shippea Hill* (29); Fl VII*b*; with sporophytes.

The material from Shippea Hill is beautifully preserved and robust. The straggling habit can be seen and there are chloroplast-like contents in the cells. Twelve capsules were found. The wood peat from which the material was extracted clearly indicates the habitat, a moist base-rich fen woodland, a habitat where the species can flourish today. *E. speciosum* is sparse in Scotland and Ireland but more widespread in England and Wales.

There is a Pliocene record from Reuver (table 10) and a Post-glacial one from Moravia (Pilous 1968).

Eurhynchium riparioides (Hedw.) Rich.

Wilsford Shaft (8); Fl VII*b**; Bronze Age ar; Dimbleby 1967 pc. *Low Wood* (69); Fl only; Pennington 1943 as *E. rusciforme* Milde. !*Seathwaite Tarn* (69); Fl VIII. *Cowdenglen* (76); Fl? only; Mahony 1869 as *Hypnum ruscifolium*. *Fort William* (97); Fl V or VI*; Dixon 1910 as *E. rusciforme* (Neck) Milde. !*Loch Cuithir* (104); Fl IV; Vasari and Vasari 1968.

This species occurs throughout the British Isles. It grows attached to rock or wood in running water. Hence there is no surprise in recovering fragments from limnic deposits. The range, wide in the Northern Hemisphere, is limited northwards in Fennoscandia. There are few subfossils from elsewhere, two examples being Federsen Wierde, Germany (Iron Age; Körber-Grohne 1967), and Ripetta, Rome (Clerici 1892).

Eurhynchium confertum (Dicks.) Milde

Minnis Bay (15); Fl VII*b**; Bronze Age ar; t; Conolly 1941. *Brigg* (54); Fl VII*b**; Bronze Age ar; t; Sheppard 1910 as *Hypnum confertum* (Dicks.) Braithw.

The tentativeness of both these Flandrian records may be ascribed to lack of easy distinction in the subfossil state between this species and others of the genus and also possibly *Brachythecium velutinum*.

E. confertum, unrecorded from a few of the northernmost counties of Britain and four in central Ireland, is a shade-tolerant species of rocks and tree bases. The range includes Europe, Macaronesia, North Africa, the Caucasus and China. There is a tentative Miocene record from Arjuzanx, France (table 9).

Entodontaceae
Entodon concinnus (De Not.) Paris
!*Galley Hill* (31); Ip only.

The solitary fragment is large and well preserved. Something of the habit of the species can be seen as well as the characteristic areolation of the leaves. *E. concinnus* has a widespread but scattered distribution in the British Isles where it favours calcicolous grassland as it does throughout Europe. In Fennoscandia the range is markedly disjunct (map in Knusentsjerna 1945). This range includes Asia and North America. Subfossil occurrences include four of Devensian age in Poland (Szafran 1952).

Pleurozium schreberi (Brid.) Mitt.
Weymouth (9); Fl only; Gepp 1895 as *Hypnum schreberi. West Wittering* (13); Ip only; Reid 1892. *Farnham* (17); Wo?; Oakley 1939 as *Hypnum schreberi.* !*Holme Fen* (31); Fl VII or VIII. !*Nant Ffrancon* (49); LDe only. !*Holme Pierrepont* (56); Fl VIII*; Iron Age ar; Dickson and Ransom 1968. !*Chelford* (58); EDe. *Linton Mires* (63); Fl IV or V; Raistrick and Blackburn 1932 as *Hypnum schreberi* Willd. *Malham Tarn* (64); Fl VI; Pigott and Pigott 1963. *Hutton Henry* (66); Ip IV; Beaumont *et al.* 1969. !*Low Wray Bay* (69); LDe II. *Skelsmergh Tarn* (69); LDe I and Fl IV-V; Walker 1955. *Garvel Park* (76); LDe/Fl IV*; Robertson 1881 as *Hypnum schreberi.* !*New Dry Dock* (76); IV*. !*Drymen* (86); LDe II. !*Carpow* (88); Fl VIII; Roman ar. *Loch Einich* (96); Fl V or VI; t; H. H. Birks 1969 pc. *Moine Mohr A and B* (98); Fl VII and VIII; Chesters 1931. *Canna* (104); Fl VII*b*; Flenley and Pearson 1967. *Loch Fada* (104) LDe I, II and III; Birks 1969 pc. *Gort* (H15); LAn, Ho II, III and III or IV; Jessen *et al.* 1959 as *Hylocomium schreberi.*

A calcifuge species, *P. schreberi* occurs in every vice-county of the British Isles. It is abundant in a variety of woodlands and dwarf shrub heaths in the north and west of the country. The subfossils clearly indicate a widespread range since the Late Devensian period, if not earlier.

In accordance with the extensive range in the Northern Hemisphere and South America, and with the wide amplitude, there are numerous subfossils of glacial, interglacial and Flandrian age in the rest of Europe.

Plagiotheciaceae
Only three of the seventeen British species of this family have been identified from the British Pleistocene. More are to be expected in the subfossil state. Several species are restricted to the highland zone, notably *P. piliferum* which may well be a Devensian relict.

Isopterygium pulchellum (Hedw.) Jaeg.
!*Broome Heath* (27); Wo?; t; det. S. W. Greene.

I. pulchellum, restricted to the uplands of the British Isles where it often inhabits rock crevices, reaches the highest latitudes in the Northern Hemisphere where its range is extensive.

Isopterygium is a genus seldom found in the subfossil state. There is a Middle Pleistocene record of *I. pulchellum* from Russia (Abramova and Abramov 1962) and a post-glacial record of *I. depressum* from Moravia (Pilous 1968). *Taxiphyllum geophilum* is known from the Pleistocene of Louisiana (Steere 1938).

Plagiothecium undetermined species
Gort (H15); Ho II and III; Jessen *et al.* 1959.

Subfossils of this genus are sparse. There is an undetermined species from the Russian Pliocene (table 10) and the Polish Pleistocene has yielded *P. laetum, P. succulentum* and *P. sylvaticum. P. laetum* and *P. denticulatum* are known from the Moravian Post-glacial (Pilous 1968).

Plagiothecium denticulatum (Hedw.) B., S. & G.
Cromwell Bottom (63); Fl VI and VII*b* or VIII; Bartley 1964. *Burtree Lane* (66); Fl VI; Bellamy *et al.* 1966. *Garvel Park* (77); LDe/Fl IV*; Robertson 1881.

These records, particularly that from Garvel Park, are perhaps in need of revision in view of modern taxonomic advances.

Plagiothecium undulatum (Hedw.) B., S. & G.
Brigg (54); Fl VII*b**; Bronze Age ar; Sheppard 1910 as *P. undulatum* (L.) B. & S. *Cromwell Bottom* (63); Fl VII or VIII; Bartley 1964. *Fort William* (97); Fl V or VI*; Dixon 1910 as *P. undulatum* B. & S. *Loch Fada* (104); LDe II; Birks 1909 pc.

This calcifuge species has been recorded from almost throughout the British Isles where in the north and west it is common in woodland and in more open montane situations. It is easily recognised by its large size and undulate leaves.

In Fennoscandia the range is both western and southern (fig. 72). Apart from Europe *P. undulatum* occurs in Turkey and Pacific North America. Three of the subfossils are Flandrian in age. Much more interest attaches to the zone II record from Skye which establishes the Late Devensian presence of this species in western Britain. There are no subfossils known from outside the British Isles.

Hypnaceae
Hypnum undetermined species
Dogger Bank; Fl IV or V; Whitehead 1921. *Hawks Tor* (2); LDe II; Conolly *et al.* 1950. *Amesbury* (8); Fl VII*b**; Bronze Age ar; Newall 1931. !*Lowestoft* (25); An?; West and Wilson 1968. *Methwold Fen* (28); Fl VI; Godwin *et al.* 1934. *Skelsmergh Tarn* (69); LDe II; Walker 1955. *Low Wray Bay* (69); LDe I and II; Pennington 1962. *Whitrig Bog* (81); LDe III;

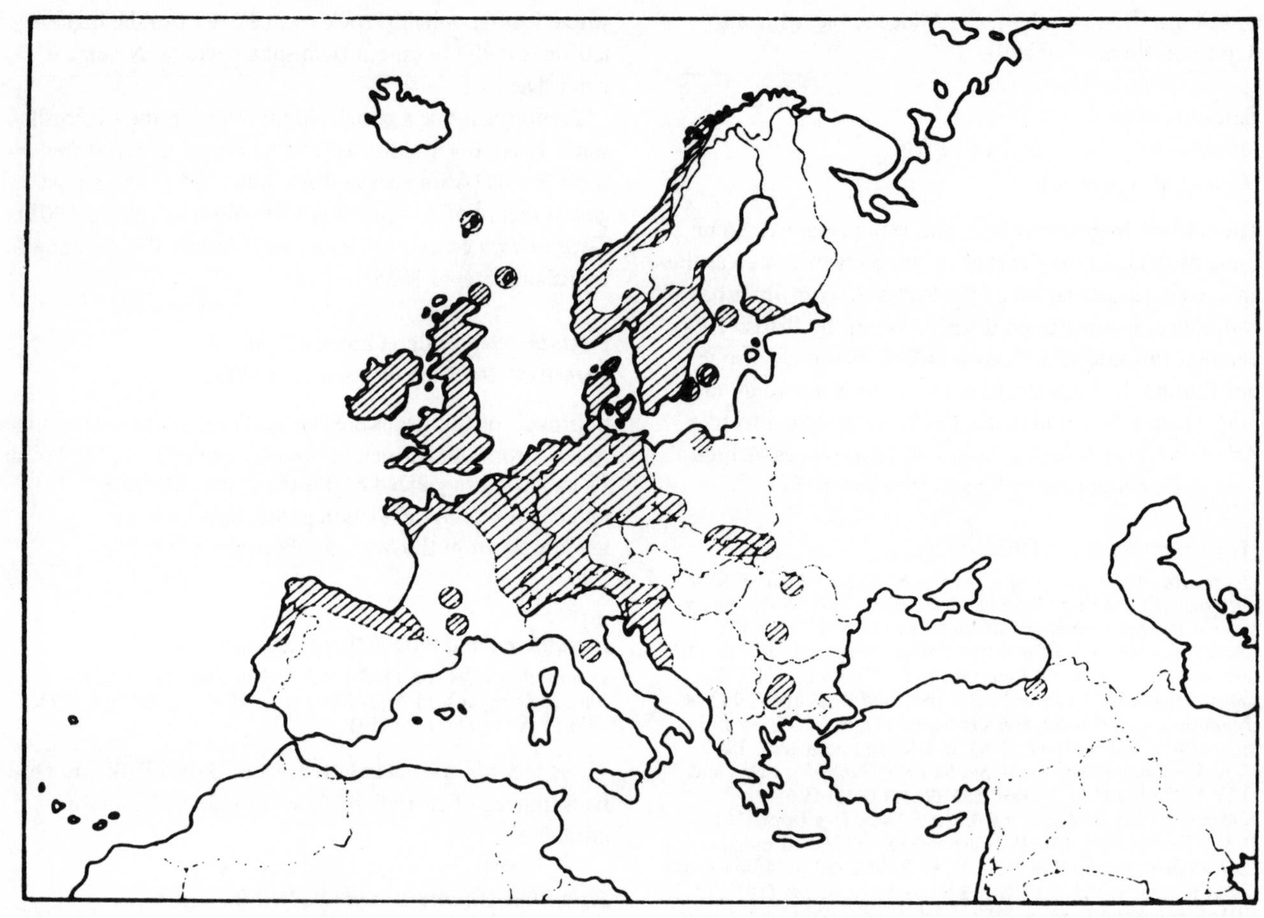

Fig. 72. European range of *Plagiothecium undulatum.* From Störmer 1969.

Mitchell 1948. *Garrall Hill* (94); LDe II; Donner 1957. !*Loch Cuithir* (104); LDe II; Vasari and Vasari 1968. *Clikhimin* (112); Fl VIII*; Iron Age ar; Hamilton 1968. *Canbo* (H25); Fl VIII; Jessen 1949. *Ardlcw Inn* (H30); Fl VII; Jessen 1949.

Almost all these records are used in the antiquated sense of Dixon (1924) who lumped *Calliergon, Campylium, Drepanocladus, Hygrohypnum, Hypnum* and *Scorpidium.* They are all but useless. Most probably they refer to species of *Calliergon* and *Drepanocladus*, though other species may be involved, even species outside the Hypnaceae.

Hypnum cupressiforme Hedw.

!*Meare* (6); Fl VII*b* or VIII. *Weymouth* (9); Fl only; Gepp 1895 as *H. cupressiforme* L. *West Wittering* (13); Ip only; Reid 1892. *Hitchin* (20); Ho only; Reid 1897. *Hoxne* (25); Ho III; t; Reid 1896 as *Stereodon cupressiforme* Brid. *Happisburgh* (27); Pa; t. !*Wretton* (28); Ip II. !*Soham Lode* (29); Fl VII*b*. !*Shippea Hill* (29); Fl VII*b*. !*Sidgwick Avenue* (29); EDe; Dickson 1963. *Trundle Mere* (31); Fl VII*b* and VIII; Vishnu-Mittre 1959. !*Aust* (34); Fl VII*a*. !*Bryn-y-Mor* (48); Fl VII*a*. *Yorkshire Pennines*; Fl VII*a* or VIII*; Burrel 1924. *Leeds* (64); Fl VI/VII/VIII*; Raistrick and Woodhead 1930. *Hutton Henry* (66); Ip IV; Beaumont *et al.* 1969. *Russland* (69); Fl only; Dickinson 1968 pc. !*Low Wray Bay* (69); LDe II; also var. *lacunosum*; t. !*Blelham Tarn* (69); LDe II. !*Kirkmichael 1 and 3a* (71); LDe I; Dickson *et al.* 1970. *Garvel Park* (76); LDe/Fl IV*; Robertson 1881. *Hailes* or *Corstorphine* (83); LDe only; Bennie 1894*a* as *H. cupressiforme* L. !*Carpow* (88); Fl VIII; Roman ar. *Fort William* (97); Fl V or VI*; Dixon 1910 as *H. cupressiforme* L. *Moine Mohr A and B* (98); Fl VII and VIII; Chesters 1931. *Canna* (104); Fl VII*b*; t; Flenley and Pearson 1967. !*Loch Cuithir* (104); Fl V and VI and VII*a*; Vasari and Vasari 1968. *Loch Meodal* (104); LDe II; Birks 1969 pc. !*Loch Droma* (105); LDe I; Kirk and Godwin 1963. !*Baggotstown* (H8); Ho III; Dickson 1964. *Gort* (H15); Ho II or III and III or IV; Jessen *et al.* 1959. !*Lissue* (H39); Fl VIII; Celtic ar; Mitchell 1951.

This cosmopolitan, highly polymorphic species is known throughout the British Isles on a great variety of substrata in both open and woodland situations. There are Hoxnian, Ipswichian and numerous Flandrian records as well as a wide scatter of Devensian records proving the extensive distribution relates to Late Devensian if not earlier times.

Hypnum revolutum (Mitt.) Lindb.

!*Drumurcher* (H32); LDe III.

Only a very small quantity of this species was recovered from the zone III layers of Drumurcher where the richest

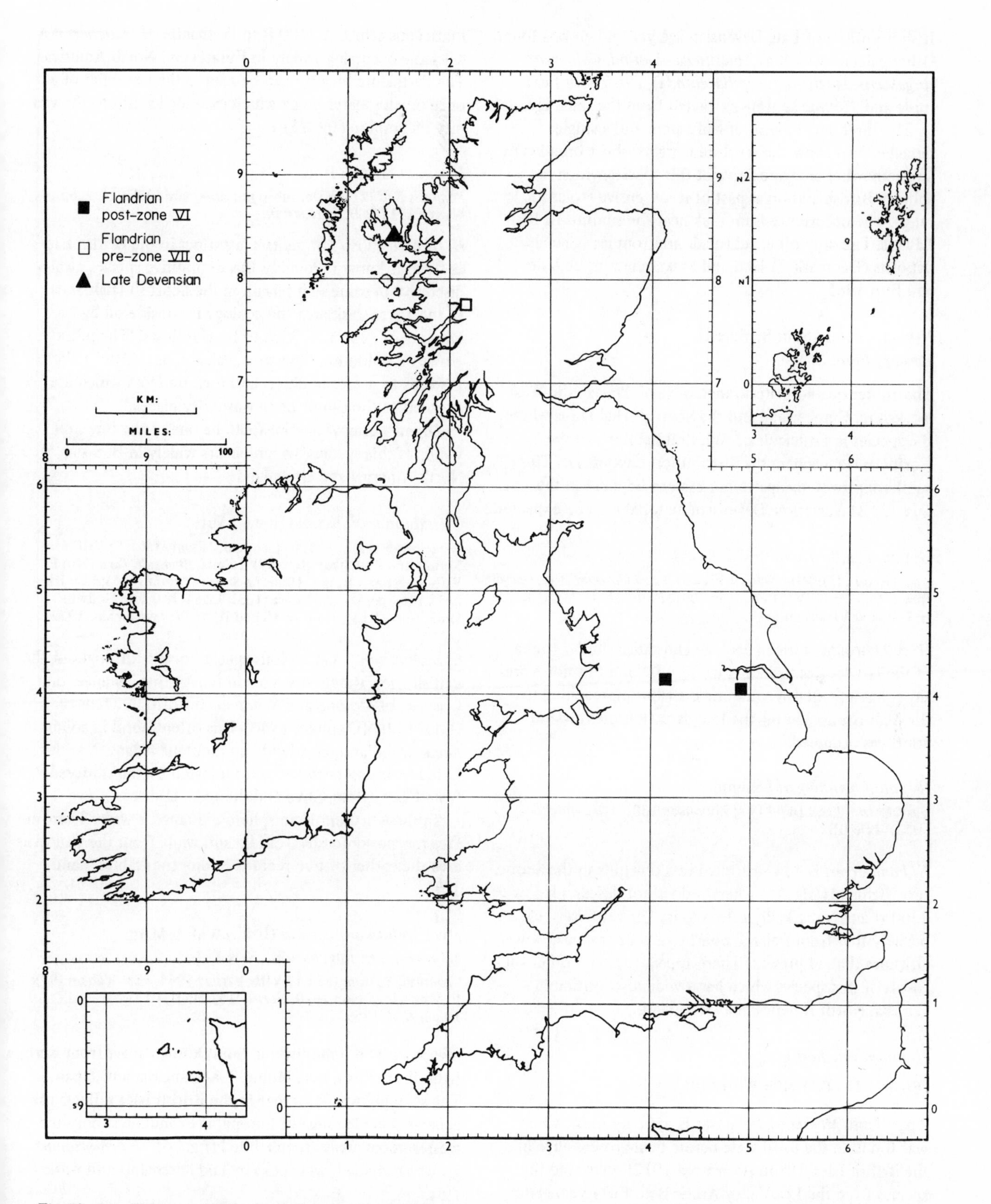

Fig. 73. Late Pleistocene subfossils of *Plagiothecium undulatum*.

Irish bryoflora of Late Devensian age yet known was found. Other calcicoles such as *Abietinella abietina, Calliergon turgescens, Distichium capillaceum* (t), *Ditrichum flexicaule* and *Timmia* sp were extracted from the same layers.

The three small pieces of leafy stem and a single detached leaf show the revolute margins, short broad cells and somewhat elongate nerve of this species which has a solitary British station as part of its extensive Holarctic range. The occurrence is on rock near the summit of Ben Lawers. The only other subfossils are from last glacial deposits (Denmark, Poland and Saskatchewan; de Vries and Bird 1965).

Hypnum hamulosum B., S. & G.

!*Broome Heath* (27); Wo.

The authenticated British range is North Wales, Mid-West Yorkshire, Cumberland and the Scottish Highlands where the species is a calcicole of rupestral habitats. In the Northern Hemisphere the distribution is extensive. The subfossil proves occupation of eastern England in the second last glaciation. Devensian material is to be expected.

Hypnum callichroum Brid.

Angel Road (21); MDe; Warren 1912 as *H. callichroum* Brid. *Garvel Park* (76); LDe Fl IV*; Robertson 1881. *Roundstone 2* (H16); LDe II, Fl V and VI; Jessen 1949.

H. callichroum, with an ecology and range similar to that of the last species including western Ireland but not Yorkshire, has an extensive Alarctic distribution reaching to the high Arctic. During the last glaciation the lowland zone was occupied.

Hypnum bambergeri Schimp.

Garvel Park (76); LDe/Fl IV*; Robertson 1881. !*Drumurcher* (H32); LDe III.

H. bambergeri is now restricted to a few hills in the Scottish Highlands (fig. 74) where it does not descend below 2300 ft and is markedly calcicolous. The subfossils, especially that from Ireland, point to a wider range at lower altitudes than at present. There appear to be no other subfossils of this species which has a wide distribution in Eurasia, North America and Greenland.

Hypnum vaucheri Lesq.

!*Ponder's End* (21); MDe; Warren 1912.

Apart from *Meesia tristicha* this is the only moss which was found in the fossil state before being found alive in the British Isles. Dixon (in Warren 1912) recovered the species from the Lea Valley Arctic Bed. Fifty years later Perry and Fitzgerald (1963) collected the plant from micaceous schist at 2400 ft in Perthshire. *H. vaucheri* is a calcicole occurring widely in Eurasia and North America. The Perthshire locality can be seen as the last relict of a once much greater range which extended south to the vicinity of London (fig. 75).

Hypnum ravaudi Boul.

Ponder's End (21); MDe; subsp *fastiatum* (Brid.) Wijk & Marg.; Warren 1912 as *H. fastigiatum.*

H. ravaudi subsp *fastigiatum* is extinct in the British Isles as far as is known. Possibly this diminutive species awaits discovery in some rich terrain in the Scottish Highlands. In the Torneträsk area the ecology is considered by Mårtensson (1956, p. 306) to be as follows. 'The moss grows on dolomite, limestone, calcareous schists or slate. It seems to prefer boulders and rock on faces which are exposed to the sun or lie in warm dry places.'

Dixon's discovery appears to be one of the few subfossils of this species, which occurs widely in Eurasia, North America and Greenland.

Ctenidium molluscum (Hedw.) Mitt.

!*Winchester* (12); Fl VIII*; Saxon ar. !*Bunny* (56); Fl VIII*; Roman ar. !*Low Wray Bay* (69); LDe II. *Ehenside Tarn* (70); Fl VII*b* or VIII; t; Walker 1966. *Loch Cill Chriosd* (104); LDe III; Birks 1969 pc. !*Loch Droma* (105); LDe I; Kirk and Godwin 1963. *Gort* (H15); Ho II or III and III or IV; Jessen *et al.* 1959.

C. molluscum, a calcicolous species, occurs throughout the British Isles. Rarely it occurs in fens as, for instance, on tussocks of *Molinia* and *Schoenus* in eutrophic fens of Oxfordshire (Clapham 1940). It is often found in soil in *Fagus* woodland, grassland and montane ledges. A wide Late Devensian occurrence is indicated by the subfossils from Easter Ross, Skye and the Lake District.

In the Northern Hemisphere the range is wide. There are Pleistocene subfossils from Poland while from the Caucasus area the earlier of two records is Miocene (tables 9 and 10).

Hyocomium armoricum (Brid.) Wijk & Marg.
[*Hyocomium flagellare* B., S. & G.]

Cromwell Bottom (63); Fl VII*b*; Bartley 1964. *Fort William* (97); Fl V or VI*; Dixon 1910. *Gort* (H15); Ho II, III and III or IV; Jessen *et al.* 1959.

Hyocomium is a monotypic genus known apart from Europe from the Azores, Asia Minor, Transcaucasia and Japan. The strongly western range in the British Isles reflects the general oceanic range in Europe; in Scandinavia only the westernmost areas are inhabited (fig. 76). *H. armoricum* grows on shaded, wet rocks by and in streams and waterfalls.

If the species invaded Britain in the Flandrian after

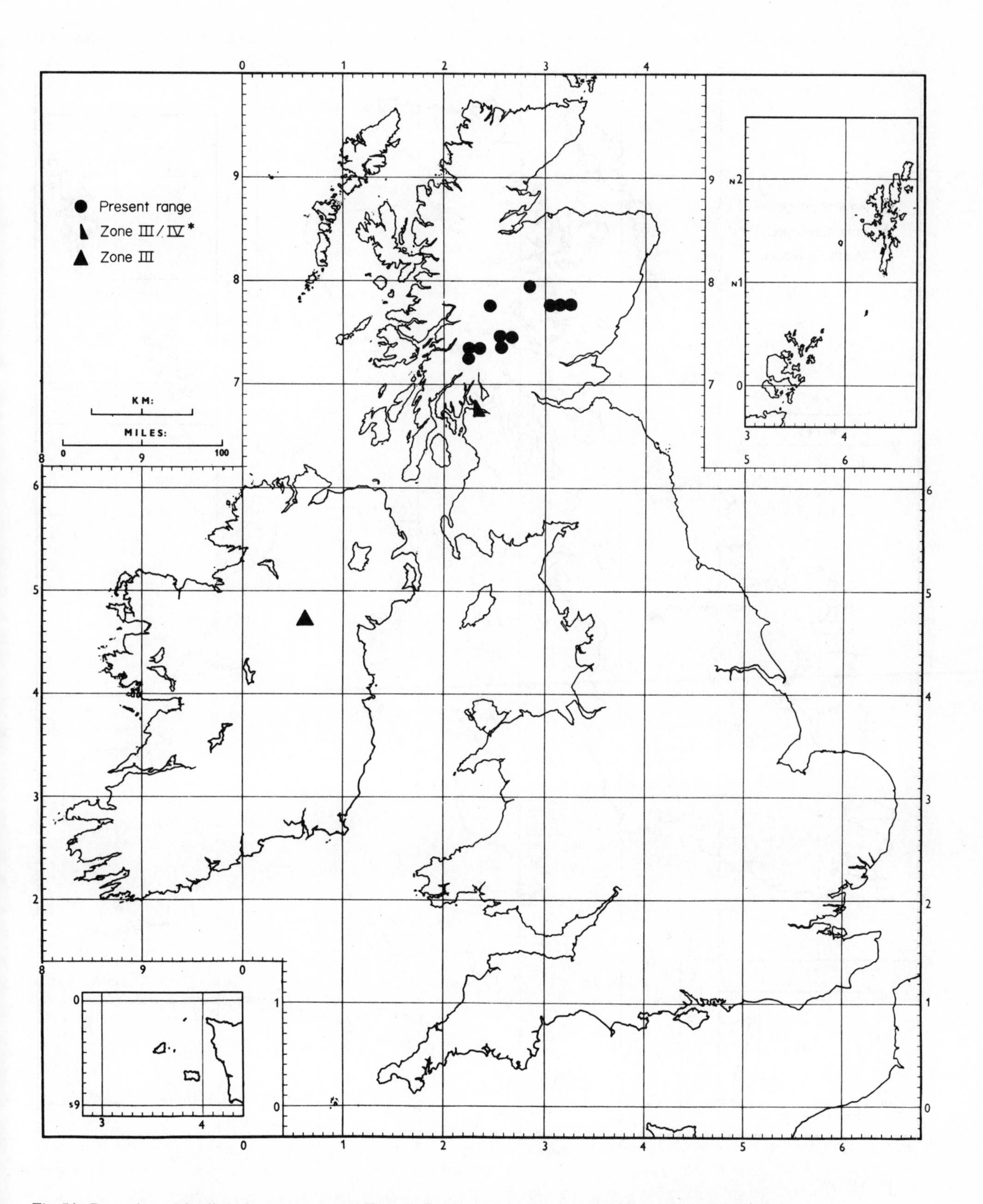

Fig. 74. Devensian subfossils and present range of *Hypnum bambergeri*. Partly after H. H. Birks, *Trans. Br. bryol. Soc.* 1969.

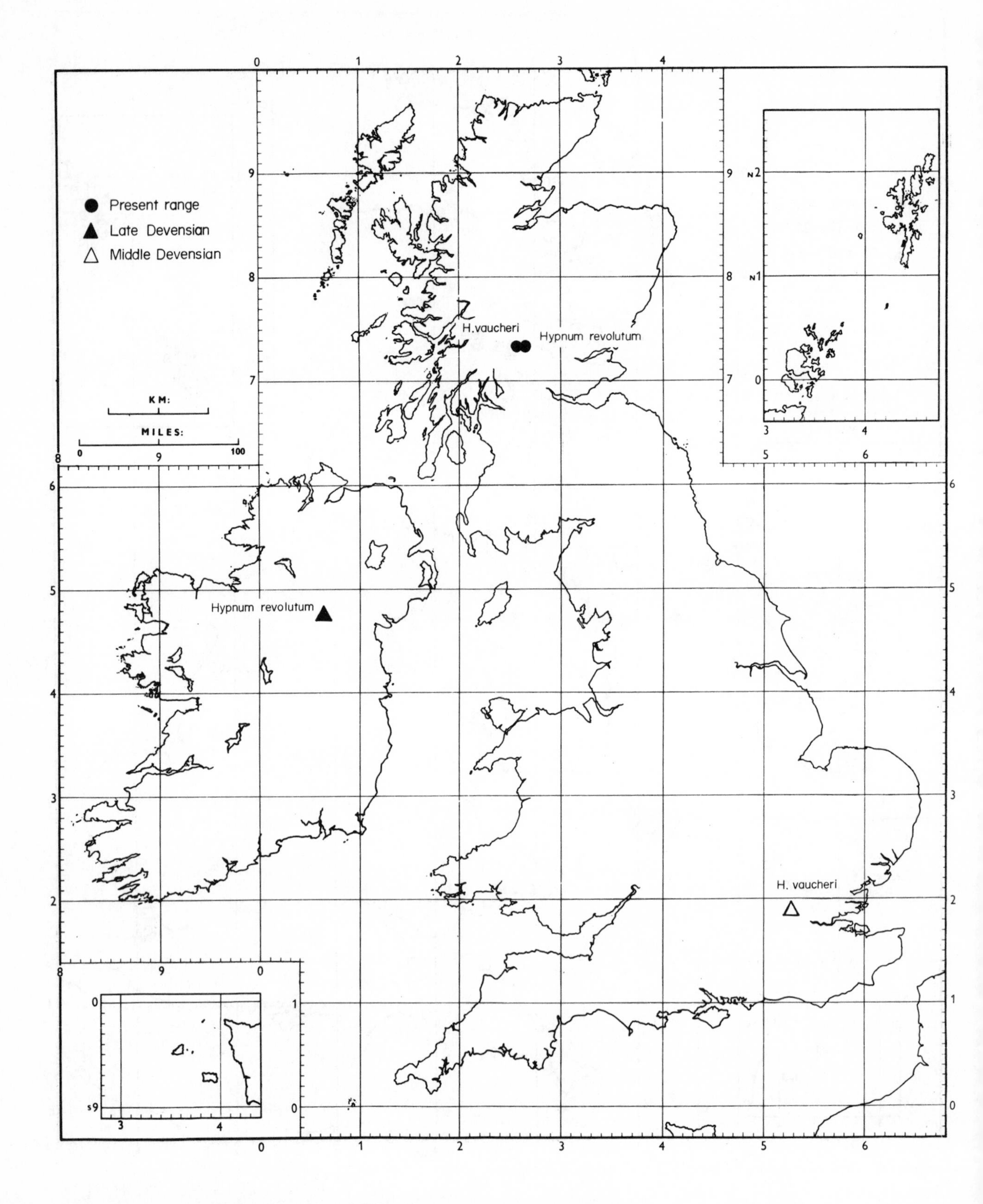

Fig. 75. Devensian subfossils and present range of *Hypnum revolutum* and *H. vaucheri*.

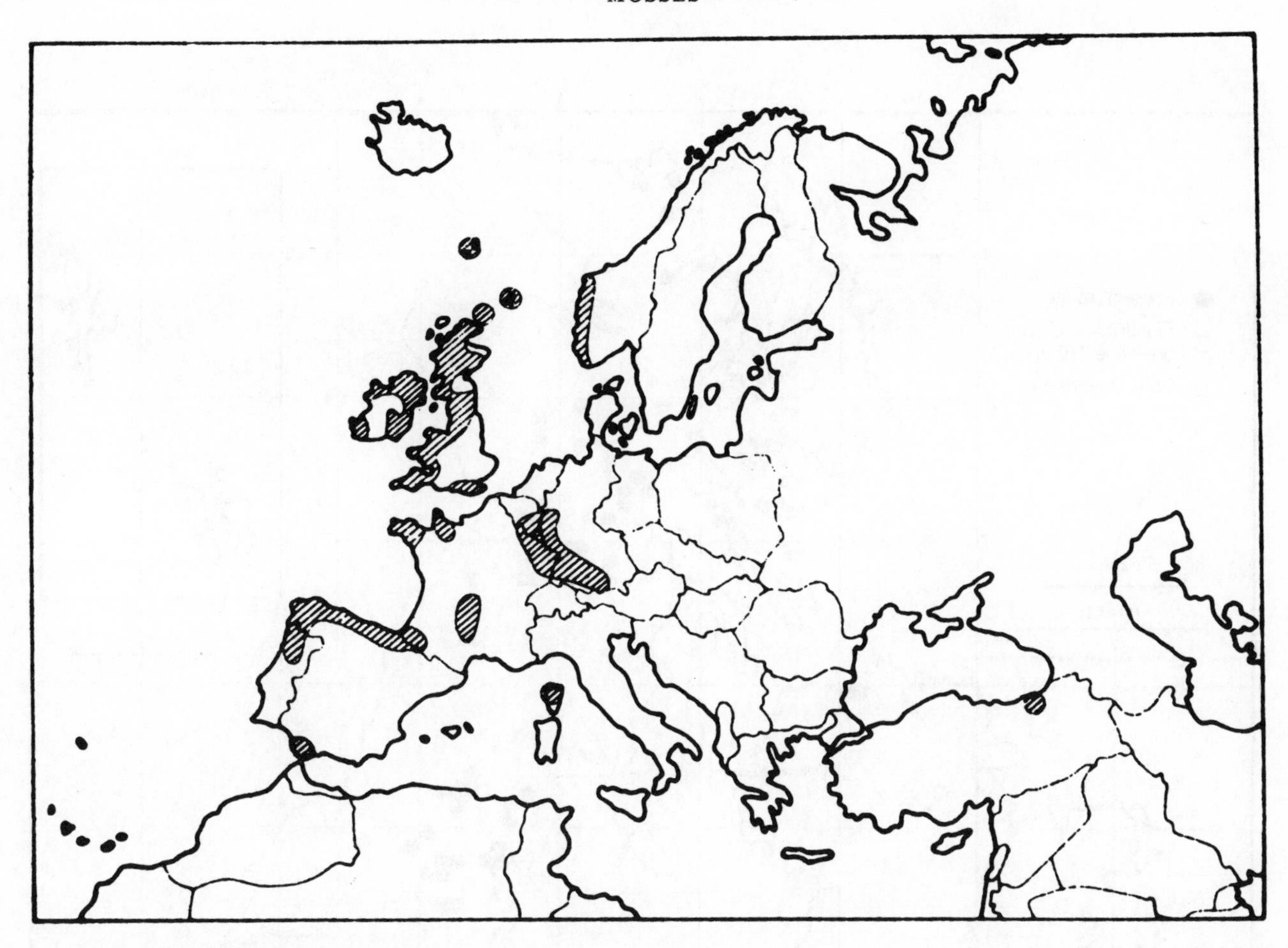

Fig. 76. European range of *Hyocomium armoricum.* From Störmer 1969.

absence in the Devensian, the Fort William record points to Boreal time for the arrival. Occupation of western Ireland for much of the Hoxnian (zones II and III and possibly IV) is proved by the Gort subfossils. The only other subfossil is of Flandrian age from Belle-Île-en-Mer, off the coast of Brittany (Gadeceau 1919).

Rhytidium rugosum (Sull.) Kindb.

!*Mochras* (48); De? only. !*Nant Ffrancon* (49); LDe only. !*Whitrig Bog* (81); LDe III. *Hailes* or *Corstorphine* (83); LDe only; Bennie 1894*a* as *Hypnum rugosum* Ehrh. !*Loch Fada* (104); Fl V; Vasari and Vasari 1968.

From the standpoint of phytogeography *Rhytidium rugosum* is one of the most interesting British plants. The scattered mostly montane localities in the highland zone combined with the lowland ones in the Breckland of East Anglia make a most unusual range. Parallels are hard to find; *Homalothecium nitens* and *Cinclidium stygium* have somewhat similar ranges but different ecologies. The great rarity in Ireland where *Rhytidium* is known only from sand dunes in Co. Derry adds to the chorological distinctiveness (fig. 77).

The determination of *Rhytidium* is an easy matter; the rugose leaves with projecting cell ends, strong single nerve, recurved margins and large groups of incrassate, porose angular cells are easily recognised when the subfossils are well preserved as are all seen by the author with one exception. The single fragment from Loch Fada is in poor condition without an intact leaf. However, the lower parts of the leaves suffice.

In the European literature *Rhytidium* is most often referred to as calcicolous. In Britain it is certainly very often on base-rich substrata such as calcareous sand, limestone and micaceous schist. It is primarily a species of dry grasslands. Indeed Herzog (1926) categorises it as a steppe species. He states (p. 175) '*Rhythidium* ist ein sehr charactenstischer Xerophyt von panboreal Verbreitung, mit Vorliebe für die Steppenformationen und Trockengehölze mittlerer Breiten, wo es zuweilen Massvegetation bildet.'

In light requirement it is heliophilous; in woodland habitats it is usually in very open situations as in pine woods on south-facing slopes of Gudbransdalen, Norway (Kleiven 1959).

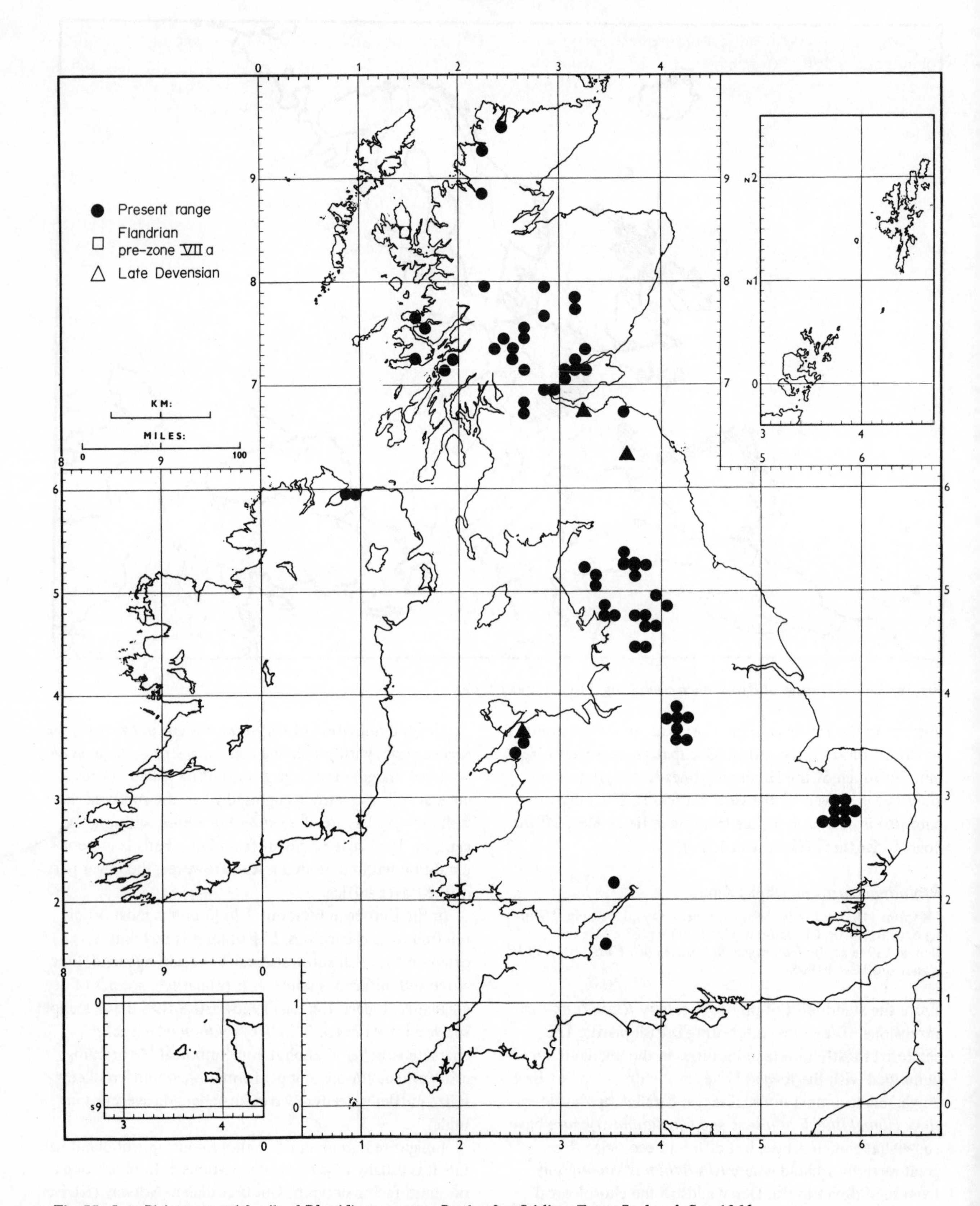

Fig. 77. Late Pleistocene subfossils of *Rhytidium rugosum*. Partly after Stirling, *Trans. Br. bryol. Soc.* 1966.

However, in marked contrast to the preference for calcicolous grasslands in Europe stands the occurrence in communities colonising the outwash of the Muldrow Glacier in Alaska. Here Viereck (1966) found *Rhytidium* at all stages of the succession from pioneer to low shrub-sedge tussock-moss tundra. *Rhytidium* reached its great frequency in the late shrub stage in which *Hylocomium splendens* and *Pleurozium schreberi* are the dominant mosses and the pH of the moss layer is 4.5.

The range is very extensive in Eurasia and North America, south to Morocco and Guatemala and north to Novaya Zemlya.

The distribution and ecology of *Rhytidium* in Fennoscandia have been the subject of much comment (Kotilainen 1929; Vaarama 1938; Albertson 1940; Tuomikoski 1946; Hallberg 1959). Albertson's map, now out-dated, shows the principal features: scattered in the mountains, rare in the south (Oslofjord, southern Sweden including Oland but not Gotland) and very rare in Finland. Hallberg's discussion of the occurrences in Bohuslän (west coast of southern Sweden) is interesting because the idea is advanced that invasion by spores, travelling from southern Norway where most records of sporophytes have been made in Fennoscandia, led to establishment.

Rhytidium is sterile in Britain as it is over much of its range, including most of Europe. Nyholm (1965) describes the Fennoscandian occurrence of sporophytes as rare. There have been numerous gatherings of sporophytes in the mountains of southeastern Siberia (Savicz 1928) where in the Lake Baikal region capsules occur on plants from all vegetational zones (Bardunov 1965).

The Late Devensian subfossils speak for an extensive range in the highland zone of Britain and, moreover, a range less discontinuous than now exists. But the major disjunction concerns the Breckland localities. It is hard to believe that they are not relict. If they too pertain to a more continuous glacial range one must envisage that the Breckland supported at least locally very open forest or even treeless areas throughout the Flandrian.

The Loch Fada record is doubly interesting. It is Flandrian in age (zone V) and proves past occurrence on Skye where the species has not been found in the living state. The disappearance is enigmatic; the basaltic ridge of Trotternish would seem to provide suitable habitats.

Subfossils have been recovered from glacial deposits in Russia, Poland, Sweden and Denmark. There is a Middle Pleistocene record and two Pliocene ones from Russia (table 10).

Herzog (1926, pp. 175-6) believed that the distribution could only be explained in one of two ways: either (*a*) a greater reproductive capacity in the past, or (*b*) a step-by-step spread in the Holarctic when Eurasia and North America were still united. Or, one might argue, a combination of both hypotheses, neither of which is likely ever to be proved; the Pliocene records are consistent with (*b*), but spores which, according to Nyholm (1965), are finely papillose are most unlikely ever to be recognised by palynologists.

In apparent ignorance of the Tertiary subfossils which might be taken to support his case, Pospisil (1968) believes that *Rhytidium* and also *Entodon concinnus, Camptothecium lutescens* and *Thuidium abietinum* are preglacial relicts in Czechoslovakia.

Rhytidiadelphus triquetrus (Hedw.) Warnst.

!*Happisburgh* (27); Ba or earlier. ***Woodwalton Fen*** (31); Fl VII*b*; Godwin and Clifford 1938 as *Hylocomium triquetrum*. ***Brigg*** (54); Fl VII*b*; Bronze Age ar; Atkinson 1887 as *Hypnum triquetrum*. *Castle Eden* (66); EP; Reid 1920 as *Hylocomium triquetrum* Linn. !*Low Water* (69); Fl only. *Garvel Park* (76); LDe/Fl IV*; Robertson 1881 as *Hylocomium triquetrum*. !*Knocknacran* (H32); Fl V.

In the living state the three British species of *Rhytidiadelphus* are very different in habit. Seldom is this revealed by subfossils. However, the species also differ in leaf structure. *R. triquetrus* is particularly distinct in both the shape and spinulose papillae of the leaves. *R. loreus* and *R. squarrosus* can be separated on the basal cells of the leaves.

R. triquetrus occurs throughout the British Isles in a variety of habitats including woodland, grassland and dunes. In the Scandinavian mountains where the species has a marked calcicolous tendency meadow birch forests and *Dryas* heath may be inhabited. The range is widespread in the Northern Hemisphere. The predominance of the subfossils from Flandrian and interglacial deposits is in agreement with the present mainly temperate range and woodland ecology.

Rhytidiadelphus squarrosus (Hedw.) Warnst.

Hoxne (25); Ho III or EWo; Reid 1896 as *Hylocomium squarrosum* Bruch & Schimp. !*Little Paxton* (31); Fl VIII*; Saxon ar. !*Holme Pierrepont* (56); Fl VIII*; Iron Age ar; Dickson and Ransom 1968. !*Chelford* (58); EDe. *Bridgewater Street* (59); Fl VIII*; Roman ar; Roeder 1899 as *Hylocomium squarrosum*. *Yorkshire Pennines*; Fl VII/VIII*; Burrell 1924. *Halton* (67); Fl VIII*; Roman ar; Simpson and Richmond 1937 as *Hylocomium squarrosum* B. & S. !*Seathwaite Tarn* (69); Fl VIII. !*Kirkmichael 1* (71); LDe I; Dickson *et al.* 1970. *Garvel Park* (76); LDe/Fl IV*; Robertson 1881 as *Hylocomium squarrosum*. !*New Dry Dock* (76); LDe only. !*Carpow* (88); Fl VIII; Roman ar. !*Garral Hill* (94); LDe II. *Fort William* (97); Fl V or VI*; Dixon 1910 as *Hylocomium squarrosum* B. & S. *Moine Mohr A and B* (98); Fl VII or VIII; Chesters 1931 as *Hylocomium squarrosum*. *Loch Fada* (104); LDe I; Birks 1969 pc. !*Winetavern Street* and *High Street* (H21); Medieval ar. !*Moville* (H34); Fl only; McMillan 1957. !*Lissue* (H39); Fl VIII; Celtic ar; Mitchell 1951.

R. squarrosus occurs throughout the British Isles where it most often occurs in anthropogenic grasslands such as lawns and *Agrostis-Festuca* communities of the lower slopes of mountains. The bulk of the subfossils are Flandrian or interglacial. However, the records from Chelford, the Greenock deposits, the Isle of Man, Garral Hill and Loch Fada point to extensive Devensian occurrence. *R. squarrosus* reaches southern Spitsbergen; it extends throughout much of the Northern Hemisphere.

Rhytidiadelphus loreus (Hedw.) Warnst.

Yorkshire Pennines; Fl VII or VIII*; Burrell 1924. !*Seathwaite Tarn* (69); Fl VIII. *Fort William* (97); Fl V or VI*; Dixon 1910 as *Hylocomium loreum* B. & S. *Moine Mohr B* (98); Fl VII and VII/VIII; Chesters 1931 as *Hylocomium loreum*. !*Loch Cuithir* (104); Fl V, VI, VII*b*; Vasari and Vasari 1968. !*Loch Fada* (104); Fl VIII; Vasari and Vasari 1968. *Clikhimin* (112); Fl VIII*; Iron Age ar; Hamilton 1968 as *Hylocomium loreum*. !*Moville* (H34); Fl only; McMillan 1957.

R. loreus, unrecorded from a few, mainly eastern vice-counties, has a markedly western range in the British Isles. A calcifuge, it is common in woodland in the highland zone and ascends above the tree-line, growing on ledges and in late snow-bed vegetation. In Fennoscandia there is a markedly southern and western tendency (fig. 78); *R. loreus* occurs in Europe, Macaronesia and North America. The subfossils are entirely Flandrian, mostly from the Atlantic or later periods. However, the Loch Cuithir and Fort William records prove Boreal occurrence in western Scotland.

Hylocomium undetermined species

Aintree (59); Fl only; Travis 1909. *Low Wray Bay* (69); LDe II; Pennington 1962. *Cushendun* (H39); Fl VI; Jessen 1949.

H. splendens is probably the species represented in all three cases.

Hylocomium brevirostre (Brid.) B., S. & G.

Hitchin (20); Ho only; t; Reid 1897. !*Russland Moss* (69); Fl V and VI; Dickinson 1968 pc. *Garvel Park* (76); LDe/Fl IV*; Robertson 1881. *Fort William* (97); Fl V or VI*; Dixon 1910 as *H. brevirostre* B. & G. *River Bann*; Fl IV-VII*a**; Mesolithic ar; Knowles 1912.

H. brevirostre is frequent only in the north and west of the British Isles where it often grows on boulders and also tree

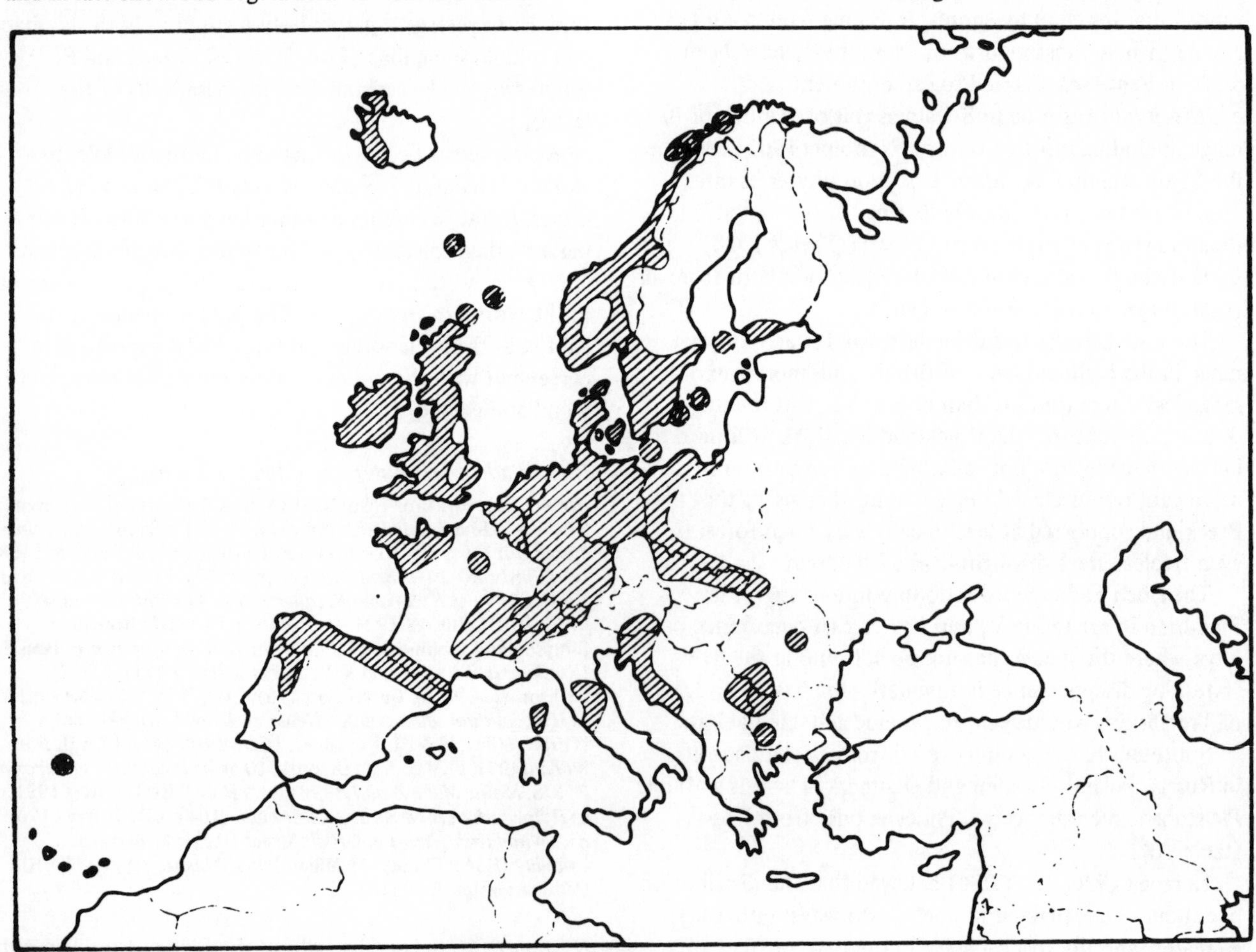

Fig. 78. European range of *Rhytidiadelphus loreus*. From Störmer 1969.

bases in woodland. It is classed as calcicolous by McVean and Ratcliffe (1962). In Fennoscandia the range is southern and Lye (1967) considers the Norwegian occurrence as hyperoceanic. Outside Europe (fig. 79) the species is known from North Africa, eastern Asia and North America. There are Flandrian records from Switzerland (Neuweiler 1905) and Belgium (Stockmans and Vanhoorne 1954).

The Mesolithic record from the River Bann is a unique discovery. A large mass of the moss was found in association with a flint flake in such a position that it is thought to have been a protection for the user's hand (plate 21).

Hylocomium umbratum (Brid.) B., S. & G.

!*Seathwaite Tarn* (69); Fl VIII.

The single fragment is known by the slender tapering habit as well as leaf shape and serration. *H. umbratum* is restricted to the north and west of the British Isles; it is rare in Ireland. The characteristic habitat is moist rocks in montane woodland. Above the tree line habitats are in sheltered positions such as block scree above 3000 ft on Ben Lawers (Watson 1925). The range extends eastwards to the Urals and also includes North Africa, eastern Asia and North America.

Hylocomium splendens (Hedw.) B., S. & G.

!*Honiton* (3); Ip. !*Lowham* (6); Fl VIII*; Roman ar. *Wilsford Shaft* (8); Bronze Age ar; Dimbleby 1967 pc. *Verulamium* (20); Fl VIII*; Roman ar; Wallace 1956. *Elan Valley Bog* (46); Fl VII*a*; VII*b*; Moore and Chater 1969. !*Nant Ffrancon* (49); LDe only. *Brigg* (54); Fl VII*b**; Bronze Age ar; t; Sheppard 1910 as ?*Hylocomium proliferum* (L.) Lindb. *Holme Pierrepont* (56); Fl VIII*; Iron Age ar; Dickson and Ransom 1968. *Bridgewater Street* (59); Fl VIII*; Roman ar; Roeder 1899; Burrell 1924. !*Burbage Brook* (63); LDe II. !*Widdybank Fell* (66); Fl VII; Hewetson 1967 pc. *Moor House* (69); Fl VII and/or VIII; Johnson 1960 pc. !*Foolmire Sike* (69); Fl VII*a* and VII*b*. !*Russland Moss* (69); Fl VI and VII; Dickinson 1968 pc. !*Low Wray Bay* (69); LDe II. !*Goat's Water* (69); LDe; Pennington 1964. !*Seathwaite Tarn* (69); Fl VIII; Pennington 1964. *Low Wood* (69); Fl only; Pennington 1943. !*Kirkmichael 1 and 3a* (71); LDe I and III; Dickson *et al.* 1970. *Lochlee Crannog* (75); Fl VIII*; Iron Age? ar; Munro 1879. *Cowdenglen* (76); Fl? only; Mahony 1869 as *Hypnum splendens*. *Garvel Park* (76); LDe/Fl IV*; Robertson 1881. !*New Dry Dock* (76); LDe only. !*Methilhill* (85); Fl VII*b**; Bronze Age ar. !*Carpow* (88); Fl VIII Roman ar. !*Dalreach* (89); Fl only. !*Loch Cuithir* (104); LDe III-IV; Fl VII*a*; Vasari and Vasari 1968. !*Loch Fada* (104); LDe III-IV; Fl V and VIII; Vasari and Vasari 1968;

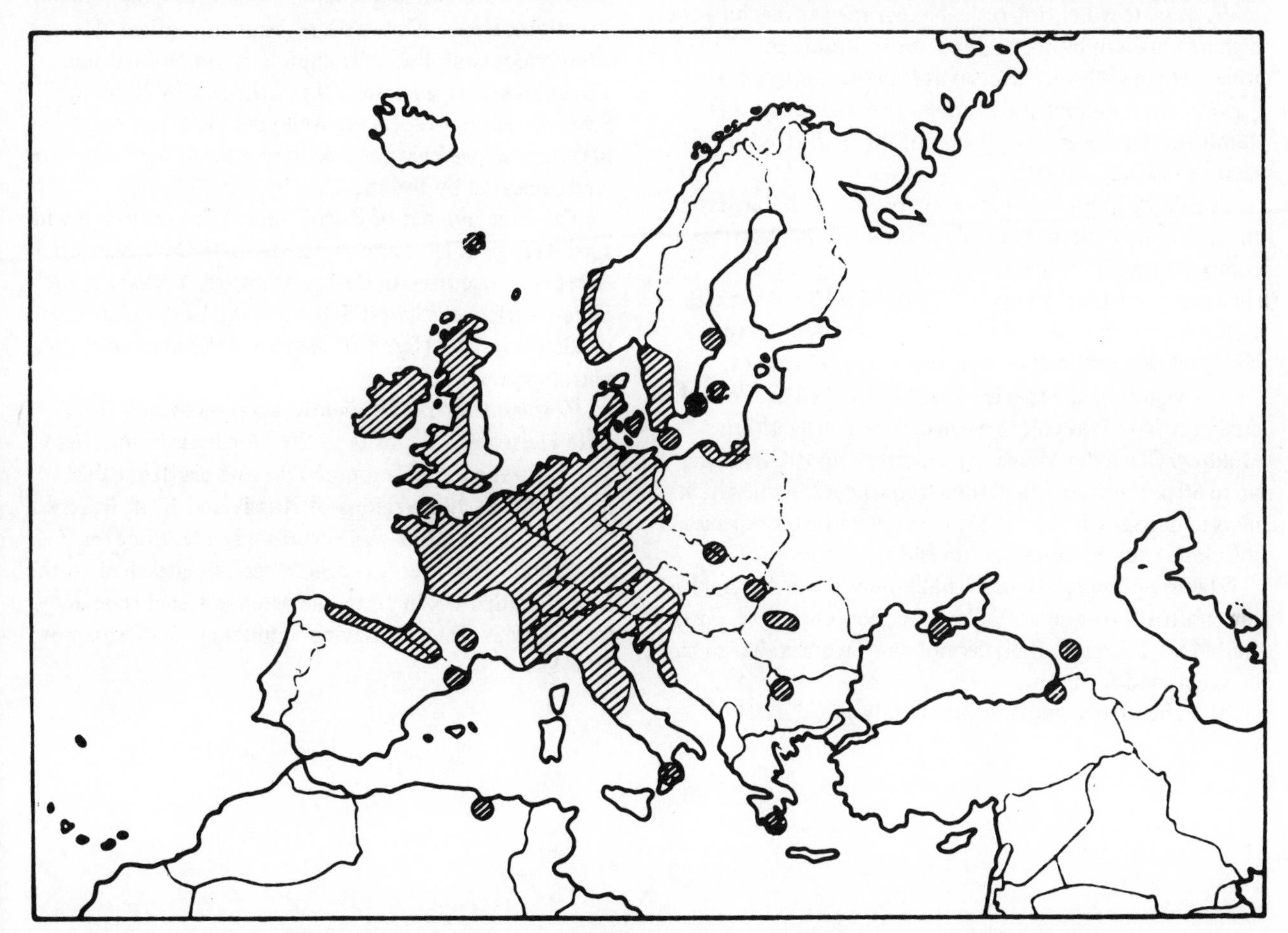

Fig. 79. European range of *Hylocomium brevirostre*. From Störmer 1969.

LDe I and II; Birks 1969 pc. *Loch Meodal* (104); LDe I and II; Birks 1969 pc. *Loch Mealt* (104); LDe I, II, III and Fl IV; Birks 1969 pc. *Loch Cill Chriosd* (104); LDe I and III; Birks 1969 pc. *Lochan Coir 'a' Ghobhainn* (104); LDe II and III; Birks 1969 pc. *Loch Droma* (105); LDe I; Kirk and Godwin 1963. !*Loch Maree* (105); Fl V or VI and VII*b* or VIII; H. H. Birks 1969 pc. !*Fugla Ness* (112); Ho F.2; Birks and Ransom 1969. !*Ballydermot Bog* (H19); Fl VIII. !*Treanscrabbagh* (H28); Fl VII*a*. !*Newtonbabe* (H31); LDe II. !*Moville* (H34); Fl only; McMillan 1957.

H. splendens is among the most commonly encountered species in British Late Pleistocene deposits. Its absence as yet from pre-Ipswichian strata probably has no significance apart from paucity of investigation. Frequent fossilisation and recognition have been assured by the wide ecological amplitude and range of this robust species.

Some species from Loch Droma and the Lake District sediments are several centimetres long, even showing the characteristic arched branching. Detached branches and leaves are often recovered and readily determinable. British glacial fossils of sufficient size and preservation to allow judgment are not referable to the high latitude taxon *H. alaskanum*, variously given specific or varietal rank or dismissed as an environmental modification.

H. splendens is a Holarctic, bipolar species reaching the highest northern latitudes. It occurs profusely in coniferous forest where with such species as *Dicranum polysetum, Pleurozium schreberi* and *Ptilium crista-castrensis* it is a dominant. However, few, if any, of the British subfossils relate to such a habitat.

Subalpine birch forest is another highly suitable habitat. In less luxuriant state but still common, *H. splendens* is a component of low-alpine *Dryas* heath rich in mosses (Mårtensson 1956). Many of the glacial subfossils might well relate to this type of vegetation; Low Wray Bay, Kirkmichael and Loch Droma are obvious examples.

The vigour of the species is well illustrated by Viereck's study (1966) of the colonisation of gravel outwash from the Muldrow Glacier in Alaska; it is an early colonist, reaching up to 50 per cent cover and total frequency. It forms thick mats under *Salix* bushes and remains an important species in the climax tundra of low shrub and sedge tussock.

The dry growing stages of mires are yet another habitat; the subfossils from Elan Valley Bog, Foolmire Sike, Russland Moss, Dalreach, Ballydermot and Treanscrabbagh can be accounted for in this way.

The abundance in many communities of the British Isles is revealed by McVean and Ratcliffe's photosociological analyses (1962); such diverse *noda* as *Pinetum Hylocomieto-Vaccinetum, Salix lapponum-Luzula sylvatica* and *Nardetum subalpinum* serve as examples. In the lowland zone the range of habitats includes calcicolous grassland and *Fagus* woodland.

No less than nine archaeological sites have yielded remains of *H. splendens*; four from Roman times, three from the Bronze Age and two possibly from the Iron Age. In some cases, for instance Carpow where a moss assemblage of some twenty taxa was extracted from the fort ditch, there is no need to suppose anything but local growth. However, in other cases the speculation is tempting that the occurrences represent fragments of man's rubbish. The species is both common and robust and so it is at once readily available and suitable as packing and stuffing material. The Brigg and Holme Pierrepont subfossils formed part of the caulking of monoxylous canoes. Remains of *H. splendens* have been recovered from three Bronze Age burials in Denmark (Iversen 1939).

Fruiting is rather scarce in *H. splendens* and perhaps most often seen on large plants growing in moist, sheltered situations. Hence it is worth reporting that the subfossils from Widdybank Fell bear abundant gametangial buds. The subfossils from zone VII*a* of the nearby Foolmire Sike, though not especially well preserved, represent luxuriant specimens; perhaps they grew in the mire woodland indicated by the stratigraphy.

The large number of British subfossils confirms a widespread range at least as early as the Late Devensian when the species occurred in Caernarvonshire, Yorkshire, the Lake District, northwest Scotland, the Isle of Man and northeast Ireland (fig. 80); such an assessment conforms with the present ecology.

H. splendens has frequently been recovered from Late Pleistocene deposits in the Northern Hemisphere. The earliest subfossils are of Pliocene age from the Kama and Bashkir regions of Russia and from Poland (table 10); the latter was surprisingly identified as *H. alaskanum*. However, no doubt need be attached to the record of this taxon from the sediment enclosing a mammoth in Taimyr (Savicz-Ljubitzkaja and Abramova 1954).

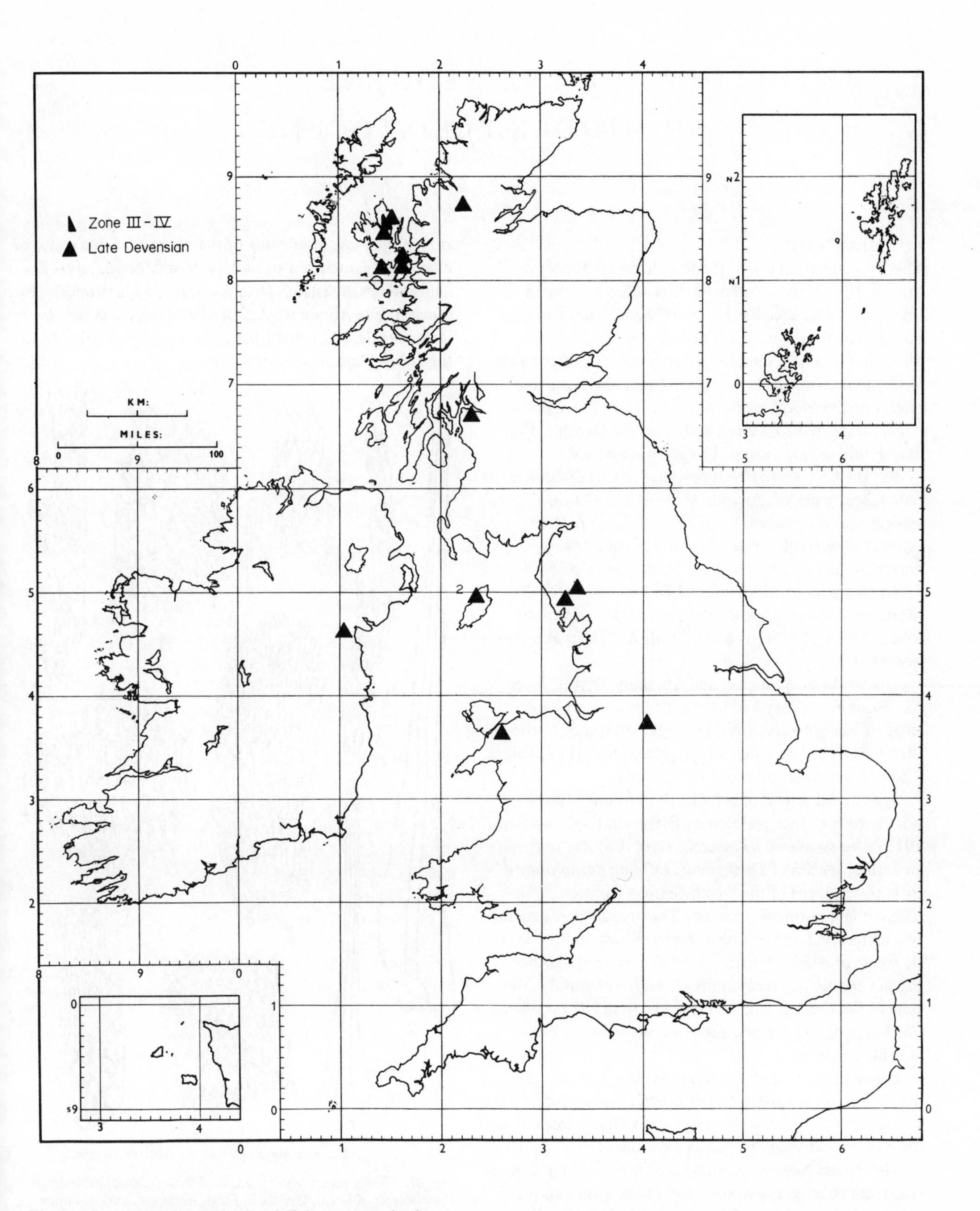

Fig. 80. Late Devensian subfossils of *Hylocomium splendens.*

6 BRYOFLORAS OF THE NEOGENE, EARLY AND MIDDLE PLEISTOCENE

6.1 NEOGENE

The Neogene bryoflora of the British Isles is almost unknown. Nevertheless the sparse discoveries are of considerable bryogeographical interest. Apart from spores of *Sphagnum* species, the plant beds formed at about the Miocene-Pliocene transition in Derbyshire (Boulter 1971) have yielded a single moss fragment probably referable to *Hypnodendron*, a genus of some thirty-eight species largely restricted to southeastern Asia and Oceania. Full description of this fossil will be given elsewhere.

For the bryogeographical purposes of this discussion only recent work on Eurasian Miocene and Pliocene assemblages is considered. Earlier work is ignored; the value of nineteenth-century studies is hard to gauge, especially that on bryophytes in amber. See Czeczott (1961) for a summarising list and Dixon (1927) for dismissal or necessarily vague re-identifications of various fossils. Steere (1942) has listed North American Miocene bryophytes.

The works of Abramova and Abramov, Dixon, Jovet-Ast, Szafran and Weyland have added greatly to knowledge of Neogene bryophytes and give substance to the theories of Tertiary relicts in the present bryoflora. Tables 9 and 10 are compilations of their results.

Spectacular alterations of distribution since the Neogene were first demonstrated by Dixon (in Reid and Reid 1915) who recovered a *Pinnatella* (sect. *Urocladium*) and an extinct species of *Trachycystis* (*Mnium antiquorum*) from the Pliocene of the Dutch-German frontier. *Pinnatella* is a large tropical genus and *Trachycystis* has a mere two living species from China, Japan, Korea and the Pacific fringe of Alaska. Szafran (1949*a*) described another extinct species of *Trachycystis* (*T. szaferi*) from the Miocene of Southern Poland and later recovered *T. flagellaris*, one of the extant species, also from the Miocene of Southern Poland.

Hence in the case of *Trachycystis* there has been a restriction to the eastern end of Eurasia (apart from Alaska) whereas in the Miocene the genus was well represented at the western extremity of this vast continent.

The richest Neogene assemblages are those from Russia: that from Duab at the western end of the Caucasus is unsurpassed (Abramova and Abramov 1959). *Echinodium*, among the most interesting of the discoveries, is a genus of nine species restricted to southeastern Australia, New Zealand and nearby islands (five species) and Macaronesia (four species). The Abramovs described an extinct species, *E. savicziae* (fig. 81), which was subsequently recorded from the Polish Miocene (Szafran 1964).

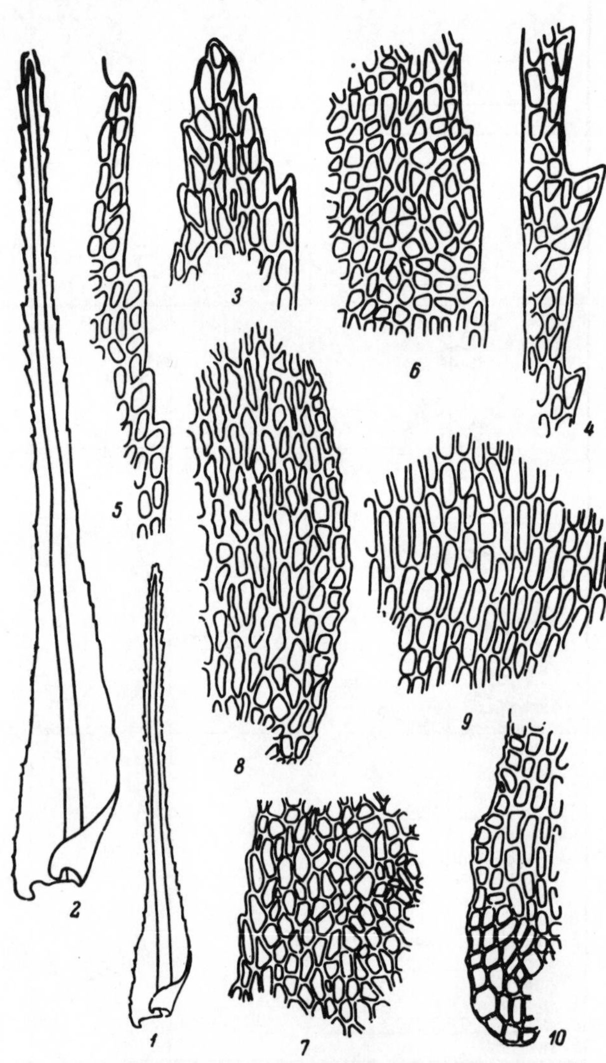

Echinodium Savicziae **A. Abr. et I. Abr.:**

Fig. 81. Finely preserved remains of *Echinodium savicziae* from the Pliocene of Duab, Caucasus. From Abramova and Abramov 1959*a*.

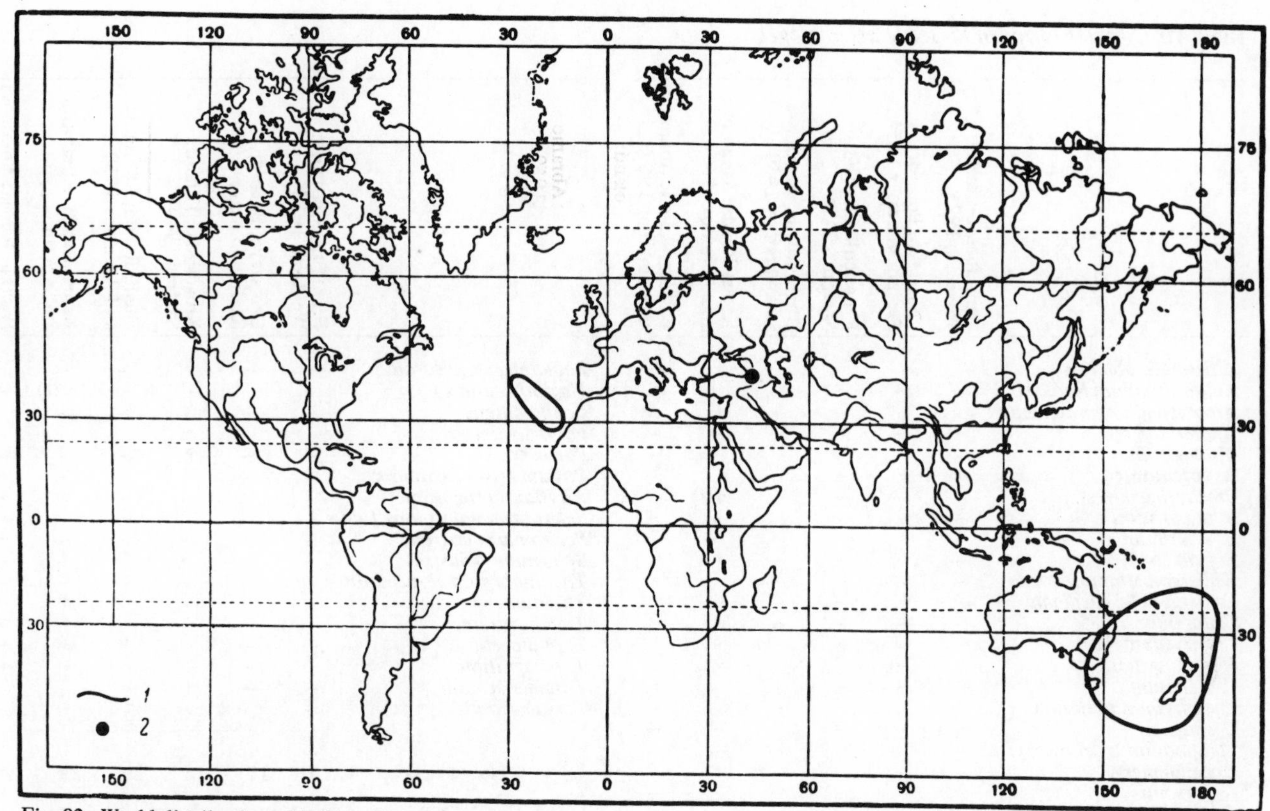

Fig. 82. World distribution of *Echinodium*. After Abramova and Abramov 1959*a*. 1 Present range. 2 Pliocene record from Duab. There is also a Miocene record from Poland.

Table 9 *Some European Miocene assemblages*

	Arjuzanx, France; Jovet-Ast and Huard 1966	Stare Gliwice, Poland; Szafran 1958	Zatoka Gdowska, Poland; Szafran 1964	Pshekha, Russia; Abramova and Abramov 1965
Anomodon longifolius	–	–	+	–
Brachythecium sp	–	–	+	–
B. rutabulum	+	–	–	–
B. sect. *Salebrosa*	–	+	–	–
B. velutinum	+	+	–	–
Cirriphyllum piliferum	+	–	+	–
Claopodium sp	–	+	–	–
C. whippleanum	+	–	–	–
Ctenidium molluscum	–	–	–	+
**Echinodium savicziae*	–	–	+	–
Eriodon sp	–	+	–	–
Eurhynchium sp	–	–	+	–
E. confertum	+	–	–	–
E. pulchellum	–	–	+	–
E. speciosum (t)	+	–	–	–
E. stokesii	+	–	–	–
E. swartzii	+	–	+	–
Heterocladium squarrosulum	–	–	+	–
Homalia lusitanica	+	–	–	–
H. trichomanoides	+	–	–	–
Isothecium myosuroides	–	–	–	+
Macrothamnium sp (t)	–	+	–	–
Neckera sp	–	–	+	–
N. complanata	+	+	–	+
Palamocladium sp.	–	+	–	–
P. euchloron	–	–	–	+
Papillaria sect. *Eupapillaria*	+	–	–	–
Pseudoscleropodium purum	+	–	–	–
Sphagnum subg. *Inophloea*	–	+	–	–
Thamnobryum alopecurum	–	–	+	–
T. sect. *Parathamnium*	–	+	–	–
**Thamnites marginatus*	+	–	–	–
Thuidium delicatulum	+	–	–	–
T. tamariscinum	+	–	–	–
Trachycystis flagellaris	–	+	–	–
No. of taxa	16	10	10	4

Total no. of taxa 37

Hypnodendron sp *Sphagnum* spp	Derbyshire, England Boulter 1969 pc
**Trachycystis szaferi*	Domanski Wierch, Poland: Szafran 1949*a*

*Extinct taxon

Table 10 *Some European Pliocene assemblages*

	Frankfurt, Germany; Weyland, 1925	Czorsztyn, Poland; Szafran in Szafer 1954	Kroscienko, Poland; Szafran 1949	Duab, Russia; Abramova and Abramov 1959	Kama, Russia; Abramova and Abramov 1962	Bashkir, Russia; Abramova and Abramov 1967
Abietinella abietina	–	–	–	–	+	+
Actinothuidium hookeri	–	–	–	–	–	+
Amblystegium juratzkanum	–	–	–	–	+	–
Anomodon sp	–	–	+	–	–	–
A. longifolius	–	–	–	+	–	–
A. viticulosus	+	+	–	+	–	–
Brachythecium sp	+	–	–	–	+(t)	–
B. salebrosum	–	–	–	+	–	–
B. velutinum	–	–	–	+	–	–
Bryum sp	–	–	–	–	+	–
Calliergon stramineum	–	+	–	–	–	+
Calliergonella cuspidata	–	+	–	–	–	+
Campylium sp	–	–	–	–	+	–
C. chrysophyllum	–	–	–	–	–	+
C. sommerfeltii	–	–	–	–	+	+
C. stellatum	–	–	–	–	–	+
Cirriphyllum piliferum	–	–	–	+	–	–
C. vaucheri	+(t)	–	–	–	–	–
**Claopodium kolakowskyi*	–	–	–	+	–	–
C. pellucinerve	–	–	–	–	+	–
**C. personii*	–	–	–	+	–	–
C. whippleanum	–	–	–	+	–	–
Clasmatodon parvulus	–	–	–	+	–	–
Cratoneuron filicinum	–	+	–	–	+	+
Ctenidium molluscum	–	–	–	+	–	–
Dichelyma capillaceum	–	+	–	–	–	–
Drepanocladus sp	–	–	–	–	+	–
D. aduncus	–	+	–	–	–	+
D. sendtneri	–	+	–	–	–	+
D. vernicosus	–	–	+	–	–	–
**Echinodium savicziae*	–	–	–	+	–	–
Ectropothecium sp	–	–	+	–	–	–
Entodon orthocarpus	–	–	–	–	–	+
Eurhynchium sp or spp	+	–	–	+	+	–
E. angustirete	–	–	–	+	–	–
E. meridionale	–	–	+	+	–	–
E. pulchellum	–	–	–	+	+	+
E. riparioides	–	–	–	+	–	–
E. stokesii	–	–	–	–	+	–
E. swartzii	–	–	–	–	+	–
Gollania berthelotiana	–	–	–	+	–	–
Heterocladium sp	+(t)	–	–	–	–	–
H. heteropterum	–	–	–	+	–	–
Homalia lusitanica	–	–	–	+	–	–
H. subarcuata	–	–	+	–	–	–
H. trichomanoides	–	+	+	+	–	+
Homalothecium nitens	–	–	–	–	–	+
H. sericeum	–	–	–	+	–	–
Hylocomium pyrenaicum	–	–	–	+(t)	–	–
H. splendens	–	–	+	–	+	+
Hypnum sp	–	–	–	+	–	–
H. cupressiforme	–	–	–	–	–	+
H. lindbergii	–	–	–	–	–	+
Isothecium myurum	–	–	–	+	–	–
Leskeella tectorum	–	–	–	–	–	+
Mnium orthorhynchum	–	–	+	–	–	–
Neckera sp	–	+	–	–	–	–
N. besseri	–	–	–	+	–	–
N. complanata	+(t)	–	+	–	+	–
N. crispa	–	–	+	+	–	–
N. pennata	–	–	+	–	–	–
Palamocladium euchloron	–	–	+	+	–	+
Pinnatella alopecuroides	–	–	+	–	–	–
Plagiothecium sp	–	–	–	–	+(t)	–
Pohlia nutans	–	–	+	–	–	–
Polytrichum sp	–	–	+	–	–	–
Pottia sp	–	–	+	–	–	–
Ptilium crista-castrensis	–	–	–	–	–	+
Rhytidium rugosum	–	–	–	–	+	+
**Sciaromiadelphus longifolius*	–	–	–	–	–	+
**Sciaromium laxirete*	–	–	–	–	–	+
Sphagnum palustre	–	+	–	–	–	–
Thamnobryum alopecurum	+	–	–	+	–	–
Thuidium sp	–	–	–	–	+	–
T. delicatulum	–	+	+	–	–	–
T. philibertii	–	–	–	+	+	+
T. recognitum	–	–	–	–	+	–
T. tamariscinum	–	–	+	–	–	–
Tortula ruralis	–	–	–	–	+	–
No. of taxa	7	11	18	29	22	24

The deposit at Reuver on the Dutch-German border yielded *Drepanocladus* sp, *Pinnatella* sect. *Urocladium* and **Trachycystis antiquorum.*

Total no. of taxa : 81.

*Extinct taxon.

Also recovered from Duab was *Gollania berthelotiana*, a Macaronesian endemic species (fig. 83) of a genus occurring in the Old World tropics, eastern Asia and Alaska. Equally interesting are three species of *Claopodium*, two of which are extinct (*C. kolakowskyi* and *C. personii*) while the third is *C. whippleanum*, a species with the remarkable range of western North America, Hawaii, Spain and Portugal (Nogouchi 1964). The last-named has also been recorded from the Miocene deposit at Arjuzanx, France (Jovet-Ast and Huard 1966).

The significance of these fossils for the understanding of the history of the oceanic species, often considered Tertiary relicts, in the British Isles is discussed on p. 205.

An affinity with tropical regions, especially southeastern Asia, is well marked in the Neogene assemblages by such taxa as *Eriodon* sp, *Macrothamnium* sp, *Papillaria* sect. *Eupapillaria*, *Thamnium* sect. *Parathamnium* and *Hypnodendron* sp (in the Miocene) and *Ectropothecium* sp, *Homalia subarcuata* and *Pinnatella alopecuroides* (in the Pliocene).

In marked contrast to these species just mentioned and

Table 11 *Present ranges of some Neogene taxa*

Actinothuidium hookeri	Small genus; S.E. Asia
Claopodium spp	E. and S.E. Asia, Oceania, W. North America, Central America, Portugal
Clasmatodon parvulus	Germany, North America
Echinodium savicziae	See fig. 83
Ectropothecium sp	Large genus; Old and New World tropics, Oceania
Eriodon sp	Small genus; Central and South America, New Zealand
Gollania berthelotiana	Macaronesia
Homalia subarcuata	S.E. Asia
Macrothamnium sp	S.E. Asia, Oceania
Hypnodendron sp	Large genus; mostly S.E. Asia and Oceania, also Australia, New Zealand, South America
Palamocladium euchloron	Caucasus, S.W. Asia; large genus in Old and new World tropics, Oceania
Papillaria sect. *Eupapillaria*	Large genus; Old and New World tropics, Oceania, Australia, New Zealand
Pinnatella alopecuroides	S.E. Asia; large genus in Old and New World tropics, Oceania, Australia
Sciaromium spp	Large genus; South Africa, S.E. Asia, Oceania, Australia, North and South America
Thamnium sect. *Parathamnium*	S.E. Asia
Trachycystis spp	Small genus; E. Asia, Alaska

also the others listed in tables 9 and 10 stand species which can flourish in northern regions (north of 60° N for this context). In the Miocene lists (especially that from Aldan) there are few: *Homalothecium nitens, Ctenidium molluscum, Drepanocladus sendtneri, Eurhynchium pulchellum* and *Ptilium cristacastrensis*. In the Pliocene there are many more (table 12).

Nevertheless, the Neogene assemblages are characterised by shade-tolerant thermophiles. Apart from the tropical-warm temperate species already mentioned, the most commonly recorded species is *Neckera complanata* (6 out of 12 assemblages). When such as *Homalia trichomanoides* (5/12), *Anomodon viticulosus* (4/12), *Cirriphyllum piliferum, Eurhynchium swartzii, Thamnobryum alopecurum* and *Thuidium delicatulum* (all 3/12) and others less commonly found are added, the dominating ecological aspect is clear; the assemblages represent forest floras, as already pointed out by Savicz-Ljubitzkaja and Abramov (1959).

Table 12 *Northern species in the Pliocene*

Abietinella abietina	*Eurhynchium pulchellum*
Calliergon stramineum	*Homalothecium nitens*
Campylium stellatum	*Hylocomium pyrenaicum* (t)
Cirriphyllum vaucheri (t)	*H. splendens*
Cratoneuron filicinum	*Pohlia nutans*
Ctenidium molluscum	*Ptilium crista-castrensis*
Dichelyma capillaceum	*Rhytidium rugosum*
Drepanocladus sendtneri	*Tortula ruralis*

Omitted from table 10 is another assemblage, in the southwestern Caucasus, studied by the Abramovs (1959). There is doubt about the age of the layers in question, the Chaudinsky Strata, which may be Late Pliocene or Early Pleistocene. The species recovered were as follows:

Anomodon attenuatus	*Homalothecium phillipeanum*
Brachythecium sp	*Isothecium myurum*
Ctenidium molluscum	*Neckera complanata*
Epipterygium tozeri	*Palamocladium euchloron*
Eurhynchium pulchellum	*Thamnobryum alopecurum*
	Thuidium philibertii (t)

This assemblage, which could well be placed with those of Pliocene age, is mainly noteworthy for the fine material of *Epipterygium* (fig. 84).

The Sphagnales extend back to the Lower Jurassic or even to the Permian if the Protosphagnales are accepted as belonging there (Lacey 1969). By the Tertiary if not earlier *Sphagnum* was widespread. However, macroscopic remains are scarce, as in the European Miocene and Pliocene (tables 9 and 10). Spore records (in the form genus *Sphagnumsporites* Raatz (*Stereisporites* Thoms. & Pf.)) are much more numerous. Jovet-Ast (1967) gives various Tertiary records.

There are Eocene records from Germany (Thomson and Pflug 1953) and Hungary (Kedves 1969). From the middle and late Tertiary of Central Europe no less than sixty-four form species (assigned to *Stereisporites*) have been described and illustrated by Krutzsch (1963*b*). Very tentatively he refers some of the form species to eight of the sections of the modern genus (*Palustria, Rigida, Squarrosa, Polyclada, Truncata, Subsecunda, Cuspidata* and *Acutifolia*). It is difficult to judge the value of these form species in terms of modern species; however, to state that by the middle Tertiary in Europe the genus was diverse seems safe.

Stuchlik (1964) has recorded five different spore types from a single Polish Miocene deposit and drawn comparisons with sections *Sphagnum, Cuspidata* and *Acutifolia*.

Apart from the fragment of *Hypnodendron* discussed on p. 170, *Sphagnum* is the only other moss genus known from the British Tertiary. The deposits in question are situated at Washing Bay, Co. Tyrone, and are considered to be middle or early Oligocene in age (Watts 1962). Johnson (1951, p. 100) described the discovery as follows:

> Scraps of Sphagnum were found in the core about a dozen times at a depth of 786-903 ft, and occasionally spores 36-40 μ in diameter were encountered. For the most part the finds consisted of a few leaf cells only,

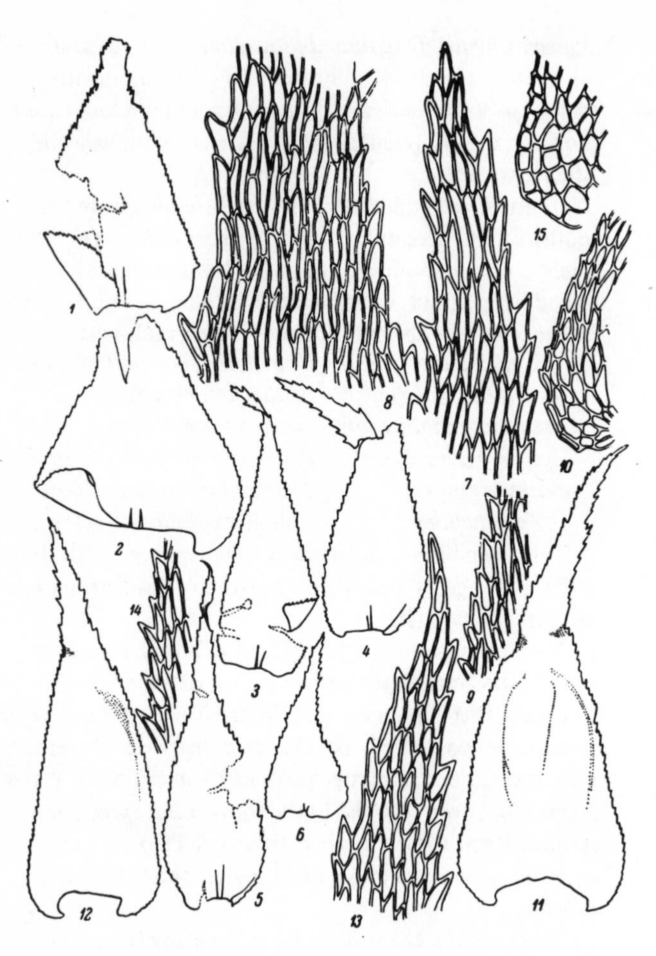

Fig. 83. Finely preserved remains of *Gollania berthelotiana* from the Pliocene of Duab, Caucasus. From Abramova and Abramov 1959*a*.

Fig. 84. Finely preserved remains of *Epipterygium tozeri* from the Pliocene or Early Pleistocene of southwestern Caucasus. From Abramova and Abramov 1959*b*.

but in one or two cases whole leaves were obtained. A complete leaf and portion of the same at higher magnification are shown in figs. 1 and 2. The characteristic leaf structure of the genus is clear.

There was sufficient difference in structure between the various leaf scraps observed to indicate that more than one species of Sphagnum was present in the core. In some specimens the walls of the hyaline cells were lined with a comb-like fringe, a feature present in one or two recent species. In one specimen in which this character appeared the surface or periclinal wall of the hyaline cell shows pores and trabeculae in addition to the marginal teeth.

The material available was inadequate for a specific identification . . .

The plant beds, already mentioned, thought to belong to the Miocene-Pliocene boundary in Derbyshire, have yielded spores of eleven form species of *Stereisporites* (Boulter 1971). The Pliocene deposits of the Netherlands have abundant spores which extend into the Lower Pleistocene (Zagwijn 1960).

These numerous Palaeogene and especially Neogene records have rendered baseless the speculation by Gams (1932) that *Sphagnum* may not have occurred in pre-Pleistocene Europe and also that by Szafran (1949) that immigration took place in the Neogene.

Finally, mention must be made of the palynological studies of Nagy (1968) on the Neogene of Hungary. She has described six taxa of Anthocerotaceae, two of *Ricciaesporites*, and one each of *Encalyptasporites* and *Ephemerisporites*. The similarities with modern taxa are striking and the affinities can scarcely be doubted.

6.2 EARLY PLEISTOCENE

Apart from the Cromer Forest Bed Series, only Castle Eden, Co. Durham, has produced moss remains which are considered as Early Pleistocene. The deposit was formerly considered to be Pliocene (Reid 1920). However, according to West (1968*a*, p. 298), 'The paucity of exotic species (vascular plants) and its stratigraphical position below the oldest glacial deposits in the area, suggest an Early Pleistocene age.' The nine species composing the assemblage were as follows:

Amblystegium kochii	*E. swartzii*
A. serpens	*Homalia trichomanoides*
Brachythecium sp	*Neckera complanata*
Eurhynchium praelongum	*Rhytidiadelphus triquetrus*
	Thuidium tamariscinum

However early its age, this assemblage is remarkable for its dullness. As Dixon (p. 110) stated, 'They are the ordinary mosses that one might meet in a wood or lane-side now, in any lowland part of England.' Like the bryofloras of the later temperate stages it is a shade-tolerant flora of deciduous woodland; remains of *E. swartzii* were the most frequent.

As well as Castle Eden, Dixon investigated another Early Pleistocene assemblage (Reid and Reid 1915). The locality is Tegelen in the Netherlands and the stage Tiglian (Zagwijn 1963*a*), correlated with the Ludhamian (West 1968). He recorded the following:

Cratoneuron filicinum	*Eurhynchium speciosum*
Ditrichum pusillum	*Homalothecium sericeum*
Drepanocladus capillifolius	*Lescuraea patens* (t)
	Philonotis fontana

As Dixon was at pains to point out, this assemblage is ecologically heterogeneous. *Lescuraea patens* and *Drepanocladus capillifolius* seem misplaced in a deposit formed under temperate conditions. The bulk of the mosses consisted of *Eurhynchium speciosum* which points, as Dixon stated, to a low-lying marsh or river valley. With plausibility he regarded *Pseudoleskea patens* as allochthonous. This is the kind of difficulty which Steere (1965) has emphasised.

Apart from *Sphagnum* spores there are no bryophyte remains recorded from the Early Pleistocene stages, the Ludhamian, Thurnian and Antian, known only from coastal areas of East Anglia (West 1968*a*, pp. 245-7). A very few mosses are known from the Baventian (table 13).

Sphagnum spores are well represented in the marine or estuarine shallow-water deposits of the Crag at Ludham, Norfolk (fig. 14), which represent a sequence of three temperate and two cold phases. It is not certain that the spores belong to the cold stages; the spores were transported by water from the nearby land and so some or all may have been reworked from the deposits of the warmer phases. Even if this did happen it is clear that the genus was present through much of the long period of time investigated. Low values of spores (10 per cent) were recorded in the marine sands and clays (Antian and Baventian) of Easton Bavents, Suffolk (Funnell and West 1962).

Table 13 *Mosses from the Cromer Forest Bed series*

	Early Pleistocene			Middle Pleistocene		
	Baventian or earlier	Baventian	Pastonian	Beestonian	Cromerian	Anglian
Anomodon sp	–	–	–	–	+	–
Antitrichia curtipendula	–	–	+	–	+	–
Aulacomnium palustre	–	–	+	–	–	–
Brachythecium sp	–	–	+	–	–	–
B. populeum (t)	+	–	–	–	–	–
Calliergon spp	–	+	–	–	+	+
C. giganteum	–	–	–	–	–	+
Campylium sp	–	–	–	–	–	+
Dicranum sp	–	–	+	–	–	–
Drepanocladus spp	+	+	+	+	+	+
Homalia trichomanoides	+	–	–	–	–	–
Hypnum cupressiforme	–	–	+	–	–	–
Pohlia sp	–	–	+	–	–	–
Rhytidiadelphus triquetrus	+	–	–	–	–	–
Scorpidium scorpioides	–	–	–	–	–	+
Sphagnum spp	+	+	+	+	+	+
S. imbricatum	–	–	+	–	–	+

6.3 MIDDLE PLEISTOCENE

Pastonian, Beestonian and Cromerian

The investigations of the Cromer Forest Bed Series, being pursued by Dr West and Mrs Wilson, are proving highly productive. Much stratigraphical, palaeoecological and phytogeographical information has accrued. However, as yet Mrs Wilson has extracted few moss species which, nevertheless, are important because very little is known of the bryogeography of the periods in question, either in the British Isles or elsewhere (table 13).

Moss remains from six sites are still under investigation. No details will be given here except that the remains, though abundant, are highly fragmentary and badly preserved and for the most part were extracted from estuarine sediments; they are allochthonous and may have travelled a long way before incorporation. Fragments of *Calliergon, Drepanocladus* and *Sphagnum* are the most abundant.

Apart from the species given in table 13, the Cromerian temperate stage has yielded three species which were extracted from the Pakefield Beds (Blake 1890), now considered as probably Cromerian (West 1970 pc). Mitten identified *Brachythecium plumosum, Drepanocladus fluitans* and *Eurhynchium swartzii*.

Duigan (1963) recorded *Sphagnum* spores at almost every one of her several sites of the Cromer Forest Bed and sometimes encountered high values, up to 20 per cent in the case of Happisburgh.

Anglian

Apart from the Late Anglian assemblages (table 14) there are only a handful of moss records referable to the Anglian glaciation. The Corton Beds at Lowestoft (West and Wilson 1968) yielded *Amblystegium* sp, *Drepanocladus* sp or spp, *Hypnum* sp or spp, *Pohlia* sp and *Sphagnum* sp. So far the recent investigations of the Cromer Forest Bed Series have revealed the following of Early Anglian age:

Calliergon giganteum — *Scorpidium scorpioides* (t)
Campylium sp — *Sphagnum imbricatum*
Drepanocladus spp

Sphagnum imbricatum is uncommon in the bryofloras of cold stages (p. 71); one may wonder if it could have been derived from the Cromerian. The remainder are unexceptional. Seventy years ago, Reid (1904), investigating the stratigraphy revealed by the construction of a well at Mundesley, extracted mosses from the 'Arctic Fresh-water Bed', now known to belong to two cold stages (West and Wilson 1968). In this case the age is Anglian. Dixon identified the following:

Amblystegium varium — *Drepanocladus capillifolius*
Calliergon richardsonii — *D. fluitans*
C. turgescens — *D. revolvens*
Campylium polygamum — *D. sendtneri*

Three species are particularly noteworthy. *Calliergon richardsonii* and *Drepanocladus capillifolius* are both extinct in the British Isles (p. 212) and *Calliergon turgescens*, representing its earliest British occurrence, is an outstanding example of a relict species (p. 142), commonly found in last glacial assemblages.

There are about twenty-two taxa of mosses from the Late Anglian layers of Hoxne, Gort and Baggotstown (table 14). The list is strikingly similar to those of the Late Devensian especially in the presence of *Distichium capillaceum* (t), *Polytrichum alpinum* and *Rhacomitrium lanuginosum* as well as several rich fen species, notably *Campylium stellatum* in all three assemblages.

Only *Brachythecium velutinum* seems anomalous. However, one must remember species such as *Cirriphyllum piliferum, Thuidium tamariscinum* and *Plagiothecium undulatum* in the Late Devensian.

Table 14 *Late Anglian mosses*

	Hoxne	Gort	Baggotstown
Amblystegium sp	–	–	+
Brachythecium sp	–	+	–
B. velutinum	–	+	–
Bryum sp	+	–	–
Calliergon cordifolium	+	–	–
C. giganteum	+	–	+
Calliergonella cuspidata	+	–	–
Campylium stellatum	+(t)	+	+
Climacium dendroides	–	+	–
Cratoneuron filicinum	+	–	–
C. commutatum	+	–	–
Distichium capillaceum (t)	–	–	+
Drepanocladus sp	+	–	–
D. revolvens	+	–	–
D. sendtneri (t)	+	–	–
Plagiomnium rugicum (t)	+	–	–
Pleurozium schreberi	–	+	–
Polytrichum alpinum	+	–	–
Rhacomitrium lanuginosum	–	+	–
Scorpidium scorpioides	+	–	–
Tortula sp	+	–	–
No. of taxa	14	6	4

Hoxnian

The three Hoxnian assemblages from southeastern England, three from western Ireland and one from the Shetland Islands give a geographical extensiveness lacking from the Ipswichian (p. 181). Well over fifty species have been recognised, the majority originating from the rich assemblage from Gort, Co. Galway, which has many points of interest. Not least is the presence of *Hyocomium armoricum* in Ho zones II, III and perhaps IV (p. 206). *Brachythecium velutinum* extends from the Late Anglian through zones I, II and III and again perhaps IV. The history of *Thuidium tamariscinum* is equally full, except for the Late Anglian. These species as well as *Anthocerus punctatus* (t), *Eurhynchium praelongum, E. schleicheri, Hypnum cupressiforme, Isothecium* spp, *Neckera* spp and *Pseudoscleropodium purum* give the flora its stamp of shade-tolerance and indicate temperateness of climate. Similar deductions can be drawn from the Hitchin, Marks Tey, Hoxne and Fugla Ness assemblages.

However, many bryophyte communities are represented in the Gort assemblage, other than those of deciduous woodland. *Hyocomium armoricum*, especially, but also such species as *Isothecium* spp, *Rhacomitrium lanuginosum, Sphagnum imbricatum* and *S. molle* point to oceanicity of climate, as does the remarkable flowering plant assemblage (p. 27).

The diversity of *Sphagnum* spp, particularly in Ho zones III and perhaps IV, indicates well developed oligotrophic, probably ombrogenous, mires which could also have supported *Aulacomnium palustre, Pleurozium schreberi, Polytrichum commune* (t) and *Rhacomitrium lanuginosum.* Rich fens (also Ho III or IV) are attested by the discovery of *Homalothecium nitens* and *Meesia uliginosa* from stratum *b* (sandy gyttja).

There remain for discussion the bryofloras from the uppermost layers at Hoxne and Baggotstown, and from the

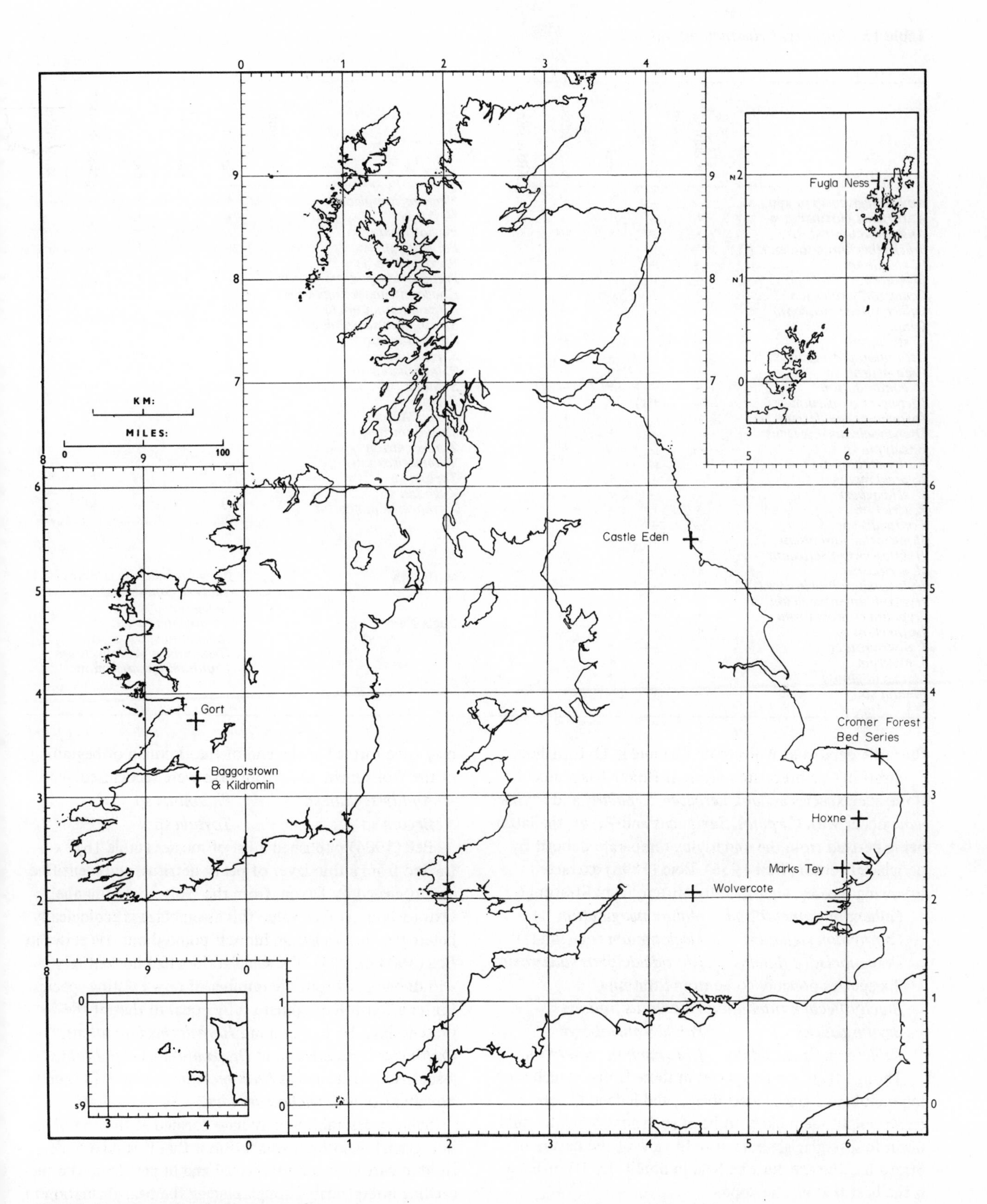

Fig. 85. Hoxnian and earlier temperate sites.

Table 15 *Important Hoxnian assemblages*

	Hitchin	Hoxne	Gort	Baggotstown	Kildromin
Amblystegium sp or spp	–	–	+	+	–
Antitrichia curtipendula	+	–	+	–	–
Aulacomnium palustre	–	–	+	–	–
Brachythecium populeum (t)	–	+	–	–	–
B. velutinum	–	–	+	–	–
Bryum spp	–	+	–	+	+
Calliergon giganteum	–	–	–	+	+
Calliergonella cuspidata	–	+	–	–	–
Campylium elodes	–	–	+	+	–
C. stellatum	–	–	+	+	+
Ceratodon purpureus	–	–	–	+	–
Cratoneuron sp	–	–	+	–	–
C. commutatum	–	–	+	–	–
Dicranum scoparium	–	+(t)	+	–	–
Distichium capillaceum (t)	–	–	–	+	+
Drepanocladus revolvens	–	–	+	–	–
Encalypta sp	–	–	–	+	–
Eurhynchium sp	–	+	–	–	–
E. praelongum	–	–	+	–	–
E. schleicheri	–	–	+	–	–
E. striatum	–	+	–	–	–
Fontinalis sp	–	–	–	+	+
Homalothecium nitens	–	–	+	–	–
H. lutescens or *sericeum*	–	–	–	+	–
H. sericeum	+	+	–	–	–
Hylocomium brevirostre (t)	+	–	–	–	–
Hyocomium armoricum	–	–	+	–	–
Hypnum cupressiforme	+	+	+	+	–
Isothecium sp	+	–	–	–	–
I. myosuroides	–	–	+	–	–
I. myurum	–	–	+	–	–
Meesia uliginosa	–	–	+	–	–
Mnium spp	–	+	–	+	–
Neckera complanata	+	+	+	–	–
N. crispa	–	–	+	–	–
Plagiothecium sp	–	–	+	–	–
Polytrichum sp	–	–	+	–	–
P. commune (t)	–	–	+	–	–
Pseudoscleropodium purum	–	–	+	–	–
Rhacomitrium lanuginosum	–	–	+	–	–
Rhizomnium punctatum	–	+	–	–	–
Scorpidium scorpioides	–	–	–	+	+
Sphagnum spp	–	+	+	+	+
S. imbricatum	–	–	+	–	–
S. magellanicum	–	–	+	–	–
S. molle	–	–	+	–	–
S. palustre	–	–	+	–	–
S. papillosum	–	–	+	–	–
Thuidium sp	–	–	+	–	–
T. recognitum	–	–	+	–	–
T. tamariscinum	–	–	+	–	–
Tortula sp	–	–	–	+	+
Zygodon sp	+	–	–	–	–
Anthoceros punctatus	–	–	+	+	–
	7	12	35	16	8

Marks Tey	*Brachythecium velutinum* (t) *Neckera complanata* *Sphagnum* spp
Fugla Ness	*Anthoceros punctatus* *Sphagnum* spp *Thamnobryum alopecurum* *Thuidium tamariscinum*

fluviatile deposit at Wolvercote Channel in Oxfordshire.

Stratum *C*, a brecciated layer, at Hoxne has a mixed flora; such species as *Salix herbacea, S. polaris* and *Betula nana* occur with *Carpinus, Sambucus* and *Taxas*, the latter being derived from the underlying temperate deposit by periglacial action (West 1956). Reid (1896) extracted the following mosses, identified by Mitten, from Stratum *C*:

Calliergon sarmentosum	*Mnium marginatum* (t)
Campylium stellatum	*Plagiomnium rugicum* (t)
Drepanocladus fluitans	*Rhytidiadelphus squarrosus*

Six species possibly come from Stratum *C*:

Brachythecium rutabulum	*Philonotus fontana*
Bryum pallens	*Pohlia wahlenbergii*
Calliergonella cuspidata	*Rhizomnium punctatum*

Though there are no species in these twelve which seem impossible in a glacial assemblage, and indeed *C. sarmentosum* might seem likely to be of such an origin, all could occur in interglacial strata also. In view of the nature of Stratum *C* the age must be held in doubt. Ho III or EWo is the best that can be done.

Stratum *H* (sandy silt) from Baggotstown (Watts 1964) may have formed at the end of the Hoxnian or beginning of the Wolstonian. Only four taxa were recognised:

Amblystegium sp	*Philonotis* sp
Bryum sp	*Tortula* sp

Bell (1904) published a list of mosses (table 16) extracted from a thin layer of peaty detritus, and identified and discussed by Dixon, from the Wolvercote Channel, Oxfordshire. At face value this assemblage is ecologically heterogeneous, as Dixon himself pointed out. He states in Bell (1904, p. 124) 'The Glacial flora had not entirely withdrawn . . . From the number of now-existing species I infer a warmth of climate fully equal to that of the present day.' He had in mind *Homalothecium nitens*, *Cratoneuron decipiens* and *Drepanocladus capillifolius* in association with such as *Eurhynchium swartzii, E. speciosum* and *Homalia trichomanoides*.

Such an assemblage may have formed at the end of an interglacial. Another explanation is that it is mixed, derived in part from an interglacial and in part from the succeeding interglacial. Though placing the peat formation in the Hoxnian, Bishop (1958) concluded that the Wolvercote

Table 16 *Mosses from Wolvercote Channel*

Amblystegium kochii	*D. exannulatus*
A. serpens	*D. fluitans*
Aulacomnium palustre	*D. lycopodioides* (t)
Brachythecium glareosum	*D. revolvens*
B. rutabulum	*D. sendtneri*
Calliergon cordifolium	*Eurhynchium speciosum*
C. giganteum	*E. swartzii*
C. stramineum	*Homalia trichomanoides*
Calliergonella cuspidata	*Homalothecium nitens*
Campylium chrysophyllum	*Philonotis fontana*
Climacium dendroides	*Plagiomnium affine*
Cratoneuron decipiens	*P. rostratum*
C. filicinum	*Pohlia nutans*
Dichodontium pellucidum	*P. wahlenbergii*
Drepanocladus aduncus	*Thuidium recognitum*
D. capillifolius	

Channel formed in the latter part of the Hoxnian and early part of the Wolstonian. The moss assemblage is best treated as Late Hoxnian or Early Wolstonian.

In the Hoxnian interglacial *Sphagnum* occurred at Hoxne itself (West 1956) where *S.* cf. *rubellum* and sections *Acutifolia* and *Sphagnum* were recorded from stratum *C* (EWo) which contained spores through its depth. Turner (1970) recorded spores mostly in low values in all zones of the interglacial deposit at Marks Tey, Essex, and also from the underlying Anglian and overlying Wolstonian deposits; macroscopic remains of sub-genera *Inophloea* and *Litophloea* were recovered from Ho zones III and IV and of the latter from zone EWo. At Gort, where spores were found in low values almost throughout, *S. molle, S. imbricatum, S. palustre* and *S.* cf. *magellanicum* were identified (p. 176). At other Irish sites of the same age, spores are much more poorly represented (Watts 1959*a*, 1964, 1967), as they usually are in English sites (for example Kirmington, Lincolnshire (Watts 1959*b*), and Hatfield, Hertfordshire (Sparks *et al.* 1969)).

7 BRYOFLORAS OF THE LATE PLEISTOCENE

7.1 WOLSTONIAN

There are few moss records, almost all from southeastern England, which can be assigned to the second last glaciation with any certainty.

The Early Wolstonian layers (as defined palynologically) of the deposit at Marks Tey (Turner 1970) yielded the four taxa *Calliergon giganteum, Drepanocladus* sp, *Neckera complanata* and *Sphagnum* subg *Inophloea*. Ecologically, the *Neckera* is out of place; all the other numerous records come from temperate stages. Much of the topmost deposits at Marks Tey is brecciated, hence the *Neckera* may be derived from the underlying Ho III deposit from which it was also extracted.

Low values of *Sphagnum* spores have been recorded from a silty channel of Early Wolstonian age in the Babbington-Lillington gravels at Brandon, Warwickshire (Kelly 1968).

River terraces in Oxfordshire have revealed two moss floras, one of which pertains with more safety to the Wolstonian than the other. The doubtful one is that from the Wolvercote Channel, discussed on p. 178, which may belong at least in part to the Early Wolstonian. Less doubt attaches to the assemblage from the basal gravels of the Summertown-Radley terrace near Dorchester (Duigan 1955). Bishop (1958) has assigned these gravels to the Late Wolstonian. Proctor in Duigan (1955) recognised the following:

Bryum sp	*Drepanocladus capillifolius*
Calliergon cordifolium	*D. revolvens*
C. stramineum	*D. sendtneri*
Campylium polygamum (t)	*Philonotis calcarea*
	Plagiomnium rugicum (t)
Cratoneuron filicinum	*Pohlia* sp

Apart from *Drepanocladus capillifolius*, well known from the Devensian, especially Middle Devensian deposits, there is nothing in the assemblage to indicate glacial conditions.

Yet another river terrace assemblage of doubtful (probably Wolstonian) age is that from Farnham in the valley of the River Wey, Surrey (Oakley *et al.* 1939). Dixon identified the following:

Ablystegium serpens	*Homalothecium nitens*
Calliergon turgescens	*Hygrohypnum molle*
Pleurozium schreberi	

Table 17 *Bryophytes from Broome Heath*

Bryum sp or spp	*Hypnum hamulosum*
Brachythecium sp	*Isopterygium pulchellum* (t)
Barbula sp	*Oncophorus virens*
Campylium stellatum	*Philonotis* sp
Calliergon giganteum	*Pohlia* sp
C. turgescens	*Rhacomitrium* sp
Calliergonella cuspidata	*Scorpidium scorpioides*
Cinclidium stygium	*Timmia* sp
Cratoneuron filicinum var. *curvicaule*	*Tortula* sp
Distichium capillaceum (t)	Foliose liverwort
Drepanocladus exannulatus var. *rotae*	Capsule
D. lycopodioides (t)	
D. sendtneri (t)	
D. capillifolius	

A rich assemblage (table 17) has been recovered from sediments dredged from a gravel pit at Broome Heath, Norfolk. The geological provenance has not been studied in detail. However, the geographical position of the pit and stratigraphy of the region indicate the Wolstonian glaciation (Sparks and West 1968). A radiocarbon sample gave a date of more than 40,000 years.

The twenty species speak of open vegetation reflecting a cold climate. Several species point to rich fens:

Calliergon giganteum	*Cinclidium stygium*
C. turgescens	*Drepanocladus capillifolius*
Campylium stellatum	*Scorpidium scorpioides*

More indicate saxicolous/terricolous communities:

Cratoneuron filicinum var. *curvicaule*	*Isopterygium pulchellum* (t)
	Rhacomitrium sp
Distichium capillaceum (t)	*Timmia* sp
	Tortula sp
Hypnum hamulosum	

The Broome Heath subfossils are very similar to those from various Devensian localities. Analogous bryogeographical changes have occurred, typified by *Cratoneuron filicinium* var. *curvicaule*, now restricted to Mid Perth and North Ebudes.

The Ipswichian deposits at Ilford, Essex (West *et al.* 1964) are superimposed on Late Wolstonian strata (as defined palynologically) which yielded the following:

Amblystegium sp	*Dicranella* sp
Barbula hornschuchiana	*Drepanocladus* spp
Brachythecium albicans	*D. exannulatus*

These are the only records certainly originating from the Late glacial stage of the second last glaciation.

7.2 IPSWICHIAN

The scanty data from the last interglacial allow little comment, either ecological or bryogeographical. Not only are the assemblages few in number and poor in species but their restriction to the south and east of England (fig. 86), in contrast to those of the previous interglacial, leaves the bryoflora of the west unknown.

Table 18 *Important Ipswichian assemblages*

	West Wittering	Selsey	Histon Road	Wretton	Hutton Henry
Amblystegium kochii	–	–	+	–	–
A. riparium	–	–	+	–	–
A. varium (t)	–	–	–	+	–
Anomodon viticulosus	–	–	–	+	–
Antitrichia curtipendula	–	–	–	+	–
Aulacomnium palustre	–	–	–	–	+
Brachythecium populeum	–	+	–	–	–
B. rutabulum (t)	–	–	+	–	–
Calliergon cordifolium	–	–	+	–	–
Calliergonella cuspidata	–	–	–	+	–
Ceratodon purpureùs (t)	–	–	–	+	–
Cratoneuron filicinum	–	–	+	+	–
Drepanocladus spp	–	–	+	–	–
D. aduncus	+	–	+	–	–
D. exannulatus (t)	–	–	+	–	–
D. fluitans	+	–	–	–	–
Eurhynchium sp	–	–	+	–	–
E. praelongum	+	+	+	–	–
E. speciosum	+	–	–	–	–
E. striatum	–	–	–	+	–
E. swartzii	–	–	–	+	–
Homalia trichomanoides	–	–	–	+	–
Homalothecium nitens	–	–	+	–	–
H. sericeum	+	+	–	+	–
Hypnum cupressiforme	+	+	–	+	+
Isothecium myurum	–	–	–	+	–
Leucodon sciruoides	–	+	–	–	–
Neckera complanata	+	–	–	+	–
Pleurozium schreberi	+	–	–	–	+
Spagnum spp	–	–	+	+	+
S. flexuosum (t)	–	–	–	–	+
S. imbricatum	–	–	–	–	+
S. magellanicum	–	–	–	–	+
S. rubellum (t)	–	–	–	–	+
No. of taxa	8	5	12	14	8

Amblystegium riparium, Calliergon stramineum and *Drepanocladus fluitans* were recorded from Trafalgar Square and *Brachythecium velutinum* (t) and *Drepanocladus* spp from Ilford.

There are no records from Ip zone I but by Ip zone II there was established a shade-tolerant bryoflora composed of species indicating the deciduous woodland also revealed by pollen analysis (fig. 16). The Wretton zone II flora is the principal basis of that statement:

Amblystegium varium (t)	*Homalia trichomanoides*
Anomodon viticulosus	*Homalothecium sericeum*
Antitrichia curtipendula	*Hypnum cupressiforme*
Calliergonella cuspidata	*Isothecium myurum*
Ceratodon purpureus (t)	*Neckera complanata*
Eurhynchium striatum	*Sphagnum* sp
E. swartzii	

In addition, the assemblages from Selsey, West Wittering and Histon Road are similarly of woodland origin, though there are also indicators of fen at the latter two. The presence of *Homalothecium nitens* (Ip zone III and IV) at Histon Road is worthy of note. It parallels such species as *Meesia longiseta* and *Helodium blandowii* in the last interglacial of Denmark (Hartz 1909). With *Meesia uliginosa* it occurred at about the same period in the Hoxnian interglacial at Gort.

The British data can be well supplemented by reference to the assemblage from Ejstrup, in Southern Jutland, which was investigated by Hartz (1909) before the advent of pollen analysis. However, the Eemian (last interglacial) age is well established (Hansen 1965).

Paralleling the Wretton species, the common feature of the assemblage, certainly one of the richest of its age (table 19), is shade tolerance; it is a deciduous woodland bryoflora. Again like the Wretton assemblage, many of the species point to base-rich soils but there are calcifuges also. A few species, such as *Homalothecium*

Table 19 *Bryophytes from the last interglacial at Ejstrup, Denmark*

**Anomodon viticulosus*	*Isothecium myosuroides*
**Antitrichia curtipendula*	**I. myurum*
Brachythecium populeum	*Marchantia polymorpha*
B. rutabulum	**Neckera complanata*
B. velutinum	*N. crispa*
Calliergon cordifolium	*Plagiomnium cuspidatum*
**Calliergonella cuspidata*	*P. undulatum*
**Ceratodon purpurens* (t)	*Pleurozium schreberi*
Dicranum bonjeanii	*Polytrichum formosum*
D. scoparium	*Pseudoscleropodium purum*
Eurhynchium praelongum	*Rhizomnium pseudopunctatum*
**E. striatum*	*R. punctatum*
**E. swartzii*	*Rhytidiadelphus squarrosus*
Homalothecium lutescens	*R. triquetrus*
**H. sericeum*	*Thamnobryum alopecurum*
Hylocomium splendens	*Thuidium tamariscinum*
Hypnum cupressiforme	

*Also in the Wretton assemblage

lutescens and *Rhizomnium pseudopunctatum*, are not typically forest plants.

The recently published moss assemblage extracted from freshwater deposits now submarine at Heligoland (Behre 1970) can also be compared with Wretton. Koppe identified the following from Ip zone III (Behre's zone V).

**Antitrichia curtipendula*	**Homalothecium sericeum*
Brachythecium velutinum	*Isothecium myosuroides*
**Calliergon cuspidata*	**Neckera complanata*
Campylium stellatum	*Pylaisia polyantha*
Ctenidium molluscum	*Sphagnum* subg *Inophloea*
**Eurhynchium swartzii*	*Thuidium tamariscinum*
Homalia trichomanoides	

*Wretton zone II

With one outstanding exception, *Sphagnum* has proved to be rare or absent from British Ipswichian deposits so far investigated. For example, at Wretton, Norfolk (Sparks and West 1970), Histon Road, Cambridge (Walker 1953), and Aveley, Essex (West 1969), the spore values are 2 per cent or less. At Wortwell, Norfolk (Sparks and West 1968), and Trafalgar Square, London (Franks, 1960), the genus is absent. In contrast, erratic lumps of peat in Devensian till at Hutton Henry, Co. Durham, have yielded *S. imbricatum, S. magellanicum, S.* cf. *rubellum* and *S.* cf. *flexuosum* (Beaumont *et al.* 1969). The genus is much better represented in the rest of northwestern Europe where spores may occur abundantly (Benda and Schneekloth 1965).

The Hutton Henry erratics were largely composed of *Sphagnum* spp and other species such as *Calluna vulgaris* and *Aulacomnium palustre*. From this Beaumont *et al.* (1969, p. 803) conclude 'that throughout much of the interglacial conditions in the eastern part of Co. Durham were suitable for the development of ombrogenous peat locally in a landscape that was otherwise well forested'.

7.3 DEVENSIAN

Moss remains have been extracted and identified from a great many Devensian sites scattered very widely in the British Isles (figs. 25 and 26) from the far northwest of Scotland to southeastern England and western Ireland. Most Late Devensian sites including all the richest come from the highland zone and few from the lowland zone. The converse is true of Middle and Early Devensian localities (fig. 26).

At least 116 species of mosses and three of liverworts are known from Devensian deposits; primarily they derive from Late Devensian sites which number about sixty-two, far exceeding those of Middle (five) and Early Devensian age (three).

This considerable body of data gives great insight into the bryoflora of the last glaciation, particularly the final 3000 years.

Early Devensian

As yet, knowledge of the first 10,000 years of the last glaciation is based entirely on three small assemblages (table 20) which, however, are not without significance. Apart from the spores from Wretton (West 1969 pc), *Sphagnum* is abundant both as spores and macroscopic subfossils in the Chelford deposit which also contained *Helodium blandowii* and *Pseudobryum cinclidioides*; after *Sphagnum*, the last-named and *Calliergon cordifolium* were the commonest taxa.

Table 20 *Early Devensian assemblages*

	Sidgwick Avenue	Wretton	Chelford
Amblystegium kochii (t)	+	–	–
Antitrichia curtipendula	–	+	–
Aulacomnium palustre	–	–	+
A. turgidum	–	+	–
Bryum sp or spp	–	+	–
Calliergon cordifolium	–	–	+
C. giganteum	+	+	–
C. turgescens	+	–	–
Campylium stellatum	+	+	–
C. polygamum (t)	–	+	–
Cratoneuron commutatum var. *falcatum*	+	–	–
C. filicinum	–	+	–
Drepanocladus capillifolius	–	+	–
D. exannulatus var. *rotae* (t)	+	–	–
D. revolvens	+	–	–
D. uncinatus	–	–	+
Encalypta sp (t)	–	+	–
Helodium blandowii	–	–	+
Homalothecium lutescens	+	–	–
Hypnum cupressiforme	+	–	–
Plagiomnium rugicum (t)	–	+	–
Pleurozium schreberi	–	–	+
Pohlia wahlenbergii (t)	–	+	–
Polytrichum sect *Juniperina*	–	+	+
Pseudobryum cinclidioides	–	–	+
Rhytidiadelphus squarrosus (t)	–	–	+
Scorpidium scorpioides	+	–	–
Sphagnum subg *Litophloea*	–	–	+
No. of taxa	10	12	9

The calcifuge nature of the Chelford assemblage, emphasised not merely by *Sphagnum* but also by *Pleurozium schreberi* and *Polytrichum* sect *Juniperina*, contrasts with the other two assemblages composed almost entirely of species of base-rich soils or rich fens, although *Polytrichum* sect *Juniperina* also occurred at Wretton. Most cogent in this respect from Sidgwick Avenue are *Campylium stellatum* (the commonest species) and particularly *Calliergon turgescens* and *Homalothecium lutescens*, and from Wretton

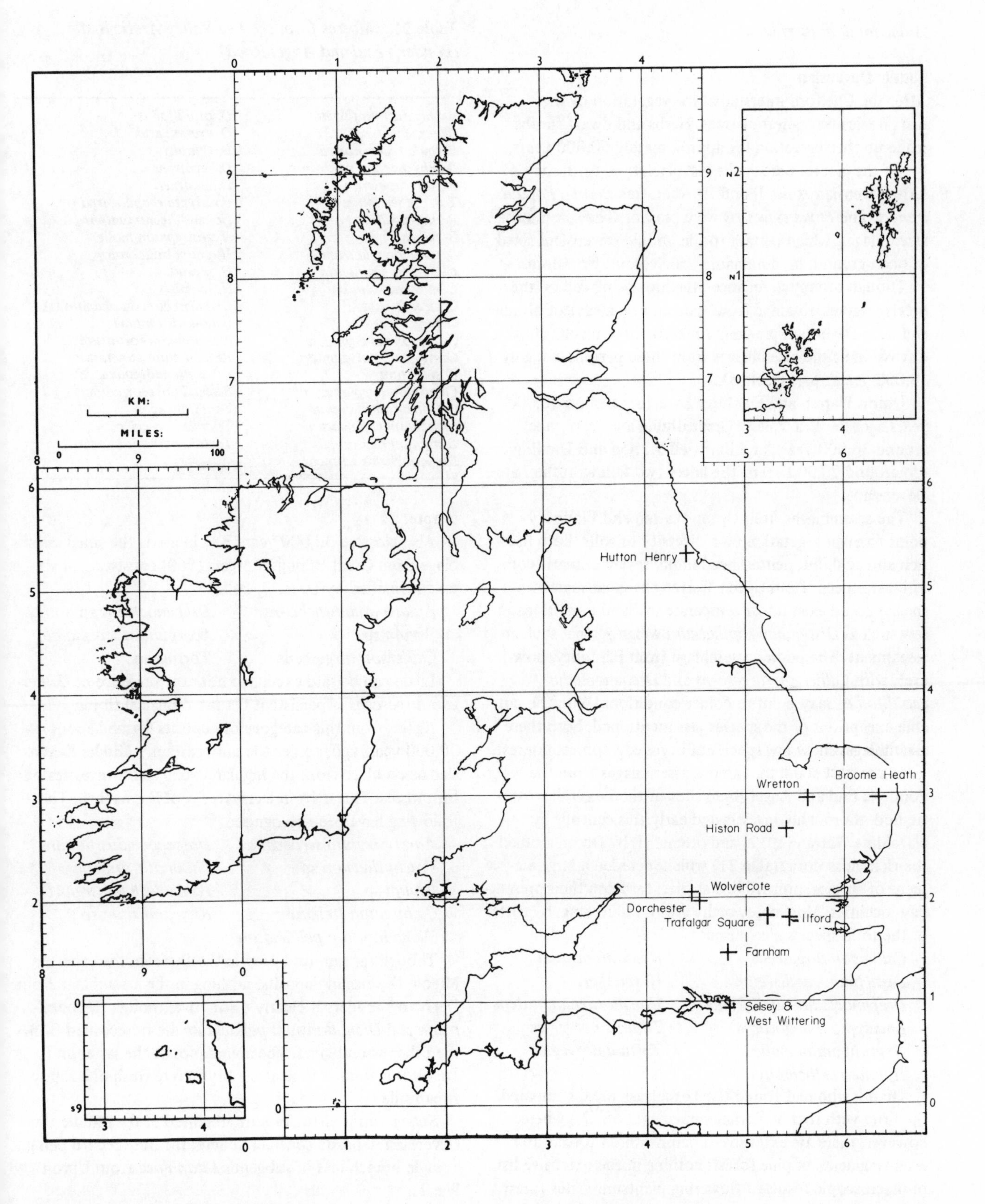

Fig. 86. Ipswichian and Wolstonian sites.

Aulacomnium turgidum.

Middle Devensian

After the Chelford interstadial the vegetation of the British Isles was bereft of trees. Herbs and dwarf shrubs made up the vegetation for approximately 50,000 years, by far the greater part of the last glaciation, until the Late Devensian zones II and III when tree *Betula, Populus tremula* and *Pinus sylvestris* were present. The seven moss assemblages, which pertain to the Middle Devensian, speak of open-ground, predominantly calcicolous, bryofloras.

Though too much reliance must not be placed on the precise figures obtained from radiocarbon assays of Middle and Late Devensian pre-zone I material, the results allow discussion of the assemblages from three periods, roughly 10,000 years apart (table 3).

Upton Warren and Fladbury have ages around 40,000 years ago, the Lea Valley, Great Billing and Derryvree around 30,000 years, and Barnwell Station and Dimlington around 20,000 years, the latter two falling in the Late Devensian.

The assemblages from Upton Warren and Fladbury point to open vegetation on a diversity of soils, both base-rich and acid, but neither need imply severe climatic conditions. Indeed, from Upton Warren the great majority of species could exist under temperate lowland conditions. A few such as *Distichium capillaceum* weigh against such an assessment. The poorer assemblage from Fladbury, however, with *Calliergon turgescens* and *Drepanocladus capillifolius*, may point to colder conditions. With the notable exceptions of the species just mentioned, both these assemblages show few species of bryogeographical interest.

In contrast stand the famous assemblages from the Ponder's End and Angel Road sites of the Lea Valley Arctic Bed, thoroughly investigated early this century by Hazeldine Warren (1912) and others. H. N. Dixon studied the rich moss flora (table 21) which included a large element of species, primarily calcicoles, far from their present-day localities. Most noteworthy are the following, a quarter of the total species recognised:

Calliergon turgescens	*Hypnum ravaudi*
Distichium capillaceum	*H. vaucheri*
Drepanocladus capillifolius	*Homalothecium nitens*
Encalypta rhabdocarpa	*Timmia norvegica*
Hygrohypnum molle	*Tortula norvegica*
Hypnum callichroum	

Dixon, who had visited the Torneträsk area, compared the flora with that of Tornelappmark (at 68° N) where, however, there are extensive subalpine birch forests and even fragments of pine forest; nothing in the extensive list of macroscopic fossils of flowering plants indicates forest. A few of the mosses are 'southern' species as defined in chapter 1.

Table 21 *Mosses from the Lea Valley Arctic Bed (Ponder's End and Angel Road)*

Amblystegium fluviatile	*D. capillifolius*
A. serpens	*D. exannulatus*
Barbula recurvirostra	*D. fluitans*
Brachythecium mildeanum	*D. revolvens*
Bryum capillare	*D. sendtneri*
T *B. creberrimum*	*Encalypta rhabdocarpa*
B. intermedium (t)	*Homalothecium nitens*
B. pallens	*Hygrohypnum molle*
B. pseudotriquetrum	*Hypnum callichroum*
Calliergon giganteum	*H. ravaudi*
C. sarmentosum (t)	*H. vaucheri*
C. stramineum	*Orthotrichum diaphanum* (t)
C. turgescens	*Philonotis fontana*
Calliergonella cuspidata	*Plagiomnium rostratum*
Campylium polygamum	*Rhizomnium punctatum* or *pseudopunctatum*
C. stellatum	*Scorpidium scorpioides*
Ceratodon purpureus (t)	*Thuidium* sp
Cratoneuron filicinum	*Timmia norvegica*
Distichium capillaceum	*Tortula norvegica*
Ditrichum flexicaule	
Drepanocladus aduncus	

Also close to 30,000 years B.P. in age is the small assemblage from Great Billing (Morgan 1969) consisting of six taxa identified by Dr F. G. Bell.

Aulacomnium palustre	*Drepanocladus* sp
Bryum sp	*Scorpidium scorpioides*
Calliergon turgescens	*Tortula* sp

Little can be said except to note the presence of *Calliergon turgescens*, a persistent feature of glacial floras.

Again, from this category of deposits formed about 30,000 years ago, comes the only certainly Middle Devensian assemblage from the highland zone so far investigated for mosses. This is from Derryvree, Co. Fermanagh. The following have been recognised:

Amblystegium serpens	*Drepanocladus fluitans*
Brachythecium sp	*Philonotis fontana*
Bryum sp	*Pohlia wahlenbergii* (t)
Campylium stellatum	*Rhacomitrium* sp
Dichodontium pellucidum	

Though far removed geographically from the other Middle Devensian deposits, nothing in the assemblage from Derryvree renders it clearly distinctive, though *Rhacomitrium* and *Dichodontium pellucidum* are unrecorded from the other assemblages. The abundance of the latter and *Philonotis fontana* indicates a bryophyte flush or a sandy streamside.

Sphagnum is virtually unrepresented in the Middle Devensian deposits considered here; the only record being a single branch leaf of subgenus *Litophloea* from Upton Warren.

Late Devensian

At Barnwell Station some 20,000 years ago grew the following, amongst others:

Abietinella abietina	*Drepanocladus revolvens*
Bryum sp	*Homalothecium lutescens*
Calliergon giganteum	*Scorpidium scorpioides*
Campylium stellatum	*Tortula* sp
Cratoneuron filicinum	

In a completely calcicolous flora, the majority are rich fen species which are a characteristic feature of most of the Middle Devensian assemblages.

Poorest of all in species is the Dimlington assemblage consisting of hundreds of fragments of *Pohlia wahlenbergii* var. *glacialis* (t), a variety characteristic of the sides of cold streams high in the mountains.

The thirteen taxa from zone I of Colney Heath constitute the richest of the few Late Devensian assemblages yet recovered from the lowland zone. With *Calliergon giganteum, Calliergonella cuspidata* and *Drepanocladus exannulatus* most abundant in fragments, the assemblage is overwhelmingly derived from a rich fen, as is so often the case with Devensian material. And, moreover, the two terricolous species recovered, *Abietinella abietina* and *Homalothecium lutescens*, are both heliophilous calcicoles.

There is little, if any, distinction from the Middle Devensian assemblages, especially those from Barnwell Station and the Lea Valley.

By way of geographical contrast the assemblage from Loch Droma, some 800 years younger than that from Colney Heath, is the richest zone I bryoflora from the highland zone. In the presence of *Hylocomium splendens* and *Rhacomitrium lanuginosum*, the two commonest species, and four species of *Polytrichum*, the assemblage differs strikingly from that of Colney Heath. A greater diversity of bryophyte vegetation is indicated: snowbeds, rich fens, rocks, flushes and a range of soil types from base-poor to base-rich.

Tortella tortuosa, Rhacomitrium fasciculare, Hypnum cupressiforme and *Homalothecium lutescens* or *sericeum* were the most frequently encountered species in the zone II assemblage from Low Wray Bay, Lake Windermere, which also included *Hylocomium, Rhacomitrium* and *Polytrichum* as well as *Antitrichia*.

These four last-named genera occur again and again in Late Devensian deposits in the western parts of the British Isles as at zone III of Kirkmichael, Isle of Man, where the most represented species were *Meesia tristicha, Cratoneuron commutatum, Climacium dendroides* and *Antitrichia curtipendula*. Again the rich assemblage points to a variety of habitats including mires with *Sphagnum*, a somewhat unusual occurrence in glacial deposits.

Macroscopic remains of *Sphagnum* are unrecorded from most Late Devensian sites and where they have been found they are sparse, never abundant.

Perhaps the outstanding zone III bryoflora is that from Drumurcher, Co. Monaghan. Of the twenty-six taxa recovered from zone III, twenty did not occur in the zone IV stratum. Of these twenty, thirteen listed below are now of restricted or very restricted montane distribution or are of otherwise disjunct range.

Abietinella abietina	*Hypnum bambergeri*
Bartramia ithyphylla	*H. revolutum*
Brachythecium glaciale (t)	*Plagiopus oederi*
Calliergon turgescens	*Polytrichum alpinum*
Distichium sp	*P. urnigerum*
Ditrichum flexicaule	*Timmia* sp
Drepanocladus uncinatus	

This is a markedly calcicolous and saxicolous bryoflora; *Cratoneuron filicinum* was the most abundant species.

The abundant data from these assemblages and many others undiscussed point to a Late Devensian bryoflora rich in species of diverse bryogeographical groups. This can be finally and well exemplified by a study, still in progress, of the mosses from a Late Devensian core covering zones I to III from Nant Ffrancon in Snowdonia. The complete tally will number some fifty species of which the following are among the most significant.

Antitrichia curtipendula	*Pleurozium schreberi*
Calliergon turgescens	*Polytrichum alpinum*
Fissidens cristatus (t)	*Rhacomitrium heterostichum*
F. osmundoides (t)	*R. lanuginosum*
Hylocomium splendens	*Rhytidium rugosum*
Oligotrichum hercynicum	*Sphagnum imbricatum*
Philonotis seriata	

7.4 FLANDRIAN

The investigation of more than seventy Flandrian deposits (excluding *Sphagnum* peats) in Britain and a much smaller number in Ireland has revealed bryophyte remains totalling some 160 mosses and twelve liverworts. Figs. 22, 23 and 24 show the wide distribution of these sites in Britain and the paucity in Ireland.

The Flandrian assemblages are readily divisible into three categories according to provenance, each category yielding different information.

(1) Fluviatile and lacustrine assemblages (fig. 22): Water-transported material from river gravels or in-washed layers from lakes may be species-rich and derived from various habitats. Fort William is an outstanding example, as is Seathwaite Tarn, and Windmillcroft and Cromwell Bottom less so.

(2) Mire assemblages (fig. 23): Even eutrophic peats are species-poor by comparison with allochthonous assemblages but bryogeographically may be of great interest.

Table 22 *Devensian and/or Flandrian zone IV species now with restricted disjunct ranges or primarily species of the montane areas of the British Isles*

**Abietinella abietina*	**Lescuraea patens*
**Andreaea rupestris*	*Meesia tristicha*
Antitrichia curtipendula	**M. uliginosa*
**Aulacomnium turgidum*	**Mnium orthorynchum*
**Barbula icmadophila*	**Oligotrichum hercynicum*
**B. spadicea*	**Onoophorus virens*
Blindia acuta	*Paludella squarrosa*
Brachythecium glaciale	**Philonotis seriata*
T**B. reflexum*	**Plagiopus oederi*
Calliergon sarmentosum	**Pohlia wahlenbergii* var. *glacialis* (t)
**C. turgescens*	*Polytrichum alpinum*
Cinclidium stygium	**P. norvegicum*
**Cratoneuron decipiens*	*P. urnigerum*
**Dicranum elongatum*	*Pseudobryum cinclidioides*
Distichium capillaceum	**Rhacomitrium fasciculare*
T**D. inclinatum*	*R. heterostichum*
Drepanocladus uncinatus	*R. lanuginosum*
**D. vernicosus*	*Rhizomnium pseudopunctatum*
**Encalypta rhabdocarpa*	*Rhytidium rugosum*
Fissidens osmundoides	*Sphagnum imbricatum*
Helodium blandowii	**Splachnum sphaericum*
Homalothecium nitens	**Timmia norvegica*
**Hygrohypnum molle*	**Tortella fragilis*
**Hypnum bambergeri*	**Tortula norvegica*
**H. revolutum*	
**H. vaucheri*	

Devensian species now extinct in the British Isles

**Drepanocladus capillifolius* **Hypnum ravaudi*

*Species unrecorded from post-zone IV deposits.

Especially noteworthy bryofloras have been discovered at more than twenty sites. Malham Tarn Moss, the Norfolk Broads, the Somerset Levels, Burtree Lane, Dun Moss, Fallahogy and the bogs of Co. Kildare are among the most significant.

(3) Archaeological assemblages (fig. 24): More bryogeographical and ecological significance attaches to these than at first thought one might imagine. Primarily they are of interest in revealing the past uses of mosses. The best examples are caulking for boats, North Ferriby, Brigg (Bronze Age); wells and other shafts, Wilsford (Bronze Age), Barnsley Park and Bunny (Roman); ditches, Carpow (Roman), Winchester (Saxon); pits, Winetavern Street, Dublin (Viking); and sods, Newgrange (Neolithic).

Forests

There are no assemblages certainly derived from the forests of *Betula* and *Pinus* which grew in zone IV and later times and knowledge of the deciduous forest bryoflora, especially of zones V, VI and VII*a*, is scanty and may remain so. The only deposits, apart from peat formed in fen woodland, which yield species likely to have grown in the closed portions of the forests are water-transported. Such allochthonous accumulations are liable to contain stream-side species and others intolerant of the shade of closed canopy deciduous forests.

The assemblage, imprecisely dated (zone V or VI), from the raised beach at Fort William, is by far the richest of Flandrian age. Dixon (1910), in discussing the provenance of the thirty-six taxa in table 23, recognised that the several habitats represented could be explained by envisaging a stream flowing through a rocky woodland. Species which occur frequently or exclusively in interglacial and Flandrian deposits are *Antitrichia curtipendula,* which made up the bulk of the material, *Eurhynchium* spp, *Hylocomium brevirostre, Isothecium* spp, *Neckera* spp and *Thuidium* spp. *Dicranum scottianum, Hyocomium armoricum* and *Plagiothecium undulatum* are considered in the discussion on oceanic species.

Table 23 *Mosses from Fort William*

Andreaea rothii	*Hygrohypnum luridum*
Antitrichia curtipendula	*Hylocomium brevirostre*
Blindia acuta	*Hyocomium armoricum*
Brachythecium plumosum	*Hypnum cupressiforme*
B. rutabulum	*Isothecium myosuroides*
Bryum pseudotriquetrum	*I. myurum*
Calliergonella cuspidata	*Mnium hornum*
Cratoneuron commutatum var. *falcatum*	*Neckera complanata*
	N. pumila
Dichodontrium pellucidum	*Plagiothecium undulatum*
Dicranum bonjeanii	*Rhacomitrium canescens*
D. scottianum	*R. heterostichum*
Drepanocladus fluitans	*Rhytidiadelphus loreus*
D. uncinatus	*R. squarrosus*
Eurhynchium praelongum	*Sphagnum fimbriatum* or *S. girgensohnii*
E. riparioides	
E. striatum	*Thuidium delicatulum*
Fissidens osmundoides	*T. philiberti*
Grimmia sp	*T. tamariscinum*

Another fluviatile sequence which proved productive of moss remains was discovered in a gravel pit at Cromwell Bottom, Elland, near Leeds. Bartley (1964) analysed five organic accumulations sealed above and below by water-borne sands and gravels. Remains of *Betula, Cornus sanguinea, Corylus, Quercus* and *Salix* were recovered from deposit 1, referred to zone VI, as well as *Antitrichia curtipendula, Eurhynchium praelongum* (t) and *Plagiothecium denticulatum.*

The mosses extracted from 'lagoon silt and peat' (Jessen 1949, p. 138) at Cushenden, Co. Antrim, are also of zone VI age. They give the only insight of Irish forest bryoflora of the first half of the Flandrian. Hesselbo identified the following:

Eurhynchium sp	*Neckera complanata*
Hylocomium sp	*Thamnobryum alopecurum*
Thuidium tamariscinum	

The zone VII*a* bryoflora from terrestrial as distinct from wetland habitats, is very little known. There are no species-rich assemblages and, apart from scattered records, there are only three assemblages to discuss. Loch Cuithir, Isle of Skye, produced *Antitrichia curtipendula, Homalothecium lutescens* or *sericeum, Hylocomium splendens* and *Hypnum cupressiforme*. Buried organic layers at Aust, Gloucestershire, and Bryn-y-Mor, Merioneth, each yielded a few species pointing to woodland. Both gave *Eurhynchium praelongum* and *Hypnum cupressiforme*, with *Antitrichia curtipendula* and *Ulota* sp only in the former and *Dicranum scoparium* and *Neckera crispa* only in the latter.

Zone VII*b* is better endowed with terrestrial moss assemblages. Cromwell Bottom again is pertinent. Deposit 3 yielded remains of *Betula, Quercus, Pteridium* and *Ranunculus bulbosus* and the following mosses:

Dicranum majus	*Mnium hornum*
Eurhynchium praelongum (t)	*Thuidium tamariscinum*
Hyocomium armoricum	

Deposit 5, which may belong to either zone VII*b* or VIII, though more likely the former, gave a more diverse flora.

Amblystegium riparium	*Neckera complanata*
Antitrichia curtipendula	*Plagiothecium denticulatum*
Dichodontium pellucidum	*Rhizomnium punctatum*
Homalothecium sericeum	
Mnium hornum	

Associated tree and shrub remains were *Alnus, Betula, Cornus sanguinea, Fraxinus, Ilex* and *Pinus*.

As another instance of the Flandrian forest bryoflora mention may be made of the imprecisely dated assemblage from Windmillcroft, central Glasgow (Appendix 1). Though riverbank species were well represented in the deposits, which were organic layers in fluviatile gravels, very similar to the Cromwell Bottom situation, a striking component is the woodland one.

Homalia trichomanoides	*N. crispa*
Isothecium myosuroides	*N. pumila*
Neckera complanata (t)	*Thuidium tamariscinum*

The zone VII*b* assemblage from Shippea Hill represents the ground and trunk floras of a fen woodland comprising several shrubs and trees, notably *Tilia platyphyllos*. *Neckera complanata, Homalothecium sericeum* and perhaps also *Hypnum cupressiforme* may well have been corticolous with *Cratoneuron filicinum, Eurhynchium speciosum, E. swartzii* and *Thamnobryum alopecurum* in the more or less wet ground layer.

There may only be one instance of epiphytic mosses preserved *in situ*. In the H. N. Dixon collection lodged in the British Museum (Natural History) there is a packet labelled 'mosses from trunk of oak tree in Pleistocene peat bed, Royal Albert Dock, Essex, 1914'. The three mosses in the packet are *Homalothecium sericeum, Neckera complanata* and *Thamnobryum alopecurum*, all of which were recovered from the similar deposit at Shippea Hill just discussed. This peat bed in the Royal Albert Dock formed sometime in zone VII.

Table 24 *Caulking bryophytes from the Brigg dugout canoe*

Brachythecium rutabulum	*Thuidium tamariscinum*
Bryum sp	T *Calypogeia trichomanis*
Distichium capillaceum	*Diplophyllum albicans*
Eurhynchium confertum	*Lepidozia reptans*
E. striatum	*Lophocolea cuspidata*
Fissidens bryoides	*Lunularia cruciata*
Hylocomium splendens (t)	*Metzgeria furcata*
Mnium hornum	*Pellia epiphylla*
Plagiothecium undulatum	*Plagiochila asplenioides*
Rhizomnium punctatum	*Riccardia* sp
Rhytidiadelphus triquetrus	*Scapania* sp

Archaeology has provided several assemblages primarily connected with the forest environment. Two examples must suffice. The Bronze Age dugout canoe from Brigg in Lincolnshire was an exceptional discovery. No less than twenty-two bryophytes including ten liverworts had been used to plug the seams of the stern board and a patch (table 24). Only *Distichium capillaceum* does not fit the assessment that the bryophytes were collected from a moist, shaded river bank. One does not know where the river bank was located; it need not have been in Lincolnshire or, indeed, anywhere in the lowland zone.

The other archaeological example is Winetavern Street, Dublin, of Medieval age. Pit 1 yielded thirteen bryophytes.

Calliergonella cuspidata	*N. crispa*
Cirriphyllum piliferum	*Plagiochila asplenioides*
Dicranum scoparium	*Plagiomnium affine*
Eurhynchium praelongum	*Rhytidiadelphus squarrosus*
E. striatum	*Thamnobryum alopecurum*
Fontinalis antipyretica	*Thuidium tamariscinum*
Neckera complanata	

These bryophytes were, of course, not *in situ* but had been collected for some specific use. With the exception of the aquatic *Fontinalis*, the assemblage has an ecological consistency; the species were gathered from partially or deeply shaded soil, trunks and rocks in forest with a base-rich substratum.

To finish this section on forest bryofloras mention may be made of the assemblage from Seathwaite Tarn, in the southwestern fells of the English Lake District, one of the youngest but also one of the ecologically most significant bryofloras. Organic layers washed into the tarn in the first few centuries A.D. contained nineteen taxa of mosses.

On a brief survey of the present bryoflora of the drainage basin of the tarn the author failed to find seven of the species from the wash-in layers. These are as follows:

Calliergonella cuspidata | *Hylocomium umbratum*
Dichodontium pellucidum | *Plagiomnium rugicum*
Eurhynchium riparioides | *Thamnobryum alopecurum*
Fontinalis sp

A further four species from the core are not at present significant components of the vegetation of the hillsides round the tarn. These are *Breutelia chrysocoma, Hylocomium splendens, Rhytidiadelphus loreus* and *R. squarrosus*.

The wash-in assemblage cannot be accounted for in terms of the present bryoflora of the streams and hillsides. Reasons for this are discussed on p. 195; human destruction of forest may be the major factor.

Grasslands, heaths and other habitats

The small assemblage from Newgrange, Co. Meath, is unusual, indicating a quite different environment. *Brachythecium rutabulum* (the major component) and *Plagiomnium undulatum* were recovered from the sods forming a wall, a part of the construction of the vast Neolithic burial mound. One can envisage a damp, weedy turf as the habitat; Professor Mitchell found *Ranunculus repens, Taraxacum officinale, Stellaria* sp and *Cerastium* sp to be the most common flowering plants.

Moss remains unequivocally derived from grasslands such as those from Newgrange are scarce. A less convincing example is the assemblage from the Bronze Age site at Wilsford, near Amesbury, Wiltshire. Mr E. C. Wallace identified the following:

Eurhynchium swartzii *Hylocomium splendens* *Neckera complanata*	frequent
Calliergonella cuspidata *Eurhynchium praelongum* *Homalothecium lutescens*	scarce
Brachythecium glareosum (t) *Eurhynchium riparioides* *Thamnobryum alopecurum*	rare

The provenance of these mosses is extraordinary; they were found with much woody and herbaceous debris at the bottom of a shaft 30 m deep under a pond barrow (Ashbee 1963). For all the species except the two last-named an adequate interpretation is derivation from calcicolous grassland on the surrounding chalk. Though the first five all flourish in forest, *Homalothecium lutescens* is not a woodland plant.

While the Wilsford bryoflora may be derived, at least in part, from a grassland, not all assemblages fossilised in man-made diggings need necessarily have external provenances. The assemblage from the Roman well at Bunny may indeed have a source outside the well, but that from Barnsley Park is different. Here most if not all of the species could have grown on the inside damp stonework of the wall. Indeed the growth form of *Thamnobryum* (plate 9) and *Eurhynchium swartzii* is a strong indication of such a source. It is a very shade-tolerant assemblage.

Archaeology provides another example of a habitat little revealed by macroscopic analyses. The plants used as caulking in canoe No. 2 from Holme Pierrepont, near Nottingham, were derived from a heath, in particular a *Callunetum*. The mosses were as follows, *Hylocomium splendens* and *Pleurozium schreberi* being the most abundant:

Dicranum scoparium | *Polytrichum commune*
Hylocomium splendens | *Rhytidiadelphus squarrosus*
Hypnum cupressiforme
Pleurozium schreberi

Calluna vulgaris was most abundant of the flowering plants which also included *Carex pilulifera*.

The extensive areas of sandhills in South Lancashire are well known for the occurrence of such mosses as *Amblyodon dealbatus, Bryum mamillatum, B. neodamense, Catoscopium nigritum* and *Meesia uliginosa* (Savidge 1963). From the Liverpool area there are two subfossil assemblages which pertain to dune slacks. The better is that from Aintree (Travis 1909) where *Campylium polygamum, Drepanocladus aduncus, D. revolvens* and *Scorpidium scorpioides* were found together; the age is vague, zone VII*b* or VIII.

In the second assemblage of similarly imprecise age (Travis 1913), which was recovered from gravels at Seaforth, *Amblydon dealbatus*, if not others, may have had a dune slack origin.

Across the River Mersey at Wallasey, Cheshire, the richest dune slack assemblage was found (Travis 1922). Again the age is merely zone VII or VIII. According to Travis no such environment as indicated by the following exists at Wallasey any more.

Calliergon giganteum | *Drepanocladus aduncus*
Calliergonella cuspidata | *D. lycopodioides*
Cratoneuron filicinum | *D. revolvens*
C. commutatum var. *falcatum* | *D. sendtneri*
| *Scorpidium scorpioides*

Mires

The bryogeographical and general ecological interest of the mires of the British Isles is well known. It stems not just from the present occurrence of disjunct species and the dominance of mosses, especially *Sphagnum* species, but also from peat stratigraphy, composed very largely of *in situ*, often well preserved, remains of mosses.

Layers of *Sphagnum imbricatum* peat were recognised

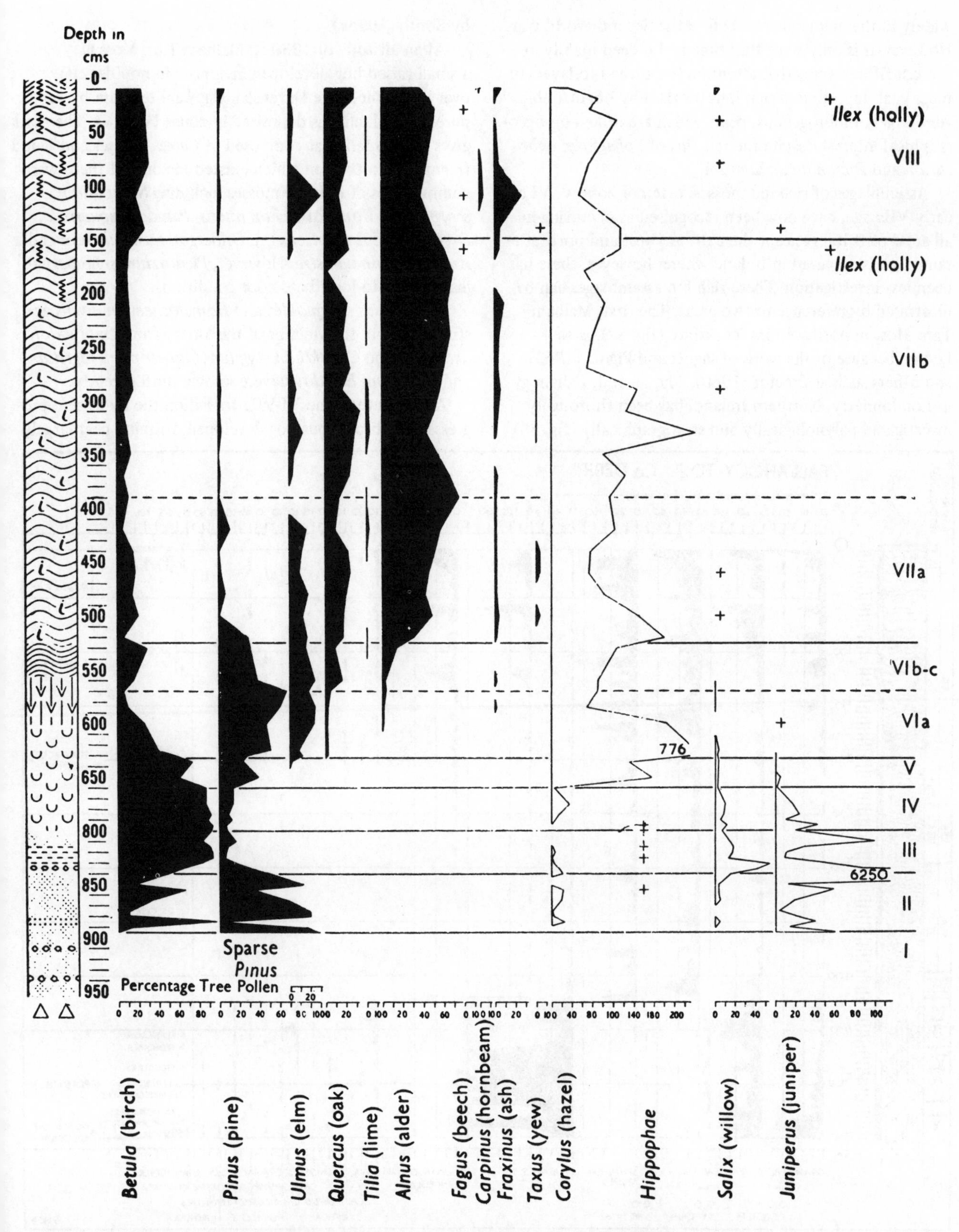

Fig. 87. Tree pollen diagram from Malham Tarn Moss, Yorkshire. From Pigott and Pigott 1963.

widely in the highland zone before the Second World War. However, it is only since that time and indeed mainly in the last fifteen years that attention has turned to layers of moss peat deposited in rich fens overlain by oligotrophic, very largely ombrogenous, peats which are of less bryogeographical interest despite the remains of *Sphagnum imbricatum* and *Dicranum undulatum*.

Assemblages of rich fen mosses, often of zone V, VI or early VII*a* age, have now been recognised as occurring in all areas of Britain except the extreme west and north; the same lack is apparent in Ireland where, however, there has been less investigation. These rich fen assemblages can be illustrated by reference to two areas. The first, Malham Tarn Moss in northwestern Yorkshire (fig. 87), is well known because of the work of Pigott and Pigott (1963) and others such as Proctor (1960). The second, Fallahogy in Londonderry, Northern Ireland, has been thoroughly investigated palynologically and stratigraphically (fig. 88) by Smith (1958*a*).

At an altitude of 1230 ft, Malham Tarn Moss is a small raised bog developed in a basin in boulder clay over limestone. Late Devensian clays are overlain by very pure marl which was deposited in zones IV to VI. The marl gives way to fen peat composed of *Carex*, mosses and wood fragments. In this fen which existed in zone VI there were communities of rich fen mosses including *Scorpidium scorpioides, Homalothecium nitens, Paludella squarrosa, Drepanocladus revolvens* (t), *Calliergon sarmentosum* and *Aulacomnium palustre*. However, *Pleurozium schreberi* at least points to local base-poor conditions.

Scorpidium scorpioides and *Homalothecium nitens* are still present in the vicinity of the Moss as are *Amblyodon dealbatus* and *Cinclidium stygium*; *Catoscopium nigritum* and *Helodium blandowii* were known until recently.

At about the zone VI-VII*a* transition the hydrosere progressed; ombrogenous bog developed, forming peat largely

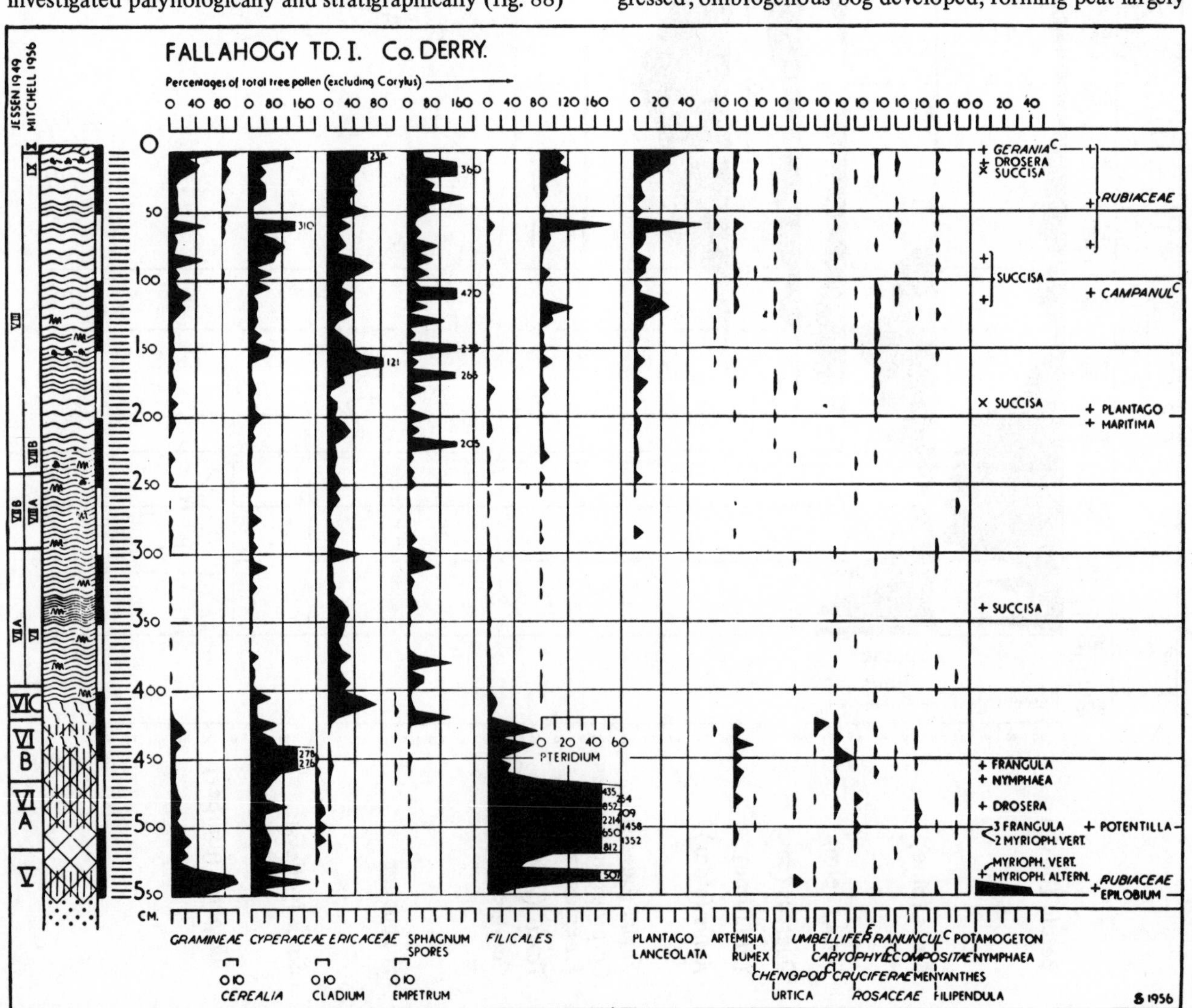

Fig. 88. Non-tree pollen diagram from Fallahogy. From Smith 1958*a*.

composed of *Eriophorum vaginatum* and *Sphagnum imbricatum* which no longer occurs on the Moss, nor does *Dicranum undulatum*, found close to the surface. Burning, grazing and marginal peat cutting have altered the vegetation of the Moss.

At an altitude of 150 ft the raised bog at Fallahogy is about one mile long and half a mile wide. It formed between drumlins of the last glaciation in the Lower Bann Valley. Over the basal coarse sand, detritus mud, with remains of reed-swamp plants, mosses and wood fragments, was deposited in zones V and VI. *Homalothecium nitens, Paludella squarrosa* and *Drepanocladus fluitans* (t) have been recorded from a layer of moss peat from zone VI. Perhaps one can envisage a transitional rich fen for this assemblage.

Like the development at Malham Tarn Moss, the fen gave way to ombrogenous bog at about the zone VI-VII*a* transition; *Sphagnum imbricatum* and *S. papillosum* are the only moss species identified from the 4 m of raised bog peat. Marginal peat cuttings have been made.

8 THE FORMATION OF BRYOPHYTE DISTRIBUTIONS

8.1 THE ADVENT OF MAN

Past uses of mosses

Sphagnum is something of an exception to the generalisation that mosses have few economic uses in the urban civilisations of modern times (Welch 1948; Ando 1957; Schofield 1969*a*). As a major component of peat it is used in horticulture and as fuel. Its use in the First World War as a wound dressing is well known.

In the past, from at least as early as Mesolithic times, mosses have been used in a variety of ways. Indeed it is likely that there were many more uses, some highly specific, than we realise.

The Mesolithic flint flake from the River Bann (plate 21) may be a unique discovery in possessing a moss padding for the user's hand. One may speculate, however, that such a use was widespread but that the united preservation of flake and handle has been an extremely rare event. Also from the Mesolithic there is an instance of a type of use which has been very common until modern times; this is the plugging of seams or cracks. The excavation of the birch platform at Star Carr (Clark 1954) revealed wads of *Holmalothecium sericeum* and *Isothecium myosuroides* between the timbers. A modern example of such a use is the plugging of cabins in Valais, Switzerland (Doignon 1954).

The Bronze Age cist grave at Methilhill, Fife, excavated by Henshall (1964), contained a highly disintegrated mass of plant debris in the vicinity of the skeleton's chest where there was also placed a sheathed dagger. The details of the excavation and subsequent botanical analysis did not reveal the reason why *Sphagnum palustre* was a major component of the plant debris and *Hylocomium splendens* a minor one. The speculation is tempting that perhaps the *Sphagnum* had been a wound dressing (Lambert 1964). However, it is more likely, perhaps, that the dagger was packed in *Sphagnum*. Barrow 85 at Amesbury offers a parallel. Here, according to Newall (1931), a dagger and a scraper were carefully packed in moss and yew leaves; the mosses were abundant *Sphagnum* and a little '*Hypnum*' and *Plagiomnium undulatum*. Mosses have also been recovered from Bronze Age graves in Denmark (Iversen 1939).

The excavation of a Bronze Age crannog at Ulrome, Holderness (Smith 1911; Varley 1968), revealed *Antitrichia curtipendula* and *Hypnum cupressiforme*, considered by Smith (1911) to have been 'possibly used for tinder for striking fire'. The lake dwellings of Switzerland and adjacent regions of Europe have yielded mosses as described by Keller (1878) who states (p. 524) that mosses were 'undoubtedly used for stopping the holes in walls of the huts and also for bedding'. On p. 532 he lists *Antitrichia curtipendula, Neckera crispa, N. complanata, Thuidium delicatulum, Anomodon viticulosus, Leucodon sciuroides*, and *Hylocomium brevirostre*. For the most part these are all species which turn up again and again in archaeological contexts.

However, it is the use of mosses as part of boat-building techniques in the Bronze Age which demands most attention. The outstanding examples are the three craft of sewn oak planks found in the muds of the Humber at North Ferriby (Wright and Wright 1947; Wright and Churchill 1965). The highly-skilled workmen who constructed these complex boats, about 50 ft long and 8½ ft broad (fig. 89), deliberately selected for caulking material *Neckera complanata*, pure but for a slight admixture of *Eurhynchium striatum*, and also *Polytrichum commune*, twisted into ropes, in the case of boat 3. The considerable quantity of *Neckera complanata* needed for the purpose makes one wonder why this species was chosen rather than any commoner moss. Even if the species were commoner then than it is today, collection in large quantities may not have been easy. No favourable physical property peculiar to *Neckera complanata* can be readily envisaged. Perhaps there was sympathetic magic involved, as in the case of *Fontinalis antipyretica* used as plugging for wooden dwellings by Nordic peoples in the belief that fire was prevented.

The 'raft' (another sewn boat; Thorpe 1887) and a monoxylous canoe from Brigg, Lincolnshire (Atkinson 1887; Sheppard 1910; Smith 1958*c*), were also caulked with bryophytes; twenty-three species of liverworts and mosses made up the caulking of the stern board and patch on the canoe which was recovered from brackish water clay of the old River Ancholme. The radiocarbon age of wood from the canoe is 2784 ± 100 years B.P. (Godwin and Willis 1961). In contrast to their predecessors at North Ferriby, the builders of the Brigg canoe may have gathered

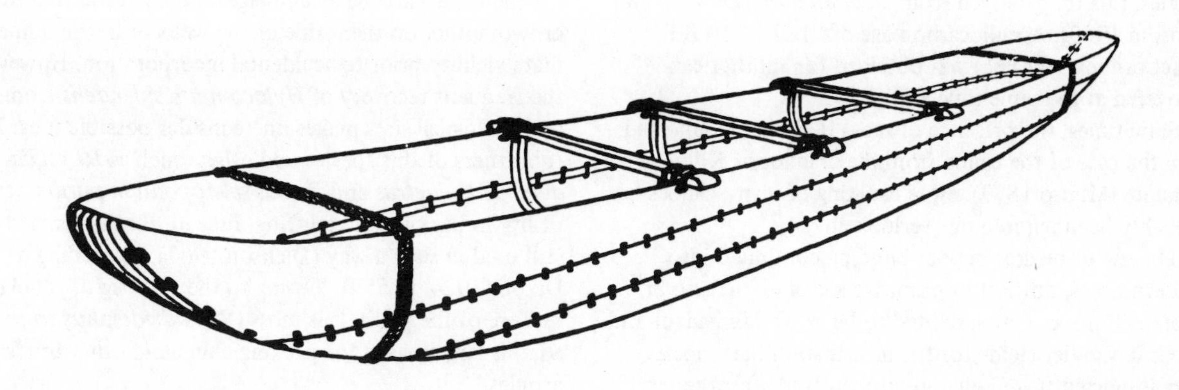

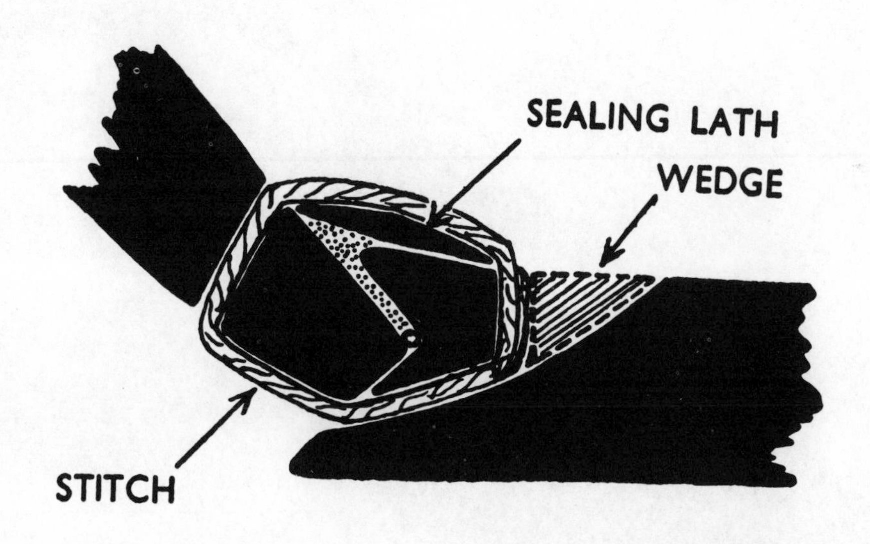

Fig. 89. The North Ferriby boats. Above, reconstruction of boat No. 1. Below, the method of caulking. The moss wads are *Neckera complanata* and the moss rope is *Polytrichum commune* (plate 21). From Wright and Churchill 1965.

the twenty-three species quite casually from a wooded river bank.

Mosses may well have been used very widely as caulking material, not just in the Bronze Age but also in the Iron Age and later times. A crack in a monoxylous canoe from Holme Pierrepoint, Nottingham, was caulked with six species, this time derived from a *Callunetum* (Dickson and Ransom 1968). A radiocarbon age of 2180 ± 110 B.P. (MacCormick 1970 pc) was obtained for another canoe recovered at the same time and place.

Sometimes, the presence of moss is all that is reported, as in the case of the canoe from the crannog at Kilmaurs, Ayrshire (Munro 1879). Moss caulking of many canoes has probably been ignored or overlooked.

The use of mosses in boat-building continued into modern times, not just in primitive societies but also in Western Europe. I am indebted to Dr W. D. Margadant and Mr G. P. van der Heide for the information that mosses were imported from Belgium into Holland after the sixteenth century for caulking carvel-built boats. As a final instance it may be noted that Hugh Miller (1854, p. 272) in the early nineteenth century, on visiting a boatyard at Gairloch, found 'the Highland builder engaged in laying a layer of dried moss, steeped in tar, along one of his seams, and learned that such had been the practice in that locality from time immemorial'.

There is no pressing need to assume that the mosses recovered from Roman wells are indications of any utility; for the most part the assemblages can be accounted for by growth either on the inside of the wells or in the immediate vicinity prior to accidental incorporation. However, the frequent recovery of *Hylocomium splendens* from archaeological sites makes one consider possible uses. The robustness of this species and others such as *Rhytidiadelphus triquetrus* and *Pseudoscleropodium purum* confers utility in packing and stuffing. Indeed, the last-named is still used in such a way (Dickson 1967). According to Dixon (1924, p. 559), 'Owing to the very rigid, yet elastic texture of the plant, this moss (*Rhytidiadelphus triquetrus*) is largely used for packing china and other brittle articles.'

Polytrichum commune has been recovered from many archaeological sites. Its size, flexibility and toughness make it suitable for brooms, ropes and weaving into small con-

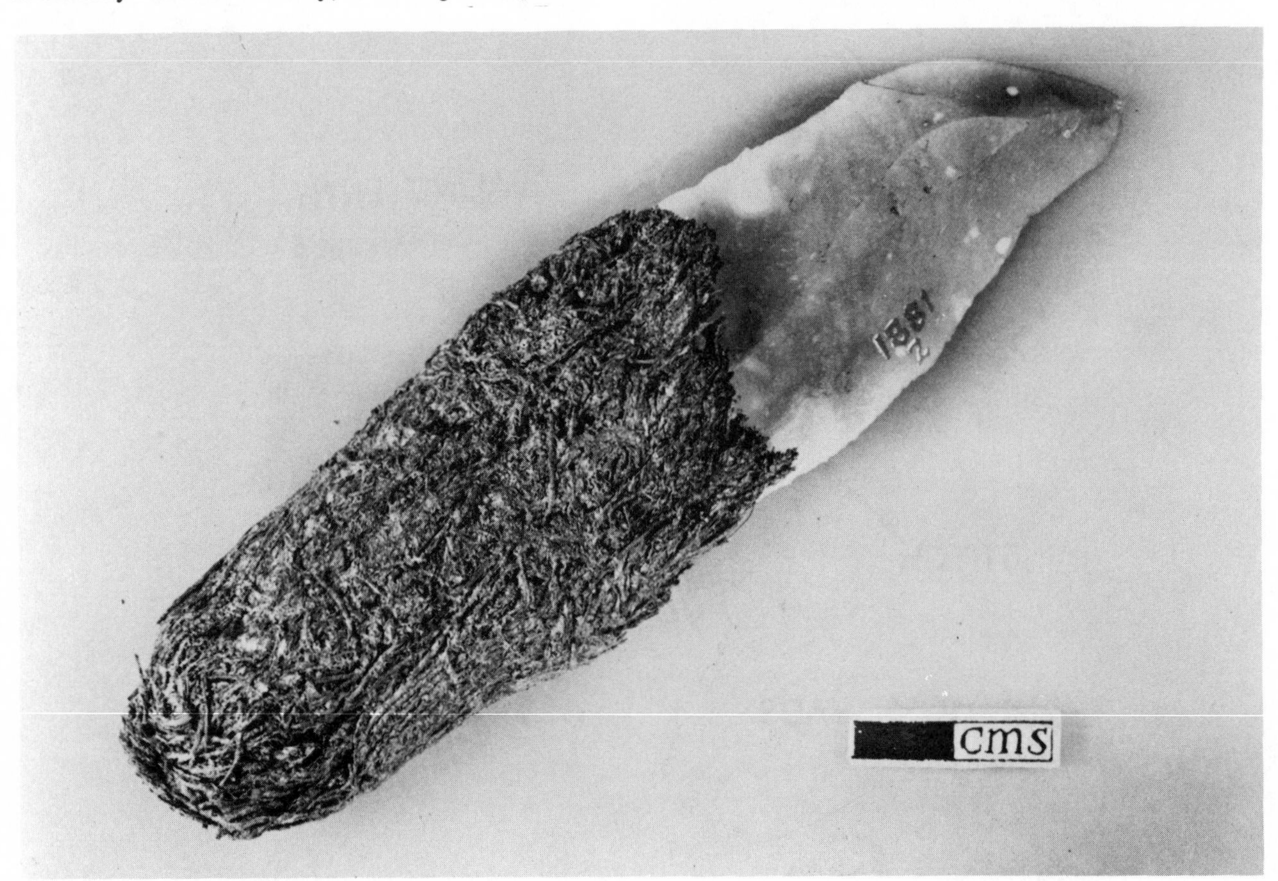

Plate 21. (*a*) Mesolithic flint flake from the River Bann. Moss handle consists of *Hylocomium brevirostre*. National Museum of Ireland.

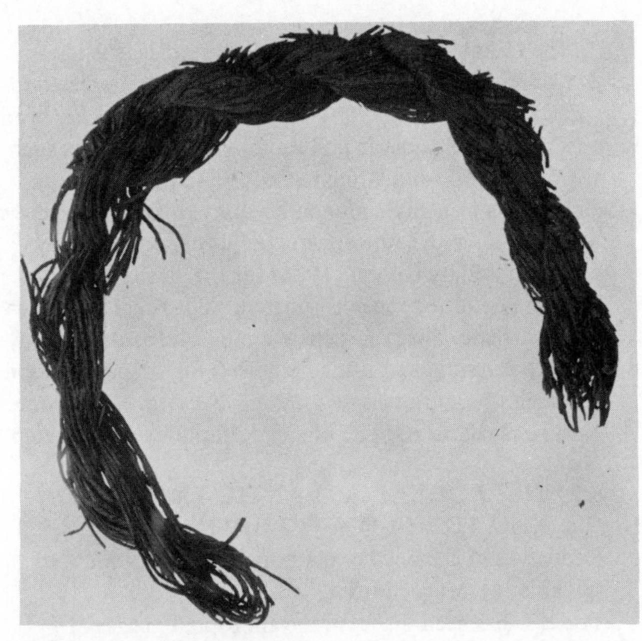

Plate 21. (*b*) Fragment of a twist rope of *Polytrichum commune* used for caulking North Ferriby Boat No. 3.

tainers (plate 21).

However much bryophytes have been exploited in the ways discussed above and in other unsuspected ways the effect on bryophyte distribution patterns must have been minimal. Perhaps *Neckera complanata* was locally depleted in Bronze Age Yorkshire! Nothing stronger can be said.

However, the advent of man as a destroyer of forests and mires has had a major effect on the formation of bryophyte ranges.

Destruction of forests

The realisation of the extent in space and time of man's influence on vegetation, particularly forest, is one of the major successes of pollen anlysis. Whatever the effects of Mesolithic folk, it is certain that Neolithic man began to clear the forests in however small and temporary a fashion at least 5000 years ago and gradually through the Bronze, Iron and historical ages the destruction became more and more effective until the present virtually deforested landscape resulted. Smith (1970) and Turner (1970) have given effective reviews of the British evidence. The maps of Scotland published by McVean and Ratcliffe (1962) showing the forest cover before and after man's intervention make the point dramatically. One can readily imagine that such a profound change in the dominant vegetation altered the bryophyte vegetation.

The anthropogenic destruction of forest by fire must have created many times a type of habitat only very rarely produced without man's intervention under the British climate. *Ceratodon purpureus, Funaria hygrometrica, Leptobryum pyriforme, Marchantia polymorpha* and others which can spread explosively on burnt ground at present, may well have been flourishing similarly for several thousand years; *Ceratodon purpureus* and *Funaria hygrometrica* certainly, and possibly the others also, were present and widespread before the forest destruction began and so were ready to take advantage of the new habitats. Nevertheless, no macroscopic subfossils have been extracted from charcoal horizons, as far as I know. However, there are claims of *Funaria* spores from forest clearance horizons.

The major effect of the reduction of forest must have been the decline of shade-tolerant and shade-demanding species. Again there is no evidence from subfossil assemblages, with the possible exception of Seathwaite Tarn. However, the fossil record gives more than a hint that the Flandrian forests had members of the Neckeraceae, notably *Neckera complanata* and *Thamnobryum alopecurum*, and *Eurhynchium* species such as *E. praelongum* and *E. striatum*, as important components of the bryophyte vegetation. With the elimination of vast tracts of forest the ecological significance of these species waned.

Possibly the most cogent subfossils from the Seathwaite Tarn core already discussed (p. 188) are *Hylocomium umbratum* and *Thamnobryum alopecurum*. Why should they and five other species be absent from the catchment area and four other species be insignificant there now? Discussing these wash-in layers Pennington (1970, pp. 72-3) envisages deforestation and worsening soil conditions taking place as a result of Brigantian occupation. The destruction of forest may well have deprived these two species of the shaded habitats they favour. However, it is difficult to dissociate the effects of forest clearance and soil impoverishment.

Destruction of mires

Brief accounts of the extinctions of *Paludella squarrosa* and *Helodium blandowii* as British species have already been given in the relevant pages of chapter 5. These are the most extreme, clear-cut cases; *Paludella squarrosa*, reduced to three localities in Yorkshire and Cheshire, and *Helodium blandowii*, reduced to four localities in the same counties, have vanished in the last hundred years. While precise details of the events at Malham Tarn and Skipwith Common may be lacking, it is certain that the elimination of these species from Britain is anthropogenic. Baker (1906, p. 533) has described the extinctions of *Paludella* as follows.

> Deep peat bogs, very rare.
> Found in Terrington North Car, near Castle Howard, by H. Ibbotson, 1842. Growing there in fair abundance in 1853. (M.B.S.) After the wet summer of 1860, a deep drain was cut through the centre of the peaty bog. In July, 1868, the bog was visited by Messrs Slater and Stabler in search of this moss. It had then almost dis-

appeared, by careful searching, they succeeded in observing a few stems of the plant. It has not been seen since, and must now be considered extinct as a British Moss until found in some other locality. The only other known British locality for this rare plant was Knutsford Moor, Cheshire, where it was found by W. Wilson in 1832. Draining operations on this locality have long since destroyed the plant.

For the same reasons *Homalothecium nitens* has become more disjunct, having disappeared almost completely from the lowland zone and from parts of the highland zone as well. The same applies to *Dicranum undulatum*. A similar explanation may apply to the final restriction of *Meesia tristicha* to Ireland and to the total disappearance of *Meesia longiseta*, unknown in the living state in the British Isles, but with two English localities in the last few thousand years.

These are the most striking examples of anthropogenic reductions because they apply to species already more or less relict in Britain for reasons unconnected with man. However, many mire plants, bryophytes and vascular plants alike of greater abundance under present British conditions have undergone numerous local extinctions and show considerable contractions, especially in the lowland zone, as a result of the wholesale devastation of many mires and the greater or lesser alteration of hundreds of others. The total and very recent elimination of the mires at Chat Moss in South Lancashire, the Fenland Margin in Huntingdonshire, the Somerset Levels, and Amberley Wild Brooks in Sussex, makes the point which need not be elaborated here. Cutting of peat for fuel, drainage operations and the accidental or intentional burning of mires have been the destructive agents.

Many species of *Sphagnum*, with the outstanding example of *S. imbricatum*, whatever the full explanation of its decline, have become less common as a result of these activities.

Other anthropogenic events

There are other ways in which the bryoflora has been affected by man's activities. However, an extensive discussion would be inappropriate in a book which deals primarily with the subfossil evidence. Two examples must suffice: the creation of a field system of agriculture and atmospheric pollution.

There are no British subfossils, and very few from anywhere relevant to this topic, of terricolous mosses such as species of *Pottia*, *Phascum* and *Ephemerum* which may well have increased as a result of agricultural practices which maintain bare soil. The spores of *Riccia* and *Anthoceros* recovered from agricultural horizons in the Netherlands are pertinent in this context, as may be the spores of the latter from Kent and elsewhere in the British Isles.

The recent decrease of corticolous species (such as *Antitrichia curtipendula, Cryphaea heteromalla, Leucodon sciuroides, Orthotrichum lyellii* and *Ulota crispa*) has been ascribed to atmospheric pollution by several authors such as Coker (1967) and Whitehouse (1964). Partial or total elimination of many common bryophytes from areas downwind of sources of atmospheric pollution, convincingly demonstrated by Gilbert (1968) for the Newcastle upon Tyne region, can be seen throughout industrialised Britain and elsewhere. That again there are no subfossils certainly relevant is hardly surprising for so recent a transformation. However, the subfossils of *Antitrichia curtipendula* from Nottinghamshire and Lincolnshire should be remembered.

8.2 THE BOREAL-ATLANTIC TRANSITION AND THE CLIMATIC OPTIMUM

Reduction of the northern-continental species and the spread of ombrogenous bog

The case has been made on the appropriate pages of chapter 5 and the relevant figures that during Late Devensian zones I-III species such as *Calliergon trifarium, Cinclidium stygium, Helodium blandowii, Homalothecium nitens, Meesia tristicha* and *Paludella squarrosa* were widespread in the British Isles. This statement applies especially to Britain but perhaps less so to Ireland, though the occurrence of *Homalothecium nitens* in Late Devensian Ireland at Roundstone and elsewhere should be noted.

Perhaps there was some restriction of these species at the Late Devensian-Flandrian transition. If the glacial habitats of *Homalothecium nitens* were of the more terrestrial rather than paludal type local eliminations may be envisaged. However, burial by mineral debris as occurred at Kirkmichael (fig. 90) may well have been highly exceptional.

There is abundant evidence that these species easily survived the vegetational changes of some 10,000 years ago. They flourished in Flandrian zones IV to VI. To recall the substantial layers of *Paludella* peat at Dun Moss, Burtree Lane, Malham Tarn Moss and Fallahogy (fig. 90) is enough to make the point.

These species are the 'subarctic glacial relicts' of various authors, the first of whom was Herzog (1926). In the light of the now numerous subfossils this designation is unsatisfactory. 'Relicts' in the southern parts of their ranges they most certainly are. However, the adjective 'glacial' is now inapposite. It is true that these species were widespread during last glacial times but the subfossil layers point to a greater abundance in the first few millennia of the Flandrian. It is difficult to coin an apt, terse designation. 'Northern-continental rich fen relicts' is more accurate but too long.

All the species under discussion survive in the British Isles, if only precariously as in the case of *Meesia tristicha*,

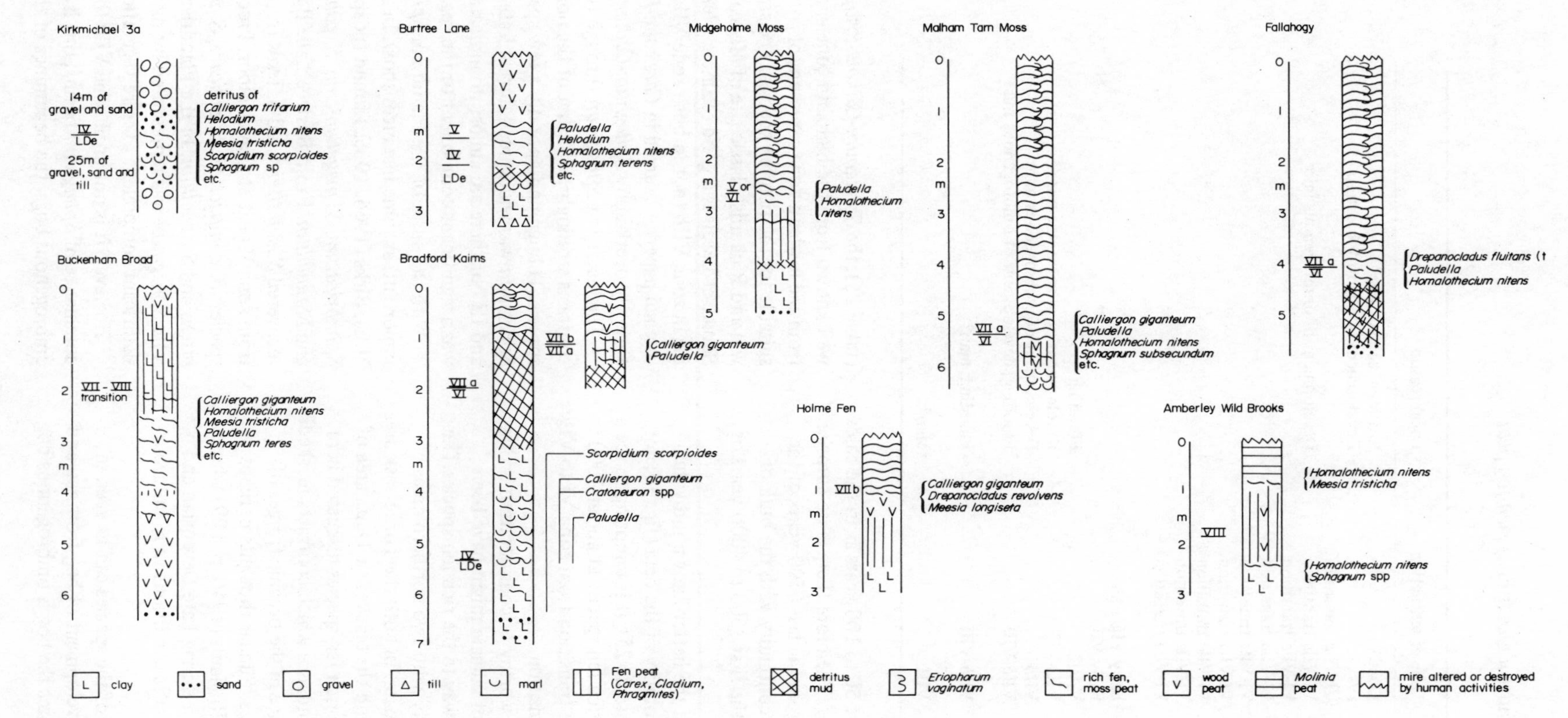

Fig. 90. Stratigraphy of some mires with rich fen peats. Modified from various authors.

Table 25 *Important Flandrian localities of rich fen peat*

Locality	Time of extinction	Local reason
Kirkmichael	LDe-IV	Erosional catastrophe?
Burtree Lane	IV-V	Development of oligotrophic mire
Wybunbury Bog	VI-VII*a* transition?	"
Malham Tarn Moss	VI-VII*a* transition	Development of ombrogenous mire
Fox Earth Gill	VI-VII*a* transition	"
Upper Valley Bog	VI-VII*a* transition	"
Midgeholme Moss	VI-VII*a* transition	"
Dun Moss	VI-VII*a* transition	"
Burreldale Moss	VI-VII*a* transition?	"
Fallahogy	VI-VII*a* transition	"
Co. Kildare (several localities)	VI-VII*a* transition?	"
Dufton Moss	Early VII*a*	"
Whixall Moss	Early VII*a*	"
Bradford Kaims	V-VI?	Flooding
	VII*a*-VII*b*	Development of oligotrophic mire
Holme Fen	VII*b*	Development of ombrogenous mire
Somerset Levels (several localities)	VII*b*-VIII	Flooding; development of ombrogenous mire
Norfolk Broads (two localities)	VII*b*-VIII	Flooding; man?
Amberley Wild Brooks	VIII	Man?

or did survive until the last 50 to 100 years as in the cases of *Paludella* and *Helodium*. Therefore there is no surprise in the scatter of records from the last 6500 years of the Flandrian, which provide continuity with the bulk of assemblages deposited in the first 3500 to 4000 years of the Flandrian.

On the basis of not just the better-known individual stratigraphies but of the totality of the data (fig. 90; not all the data are shown on table 25), it is tempting to see a marked reduction of the rich fen species at about 7000 to 7500 years ago; this is the transition from zone VI to VII*a* or the Boreal-Atlantic transition.

A speculation perhaps worthy of consideration is that an increase in oceanicity of climate might have been directly inimical in some way to the rich fen species.The author has some familiarity with two northern areas of markedly continental climate. In both the Torneträsk area of northeastern Sweden and the Mackenzie Delta area of Canada some or all of the rich fen species discussed here are abundant; it is an abundance which contrasts markedly with the scarcity or absence in the oceanic fringes of Europe and North America. Similar thoughts perhaps occurred to Godwin and Richards (1946, p. 129) whose notion, however, that even during Late Devensian and early Flandrian times climatic conditions were unfavourable, can be dismissed.

The local eliminations of the species can be seen in purely seral terms, the development of bog from fen and fen woodland. In most cases the bog is ombrogenous bog (table 25); the *Sphagnum-Calluna-Eriophorum* peats so well known from the highland zone and more sparsely from the lowland zone represent oligotrophic mires unsuitable for the rich fen species. The stratigraphy at Godwin and Richards' classic site of *Meesia tristicha* in the Somerset Levels is a good example. Here the *Meesia*, growing in zone VII*b* (not in both zone VII*a* and VII*b* as in the original paper), occurred in *Carex* fen-*Betula* wood peat abruptly overlain by *Sphagnum-Calluna-Eriophorum* peat.

The history of *Sphagnum* through the last 12,000 years contrasts strikingly with that of the northern-continental species. Though perhaps only a few species were present, *Sphagnum* was widespread during Late Devensian zones II and III but there are almost no indications that the genus was a significant component of the vegetation (p. 208).

With the onset of the Flandrian, *Sphagnum* became more important, as three instances show. In zone IV-V at Chat Moss, Birks (1964, 1965*a*) found six species: *S. cuspidatum, S. imbricatum, S. magellanicum, S. palustre, S. papillosum* and *S. tenellum*. From Burtree Lane Bellamy *et al.* (1966) recovered *S. subnitens* and *S. teres* from zone V peat, and from zone V peat at Cranes Moor, Proctor determined five species: *S. compactum, S. palustre, S. papillosum, S. subnitens* and *S. tenellum*. In the Flandrian long before zone VII*a* *Sphagnum* was represented by many species and was a significant component of mire vegetation.

However, it is not until zone VII*a* that the ecological dominance of *Sphagnum* begins; this is the spread of ombrogenous bogs, the beginnings of the blanket bogs of

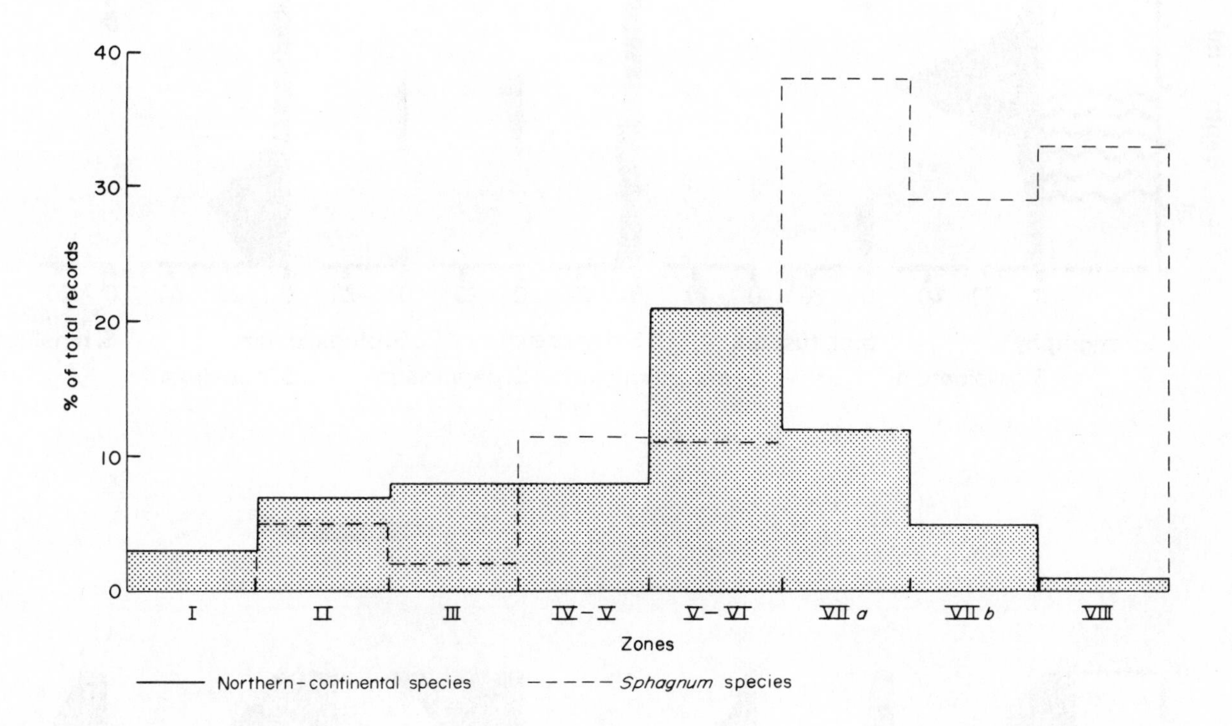

Fig. 91. Northern-continental species and *Sphagnum* zone by zone. Percentages of total records excluding spores. Records from zone III-IV are included in the zone III column. Records from zone III/IV*, IV, IV-V and V are aggregated in the zone IV-V column. Records from zone V-VI and VI are aggregated in the zone V-VI column. Records imprecisely dated within zones I to III and zones VII*a* to VIII are excluded.

the uplands and the massive development of the raised bogs in the valleys.

Therefore, if this initiation of ombrogenous bog is seen in climatic terms, a response to increasing wetness about one third of the way through the Flandrian, the reason for the decline of the northern-continental species is clear; a general tendency for the mires to become more oligotrophic greatly reduced the number of available habitats.

It is to be regretted that there are no radiocarbon dates relating directly to this matter, with the sole exception of Blackriver Bog *B*, Hammond's site in Co. Kildare, where the rich fen peat has an age of 7500 B.P.

The carbon dating of suitable profiles with meticulously examined stratigraphies is greatly to be desired.

In concluding this section mention may be made of the frequency curves for *Sphagnum* spores which are shown in pollen diagrams from the ombrogenous peats of zones VII and VIII. Fig. 88 shows a good example from the raised bog peat at Fallahogy; the *Sphagnum* curve fluctuates very markedly from almost zero to values of more than 400 per cent. This is quite typical. However, according to Dr A. G. Smith differential destruction during the chemical preparation of the Fallahogy samples may partly account for the fluctuations.

It has to be realised that such a *Sphagnum* curve is the summation of spores produced by more than one species, possibly several species. The various mire communities in and around a raised bog may support up to fifteen species of *Sphagnum*, each with particular ecological requirements. Tallis (1964) has clearly demonstrated that *Sphagnum* spore peaks can be separated into various components. At his site at Wesseden II, in the Southern Pennines (fig. 92), the lower peak is composed of *S. acutifolium* and the upper mostly of *S. cuspidatum* and to a lesser extent of *S. papillosum*. At the nearby Dean Head Hill he recognised eight species (fig. 92). He concludes (p. 350) 'A single change in environmental factors cannot account for the separate spore peaks, and in the absence of experimental work on the conditions governing fruiting in *Sphagnum* it would perhaps be unwise to speculate on possible causes but the data do show that fluctuations in

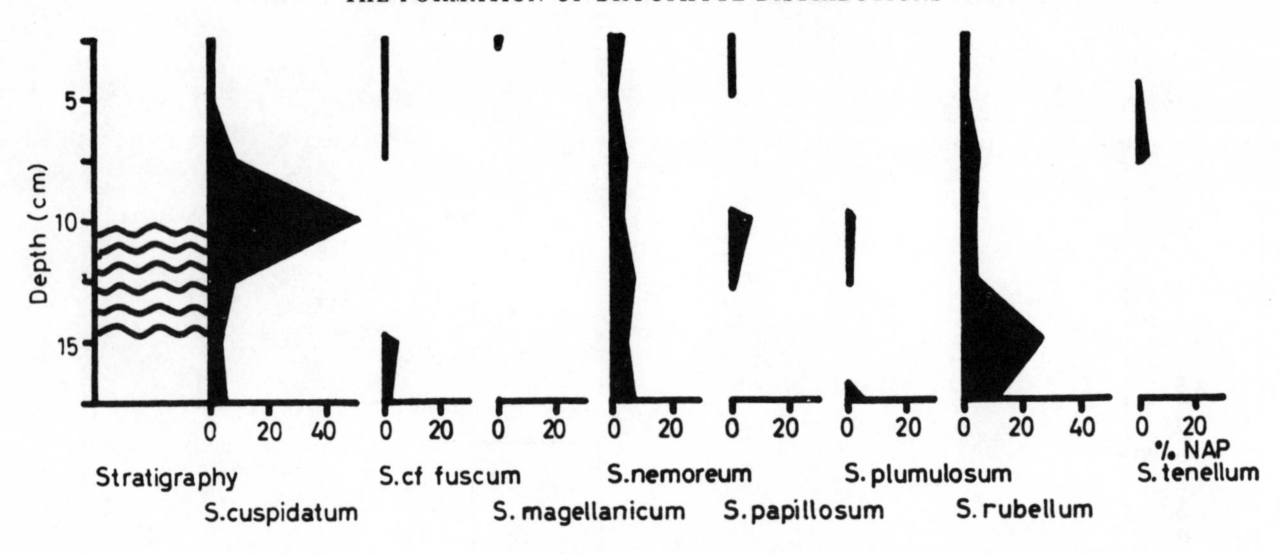

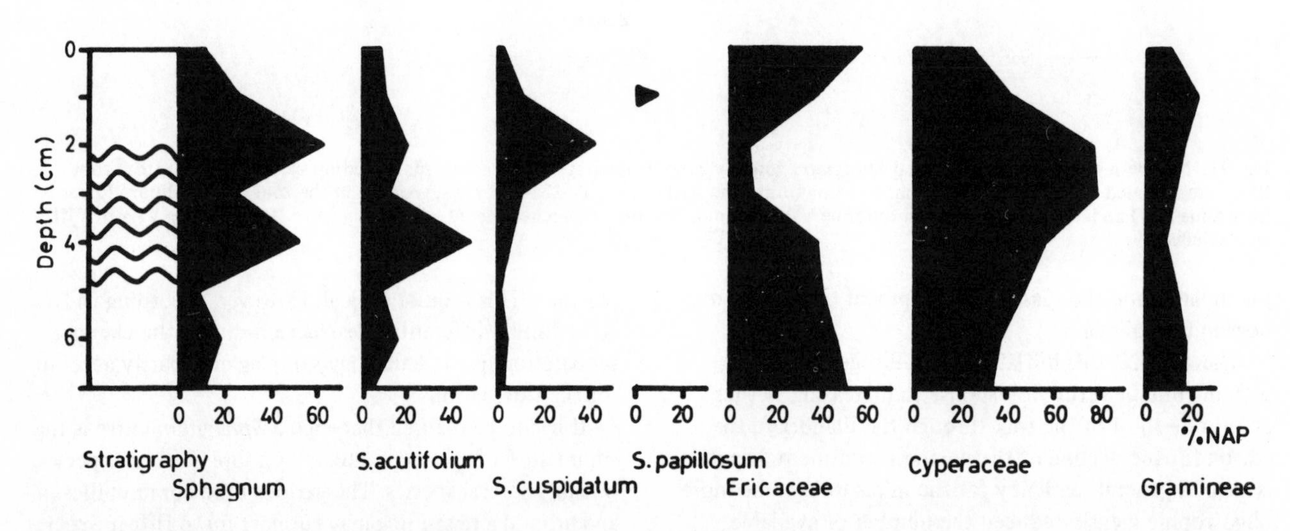

Fig. 92. *Sphagnum* spore diagrams. Top, spore diagram for various *Sphagnum* species at Dean Head Hill. Bottom, diagram across the uppermost *Sphagnum* band at Wessenden II. From Tallis 1964.

Sphagnum spore counts in southern Pennine peats are not random, and may indeed have great ecological significance.'

The point made by Tallis, that little is known of the factors controlling sporophyte production, is worth stressing. Even worthwhile field observations of sporophyte frequency appear to be lacking. In the British Isles some species fruit freely, such as *S. fimbriatum* and *S. tenellum*, others much less frequently and more sporadically.

Tallis (1964) and others have shown that layers of unhumified *Sphagnum* peat give higher spore counts than humified layers and some authors, notably Nichols (1967, 1969), have used these changes in frequency linked to stratigraphy to make far-reaching climatic deductions. However, the multiple nature of the *Sphagnum* curve and current ignorance of fruiting behaviour hinder any easy acceptance of these conclusions.

Northward spread of thermophiles

In his 1932 review (p. 317) Gams stated 'During the warm postglacial period, southern and oceanic species reached higher latitudes and altitudes than at present.' This seems likely enough. However, there are no bryophyte parallels

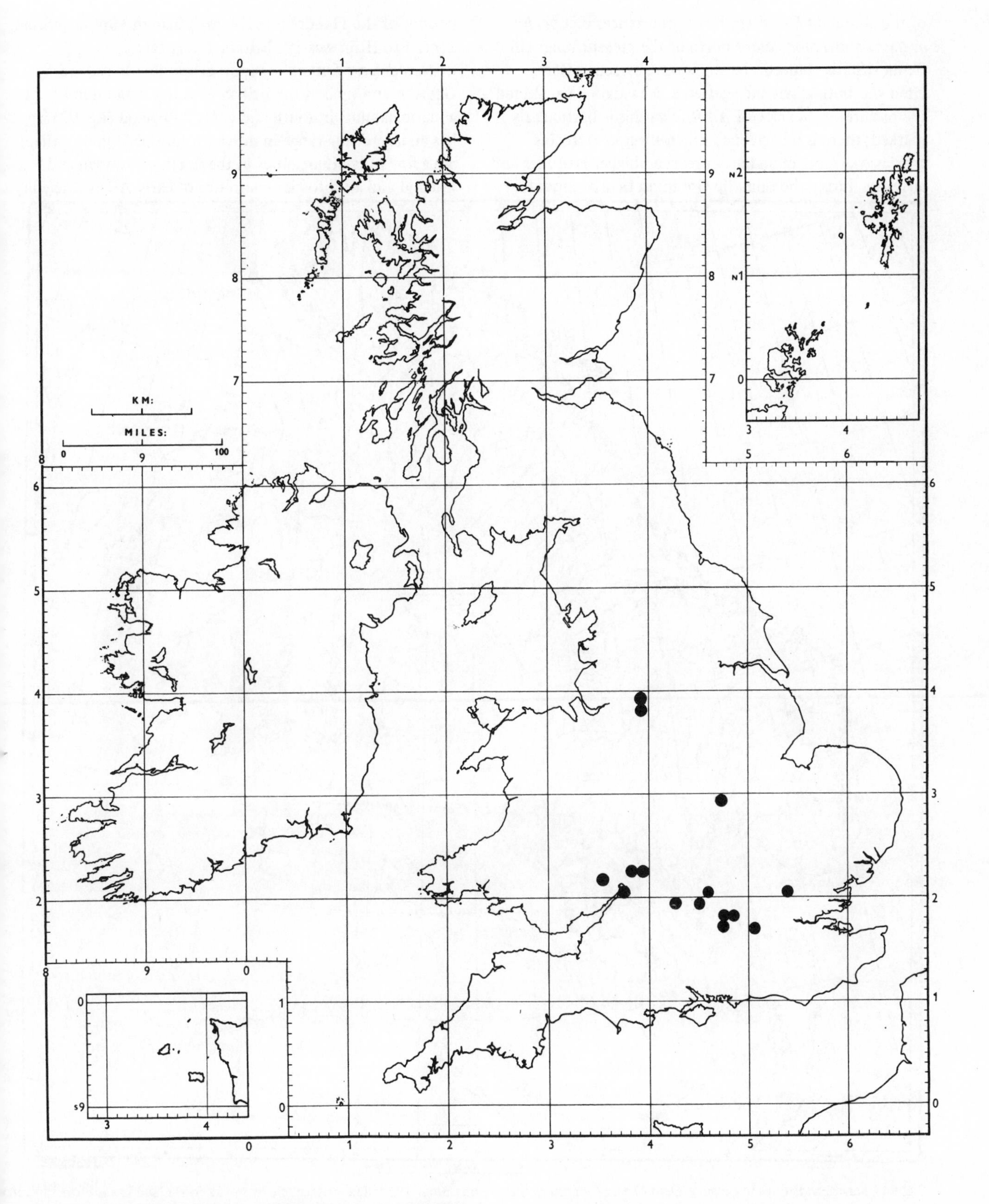

Fig. 93. British range of *Octodiceras fontanum*. From Smith 1963.

of the abundant Flandrian fossil occurrences of *Corylus avellana* and *Trapa natans* north of the present ranges in Fennoscandia. Indeed, the supporting evidence Gams cited was both slight and scattered. New data have altered the picture. In the case of *Antitrichia* this is particularly marked; there is no evidence which suggests that this species was more abundant or reached higher latitudes and altitudes during the Climatic Optimum than at any other periods of the Flandrian or, indeed, than in Late Devensian zones I to III in western Britain at any rate.

Octodiceras fontanum, another species mentioned by Gams, is confined in the British Isles to a small number of aquatic habitats in south and central England (fig. 93). It has an interesting range in northern Europe (fig. 94); there are a few isolated localities in the north of Sweden and Finland and a subfossil occurrence of early Atlantic age in

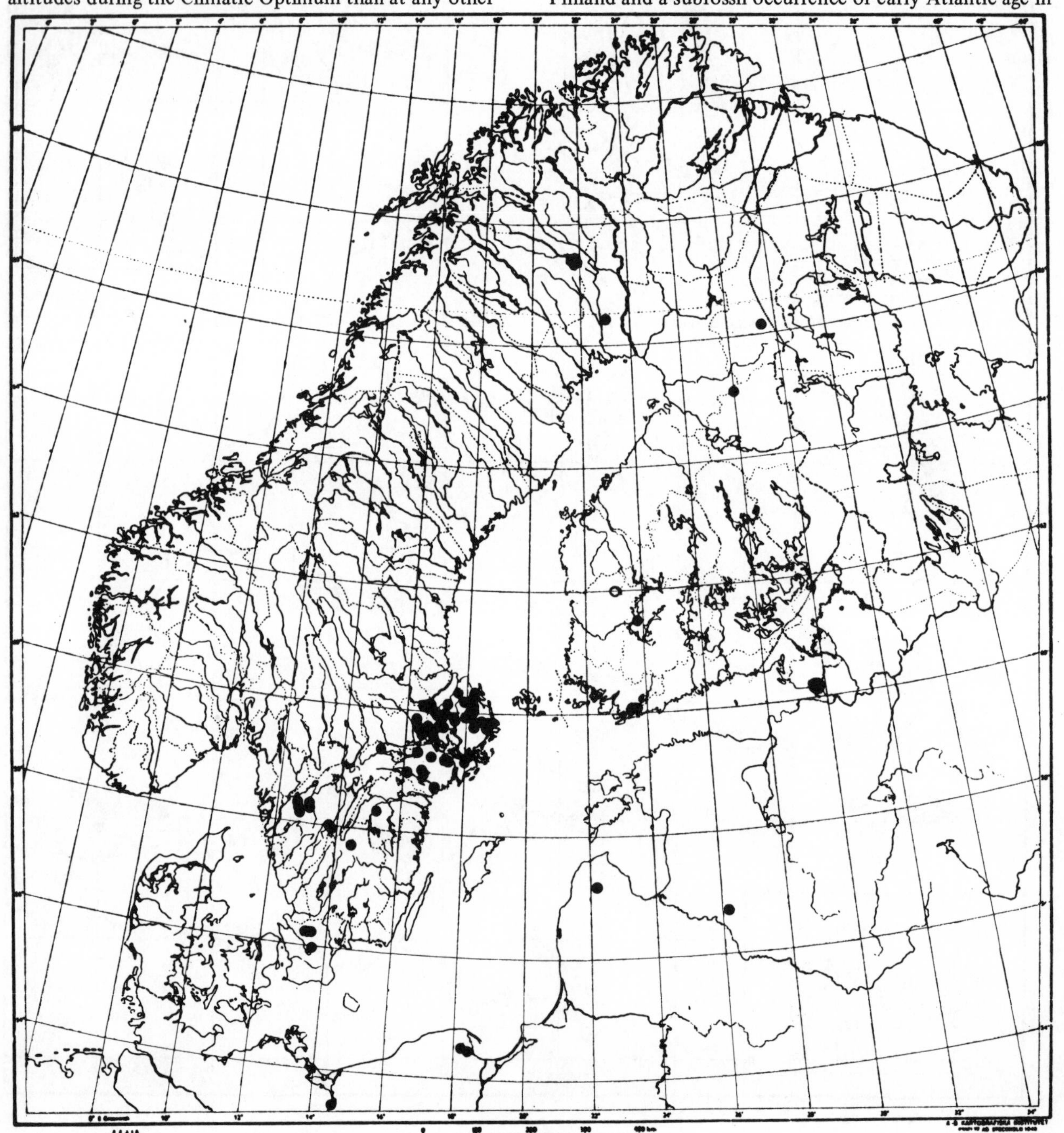

Fig. 94. Northern European range of *Octodiceras fontanum*. After Lohammar 1954. The open circle in southern Finland is a subfossil locality

southern Finland. The northern stations are regarded as relict by Lohammar (1954), who envisages restriction since the Climatic Optimum as a result of edaphic and limnological changes which he does not specify.

Except by implication for the oceanic species as a group, there are no claims in the literature that any present British patterns can be explained in terms of the Climatic Optimum and subsequent deterioration. Apart from *Octodiceras fontanum*, there are at least three further examples from Fennoscandia.

Isothecium striatulum in Fennoscandia has been mapped by Störmer (1955) who believes that the isolated Norwegian occurrences are relict from 'the warmer climatic period after the last glaciation'. There is nothing inherently impossible about such an interpretation. However, there are no supporting fossils and, moreover, Störmer does not interpret the gap between the Scandinavian localities and the northern limit in Germany as a result of extinction but merely as a lack of suitable habitats.

In his fine treatise on *Mosses with a Western and Southern Distribution in Norway*, Störmer (1969) speculates on the history of many species including *Pterogonium gracile* and *Breutelia chrysocoma*, both of which he sees as immigrants to Norway during the Climatic Optimum. For *Breutelia* he states (p. 213) 'It may have immigrated to our south coast during the post-glacial warmth period, perhaps as early as the late Boreal period, when the still existing "land bridge" between England and Jutland made immigration easier than in the following periods.'

8.3 THE DEVENSIAN-FLANDRIAN TRANSITION

Bryogeographical changes after the Late Devensian: eclipse of the arctic-alpines

The bryoflora of the Late Devensian was species-rich. As implied below by the discussions of ubiquitous and southern species, it was varied in present geographical elements. The outstanding feature of the Late Devensian bryoflora is the large proportion of species (41 spp or c. 34 per cent) now of restricted disjunct or montane ranges in the British Isles (table 22). Moreover, if one thinks of the Devensian as a whole, the number of phytogeographically interesting species is augmented; the majority of species in table 22 can be considered arctic-alpines. More species will be added as investigation proceeds, and it is likely that the two species now extinct in the British Isles represent only a small pro-

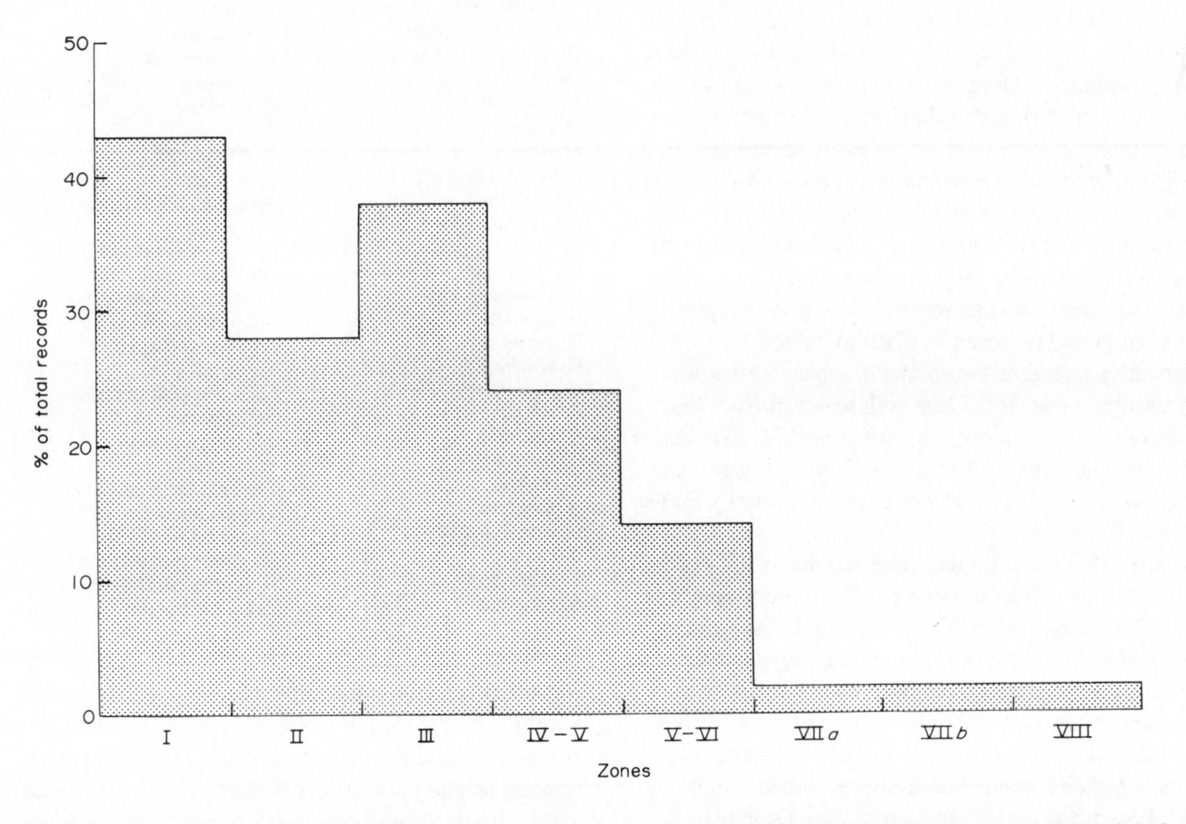

Fig. 95. Arctic-alpine and other open ground species zone by zone. Species such as *Abietinella abietina, Ditrichum flexicaule, Rhacomitrium* spp and *Rhytidium rugosum* are included. See the legend of fig. 90 for further explanation.

portion of the exotic species which occurred during the last glaciation.

No less than thirty-four of the fifty-three taxa in table 22 are absent from post-zone IV Flandrian subfossil assemblages; the great majority of the thirty-four are arctic-alpines. It is clear that the species in this category, which at least in some if not in many cases were widespread and abundant in the Late Devensian, suffered a great decline at the Devensian-Flandrian transition (fig. 95).

The arctic-alpines are Late Devensian relicts. Perhaps the best examples as judged by known Devensian occurrence far from the present restricted British ranges are as follows:

Aulacomnium turgidum	*Hypnum vaucheri*
Calliergon turgescens	*Lescuraea patens*
Dicranum elongatum	*Philonotis seriata*
Hygrohypnum molle	*Polytrichum norvegicum*
Hypnum bambergeri	*Timmia norvegica*
H. revolutum	*Tortula norvegica*

Calliergon turgescens, Hypnum bambergeri and *H. revolutum* have vacated Ireland since the Late Devensian and are now restricted to the central Scottish Highlands.

Possible Devensian relict species

Though unknown from British Devensian deposits there are many species in the present British bryoflora which may be deemed relicts of a wider range in the last glaciation. Table 26 lists over 100 species which seem obvious candidates to the author. More could be added. Some have better claims to such status than others. The asterisks in table 26 indicate some of the most easily acceptable. For example, *Ctenidium procerrimum*, a species which lacks sporophytes throughout its entire range, has a very limited British range (Mid Perth and Easterness) on high-altitude base-rich rock. It seems more than reasonable to consider this species a relict, an assessment which gains support from a last glacial occurrence, albeit in Poland.

Similarly, several of the species in table 26 are known from European last glacial sites well removed from the present ranges. Examples are *Bryum weigelii* (t, Germany), *Cirriphyllum cirrosum* (Denmark, Poland), *Conostomum tetragonum* (Poland), *Encalypta alpina* (Denmark), *Pohlia cruda* and *P. cucullata* (both Denmark).

On the other hand, doubts attach to the validity of inclusion of some species in table 26. The present ranges of such as *Brachythecium reflexum, Heterocladium dimorphum, Hylocomium pyrenaicum* and *Pterygynandrum filiforme* may well be reduced from a greater spread at some stage of the Late Plesitocene, but not necessarily the Devensian. One might imagine common occurrence in the subalpine *Betula* forests extensively developed in the British Isles in the early Flandrian. A Late Devensian occurrence of *Ptilium*, at any rate in southern

Table 26 *Possible Devensian relict species*

Amblystegiella confervoides
A. sprucei
Amphidium lapponicum
A. mougeotii
Andreaea nivalis
Anoectangium aestivum
Anomobryum concinnatum
A. filiforme
*Aongstroemia longipes**
Arctoa fulvella
Barbula ferruginascens
B. rufa
*Blindia caespiticia**
*Brachythecium erythrorrhizon**
B. reflexum
B. starkei
*Bryum arcticum**
B. calophyllum
B. muelenbeckii
B. neodamense
B. rufifolium
B. schleicheri
B. tortifolium
B. weigelii
*Campylium halleri**
*Cirriphyllum cirrosum**
Conostomum tetragonum
*Ctenidium procerrimum**
Desmatodon cernuus
D. leucostoma
Dicranella grevilleana
Dicranoweisia crispula
Dicranum blytii
D. falcatum
D. glaciale
D. starkei
Ditrichum lineare
D. zonatum
*Encalypta alpina**
E. ciliata
*Eurhynchium pulchellum**
Grimmia agazzisii
G. alpestris
G. anodon
G. atrofusca
G. doniana
G. elatior
G. elongata
G. funalis
G. incurva
G. montana
G. muelenbeckii
G. ovalis
G. trichodon
G. torquata
G. unicolor
*Haplodon wormkjoldii**
Heterocladium dimorphum
Homomallium incurvatum
Hygrohypnum dilatatum
H. smithii
Hylocomium pyrenaicum
Isoterygium mullerianum
*Lescuraea saxicola**
L. incurvata
*Mnium lycopodioides**
M. marginatum
M. spinosum
Myrinia pulvinata
Myurella julacea
M. tenerrima
Oncophorus wahlenbergii
Orthothecium rufescens
Paraleucobryum longifolium
Plagiobryum demissum
P. zierii
*Plagiomnium medium**
Plagiothecum laetum
P. piliferum
P. platyphyllum
P. roseanum
P. striatellum
Pohlia acuminata
P. bulbifera
P. cruda
P. cucullata
P. drummondii
P. elongata
P. gracilis
P. ludwigii
P. polymorpha
P. proligera
Pseudoleskeela catenulata
*P. nervosa**
Pterygynandrum filiforme
*Ptychodium plicatum**
*Saelania glaucescens**
*Seligeria oelandica**
Sphagnum lindbergii
S. riparium
Splachnum vasculosum
*Stegonia latifolia**
Tayloria lingulata
T. longicolla
Tetraplodon angustatus
*Timmia austriaca**
Trematodon ambiguus
*Trochobryum carniolicum**
*Weissia wimmerana**
Zygodon gracilis

Britain, is far from impossible. However, I have looked without success for remains of this species in the sediments from an Alleröd pine forest (as proved by macroscopic remains of pine) at Helvoirt in the Netherlands (Polak 1963). It was a similar disappointment when the investigation of the moss assemblage from the Chelford coniferous forest failed to recover the species. Nevertheless, the last glacial occurrence of the species in Poland was demonstrated many years ago by Zmuda (1914).

Catoscopium nigritum (fig. 5) and *Sphagnum warnstorfii* might both have been added to table 26. However, they are better placed among the rich fen relicts. The former has a disjunct range in the highland zone where it inhabits flushes and dune slacks. There are two discoveries of subfossils from outside the British Isles; Jasnowski (in Tolpa 1961) recovered the species from a rich fen assemblage of last interglacial age in Poland. The other comes from North America (Miller 1969).

There are no subfossils of *Sphagnum warnstorfii* from the British Isles. On grounds of present ecology, it can be expected to occur in early Flandrian fen peats associated with *Homalothecium nitens* and *Paludella squarrosa*; it is a component of minerotrophic fen peats in northern Finland (Ruuhijarvi 1963).

Establishment of ubiquitous ranges in the Flandrian: the spread of forest species

Pleistocene assemblages have yielded much information pertinent to the understanding of the history of species which at present are among the commonest in the British bryoflora. It is possible to give minimal ages for the establishment of widespread occurrence of many of the 85 ubiquitous species defined here as those known from all 152 vice-counties or not less than 140.

Only 18 of these 85 are unknown in the subfossil state; for the most part the absentees have habitats unconducive to fossilisation (p. 41). Of the 67 ubiquitous species known as subfossils 26 listed in table 27 seemingly became widespread only in the Flandrian.

Table 27 *Ubiquitous species unknown from Devensian deposits. Earliest Flandrian record in brackets*

Atrichum undulatum (VI to VIII)	*Isothecium myosuroides* (IV to V)
Brachythecium rutabulum (V or VI)	*I. myurum* (V or VI)
T *Campylopus pyriformis* (VII*a*)	*Mnium hornum* (V or VI)
Dicranella heteromalla (VII or VIII)	*Neckera complanata* (V or VI)
D. varia (VII*b*)	*Plagiomnium undulatum* (VI)
Eurhynchium confertum (VII*b*)	*Pohlia nutans* (IV or V)
E. praelongum (VI)	*Polytrichum formosum* (V)
E. riparioides (IV)	*Rhacomitrium canescens* (V or VI)
E. striatum (IV to VI)	*Rhytidiadelphus loreus* (V)
E. swartzii (VII*b*)	*Sphagnum flexuosum* (VII*a*)
Fissidens bryoides (VII*b*)	*S. subnitens* (V)
Funaria hygrometrica (VI)	*Thamnobryum alopecurum* (VI)
Homalia trichomanoides (VI to VIII)	

It is hard to believe that some of these species were not widespread earlier than the Flandrian zones indicated. For instance the *Dicranella* spp, *Eurhynchium confertum* and *E. swartzii* and *Fissidens bryoides* may well have occurred widely long before zone VII*b* and *Rhacomitrium canescens* could well turn up in Devensian deposits.

However, many species in the list do make ecological and chorological sense. Thus the several species with earliest records from zone IV, V or VI are among those which on grounds of present tolerances might not be expected as glacial species but which are readily acceptable as having become widespread with the climatic amelioration. These remarks apply most easily to *Eurhynchium praelongum*, *E. striatum*, *Homalia trichomanoides*, *Isothecium* spp, *Mnium hornum*, *Neckera complanata*, *Plagiomnium undulatum* and *Thamnobryum alopecurum* (fig. 59) all of which can be regarded as primarily species of temperate range and forest ecology.

Many other species, while not in the ubiquitous category, are absent from Devensian assemblages but known from those of Flandrian age. Most readily placed with the species mentioned in the last paragraph are *Brachythecium velutinum*, *Neckera crispa*, *N. pumila*, *Ulota* spp and *Zygodon viridissimus*. There are also oceanic species (*Breutelia chrysocoma*, *Dicranum scottianum*, *Hylocomium umbratum*, *Hyocomium armoricum*) and eleven species of *Sphagnum*.

Oceanic species: Late Devensian immigrants?

That the extreme oceanic species with highly disjunct ranges are Tertiary relicts is an old theory originally erected on the basis of no subfossil evidence. Modern work on Miocene and Pliocene assemblages (p. 170) gives substance to this interpretation and more support may be forthcoming as work progresses. It seems more than plausible that the spatial disruption of these taxa can be related to climatic-vegetational changes of the Neogene-Pleistocene transition. It is a reasonable assumption that there was a rich oceanic bryoflora in Europe at the beginning of the Pleistocene; what has happened to the component species, diverse in ecological requirements, in two million or more years of major climatic fluctuations?

Subfossil assemblages may go a long way towards answering this question. However, though not as meagre as made out by Ratcliffe (1968), the hard facts of the subfossils are still sparse and lead to few firm conclusions.

One interpretation of the oceanic species in the British Isles is that they are interglacial relicts; Gams (1952) is the principal advocate. A second theory, perhaps not entirely exclusive of the first, is that the oceanic bryoflora immigrated in the Flandrian. The advocates are Greig-Smith (1950) and Richards (1954).

For the interglacial relict theory the moss and other plant remains from Gort are crucial. The occurrences of taxa such as *Hyocomium armoricum, Sphagnum molle, Rhododendron ponticum, Erica ciliaris, Doboecia cantabrica* and the Hymenophyllaceae prove the establishment of an oceanic flora, perhaps even richer than that of the present, during an interglacial period. Much more is implied; the British oceanic bryoflora cannot be seen only in terms of the Flandrian. Each interglacial may well have seen a resurgence of the oceanic flora.

It has to be remembered that the Gort data are now known to refer to *two* interglacials ago and not to the last interglacial as stated by Gams (1952). To claim the species as interglacial relicts needs proof of survival in the intervening time, especially the two glacial periods. The subfossils are lacking as yet.

Could any of the oceanic species have survived the rigours of the glacial climates? Species capable of growing in areas of late snow lie at present spring to mind (Ratcliffe 1968), species such as *Anastrophyllum donianum, A. joergensenii, Plagiochila carringtonii, Pleurozia purpurea, Scapania nimbosa, S. ornithopodioides* and others, all Northern Atlantic species of Ratcliffe (1968). One may doubt, however, whether *Myurium hebridarum* could have survived the glaciation on the island of Rhum, as envisaged by Heslop Harrison (1948).

For many species surival through the long period of the last glaciation seems very unlikely. The period of maximum ice extension when a great many of the present localities were occupied by ice may have been short-lived but it was preceded by long spells of periglacial climates with very cold dry winters. The occurrences of fossil ice wedges in western Britain have to be recalled. It may seem impossible that species such as *Colura calyptrifolia, Cyclodictyon laetevirens, Daltonia spachnoides, Dumortiera hirsuta* and *Telaranea sejuncta* could tolerate periglacial climates.

Nevertheless, in this context Steere's discoveries at Umiat in the north slope of Alaska are perhaps pertinent. Here in an area of severe continental climate with active ice wedge formation occur the genera *Herberta, Neckera, Radula* and *Tetraphis*, among others (1965).

The Flandrian theory assumes that the glacial climate was too severe for the survival of the oceanic bryoflora which must have occurred under some more genial climate and, following the warming of the Flandrian, immigration took place. The warm, wet Atlantic period saw the greatest expansion of the oceanic bryoflora which is relict from that time.

The few Flandrian records of oceanic species, as categorised by Ratcliffe (1968) (table 28), prove occurrence before zone VII*a*. Here the Fort William assemblage is the significant one; somewhat vaguely dated as zone V or VI, it contained not only *Dicranum scottianum, Hyocomium armoricum* and *Thuidium delicatulum* but also *Antitrichia curtipendula, Hylocomium brevirostre, Isothecium myosuroides, Plagiothecium undulatum* and *Rhytidiadelphus loreus*, which, though undiscussed by Ratcliffe, are apposite because of their markedly western ranges in the European context. Therefore, if the Fort William dating is accepted, it is clear that the oceanic species occurred before the Atlantic period, though not necessarily long before.

Unless one regards the onset of ombrogenous peat formation as indirect proof, the subfossils give no hint of a spread in the Atlantic period, nor is any proof likely to be forthcoming. Mire profiles are not liable to yield the oceanic species of most strongly disjunct range and the in-washed assemblages from lakes, such as Seathwaite Tarn, are briefly episodic and can give no continuous sequence.

However, the data which seem most significant to the author are the widespread discoveries of *Antitrichia curtipendula* along the western seaboard of Britain in Late Devensian zones I to III, the similar but lesser evidence for *Sphagnum imbricatum* and, not least, the zone II occurrence of *Plagiothecium undulatum* on the Isle of Skye. The last thousand or two years of the Devensian were evidently suitable for these species which are markedly western in their European ranges. Though the British distributions may well be more western as a result of anthropogenic influences and though the pattern of *Antitrichia* has enigmatic features, the extrapolation is tempting that more demanding species, oceanic species as usually defined and not merely snow-bed species, found amenable habitats, if perhaps only very locally in the extreme west, during the last few thousand years of the last glaciation. Perhaps an even more abrupt west-east gradient of oceanicity than presently exists needs to be envisaged and perhaps one can envisage gorges in western areas providing suitable mild, moist habitats, as Billings and Anderson (1966) imagine was the case in the Southern Appalachians.

Until more relevant subfossils are discovered speculation is predominant. One can readily agree with Ratcliffe (1968, p. 401) when he states 'My own attitude to this Quaternary question is that while the divergent views each appear to have some validity, none of them is an adequate explanation alone, and that the truth lies in a combination of them all.'

8.4 THE DEVENSIAN

Per-glacial survival

The ecological dominance of mosses and lichens in areas of high altitude and latitude is well known and needs no

Table 28 *British History of Oceanic Species*

	Ratcliffe's elements	Geological Stages		
		pre-De	De	Fl
Breutelia chrysocoma	Sub-Atlantic	–	–	VIII
Campylopus flexuosus	Sub-Atlantic	–	–	VIII
Dicranum scottianum	Widespread Atlantic	–	–	V or VI*
Fontinalis squamosa	Sub-Atlantic	–	3 all LDeI	–
Hylocomium umbratum	Western British	–	–	VIII
Hylocomium armoricum	Sub-Atlantic	Ho II, III, III or IV	–	V or VI, VII*a*
Hypnum callichroum	Western British	–	MDe	–
T *Rhacomitrium ellipticum*	Sub-Atlantic	Record discounted see p.		
Sphagnum imbricatum	Western British	Ho III; Ip III	LDe II; 2LDe unzoned	Numerous beginning IV-V
Thuidium delicatulum	Sub-Atlantic	–	LDe II, III	V, V or VI, VII*b* or VIII, 3VIII
	Unclassified by Ratcliffe			
Antitrichia curtipendula	–	Pa; Cr; Ho II; Ip II and III	EDe?; LDe Numerous beginning I	Numerous beginning IV or V
Hylocomium brevirostre	–	Ho unzoned	–	IV*; 2V-VI; IV-VII*a**
Isothecium myosuroides	–	Ho III		IV, V or VI*; VII-VIII
Plagiothecium undulatum	–	–	LDe II	V or VI*; VII*b*, VII or VIII
Rhytidiadelphus loreus	–	–	–	V, V or VI*, VI, VII, 2VII or VIII, 2VIII

stressing here. This ability of mosses to flourish in regions of severely cold climate leads readily to the belief that even under the most rigorous conditions of the Pleistocene glaciations the ice-free parts of the British Isles could have supported bryofloras composed of many species. Even if one imagines that at some stage or stages of the glaciations the periglacial areas were 'polar deserts', as some Dutch workers believe the landscape of Holland to have been, mosses would still have been present.

So little is known of the moss floras of the interglacials and pre-Devensian glacials that it is impossible to do more than guess what species may have been per-glacial survivors in the sense of survival through the Pleistocene. With the investigation in recent years of assemblages of Early Devensian age and particularly the rich assemblages of Middle Devensian age it has become possible to list the species which are most likely to have survived in the British Isles throughout the 60,000 or more years of the last glaciation.

Twelve species are known from deposits formed in the Early, Middle and Late Devensian (table 29). A further thirty-two species can be assessed as per-glacial survivors on the more flimsy evidence of Early and Late Devensian or Middle and Late Devensian records (table 30).

The vast majority of these forty-four species are precisely those which on grounds of present distribution and ecology could be envisaged as per-glacial survivors. To emphasise, however, the ecological and phytogeographical diversity of the species, one need mention only such as *Drepanocladus capillifolius, Calliergon stramineum, Homalothecium lutescens* and *Rhytidiadelphus squarrosus*. In table 29, nine out of twelve species belong to the Amblystegiaceae. In the glaciated areas the morainic topography allowed many accumulations of shallow water and in the times and areas of permafrost the summer melts produced very wet landscapes; this accounts for the abundant occurrence of Amblystegiaceae in Devensian floras. Species of *Calliergon, Campylium, Cratoneuron, Drepanocladus* and *Scorpidium* were abundant then just as they are in wet tundra at present.

Though it has to be remembered that the last glaciation lasted at least 60,000 years and included marked climatic fluctuations it seems more than likely that, on the basis of the above records, a rich bryoflora survived in the unglaciated areas. Further work on Early and Middle Devensian

Table 29 *Species with records from the Early, Middle and Late Devensian*

	Sidgwick Avenue	Wretton	Chelford	Upton Warren	Fladbury	Great Billing	Lea Valley	Derryvree	Barnwell	Dimlington	LDe zones I to III	Flandrian
Aulacomnium palustre	–	–	+	–	+	+	–	–	–	–	+	+
Calliergon giganteum	+	+	–	+	+	–	+	–	+	–	+	+
C. turgescens	+	–	–	–	+	+	–	–	–	–	+	–
Campylium polygamum	–	+	–	+	+	–	+	–	–	–	+	+
C. stellatum	+	+	–	+	–	–	+	+	+	–	+	+
Cratoneuron filicinum	–	+	–	+	–	–	+	–	+	–	–	–
Drepanocladus capillifolius	–	+	–	–	+	–	+	–	–	–	+	–
D. exannulatus	+	–	–	–	–	–	+	–	–	–	+	+
D. revolvens	+	–	–	+	–	–	+	–	+	–	+	+
Plagiomnium rugicum	–	+(t)	–	+	–	–	+(t)	–	–	–	+	+
Pohlia wahlenbergii	–	+(t)	–	+	–	–	–	+(t)	–	+	+	–
Scorpidium scorpioides	+	+	–	–	–	+	+	–	+	–	+	+

Table 30 *Possible per-glacial survivors*

Abietinella abietina
Amblystegium serpens
Aulacomnium turgidum
Barbula recurvirostra
Bryum pallens
B. pseudotriquetrum
Calliergon cordifolium
C. sarmentosum
C. stramineum
Calliergonella cuspidata
Ceratodon purpureus
Climacium dendroides
Cratoneuron commutatum
Dichodontium pellucidum
Dicranum scoparium
Distichium capillaceum
Ditrichum flexicaule
Drepanocladus aduncus
D. fluitans
D. uncinatus
Encalypta rhabdocarpa
Helodium blandowii
Hypnum callichroum
H. cupressiforme
Philonotis fontana
Pleurozium schreberi
Polytrichum juniperinum
Rhizomnium punctatum or *pseudopunctatum*
Rhytidiadelphus squarrosus
Tortella tortuosa
Tortula ruralis
T. norvegica

assemblages will strengthen this conclusion.

Moreover, though the Ipswichian bryoflora is little known (table 18) there is enough to claim the continuous British presence of a few species not just through the Devensian but since at least the latter part of the last interglacial. The best example is *Aulacomnium palustre* (Ip II to IV). The rest comprise *Homalothecium nitens* (Ip III and IV), *Cratoneuron filicinum* (Ip III and IV), *Hypnum cupressiforme* (Ip II and IV) and *Pleurozium schreberi* (Ip IV).

Devensian history of ***Sphagnum***

The interglacial occurrences of *Sphagnum* need little comment. Though the numbers of species is small, *Sphagnum* floras of similar diversity and distribution to the present British floras are likely suppositions. However, the history of the genus through the glacial periods, especially the last glaciation, demands fuller consideration.

The rich occurrence in Early Devensian deposits is paralleled elsewhere in northwestern Europe. *Sphagnum* peat has been recorded from the Early Weichselian deposits at Luneburg and Harksheide in northwestern Germany (Hallik 1952, 1955). Danish sites of similar age at Brörup, Nörbölling and Rodebaek have yielded very large numbers of *Sphagnum* spores and lesser numbers were reported from Herning (Andersen 1961).

Some 57,000 years ago at Chelford the climate was mild enough to support coniferous forest, a character which differentiates this interstadial from all other British Devensian deposits, except that of similar age at Wretton (West 1970 pc). Abundance of *Sphagnum* spores is a further distinguishing feature of both these deposits, correlated with the European Brörup interstadial which is also characterised by large quantities of *Sphagnum*. Dutch deposits of this age at Bruchterveld (Polak and Hamming 1963), Lunteren and Amersfoort (Zagwijn 1961) are good examples, as are those of the earlier Amersfoort interstadial at Warnssum, Moershoofd and the type site (Zagwijn 1961).

After the Chelford interstadial the almost total lack of records seems to indicate the widespread extinction or diminution of the genus in the British Isles. There are no records from Fladbury, Great Billing, the Lea Valley Arctic Bed or Barnwell Station. The single leaf from Upton Warren is the only Middle Devensian record.

As yet only three *Sphagnum* species have been identified from the Late Devensian on the basis of macroscopic fossils:

S. imbricatum: ! *Stannon* (2); LDe only. ! *Nant Ffrancon* (49); LDe only. *Roundstone 2* (H16); LDe II; Jessen *et al.* 1949.
S. papillosum: ! *Stannon* (2); LDe only. ! *Nant Ffrancon* (49); LDe only. *S. palustre*: *Loch Mealt* (104); LDe II; Birks 1969 pc.

There are several records without specific determinations, of which the following are examples:

! *Stannon* (2); LDe only. *Hawks Tor* (2); LDe II; Conolly *et al.* 1950. ! *Brook* (15); LDe II; one leaf. *Aby Grange* (54); LDe II; Suggate and West 1959. ! *Kersall Moss* (59); LDe III; a few branch leaves of subgenus *Litophloea*. ! *Kirkmichael 3a* (71); LDe III; two tiny branch fragments of subgenus *Litophloea*; Dickson *et al.* 1970. *Garvel Park* (76); LDe only; Robertson 1881. ! *Loch Kinord* (92); LDe III; one tiny branch fragment of subgenus *Litophloea*; Vasari and Vasari 1968.

Similarly, *Sphagnum* spores are rare in Late Devensian deposits such as the few following:

Hockham Mere (28); LDe II; Godwin and Tallantire 1951. *Llyn Dwythch* (49); LDe I, II and III; Seddon 1962. *Little Lochans* (74); LDe II, III; Moar 1969*a*. *Loch Mahaick* (87); LDe II; Donner 1962. *Yesnaby* (111); LDe I, II; Moar 1969*b*. *Cannons Lough* (H40); LDe II, III; Smith 1961.

These records show that in Late Devensian times the genus was widespread and represented by at least several species, but there is no evidence of extensive *Sphagnum* mires which are so marked a feature of the Flandrian. It has to be remembered, moreover, that the presence of *Sphagnum* spores in tiny values may be a result of long-distance transport. Surface samples from Spitsbergen, where sporophytes are unknown, gave values up to 4 per cent (Srodon 1960); clearly this is long-distance transport.

Absence or scarcity of *Sphagnum* is also characteristic of Late Weichselian deposits on the mainland of northwestern Europe. However, there are at least two well marked exceptions to this generalisation. Casparie and Van Ziest (1960) found consistently high values of spores through zone III in a deposit at Waskemeer in the Province of Friesland, Holland. They believe that there could have been no raised bog because the deposit is a sandy gyttja but they envisage the *Sphagnum* as having grown in the marginal water of the lake. High values of spores were found in the Late Devensian layers at Stannon (fig. 96).

In summary, the Devensian history of British *Sphagnum*, largely corroborated from the rest of northwestern Europe, appears to have been abundance in the Early Devensian followed by greatly contrasting scarcity in the Middle Devensian and increased diversity in the Late Devensian.

If the history outlined above is correct, what accounts for the striking fluctuations through the last glaciation? During the long period of the Middle Devensian was the genus wiped out? Greatly reduced abundance seems more likely.

Drastic cooling of the climate might have reduced the abundance but is unlikely to have caused extermination. If the climatic severity compared with that of present northernmost Norway, a rich *Sphagnum* flora could have existed. Lange (1969), referring principally to Troms and Finmark counties of Norway, gives thirty-two species of which the nine most abundant are as follows:

S. angustifolium	*S. riparium*
S. fuscum	*S. rusowii*
S. girgensohnii	*S. squarrosum*
S. lindbergii	*S. teres*
	S. warnstorfii

If one imagines the climate having been as severe as that of a high Arctic region, a comparison can be made with Spitsbergen. Rönning (1961) accepts nine species for the present flora of Spitsbergen. They are as follows:

S. aongstroemii	*S. girgensohnii*
S. balticum	*S. lindbergii*
S. fimbriatum	*S. squarrosum*
S. fuscum	*S. teres*
	S. warnstorfii

Only *S. squarrosum* and *S. teres* are common and five of the rest have only a few or merely one locality. All but *S. aongstroemii* are present British species. Perhaps the Middle Devensian flora was like that of Spitsbergen today: a few species, some possibly widespread but insignificant components of the vegetation, and others local and rare.

If, on further investigation, Middle Devensian deposits yield *Sphagnum* species, perhaps the most likely are *S. girgensohnii, S. fimbriatum, S. squarrosum, S. teres* and *S. warnstorfii*. They are the most frequent species not only of Spitsbergen but also of Iceland and Greenland (Lange 1963), as well as among the commonest of northernmost Norway, *S. fimbriatum* excepted. By the same criterion species of section *Sphagnum* would be unexpected: only *S. magellanicum* and *S. papillosum* occur in northernmost Norway.

The absence of *Sphagnum* from many Devensian deposits may signify nothing but unsuitable edaphic conditions. All the Middle Devensian assemblages so far investigated are markedly calcicolous, reflecting base-rich if not actually calcareous substrata, unfavourable for many species of *Sphagnum* as well as *Polytrichum*, also absent from the same deposits. However, not all *Sphagnum* species are calcifuge; one might have expected such as *S. squarrosum* or *S. warnstorfii* in the species-rich assemblages from the Lea Valley. Similarly many Late Devensian assemblages are predominantly calcicolous.

Species characteristic of the Middle or Late Devensian

The data are much too scanty to allow one to decide if any of the Middle Devensian species are characteristic of that period rather than the Early or Late Devensian. However, perhaps *Drepanocladus capillifolius* may prove to have been particuarly abundant in the Middle Devensian. There are five Devensian localities, all in central and southeastern England. Three of them belong to the Middle Devensian when perhaps this species, extinct in the British Isles and eastern in Europe as a whole, flourished more than in the preceding and later periods.

Similarly, little significance can yet be attached to the presence of numerous species in Late Devensian assemblages and not in those of the Middle Devensian except lack of investigation of deposits of the latter age. It may indeed be that in some cases there was immigration with the ameliorating climate of the Late Devensian. However, what little is known of the Middle Devensian bryoflora suggests that the great majority of the solely Late Devensian species will turn up when the now paltry number of Middle Devensian assemblages is increased.

By way of illustration, the absence of Meesiaceae from the Middle Devensian, if genuine, would be interesting bryogeographically. *Meesia tristicha* has two Late Devensian localities, *M. uliginosa* one and *Paludella* one or possibly two. However, such species as *Calliergon giganteum, C. turgescens, Homalothecium nitens* and *Scorpidium scorpioides*, which frequently occur in rich fens with members of the Meesiaceae in present vegetation, are well represented in Middle Devensian assemblages. It is difficult to see any climatic, dispersal or edaphic reasons for the absence of the Meesiaceae.

However, it is tempting to exempt a few Late Devensian species from the speculation that they will be extracted from Middle Devensian deposits in due course and to consider that the Late Devensian occurrences, in one case abundant occurrence, combined with Middle Devensian absence is ecologically meaningful.

The case of *Sphagnum* has already been discussed in this context with *Sphagnum imbricatum* as an outstanding example; the difficulty of separating the effects of climate and substrate were pointed out. The same is the case here. However, it is difficult to envisage *Plagiothecium undulatum*, known from the Late Devensian of Skye, as a Middle Devensian species. Even more cogent is *Antitrichia curtipendula*, so well represented in the western Late Devensian sites but absent from those of the east and all Middle Devensian sites (fig. 56).

The abundant Late Devensian zone I to III occurrences of *Hylocomium splendens, Polytrichum alpinum* and *Rhacomitrium lanuginosum* must also carry some weight in this argument. None is recorded from the Middle Devensian. This seems an unreal absence, however, because the ecological amplitudes and geographical distributions are too wide.

Perhaps the explanation is that they underwent great expansion in the Late Devensian in response to the ameliorating climate, and with such species as *Antitrichia curtipendula* and *Plagiothecium undulatum*, newly arrived immigrants of late zone I, flourished under the oceanic climate and edaphic conditions of the western hills.

The establishment of ubiquitous ranges in the Devensian

Of the sixty-seven ubiquitous species known as subfossils at least twenty-three, on the basis of two or more widely separated localities, can be assumed to have had very extensive distributions since the Late Devensian (table 31). The best examples are *Aulacomnium palustre* (fig. 45), *Calliergonella cuspidata, Climacium dendroides* (fig. 53) and *Cratoneuron filicinum*. In these cases the localities are numerous and situated in the lowland zone as well as the highland zone including Ireland.

Table 31 *Ubiquitous species probably widespread since the Late Devensian*

Aulacomnium palustre	*Hylocomium splendens*
Calliergonella cuspidata	*Hypnum cupressiforme*
Campylium stellatum	*Philonotis fontana*
Ceratodon purpureus	*Pleurozium schreberi*
Climacium dendroides	*Pohlia wahlenbergii*
Cratoneuron commutatum	*Polytrichum commune*
C. filicinum	*P. juniperinum*
Ctenidium molluscum	*P. piliferum*
Dicranum scoparium	*Rhizomnium punctatum*
Fissidens adianthoides	*Rhytidiadelphus squarrosus*
Fontinalis antipyretica	*Sphagnum palustre*
Homalothecium lutescens	
Ubiquitous species perhaps widespread since the Late Devensian	
Amblystegium serpens	*Fissidens cristatus*
Barbula fallax	*Grimmia apocarpa*
B. recurvirostra	*Plagiomnium rostratum*
Brachythecium rivulare	*Plagiothecium undulatum*
Bryum caespiticium (t)	*Pseudoscleropodium purum*
B. pallens	*Rhytidiadelphus triquetrus*
Cirriphyllum piliferum (zone III/IV*)	*Thuidium tamariscinum* (zone III/IV*)
Dicranum bonjeanii (t)	

There are no Irish Late Devensian records of thirteen of the twenty-three species. However, the extensive spread of most of the species in Britain including many localities in the west (eight species in the Isle of Man) makes the assumption of occurrence in Ireland seem likely; the paucity of investigation in Ireland should be borne in mind.

A further fifteen species of the eighty-five in the ubiquitous category are known from merely one Late Devensian locality (table 31). If this minimal evidence is also accepted then thirty-eight species may have had widespread ranges since Late Devensian times. For the greater majority of the thirty-eight species there are no chorological or ecological impediments attached to this assessment; they are species of wide ecological amplitude capable of adapting to Flandrian climatic-vegetational changes. However, seven species require further mention. They are as follows:

Barbula fallax — *Plagiothecium undulatum*
Cirriphyllum piliferum — *Pseudoscleropodium purum*
Fissidens cristatus — *Thuidium tamariscinum*
Homalothecium lutescens

They are all southern species as defined below. Most demanding of comment is *Plagiothecium undulatum*, a species with a markedly western range in Europe (fig. 72). Though recorded from throughout the British Isles it is much commoner in the west than in the east; in East Anglia it is a scarce species. On the basis of a single Late Devensian record from the Isle of Skye one cannot assume ubiquitous occurrence at that time, though the species may have been widespread in the western fringe of the highland zone.

There remain for discussion several species, two of which have been recorded only from the Lea Valley Arctic Bed. They are *Bryum capillare* and *Orthotrichum diaphanum* (a suspect record of a southern species). One cannot maintain that widespread ranges were sustained since the Middle Devensian, even if the few large assemblages were not restricted geographically and disparate in age (fig. 26), because much of the British Isles was occupied by ice; the very late eastern advance as far as Holderness has to be remembered. However, it may be that some, if not many, of the eighty-five species had ranges as extensive as suitable habitats allowed. *Bryum capillare* may be a case in point, as may be *Ceratodon purpureus* and *Cratoneuron filicinum*.

It is a pity that the Derryvree flora, far removed geographically from the Lea Valley Arctic Bed, but close in age, is so meagre. At least, however, some idea is gained of the Middle Devensian flora of the western British Isles. *Amblystegium serpens* and *Philonotis fontana*, the only species shared by these two assemblages, are the best examples of present ubiquitous species which may have been widespread since pre-Late Devensian times.

Other species widespread in the Late Devensian

Species which are at present ubiquitous in the British Isles were by no means the only species of extensive occurrence in the Late Devensian, as is shown in table 32 which includes only those species with three or more Late Devensian records.

In many cases one can assume extensive occurrence only in the highland zone. *Antitrichia curtipendula* is the outstanding example and *Polytrichum norvegicum* and *Sphagnum imbricatum* also demand mention.

It is likely that many of the species listed in tables 31 and 32 were not merely extensively distributed but abundant in the Late Devensian vegetation. Those species for which such an assessment is most probable are given in table 33; all have nine or more Late Devensian localities.

Table 32 *Species, other than those in the ubiquitous category, with three or more Late Devensian localities*

Abietinella abietina	*D. fluitans*
Antitrichia curtipendula	*D. uncinatus*
Calliergon cordifolium	*Helodium blandowii*
C. giganteum	*Homalothecium nitens*
C. sarmentosum	*Polytrichum alpinum*
C. turgescens	*P. norvegicum*
Dichodontium pellucidum	*P. urnigerum*
Distichium capillaceum	*Rhacomitrium fasciculare*
Ditrichum flexicaule	*R. lanuginosum*
Drepanocladus aduncus	*Scorpidium scorpioides*
D. exannulatus	*Sphagnum imbricatum*

Table 33 *Species with most Late Devensian records*

Antitrichia curtipendula	*Homalothecium nitens*
Aulacomnium palustre	*Hylocomium splendens*
Calliergon giganteum	*Hypnum cupressiforme*
Campylium stellatum	*Polytrichum alpinum*
Climacium dendroides	*Rhacomitrium lanuginosum*
Cratoneuron filicinum	*Scorpidium scorpioides*
Drepanocladus revolvens	

The striking difference between tables 31 and 32 is the presence of arctic-alpine species (e.g. *Polytrichum alpinum, Polytrichum norvegicum, Calliergon turgescens*) and species with otherwise restricted, disjunct ranges (*Abietinella abietina, Homalothecium nitens, Helodium blandowii*) in table 32 and their absence in table 31.

Southern species in the Devensian

Godwin (1956, pp. 317-18) listed some twenty-four angiosperm species known from Late Devensian assemblages, which at present have southern ranges as indicated by their restricted northward occurrence in Scandinavia. To these more can now be added such as *Bunium bulbocastanum* (t) and *Herniaria* sp, both recorded from Colney Heath (Godwin 1964). Also in this category is *Bidens cernua* from the Late Devensian deposit at Stannon (Appendix 2). These southern species are by no means exclusive to Late Devensian deposits; Bell (1969) has collated the substantial evidence from Middle Devensian assemblages.

The Middle Devensian vegetation may have had the gross appearance of present-day tundra. However, in floristic composition, as Bell has emphasised, there are wide divergencies; many species now abundant at high altitudes and latitudes occurred freely but there was a marked element of species never found in such areas today. One need mention only such as *Damasonium alisma, Diplotaxis tenuifolia* and *Onobrychis viciifolia* to make the point.

Among the mosses many species now abundant at high

Table 34 *British Devensian species with present limited northern extension in Fennoscandia, according to such authors as Nyholm (1954-69) and Mårtensson (1956)*

Amblystegium fluviatile	MDe only	*Fontinalis antipyretica*	LDe only
A. kochii	LDe only	*F. squamosa*	LDe only
A. riparium	EDe only	*Homalothecium lutescens*	EDe; LDe
Anomodon viticulosus	De?	*Hylocomium brevirostre*	LDe only
Antitrichia curtipendula	EDe; LDe	*Hypnum cupressiforme*	EDe; LDe
Barbula fallax	LDe only	*Orthotrichum diaphanum*	MDe only
B. spadicea	LDe only	*Plagiomnium affine*	LDe only
Brachythecium mildeanum	MDe only	*P. rostratum*	MDe only
Bryum creberrimum	MDe only	*Plagiothecium undulatum*	LDe only
B. intermedium	MDe only	*Pseudoscleropodium purum*	LDe only
Calliergonella cuspidata	MDe; LDe	*Sphagnum imbricatum*	LDe only
Campylium elodes	LDe only	*S. papillosum*	LDe only
Cirriphyllum piliferum	LDe/Fl IV*	*Thuidium tamariscinum*	LDe/Fl IV*
Ctenidium molluscum	LDe only	T *Tortella inclinata*	De?

latitudes were common in the Middle Devensian. Paralleling the angiosperms, however, there was a not inconsiderable representation of southern species. Thus in the Ponder's End assemblage Dixon (Warren 1912) found the following:

Amblystegium fluviatile — *Calliergonella cuspidata*
Brachythecium mildeanum — *Orthotrichum diaphanum* (t)
Plagiomnium rostratum
Bryum intermedium

These render any exact comparison with tundra impossible. Dixon compared the assemblage with the flora of the Torneträsk area, though he pointed out that some of the species do not occur there. In the light of more up-to-date information (Mårtensson 1956) the comparison seems to have lost some of its aptness. Four of the six given above do not occur there and *Calliergonella cuspidata* and *Bryum intermedium* probably do not. Moreover, Middle Devensian vegetation was treeless; the Torneträsk area has abundant *Betula* forest and stands of *Pinus*.

Whatever the physionomic aspect of Middle Devensian vegetation there is no need to envisage tundra-like vegetation for much of the British Isles in the Late Devensian zones I, II and III. Tree birches were certainly widespread and common in the lowland zone, and in much of the highland zone they occurred, as macroscopic fossils show, if only locally. By no stretch of the imagination could many of the Late Devensian southern species in table 34 be considered as tundra species. Most cogent in this respect are *Antitrichia curtipendula, Hylocomium brevirostre* and *Plagiothecium undulatum*, all western in the European context; they have been discussed above. In present and past chorological contrast stands *Homalothecium lutescens* which may have had a predominantly eastern Late Devensian occurrence (fig. 69).

Like the angiosperms, the southern mosses are a heterogeneous group which serves to emphasise the diversity of the Devensian flora and dispel the image of the last glacial flora as exclusively arctic in character.

Though the evidence is meagre it is clear that during the Wolstonian glaciation, if not earlier glaciations also, southern species were represented. The assemblage from Broome Heath yielded *Calliergonella cuspidata* among an otherwise markedly northern flora including *Oncophorus virens, Calliergon turgescens* and *Timmia* sp. The Late Wolstonian flora at Ilford contained *Barbula hornschuchiana.*

High-arctic species

Steere (1953, 1965) has drawn attention to a group of more or less circumpolar species characteristic of high latitudes (by no means exclusive to areas north of the polar tree line); one of the seventy-three species he lists is *Haplodon wormskioldii*, which is well known from the northern Pennines and the central Scottish Highlands. Figs. 43 and 44 show representative distribution patterns of species in this category.

Pleistocene subfossils of these species are almost unknown. However, the two discoveries are significant. One is the oft-quoted discovery of *Cinclidium latifolium* in Denmark. Recently Miller (1969) has recovered remains of *Aulacomnium acuminatum* from a Late Wisconsin deposit in New York. This demonstrates a northwards contraction of the same magnitude as that of *Cinclidium latifolium*.

These fossils show that at least some of the high-arctic species have been more widespread to the south in the last glaciation. Perhaps it is not impossible that subfossils will be recovered from British deposits.

Steere and others such as Crum (1965, 1966) envisage these species as survivors in the Arctic of a flora more widespread in pre-glacial times; Neogene deposits have yet to yield the proof.

Extinct species

Only a handful of living species is known to have become

extinct in the British Isles in the Pleistocene. They are *Calliergon richardsonii* **T**, *Drepanocladus capillifolius*, *Hypnum ravaudi* and *Meesia longiseta*.

The last-named is mooted as an anthropogenic extinction (p. 196) while of the others *Hypnum ravaudi* can be seen readily in terms of disappearance at the Devensian-Flandrian transition. Such may also be the explanation of the extinction of *Drepanocladus capillifolius*. In the absence of re-examination consideration of *Calliergon richardsonii* is pointless.

There are no totally extinct species described from the British Pleistocene; nor has the author found any material which unequivocally demands such treatment. However, from the Middle Devensian deposit at Earith, Huntingdonshire, Dr F. G. Bell has extracted a finely preserved *Pottia*-like subfossil which may need recognition as an extinct taxon. As stressed by Gams (1932, p. 302) few extinct species were erected early this century; nineteenth-century examples are pure misunderstandings or are based on inadequate grounds. Brizi (in Clerici 1892), for instance, giving very brief diagnoses, founded a new *Dicranum* and a new *Rhynchostegium*. No regard for the variability of mosses is indicated; the new species may be based merely on submerged forms of widespread extant species. Gams' scepticism extended to *Camptothecium woldenii* which Grout described in 1917.

About the time of Gams' judgement several extinct taxa were described. One of these, *Hygrohypnum szaferi* Podp. from the Polish Pleistocene, is the only so-called extinct taxon I have examined. I agree with Lisowski (Szafran 1961, p. 192) that the specimens are best taken as *Calliergon turgescens*.

Several extinct taxa were erected by Williams (1930*a* and *b*) and Steere (1942) from the Pleistocene of Minnesota and Iowa. They are as follows:

Calliergon aftonianum Steere	*Drepanocladus apiculatus* Steere
C. hansenae Steere	*D. minnesotensis* Williams
C. kayianum Steere	*Neocalliergon integrifolium* Williams

All need critical re-evaluation; the diagnoses and drawings by Williams, in particular, do not inspire confidence. The variability of *Calliergon* and *Drepanocladus* with environment is notorious. Again one has to contemplate that these taxa are environmental modifications.

The most recently described extinct species (1968) is *Leskea moravica* Pilous from the Post-glacial of Czechoslovakia; no account of its relationships with other species is given.

There is much less reason to question the extinct Neogene taxa described by Dixon in Reid and Reid (1915), Szafran (1949*a*) and the Abramovs. The latter in particular have given abundant grounds in descriptions and illustrations.

8.5 THE GLACIAL-INTERGLACIAL CYCLE

The vast bulk of subfossil evidence allows detailed discussion only of Flandrian and Devensian bryogeographical changes. Knowledge of the bryofloras of earlier stages of the Pleistocene is scanty. Nevertheless some cautious statements, liable to modification or abandonment as work progresses, may be made.

One can perhaps see an identical type of glacial bryoflora in the Wolstonian as in the Devensian. The assemblage from Broome Heath is the outstanding evidence; an open-ground, calcicolous, arctic-alpine bryoflora existed in the lowland zone during the second last glaciation. The very fragmentary bryofloras from earlier cold periods tell much the same story. Whether the pre-Devensian glacial bryofloras were richer in species from high latitudes, for instance, than the last glacial bryoflora, remains to be seen.

The same general conclusions apply to the interglacial bryofloras: the Flandrian and earlier temperate bryofloras seem very similar. The Gort evidence is the most striking in this context. A bryoflora of at least the same richness, including oceanic species, as that of the present is implied for the second last interglacial. Nothing in the poor data from the Ipswichian and pre-Hoxnian temperate stages leads one to suppose that the bryofloras differed in any significant respect from the Flandrian bryoflora.

There is no hint from subfossil assemblages of marked reduction in numbers of species through the Pleistocene as a consequence of cold-warm climatic fluctuations, certainly not as total extinctions nor even merely as vacation of the British Isles, though such species as *Drepanocladus capillifolius* and *Hypnum ravaudi* must be remembered.

One can readily envisage that during each of the glacial and other cold periods the bryoflora of the British Isles resembled that of the Devensian, an arctic-alpine and other open-ground species bryoflora. Then with the onset of interglacial temperate conditions the character of the bryoflora was greatly changed; forest, shade-tolerant species dominated except where suitable habitats persisted, mainly in the uplands, for the glacial species. Such a marked bryogeographical change has occurred at least three times and more likely several times. At the change from interglacial to glacial conditions the reverse took place. The forest species waned and the open-ground species spread southwards to colonise the lowlands.

8.6 THE NEOGENE-PLEISTOCENE TRANSITION

The first striking feature of Neogene assemblages is that these ancient bryofloras consist to a very large degree of species still extant. Only 8.1 per cent (3 out of 37) of the Miocene taxa listed in table 9 and 6.2 per cent (5 out of 81) of the Pliocene taxa (table 10) are now extinct. Only about 7 per cent known from the Neogene as a whole have totally disappeared. This conforms fully with the assessment by Herzog (1926) that speciation in mosses is slow. That a great many moss species are very old cannot be doubted.

Though the number of total extinctions may have been low, the second outstanding feature of the Neogene assemblages is the large number of species which have undergone huge geographical changes.

If one can think of stating that the glacial-interglacial fluctuations seem not to have impoverished the British bryoflora this is not the case for the change from Neogene to Pleistocene. Here one must rely on the European, even Eurasian, evidence. However, the extrapolation to the British bryoflora is justified; the Reuver assemblage, tiny but cogent, was close at hand. One need only recall the disappearance of *Trachycystis* spp from the western end of Eurasia. Such species as *Myurium hebridarum* (fig. 7) and the British endemic *Campylopus shawii*, whatever the details of their Pleistocene histories may be, can be interpreted as Neogene relicts.

It may well be that many species disappeared from the British Isles at the onset of the Pleistocene never to return. It is greatly to be desired that more British Neogene, especially Pliocene, material be found in order to confirm this speculation.

9 CONCLUSIONS

The macroscopic remains of mosses, the facts of historical bryogeography, reveal four periods crucial in the formation of the distributions of British bryophytes, especially those distributions of the more restricted and disjunct types.

Firstly there was the Neogene-Pleistocene transition about two to three million years ago. In the Neogene many species now of extreme oceanic range had more continuous Eurasian ranges which became disrupted as the Neogene gave way to the Pleistocene.

Secondly, there was the Devensian-Flandrian transition, about 10,000 years ago; this was the latest of many major Pleistocene climatic changes. Through the long span of the last glaciation, some 60,000 years or more, there existed in the British Isles a bryoflora which, though with a large proportion of arctic-alpine species, was diverse in geographical elements. Bryophytes, especially mosses, may well have been important components not only of the tundra-like, treeless vegetation of the Middle and Late Devensian pre-zone I vegetation but also of the vegetation of zones I to III. Among the pioneers on the fresh bare soils produced by glacial and periglacial activity were mosses including many species now restricted to the more base-rich substrata. Calcifuge species may have been poorly represented except for *Sphagnum* spp, *Polytrichum* spp and *Pleurozium schreberi* in the Late Devensian zones I to III and in the Chelford interstadial. Many species of Amblystegiaceae were abundant.

With the change from late glacial to more temperate conditions the bryophyte vegetation changed greatly. In the lowland zone and at lower altitudes in the highland zone there was very widespread elimination of arctic-alpine and other species of open ground. In the highland zone the present patterns of the now montane species began to emerge. Where forests developed, forest bryofloras arose; shade-tolerant species of Neckeraceae, *Eurhynchium* spp and some of *Brachythecium* and some of Mniaceae, all lacking in Devensian assemblages, began to flourish.

The breaking-up of the continuous ranges of the arctic-alpine and other open-ground species might be seen in purely climatic terms; an increase in temperature might in some way be directly inimical. However, competition is the explanation favoured here. The spread of the forests eliminated the essential open habitats. This is exactly the explanation put forward by Godwin (1956) and others such as Pigott and Walters (1954) to explain many of the disjunct distributions shown by vascular plants of arctic-alpine and other geographical types.

Thirdly, at the Boreal-Atlantic transition, about 7000 years ago, the northern-continental species declined, having been major components of the mires for the first 3000 years of the Flandrian. The ecological dominance of *Sphagnum* began.

Here again the explanation of highly restricted, disjunct ranges is vegetational change. The scattered localities of the northern-continental species are not the products of spore or other dispersal, nor are those of the arctic-alpine species.

Lastly, there was the emergence of man as a serious factor in the ecosystem. By clearing the forest from at least 5000 years ago and in much more recent times by damaging the mires, man has changed the bryophyte vegetation. A few species analogous to higher plant weeds may have benefited. Many others have declined, even to extinction in the case of a few in the British Isles.

APPENDIX 1 REVISED DATINGS

As a result of recent geological, palynological and radiometric advances it is possible to reconsider the ages of various assemblages from deposits investigated last century or during the first half of the present century. By far the most important reassessments pertain to the Lea Valley Arctic Bed, Fort William and Windmillcroft.

Angel Road and Ponder's End, Middlesex. Middle Devensian. C^{14} 28,000 ± 1500 B.P. Q–25
Dixon recognised some forty species of mosses (table 21) from the deposits at Angel Road and Ponder's End which form part of the complex of deposits known as the Lea Valley Arctic Bed (Warren 1912). Godwin (1960, p. 292) gives the following brief description of the fossiliferous strata: 'In the coarse gravels of the lowest and latest terrace of this wide valley, there have been recovered deeply embedded and massive rafts, often tilted at substantial angles, of finely bedded silt full of organic remains. They are so soft that they can only have been incorporated when frozen solid and even so they cannot have travelled far.'

Plant debris sieved off from the bed has a radiocarbon age of 28,000 ± 1500 B.P. The provenance of the sample is not known precisely. The locality may be Nazeing or Broxbourne (Godwin and Willis 1960).

Fort William, Inverness-shire. Fl zones V or VI*
The moss flora from Fort William described by Dixon (1910) is the largest subfossil assemblage ever reported from a British Flandrian deposit. Some thirty-six taxa are recorded.

Quoting information given by geologists of the Geological Survey, Dixon (p. 108) states 'It (the plant bed) lies just at high-water mark, being covered by one or two feet of water at high spring tides. It contains prostrate tree trunks up to one foot in diameter, and is overlain by coarse stratified gravel and sand up to 15 feet in thickness. This sand and gravel belongs to the so-called "25 foot raised beach" of Scotland.' No details are given about the tree trunks except that *Corylus* is listed.

In papers on the raised beaches of Scotland, Donner (1959, 1963) has broadly correlated the 25-ft beach with Fl zone VII*a*. He states (1962, p. 12) 'After a relative regression during the Preboreal and Boreal periods the 25-ft beach was formed during the Atlantic period, Fl zone VII*a*, and possibly already in part during the Boreal period, Fl zone VI, as indicated by the radiocarbon measurements which gave a date of 8000-5000 B.P. (Godwin and Willis 1961, 1962).'

Because of the stratigraphic position of the plant bed and the lack of information regarding the angiosperm flora it is difficult to be precise about the age of the bed. On the basis of the present data it may be judged that the bed formed at any time in Fl zone V or VI.

Windmillcroft, Glasgow. Fl zones VI/VII/VIII*
The earliest records of British subfossil mosses known to the author are given by Mahony (1868) who described plant remains in alluvial deposits found very close to the present course of the Clyde in central Glasgow.

He saw 'great trunks of trees', probably oak, and found leaves of birch, hawthorn, hazel, oak and wych-elm as well as hazel nuts and a seed of alder. Fourteen species of mosses are listed.

Brachythecium rivulare	*Mnium marginatum*
Climacium dendroides	*Neckera complanata* (t)
Cratoneuron filicinum	*N. crispa*
Drepanocladus uncinatus	*N. pumila*
Fontinalis antipyretica	*Philonotis fontana*
Homalia trichomanoides	*Polytrichum commune*
Isothecium myosuroides	*Thuidium tamariscinum*

The deposit cannot be precisely dated. To judge from the angiosperm assemblage, it may be considered that the deposit formed at any time from Fl zone VI to zone VIII.

Miscellaneous sites
There remain for mention a few sites of very imprecisely known age.

Given by Godwin as Late glacial (1956, pp. 66-7), the deposits at *Corstorphine, Hailes* (Bennie 1894*a*) and *Dronachy* (Bennie 1896) are listed in the Pleistocene Record as Late Devensian. Included in the 1894 paper is a list of twenty mosses, containing many typical Late Devensian species. Unfortunately Bennie does not differentiate between the sites, and hence the records are given in the Pleistocene Record as '*Hailes or Corstorphine*'.

The assemblages from *Faskine* (Bennie 1894*b*) and *West*

Craigneuk (Dixon 1907), thin peat layers overlain by considerable thicknesses of till, may be regarded as Devensian. They are given in the Pleistocene Record as 'Devensian? '.

Records from the Flandrian deposits at *Seaforth* (Travis and Travis 1913) and *Wallasey* (Travis 1922) are listed as Fl zone VII or VIII, as stated by Godwin (1956, pp. 149 and 158 respectively). In view of the remains of oak, hazel and alder recovered from the gravels, the Flandrian deposit near *Leeds* studied by Raistrick and Woodhead (1930) may be of zone VI or later age.

APPENDIX 2 DEVENSIAN AND FLANDRIAN MOSS ASSEMBLAGES INVESTIGATED BY J.H.DICKSON

This is a catalogue of the Devensian and Flandrian deposits which have yielded assemblages of subfossil mosses studied by the author. The sites are arranged alphabetically.

Apart from the list of mosses, for each site the basic information regarding location (including national grid reference), stratigraphy and age is given briefly, as well as reference to the subfossil angiosperm flora and to any previous work on moss remains.

Abbreviations used in this appendix for the first time are as follows:

b - bad	le - leaf or leaves
vb - very bad	num - numerous
g - good	st - stem or stems
vg - very good	

Amberley Wild Brooks, Sussex. Fl zone VIII
Aulacomnium palustre (1st; 4 mm; b), *Homalothecium nitens* (num st; 10 mm; g) and *Sphagnum* spp (num st and le) were identified by the author from a moss peat obtained by Dr D. M. Churchill, formerly of the Sub-Department of Quaternary Research, from the derelict raised bog at Amberley Wild Brooks (TQ 037145). The sample, referrable to early Fl zone VIII, came from a depth of 2.50 m, just above the transition from eutrophic *Juncus* saltmarsh peat to oligotrophic *Sphagnum* peat. Subsequently a sample collected over twenty years ago by Professor H. Godwin (depth 0.95 m; Fl zone VIII) proved to be *Meesia longiseta* (few st; 8 mm; g).

Aust, Gloucestershire. Fl zone VII*a*
Drilling operations at Aust (ST 576891) in connection with the new bridge across the Severn revealed peat at a depth of –5.75 m O.D. A sample sent to Professor H. Godwin was examined by Mrs C. A. Dickson who showed that it contained a pollen and macroscopic fossil flora indicative of Fl zone VII*a*. The following mosses were identified by the author:

1 le; b	*Antitrichia curtipendula*
1 st; 1 mm; g	*Eurhynchium praelongum* (t)
1 st; 8 mm; g	*Hypnum cupressiforme*
1 st; 3 mm; g	*Ulota* sp

Ballaugh, Isle of Man. LDe II-III and Fl IV
The organic sediments filling the kettle hole at Ballaugh (SC 349945) proved much less productive of mosses than the nearby cliff sections at Kirkmichael (p. 223). Only two species have been identified; *Polytrichum norvegicum* (zone II-III: 1 st; 3 mm; b) and *Climacium dendroides* (zone IV: 1 st; 4 mm; b).

Barnsley Park, Gloucestershire. Fl zone VIII*
Well No. 2 of the Romano-British villa at Barnsley Park, near Cirencester (Webster 1967), yielded abundant moss remains in a beautiful state of preservation. *Thamnobryum* (plate 9) was by far the most abundant. There is no doubt concerning the Roman age of the mosses, which were recovered from a black mud containing diverse organic remains, because the well was deliberately filled in with building stones.

3 st; 30 mm; b	*Anomodon viticulosus*
2 st; 40 mm; vg	*Cratoneuron filicinum*
num st; 30 mm; vg	*Eurhynchium swartzii*
1 st; 2 mm; b	*Fissidens* sect. *Bryoideum*
1 st; 5 mm; g	*Homalothecium sericeum*
1 st; 8 mm; b	*Neckera complanata*
num st; 43 mm; vg	*Thamnobryum alopecurum*
1 st; 5 mm; g	*Thuidium tamarsicinum*
1 st; 5 mm; vb	*Tortula* sp

Barnwell Station, Cambridgeshire. Late Devensian pre-zone I, C^{14} 19,500±650 B.P. Q-590
When Miss Marjorie Chandler (1921) investigated the flora of the gravel deposit at Barnwell Station (TL 472597) she recognised some eighty-six taxa of angiosperms including many of great phytogeographical importance. She also recovered moss fragments but although many of the specimens were mounted and tentatively identified nothing was published. For some years the specimens were in the possession of Professor P. W. Richards of the University College of North Wales, until 1961 when he sent them to the author. The following species have now been recognised:

8 st; 7 mm; g	*Abietinella abietina*
8 st; 8 mm; b	*Bryum* sp
6 st; 8 mm; b	*Calliergon giganteum*
num st; 15 mm; g	*Campylium stellatum*
3 st; 9 mm; vb	*Cratoneuron filicinum*
3 st; 15 mm; g	*Drepanocladus revolvens*
2 st; 5 mm; g	*Homalothecium lutescens*
num st; 10 mm; g	*Scorpidium scorpioides*
1 st; 4 mm; vb	*Tortula* sp

The deposit overlies the lowest terrace of the River Cam. Marr and Gardner (1916) describe the section as showing approximately 1.5 m of gravel and sand with several peat seams, up to 25 mm thick, which yielded the plant remains. The deposit has been destroyed; plant material washed from the seams has a radiocarbon age of 19,500 ± 650 B.P.

Bigholm Burn, Dumfriesshire. LDe zones II and III
Dr N. T. Moar (1969*a*) has recently published the pollen and macroscopic fossils of a deposit (NY 31682) exposed by the stream, Bigholm Burn. The following mosses were identified:

Zone II, silty sedge peat
7 st; 16 mm; b — *Calliergon cordifolium*
2 st; 5 mm; b — *Homalothecium nitens*

Zone III, muddy clay silt
5 st; 5 mm; b — *Drepanocladus fluitans*
2 st; 3 mm; vb — *Polytrichum alpinum*

Blelham Tarn, Lake District. LDe zones I, II and III
Blelham Tarn (NY 368006), half a mile long, lies at an altitude of 125 ft, about half a mile west of the northernmost part of Lake Windermere.

The diatom and pollen flora of the tarn was investigated by Dr G. H. Evans who gave the author some moss fragments for determination.

LDe zone II, fine detritus mud
1 st; 2 mm; vb — *Antitrichia curtipendula*
1 st; 4 mm; b — *Hypnum cupressiforme*
1 le; g — *Polytrichum alpinum*

		Zone	Sediment
1 le; b	*Antitrichia curtipendula*	Fl VII*a*, VII*b*	Brown mud
1 st; 3 mm; b	*Eurhynchium* sp	Fl VI-VII*a*	Brown mud
few le; g	*Sphagnum papillosum*	Fl IV-V	Fine detritus mud
1 st; 2 mm; g	*Zygodon viridissimus*	Fl VII	Brown mud

Very close to the north bank of the tarn (NY 365006) there is a hollow, which is probably a kettle hole. The following mosses were recovered from the Late Devensian parts of the sediments:

LDe zone III, silt
2 le; b — *Polytrichum* sp

LDe zone II, fine detritus
2 le; b — *P. alpinum*

LDe zones I and II, silt
3 st; 3 mm; vb — *Rhacomitrium* sp

Bryn-y-mor, Merioneth. Fl zone VII*a*
A sample of coarse brown detritus mud from the low sea-cliffs just north of Bryn-y-mor (SH 575011) near Towyn was submitted by Professor A. Wood of the Department of Geology, University College, Aberystwyth, to Professor H. Godwin. Mrs C. A. Dickson showed that it contained pollen and macroscopic fossils indicative of Fl zone VII*a*. The following mosses were recognised:

1 le; b — *Dicranum* sp
8 st; 6 mm; g — *Eurhynchium praelongum*
3 st; 5 mm; g — *Hypnum cupressiforme*
1 st; 6 mm; g — *Neckera crispa*

Bunny, Nottinghamshire. Fl zone VIII
The Roman well (SK 579285) at Bunny, about six miles south of Nottingham, excavated by Mr R. C. Alvey, contained a rich flora of angiosperms investigated by Mrs Wilson (1968). The following mosses, all well preserved, were identified:

1 st; 20 mm; g — *Brachythecium rutabulum*
num st; 35 mm; vg — *Calliergonella cuspidata*
4 st; 14 mm; g — *Bryum* sp
num st; 10 mm; g — *Campylium stellatum*
num st; 17 mm; g — *Cratoneuron filicinum*
4 st; 40 mm; g — *Ctenidium molluscum*
1 st; 12 mm; g — *Eurhynchium striatum* (t)
7 st; 9 mm; g — *Fissidens adianthoides*
1 st; 10 mm; g — *Isothecium myurum*
1 st; 5 mm; g — *Pseudoscleropodium purum*
1 st; 12 mm; g — *Thuidium tamariscinum*

Burnbrae Bridge, Renfrewshire. Fl zones VII/VIII*
The flora of a deposit of coarse gravels temporarily exposed near Burnbrae Bridge (NS 442638) has been investigated by Mr D. W. Brett, of the Botany Department, Bedford College. Because the gravels contain tree branches and seeds of such ruderals as *Euphorbia helioscopica, Galeopsis tetrahit* agg. or *G. speciosa* and *Polygonum lapathifolium* or *P. nodosum* (identified by Mrs C. A. Dickson) they may be considered to belong to the late Flandrian. The following mosses were identified by the author:

5 st; 18 mm; g — *Antitrichia curtipendula*
2 st; 3 mm; g — *Bryum* sp
4 st; 11 mm; g — *Neckera complanata*
1 le; g — *Polytrichum urnigerum*

Carpow, Perthshire. Fl zone VIII
As a result of the excavation carried out by Dr J. J. Wilkes of Birmingham University, nineteen taxa of mosses were recovered from a layer of organic matter found some 2 m below the inner ditch of the east defences of the Roman fort at Carpow (NO 243178) on the south side of the River Tay some eight miles southeast of Perth.

1 st; 3 mm; g — *Amblystegium riparium*
1 st; 5 mm; g — *Brachythecium rivulare*
num st; 7 mm; g — *Ceratodon purpureus*
3 st; 13 mm; b — *Dicranum scoparium*

1 st; 3 mm; b	*Drepanocladus* sp
1 st; 5 mm; g	*Eurhynchium praelongum*
1 st; 5 mm; g	*E. swartzii*
1 st; 3 mm; b	*Fissidens bryoides* (t)
11 st; 6 mm; g	*Hylocomium splendens*
7 st; 5 mm; g	*Hypnum cupressiforme*
1 st; 3 mm; g	*Isothecium myurum*
2 le; b	*Orthotrichum* sp
1 le; 4 mm; b	*Plagiomnium undulatum*
3 st; 18 mm; b	*Pleurozium schreberi*
1 le; 8 mm; g	*Polytrichum commune*
3 le; g	*Pseudoscleropodium purum*
1 st; 4 mm; b	*Rhytidiadelphus squarrosus*
1 st; 2 mm; b	*Thuidium delicatulum* or *T. philibertii*
1 le; b	*T. tamariscinum* (t)

Chelford, Cheshire. Early Devensian. C^{14} 60,800 ± 1500 B.P. CRO-1475

The deposit (SJ 810725) near Chelford, investigated by Simpson and West (1958), yielded a rich pollen flora as well as various macroscopic fossils, the most noteworthy being tree-birches, *Pinus sylvestris* and *Picea abies.* Trunks of birches and conifers were found in the detritus muds which made up the organic deposit studied. Judging from the abundance of pollen, Simpson and West (1958) consider that *Pinus sylvestris* may well have been the dominant tree of the forest which the deposit clearly indicates, with birches and spruce as lesser constituents.

Nine mosses were recovered from sites *A* and *B* of Simpson and West by Mr A. O. Chater of the Botany Department, the University of Leicester, and were identified by the author. They may be considered to have been present in and around the pool in which the muds were deposited or in the ground layer of the forest.

During April 1964 Dr R. O. Kapp of Alma College, Michigan, visited the locality and collected *Carex-Menyanthes* peat from which he extracted macroscopic remains of *Populus tremula, Juniperus communis, Sphagnum*, of which two or more species are present, and several fragmengs of *Pseudobryum cinclidioides.*

		Site *A*	Site *B*
1 le; b	*Aulacomnium palustre*	–	+
11 st; 12 mm; b	*Calliergon cordifolium*	+	+
6 st; 7 mm; g	*Drepanocladus uncinatus*	–	+
6 st; 16 mm; vg	*Helodium blandowii*	+	+
4 le; g	*Pleurozium schreberi*	+	+
1 le; b	*Polytrichum* sect. *juniperina*	+	–
1 le; g	*Pseudobryum cinclidioides*	+	–
4 le; b	*Sphagnum* subg *Litophloea*	+	–
1 le; b	*Rhytidiadelphus squarrosus* (t)	+	–

Colney Heath, Hertfordshire. LDe zone I. C^{14} 13,560 ± 210 B.P. Q-385

A great variety of plant and coleopteran remains have been recovered from peat erratics deeply embedded in gravels exposed in a pit (TL 197059) at Colney Heath, near St Albans (Godwin 1964). The gravels are tentatively considered by Godwin to be glacial outwash or torrential deposits.

The five erratics examined are described as consisting of 'black organic muds with fine to coarse vegetable detritus and lenses or layers of sand or clay' (Godwin 1964). Conifer (probably juniper) wood extracted from erratic *D* has a radiocarbon age of 13,560 ± 210 B.P. (Godwin and Willis 1960). On stratigraphic grounds it may be safely assumed that this gives a good indication of the age of all the erratics.

Mrs C. A. Dickson extracted macroscopic plant remains from all five erratics but found moss fragments in one only, that labelled *B*, which had been split into three parts for sampling. The following mosses were recognised by the author:

		0-10 cm	10-20 cm	20-29 cm
2 st; 7 mm; b	*Abietinella abietina*	+	–	–
5 st; 5 mm; b	*Bryum* sp	+	–	–
num st; 18 mm; g	*Calliergonella cuspidata*	–	+	+
num st; *c*.50 mm; g	*Calliergon giganteum*	+	+	+
4 st; 11 mm; b	*Campylium stellatum*	+	+	–
9 le; b	*Cinclidium stygium*	+	+	–
1 st; 5 mm; vb	*Climacium dendroides*	+	–	–
num st; *c*.20 mm; g	*Drepanocladus exannulatus* var. *rotae*	+	+	+
1 st; 5 mm; b	*D. revolvens*	+	–	–
1 st; 3 mm; b	*D. vernicosus*	+	–	–
9 st; 15 mm; g	*Homalothecium lutescens*	+	+	–
8 st; 6 mm; b	*Scorpidium scorpioides*	+	+	+
	Operculum	–	+	–

Derryvree, Co. Fermanagh. MDe. C^{14} 30,500 ± 1170/1030. B.P. Birm. 166

Two sets of samples were obtained from this site, the only one of Middle Devensian age from the western areas of the British Isles. The moss detritus was found in silt and sand between two tills (Colhoun *et al., Proc. Roy. Soc.* in press).

Professor G. F. Mitchell's samples

Silt near DP2

1 st; 2 mm; b	*Pohlia wahlenbergii* (t)
1 st; 1 mm; vb 1 le; b	*Rhacomitrium* sp

Silt with Pisidia

1 st; 1 mm; b	*Bryum* sp
4 st; 4 mm; b	*Campylium* sp

Dr E. A. Colhoun's sample

1 st; 1 mm; g	*Amblystegium serpens*
num st; 14 mm; g	*Brachythecium* sect. *Rutabula* or *Salebrosa*
5 st; 13 mm; g	*Campylium stellatum*
num st; 11 mm; vg	*Dichodontium pellucidum*
num st; 18 mm; g	*Drepanocladus fluitans*
num st; 19 mm; g	*Philonotis fontana*

Colhoun's material was abundant and well preserved, even

consisting of small tufts of *Dichodontium*; accordingly it may well be that the moss remains did not travel far before incorporation; they may be almost *in situ*. *Philonotis* in small quantity and a greater amount of *Drepanocladus* were dissected out of a tuft of *Dichodontium*; hence there is no doubt that these three species at least were in one moss community.

Dimlington, Yorkshire. MDe. C^{14} 18,500 ± 400 B.P. I-3372. 18,240 ± 250 B.P. Birm. 108
Mossy silt exposed in a cliff in Holderness (TA 391217) proved to contain a profusion of a single species, *Pohlia wahlenbergii* var. *glacialis* (t) (hundreds of fragments; 50 mm; vg). The radiocarbon dates demonstrate that the overlying boulder clays were deposited by an ice advance very late in the last glaciation (Penny *et al.* 1969).

Drumurcher, Co. Monaghan. LDe III, Fl IV. C^{14} 10,515 ± 195 B.P. Birm. 239
The Late Devensian and Flandrian sequence of sands and muds at Drumurcher accumulated in a depression bounded by a rocky slope of slate and a till slope. Additions to the previously known vascular plant flora (Mitchell 1953) have been made by Miss J. McCutcheon who extracted the great bulk of mosses listed in table 35.

The radiocarbon date was obtained from the grey muddy silt which layers; the underlying basal sand can be referred to LDe zone III and the brown mud to Fl zone IV.

Table 35

		Mitchell's 1950 samples pit 21-19 and 22 base of mud	1970 samples		
			Basal sand	Grey, muddy silt	Brown mud
17 mm; vb	*Abietinella abietina*	–	1	1	–
3 mm; vb	*Antitrichia curtipendula*	–	1	–	–
6 mm; vb	*Aulacomnium palustre* (t)	–	1	–	–
2 mm; g	*Bartramia ithyphylla*	–	1	–	–
4 mm; g	*Brachythecium glaciale* (t)	–	4	–	–
8 mm; b	*Bryum* spp	2*	12	14 + 46	1 + 1 le
8 mm; b	*Calliergon giganteum*	–	–	–	1
15 mm; b	*C. turgescens*	1	2	–	–
6 mm; g	*Campylium elodes*	–	–	1	–
4 mm; g	*C. stellatum*	1	1	3	–
35 mm; vb	*Cratoneuron filicinum*	2†	3	37	60
7 mm; b	*Dichodontium pellucidum*	–	2	1	–
3 mm; g	*Distichium* sp	–	4	–	–
6 mm; g	*Ditrichum flexicaule*	–	12	1	–
2 mm; b	*Drepanocladus* spp	–	3	3	1
3 mm; g	*D. uncinatus*	–	1 le	1	–
4 mm; g	*Homalothecium lutescens* or *sericeum*	–	–	–	3
5 mm; b	*Hylocomium splendens*	–	–	–	1
2 mm; g	*Hypnum bambergeri*	–	2	–	–
12 mm; g	*H. cupressiforme*	–	4 + 56	–	–
3 mm; g	*H. revolutum*	–	2 + 1 le	1	–
3 mm; b	*Philonotis* sp	–	3	–	–
12 mm; g	*P. fontana*	–	4	–	–
6 mm; vb	*Plagiomnium rugicum* (t)	1	–	1 le	1 le
5 mm; g	*Plagiopus oederi*	–	–	1	–
6 mm; b	*Pohlia* sp or spp	2	15	3	1
5 mm; b	*Polytrichum alpinum*	1	2 le	–	–
3 mm; b	*P. urnigerum*	–	2 le	1 le	–
20 mm; b	*Rhacomitrium* sp	–	2	7 + num le	–
2 mm; b	*Timmia* sp	–	1 le	–	–
13 mm; g	*Tortula* sp	1	11 le	4 le	–
	Opercula	–	–	1	2
	Unidentified stems	–	38	25	45

Figures indicate number of stems except for le = leaf.

* Also 1 stem from pit 16-12.

† From pits 16-12 and 8, not 21-9.

Table 36

	Zone:	II	III	III-IV	IV	V	VI
	Deposit:	Detritus mud	Silty clay	Detritus	Mud		Peat
num st; 25 mm; g	*Calliergon giganteum*	+	−	+	+	+	+
5 st; 20 mm; vg	*C. trifarium*	−	−	−	+	+	−
2 st; 25 mm; g	*Bryum pseudotriquetrum* (t)	−	−	−	−	+	+
1 st; 18 mm; vb	*Homalothecium nitens*	+	−	−	−	−	−
1 st; 6 mm; b	*Pleurozium schreberi*	+	−	−	−	−	−
2 st; 8 mm; b	*Rhacomitrium* sp	+	+	−	−	−	−
num st; 30 mm; vg	*Scorpidium scorpioides*	+	−	−	+	+	+
1 st; 3 mm; b	*Sphagnum* subg *Litophloea*	−	−	−	−	−	+

Drymen, Stirlingshire. LDe II, III, III-IV; Fl IV to VI
About 10 m of sediment and peat have accumulated at the eastern end of Muirpark Reservoir (NS 490923) which is situated at *c.* 700 ft, a few miles east of Drymen. The deposits were investigated by Donner and subsequently by Vasari and Vasari (1968), from whom the information contained in table 36 was obtained.

Fladbury, Worcestershire. MDe. C^{14} 38,000 ± 700 B.P. GRO 1269
These moss remains were extracted by Dr G. R. Coope, coleopterist in the Geology Department of Birmingham University, who published (1961) a detailed account of the beetle assemblage from the biogenic sediment from gravels of the No. 2 terrace of the River Avon at Fladbury. See Upton Warren (p. 231).

1 st; 8 mm; b	*Abietinella abietina*
1 st; 4 mm; b	*Aulacomnium palustre*
3 st; 4 mm; b	*Bryum* spp
4 st; 12 mm; b	*Calliergon giganteum*
1 st; 11 mm; b	*C. turgescens*
4 st; 17 mm; g	*Campylium polygamum*
2 st; 3 mm; g	*Ditrichum flexicaule*
15 st; 24 mm; g	*Drepanocladus* spp
2 st; 9 mm; g	*D. capillifolius*
7 st; 9 mm; g	*Polytrichum juniperinum*
1 st; 5 mm; b	*Tortella tortuosa*
1 st; 7 mm; b	*Tortula* sp

Garral Hill, Banffshire. LDe zone II
Donner (1957) described an organic deposit which he assigned to LDe zone II, from the southern slope of Garral Hill (NJ 444551) near Keith. The flora, as indicated by pollen analysis, contained numerous open-ground herbs, typical of the Late Devensian period.

Mr P. A. Tallantire, who visited the localities during 1963, found a deposit of closely similar stratigraphy and kindly sent samples from three horizons to the author. According to Mr Tallantire, samples *T1*, *T2* and *T3* are equivalent to the *Hypnum* peat, silty clay and silty clay-mud of Donner's deposit. Pollen analyses of samples *T1* and *T2*, carried out by Miss R. Andrew, showed spectra virtually identical with those of Donner's analysis. However, a noteworthy difference is the presence of numerous grains of *Juniperus* in the new spectra.

The author found leaves of *Juniperus* in sample *T1*. The following mosses were recognised:

		T1	*T2*	*T3*	Bulk
4 st; 12 mm; b	*Aulacomnium palustre*	−	−	+	−
2 st; 9 mm; g	*Brachythecium glaciale*	−	+	−	−
5 st; 5 mm; b	*Dicranum scoparium*	+	+	+	−
11 st; 22 mm; g	*Drepanocladus exannulatus* var. *rotae*	+	+	+	+
1 st; 5 mm; b	*Philonotis fontana* (t)	−	−	+	−
1 st; 4 mm; vb and 2 le; b	*Polytrichum alpinum*	−	+	+	−
4 st; 6 mm; b	*P. piliferum* (t)	−	+	−	−
4 st; 5 mm; rb	*Rhacomitrium* sp	+	−	−	+
1 st; 10 mm; b	*Rhytidiadelphus squarrosus*	−	−	+	−

Goat's Water, the Lake District. LDe
Goat's Water (SD 253988) is a small tarn a quarter of a mile long, lying at an altitude of 1646 ft. It is surrounded by steep slopes which to the south rise to the summit of Dow Crag at 2555 ft.

On investigating the sediments of the tarn, Mrs W. Tutin found a narrow band of moss fragments overlain by clay-mud. On the basis of pollen analysis the moss horizon is referred to the Late Devensian and the clay-mud to zone III/IV transition. Three species, recognised by the author amongst fragments extracted by Mrs Tutin, were the following:

3 st; 10 mm; vb and num le; vb	*Hylocomium splendens*
num st; 20 mm; vb	*Rhacomitrium* sp
3 st; 30 mm; vb	*Tortula* sp

Godmanchester, Huntingdonshire. Fl zone VIII*
Mr C. Green of Caister-on-Sea, Norfolk, has investigated a Roman ditch (TL 248703) in Godmanchester. Amongst the plant material recovered there were three mosses:

3 st; 9 mm; g	*Calliergonella cuspidata*
1 st; 2 mm; b	*Ceratodon purpureus* (t)
13 st; 12 mm; g	*Eurhynchium striatum*

Holme Fen, Huntingdonshire. Fl zone VII*b*

The thin layers of eutrophic to oligotrophic peats at Holme Fen contain the remains of numerous species of phytogeographical interest, some of which are no longer present in East Anglia, such as *Dicranum undulatum, Scheuchzeria palustris* and *Vaccinium vitis-idaea. Meesia longiseta* is unknown in the living state in Britain.

Because of drainage of the surrounding country, including the disappearance of Trundle and Whittlesey Meres, the peat has undergone great shrinkage and raised bog communities no longer exist (Godwin and Clifford 1938; Vishnu-Mittre 1959). Birch woodland covers the present surface.

At Site I, an excavation at one side of a large ditch (TL 203892), there is an exposure of about 0.10 m of *Sphagnum* peat above approximately 0.50 m of *Cladium* peat. Pollen analysis (Vishnu-Mittre 1959) and radio-carbon dating (Godwin and Willis 1960) have shown that the wood peat and lower parts of the *Sphagnum* peat were formed during the latter part of zone VII*b*.

The following species were extracted from the lower layers of the *Sphagnum* peat and the upper layers of the wood peat.

num st; *c*.50 mm; vg	*Calliergon giganteum*
num st; *c*.50 mm; vg	*Dicranum undulatum*
5 st; 12 mm; b	*Drepanocladus revolvens*
num st; *c*.70 mm; vg	*Meesia longiseta*
num st; *c*.50; vb	*Pleurozium schreberi*
num st; *c*.50 mm; g	*Polytrichum alpestre* (t)
num st	*Sphagnum* spp
num st; *c*.30 mm; b	*S. papillosum*

Vishnu-Mittre (1959) recovered *Dicranella varia, Pohlia nutans* (both zone VII*b*) and *Sphagnum* spp (Zones VII*a* to VIII) from Holme Fen. A little less than a mile north at Trundle Mere, he found the following species:

	Zone VII*a*	Zone VII*b*	Zone VIII
Aulacomnium palustre	–	+	–
Dicranella varia	–	+	–
Dicranum scoparium (t)	+	+	–
D. undulatum	+	+	–
Drepanocladus lycopodioides (t)	–	–	+
Hypnum cupressiforme	–	+	+
Pohlia delicatula	+	+	–
Scorpidium scorpioides	–	+	+
Sphagnum spp	+	+	+
Thuidium tamariscinum	+	–	–

Holme Pierrepont, Nottinghamshire. Fl zone VIII*. C^{14} 2180 ± 110 B.P. Birm. 132

Three monoxylous canoes and a wheel recovered from an abandoned course of the River Trent (SK 629395) have been described by MacCormick (1969). A knot-hole in the gunwale of canoe no. 2 was filled in antiquity with six species of mosses (Dickson and Ransom 1968): A sample from the gunwale of canoe no. 1 gave the date listed above (MacCormick 1970 pc).

1 st; 10 mm; b	*Dicranum scoparium*
num st; 20 mm; vg	*Hylocomium splendens*
3 st; 10 mm; b	*Hypnum cupressiforme*
num st; 30 mm; vg	*Pleurozium schreberi*
5 st; 20 mm; g	*Polytrichum commune*
5 st; 11 mm; g	*Rhytidiadelphus squarrosus*

Hooks, Holderness. LDe zone II

A block of peat from a coastal cliff section (TA 292348) at Hooks was submitted by Mr J. N. Hutchinson, of the D.S.I.R. Building Research Station, to Dr R. G. West. Pollen analysis by Miss R. Andrew established an Alleröd age for the peat which is made up almost entirely of *Helodium blandowii.* Minor constituents are *Aulacomnium palustre* and *Calliergon giganteum*. To judge from the dense mat of *Helodium* and its fine state of preservation, it is clear that the mosses were preserved *in situ.*

Kirkmichael, Isle of Man. LDe zones I, II and III

The complex deposits of gravels, detritus-muds, chalk-muds and boulder clays exposed in the cliffs on the coast at Kirkmichael have been investigated by Professor G. F. Mitchell. Macroscopic remains of angiosperms and mosses, comprising a rich flora with numerous calcicolous species (Dickson *et al.* 1970) are abundant in the detritus muds, two samples of which, plus one of chalk mud, were sent to the author by Professor Mitchell.

Site 1 - 110 m north of the mouth of Glen Ballyre (SC 314918)

The deposit approximately 3.00 m deep fills a shallow basin some 30.00 m wide. The material examined (profile I, 2.33 to 2.46 m) consisted of firm dark-brown laminated mud. The seven species recovered derive in part from zone I and in part from zone II.

Zone I

2 st; 8 mm; b	*Hylocomium splendens*
1 le; b	*Hypnum cupressiforme*
7 st; 3 mm; vb	*Rhacomitrium* sp
3 st; 10 mm; vb	*Rhytidiadelphus squarrosus*

Zone II

4 st; 4 mm; b and 2 le; g	*Antitrichia curtipendula*
4 st; 5 mm; b	*Dicranum scoparium*
1 st; 6 mm; g	*Polytrichum alpinum*

Radiocarbon dating indicates that this mud was deposited just as zone I was giving way to zone II. A sample of the material has an age of 12,150 ± 120 B.P. (GRO 1616).

Site 3a - 150 m north of Glen Wyllin (SC 311910)
Some 14.00 m of Flandrian alluvial gravels in an extensive basin overlie approximately 3.00 m of Late Devensian clays, detritus-muds, chalk-muds, sands and gravels.

14.50 to 14.58 m, thin alternating seams of sandy clay and detritus-mud. On the basis of pollen analysis and a radiocarbon age of 10,270 ± 120 B.P. (Q-673) this material can be referred to late zone III.

15.50 to 15.60 m, yellow sandy chalk-mud with brown seams. The chalk-mud is referrable to zone I.

The zone III horizon yielded twenty-three species of moss, that of zone I thirteen. In a total of thirty-one species there are five species common to both horizons.

		LDe zone I 15.50 -15.60 m	LDe zone III 14.50 -14.58 m
4 st; 3 mm; b	*Abietinella abietina*	+	–
13 st; 20 mm; g	*Antitrichia curtipendula*	+	+
4 st; 8 mm; b	*Aulacomnium palustre*	–	+
3 st; 5 mm; b	*Bryum* sp	–	+
8 st; 15 mm; b	*Calliergon cordifolium*	–	+
2 st; 7 mm; b	*C. giganteum*	–	+
2 st; 3 mm; b	*Campylium stellatum*	–	+
6 st; 5 mm; g	*Ceratodon purpureus*	+	+
2 le; b	*Cinclidium stygium*	+	–
7 st; 10 mm; vg	*Climacium dendroides*	–	+
7 st; 12 mm; b	*Cratoneuron commutatum*	–	+
5 st; 8 mm; b	*C. filicinum*	–	+
7 st; 7 mm; b	*Dicranum scoparium*	+	–
3 st; 2 mm; g	*Ditrichum flexicaule*	+	–
1 st; 6 mm; g	*Drepanocladus aduncus*	+	–
1 st; 7 mm; b	*D. revolvens*	–	+
4 st; 8 mm; g	*Helodium blandowii*	–	+
7 st; 15 mm; g	*Homalothecium lutescens* or *sericeum*	+	–
2 le; b	*H. nitens*	–	+
1 st; 11 mm; b	*Hylocomium splendens*	–	+
3 st; 4 mm; g	*Hypnum cupressiforme*	+	–
7 st; 7 mm; b	*Meesia tristicha*	–	+
4 st; 14 mm; g	*Philonotis* sp	–	+
6 st; 8 mm; vb	*Plagiomnium rugicum*	–	+
1 st; 4 mm; b	*Pohlia wahlenbergii*	–	+
1 st; 4 mm; g and 2 le; g	*Polytrichum alpinum*	+	+
2 le; g	*P. juniperinum*	–	+
5 st; 8 mm; vb	*Rhacomitrium* sp	+	+
3 st; 8 mm; b	*Scorpidium scorpioides*	+	+
7 le; b	*Sphagnum* subg *Litophloea*	–	+
10 st; 10 mm; g	*Tortula ruralis*	+	–
	Capsule	–	+

Lissue, Co. Antrim. Fl VIII*
Professor G. F. Mitchell sent the author the following mosses which he had extracted with flowering plant macroscopic fossils from ditches cut in the ninth to tenth centuries on top of a drumlin.

8 st; 15 mm; g	*Dicranum scoparium*
3 st; 17 mm; g	*Eurhynchium praelongum*
1 st; 8 mm; g	*Hypnum cupressiforme*
1 st; 7 mm; b	*Pseudoscleropodium purum*
2 st; 20 mm; g	*Rhytidiadelphus squarrosus*

Little Paxton, Huntingdonshire. Fl zone VIII*
Mr C. F. Tebbutt of St Neots has investigated a Late Saxon pit found near Little Paxton (TL 195628). A layer of peat covering the bottom of the pit yielded the following mosses:

2 st; 4 mm; g	*Antitrichia curtipendula*
3 st; 12 mm; g	*Homalothecium lutescens* or *sericeum*
1 st; 7 mm; g	*Neckera complanata*
1 st; 3 mm; vb	*Rhytidiadelphus* sp
2 le; vb	*Sphagnum* subg *Litophloea*

Loch Cuithir, Isle of Skye. LDe III, III-IV; Fl IV-VIII
Loch Cuithir (NG 474597) is a tiny loch (200 m long) at *c.* 550 ft in rugged terrain in the Trotternish peninsula. The investigations of Vasari and Vasari (1968) produced at least fifteen moss taxa (see table 37).

Loch Droma, Ross and Cromarty. LDe zone I. C^{14} 12,810 ± 155 B.P. Q-547
The highly fossiliferous glacier melt-water silts at Loch Droma (NH 270747) proved to be of outstanding interest for several reasons (Kirk and Godwin 1963). As yet, few investigations employing modern techniques have been made of Devensian sites in northernmost Scotland. Not only did the deposit produce a rich assemblage of macroscopic fossils, especially mosses which are valuable environmental indicators, but its location has special significance with regard to the distribution of Late Devensian glaciers in northwestern Scotland (Kirk and Godwin 1963).

The most organic horizon of the silt has a radiocarbon age of 12,810 ± 155 B.P. (Godwin and Willis 1911) and is referred to zone I. Nineteen mosses were recovered from this layer.

1 st; 2 mm; g	*Brachythecium glaciale*
1 st; 7 mm; g	*Bryum pseudotriquetrum* (t)
2 st; 6 mm; g	*Campylium stellatum*
2 st; 9 mm; b	*Climacium dendroides*
1 st; 2 mm; g	*Ctenidium molluscum*
12 st; 8 mm; g	*Dicranum scoparium*
3 st; 2 mm; g	*Distichium capillaceum*
num st; 7 mm; g	*Ditrichum flexicaule*
3 st; 5 mm; b	*Drepanocladus exannulatus*
2 st; 6 mm; g	*D. revolvens*
num st; 30 mm; vg	*Hylocomium splendens*
1 st; 2 mm; g	*Hypnum cupressiforme* var *lacunosum* (t)
2 st; 2 mm; g	*Pohlia* sp
3 le; b	*Polytrichum alpinum*
1 st; 3 mm; b	*P. juniperinum*
1 st; 3 mm; b	*P. norvegicum*
4 le; b	*P. urnigerum*
num st; 80 mm; vg	*Rhacomitrium lanuginosum*
6 st; 6 mm; vb and 7 le; b	*Rhizomnium pseudopunctatum*

Blindia acuta, Ctenidium molluscum, Fissidens adianthoides, Hylocomium splendens and *Rhacomitrium fasciculare* were found by Dr S. E. Durno and identified by Mr E. L. Birse of the Macauley Institute for Soil Research, Aberdeen.

Table 37

	Zone:	III	III-IV	IV	V	VI	VII*a*	VII*b*	VIII
	Deposit:	Clay	Diatom mud						Various
5 st; 10 mm; b	*Antitrichia curtipendula*	–	–	+	–	–	+	–	–
num st; 12 mm; g	*Bryum* spp	–	+	–	–	–	+	–	–
6 st; 20 mm; g	*Calliergonella cuspidata*	–	+	–	–	–	–	+	+
4 st; 7 mm; g	*Cratoneuron filicinum*	–	+	–	–	–	–	+	–
2 st; 6 mm; b	*Cinclidotus fontinaloides* (t)	–	–	–	–	–	+	–	–
5 st; 10 mm; b	*Drepanocladus revolvens*	+	–	–	–	–	–	+	+
1 st; 5 mm; g	*Eurhynchium riparioides*	–	–	+	–	–	–	–	–
6 st; 12 mm; g	*Homalothecium lutescens* or *H. sericeum*	–	–	–	–	+	+	–	–
4 st; 14 mm; vb and 2 le; g	*Hylocomium splendens*	–	+	–	–	–	+	–	–
1 st; 3 mm; b	*Hypnum* sp	+	–	–	–	–	–	–	–
3 st; 18 mm; g	*H. cupressiforme*	–	–	–	+	+	+	–	–
3 st; 17 mm; g	*Philonotis fontana*	–	–	–	–	+	+	–	–
1 st; 3 mm; g	*Pohlia* sp	–	–	+	–	–	–	–	–
3 st; 5 mm; b	*Rhytidiadelphus loreus*	–	–	–	+	+	–	+	–
1 st; 7 mm; vb	*Thuidium delicatulum*	–	–	–	–	+	–	–	–
	Calyptra	–	–	–	–	+	–	–	–

Loch Fada, Isle of Skye. LDe III, III-IV; Fl IV to VI and VIII

Loch Fada (NG 493487) is situated at *c.* 500 ft in the Trotternish peninsula. The deposits have been investigated by Vasari and Vasari (1968) and subsequently by Dr H. J. B. Birks.

Sixteen further taxa were added to those shown in table 38 by Dr Birks; these records are included in chapter 5, the outstanding ones being *Andreaea rupestris, Mnium orthorhynchum, Plagiothecium undulatum* and *Rhizomium pseudopunctatum.*

Loch Kinord, Aberdeenshire. LDe II and III; Fl VI and VII*b*

Vasari and Vasari (1968) found 520 cm of deposits at the western end of Loch Kinord (NO 435997), a small loch in western Aberdeenshire. Only a few mosses were identified: these are listed in table 39.

Loch of Park, Aberdeenshire. LDe I, II; Fl VII*a*

Loch of Park (NO 772988), another of Vasari and Vasari (1968), also in Aberdeenshire, again yielded only a handful of mosses.

	Zone:	I	II, VII*a*	
	Deposit:	Clay	Detritus mud	
1 st; 8 mm; vb	*Bryum* sp	+	–	–
1 st; 18 mm; g	*Drepanocladus exannulatus*	–	–	+
2 st; 3 mm; b	*D. revolvens*	+	–	–
num st; 26 mm; vb	*Fontinalis* sp	+	–	–
1 st; 5 mm; b	*Mnium* sp	–	+	–
1 st; 5 mm; b	*Polytrichum* sp	+	–	–
1 st; 5 mm; b	*Rhacomitrium* sp	–	+	–

Low Wray Bay, Lake Windermere. LDe zone II. C^{14} 11,878 ± 120 B.P. Q-284

The sediments of Lake Windermere are abundantly fossiliferous and in sheltered bays, such as Low Wray Bay (NY 376012), clearly show the mineral-organic-mineral stratigraphy of Late Devensian deposits (Pennington 1947 and 1962). Detritus mud, the organic deposit of zone II, has yielded many macrofossils, including abundant fragments of mosses (Pennington 1962). Nineteen taxa of mosses were extracted from what remained of cores of dark organic detritus used principally for pollen analysis and the radiocarbon sample (Godwin 1960; Godwin and Willis 1958).

6 st; 5 mm; g	*Abietinella abietina*
5 st; 10 mm; vb	*Antitrichia curtipendula*
1 st; 7 mm; b	*Calliergon cordifolium*
5 st; 12 mm; b	*Campylium stellatum*
6 st; 12 mm; b	*Cratoneuron filicinum*
1 st; 4 mm; b	*Ctenidium molluscum*
2 st; 5 mm; vb	*Dicranum scoparium* (t)
2 st; 5 mm; g	*Ditrichum flexicaule*
3 st; 5 mm; b	*Drepanocladus uncinatus*
2 st; 20 mm; vb	*Fontinalis* sp
15 st; 15 mm; g	*Homalothecium lutescens* or *sericeum*
11 st; 14 mm; g	*Hypnum cupressiforme*
2 st; 15 mm; g	var *lacunosum* (t)
1 st; 3 mm; b	*Hylocomium splendens* (t)
8 st; 10 mm; b	*Pleurozium schreberi*
2 st; 5 mm; vb	*Polytrichum alpinum*
num st; 10 mm; b	*Rhacomitrium fasciculare* (t)
11 st; 5 mm; b	*Tortella tortuosa*
1 st; 3 mm; vb	*Tortula* sp

Pennington (1962) has recorded the following species from Late Weichselian sediments in the Lake District, most species being found in the Low Wray Bay deposits:

Table 38

	Zone:	III-IV	IV	V	VI	VIII
	Deposit:	Clay mud	Detritus mud			Various
2 st; 4 mm; vb	*Bryum* sp or spp	+	–	–	–	+
1 st; 22 mm; g	*Calliergonella cuspidata*	–	–	–	–	+
2 st; 11 mm; b	*Cratoneuron commutatum*	+	–	–	–	–
1 st; 5 mm; g	*C. filicinum*	–	+	–	–	–
1 st; 6 mm; vb	*Dichodontium pellucidum*	+	–	–	–	–
1 st; 4 mm; b	*Dicranella* sp	–	–	–	–	+
num st; 30 mm; vb	*Fontinalis* sp	–	+	+	+	–
5 st; 19 mm; b	*Hylocomium splendens*	+	–	+	–	+
1 st; 2 mm; g	*Philonotis* sp	+	–	–	–	–
1 le; g	*Plagiomnium rostratum* (t)	–	+	–	–	–
1 st; 6 mm; b	*Polytrichum formosum*	–	–	–	–	+
num st; 11 mm; b	*Rhacomitrium* sp or spp	+	–	–	–	–
4 st; 10 mm; g	*Rhytidiadelphus loreus*	–	–	–	–	+
1 st; 7 mm; vb	*Rhytidium rugosum*	–	–	+	–	–
1 st; 5 mm; g	*Thuidium delicatulum*	–	–	–	–	+

Table 39

	Zone:	II	III	VI	VII*b*
	Deposit:	Detritus mud	Silt	Detritus mud	
2 st; 5 mm; g	*Calliergon cordifolium*	–	–	+	+
2 st; 5 mm; g	*C. giganteum*	–	+	–	–
5 st; 16 mm; vb	*Drepanocladus exannulatus*	+	+	–	–
1st; 2 mm; vb and 1 le	*Polytrichum* sp	+	–	–	–
1 st; 2 mm; b	*Sphagnum* subg *Litophloea*	–	+	–	–

	Zone		
	I	II	III
Abietinella abietina	–	+	–
Barbula recurvirostra	–	+	–
Bryum sp	–	+	–
B. caespiticium (t)	–	+	–
Dicranum elongatum	–	+	–
Fontinalis squarrosa	+	–	–
Hylocomium sp	–	+	–
Hypnum spp	+	–	–
Rhacomitrium fasciculare	+	+	–

Moville, Co. Donegal. Fl only
McMillan (1947) reported a 'peat' at the top of the intertidal zone, a little north of Moville; the deposit may be very recent. Professor G. F. Mitchell extracted the following:

1 st; 5 mm; g	*Dicranum scoparium*
1 st; 8 mm; g	*Hylocomium splendens*
1 st; 10 mm; g	*Pseudoscleropodium purum*
1 st; 9 mm; g	*Rhytidiadelphus loreus*
1 st; 33 mm; vg	*R. squarrosus*
1 st; 20 mm; vg	*Thuidium tamariscinum*

New Dry Dock, Renfrewshire. LDe/Fl. C^{14} 9890 ± 160 Birm. 120 and 10,560 ± 180 Birm. 121
During 1961 and 1962, the construction of a large graving dock (NS 305573) at Greenock revealed deep sections of marine deposits made up of richly fossiliferous silts and sands overlying varved clays which rest on boulder clay or Old Red Sandstone. Dr W. W. Bishop, of the Department of Geology, Bedford College, regards the deposits as identical with those described last century from sections exposed during the building of the nearby docks at Garvel Park (Robertson 1881; Scott and Steel 1883).

At the invitation of Dr Bishop, the author visited the site during June 1962 in order to collect samples for pollen and macroscopic fossil analysis. The examined exposure, now destroyed, lay on the western side of the main dock close to the landward end. The strata were as follows:

0 - 1.00 m	Made ground
1.00-1.50 m	Modern beach gravel
1.50 - *c*.7.50 m	Grey silts streaked with black. Sand in seams. Occasional small stones. Well preserved marine mollusc shells
c.7.50 - *c*.800 m	Varved clay

The top of the beach gravel is at 0.5 m below O.D.

The varved section lay several m to the east of the topmost exposure of silt and the intervening area was badly disturbed by the excavating machinery. However, there is still no reason to suppose that the above sequence is incorrect.

Some metres to the south of the varved exposure occurred a section, 0.75 m deep, bluish grey in colour and made very conspicuous by several horizons, 0.10 m apart, consisting of seaweed fragments, the most abundant of which was *Desmarestia aculeata* (Brett and Norton 1969). The relationship between this section and the varves could not be seen clearly. However, according to Dr Bishop's measurements the seaweed horizons lay just above the varved strata.

Only the uppermost 2.20 m of the grey silts of the main exposure were observed in vertical section and sampled. Though pollen can often be extracted from marine deposits, in this case no pollen was recovered from the samples. For extraction of macrofossils, two columns, side by side, 2.00 m long and made up of blocks 0.25 m long, 0.16 m wide and 0.14 m thick, were dug out of the face. The blocks of one column were split in half vertically and sieved for macroscopic fossils.

A block containing one of the seaweed horizons has also been examined. The results of these analyses are shown in the first three of the tables given below. Mrs C. A. Dickson identified the angiosperm remains.

The combined list of mosses from the two investigations totals only about twenty-seven. This represents half the total identified by Rev. J. Fergusson from the Garvel Park material collected by Robertson (1881).

It is striking that only two species, *Betula pubescens* and *Rhacomitrium lanuginosum*, are common to the upper silts and seaweed layer. The presence of such arctic-alpine species as *Aulacomnium turgidum, Oxyria digyna, Polytrichum alpinum, Salix herbacea* and *S.* cf. *reticulata* in the former but not in the latter sediments is very noticeable. Until all the samples are thoroughly investigated it would be premature to draw any firm conclusions from these differences. However, it is tempting to speculate that the two floras, at first sight quite dissimilar, indicate different climatic conditions.

Apart from stating that the vegetable matter is '. . . chiefly confined to the upper muddy sand overlying the tenacious clay, occurring in layers or rather in patches . . .', Robertson (1881, p. 20) does not give the precise provenance of the plant fossils and treats them as one sample, seven marine algae, some fifty-four mosses and several angiosperms being listed. Although he indicated (p. 14) that 'In some cases small tufts of the finer seaweeds are met with deeper down in the sandy mud', he does not appear to have found definite seaweed horizons. Fergusson's bryophyte list includes such species of frequent occurrence in woodland as *Cirriphyllum piliferum* and *Thuidium tamariscinum* together with such arctic-alpine species as *Hypnum bambergeri* and *Oncophorus virens.* Again one is tempted to wonder if two floras are represented.

The radiocarbon dates, obtained from mollusc shells (Birm. 120 at 1.5 m below O.D. and Birm. 121 at 3.5 m below O.D.), place the fossilisation of the mosses from the upper silts (table 40) at about the end of the last glaciation and beginning of the Flandrian (Bishop and Dickson 1970). In chapter 5 these species are listed as 'LDe III/Fl IV*', as are Fergusson's determinations. The species from the seaweed layers are given as 'LDe only'.

Amblystegium sp (t)
Antitrichia curtipendula
Aulacomnium palustre
Brachythecium albicans
B. populeum
B. reflexum
B. rivulare
B. salebrosum (t)
Bryum pallens
B. pseudotriquetrum
Calliergon giganteum
C. sarmentosum
Campylium elodes
C. stellatum
Cirriphyllum piliferum
Climacium dendroides
Cratoneuron commutatum var. *falcatum*
C. decipiens
C. filicinum
Dichodontium pellucidum
Dicranum scoparium
Distichium inclinatum
Drepanocladus aduncus
D. fluitans
D. revolvens
T *Gymnostomum* (sp?)
Homalothecium lutescens
H. nitens
Hygrohypnum molle
Hylocomium brevirostre
H. splendens
Hypnum bambergeri
H. callichroum
H. cupressiforme
Oncophorus virens
Paludella squarrosa
Philonotis fontana
T *Plagiomnium affine*
Plagiothecium denticulatum
Pleurozium schreberi
Pohlia wahlenbergii
Polytrichum alpestre
P. alpinum
P. juniperinum
Rhacomitrium lanuginosum
Rhizomnium pseudopunctatum
Rhytidiadelphus squarrosus
R. triquetrus
Sphagnum sp or spp
Thuidium tamariscinum
T *Tortella fragilis*
T. tortuosa
Tortula norvegica
T *Trichostomum* (sp?)

Assemblage from seaweed layers at *c.* 6.0 m below O.D.

3 st; 10 mm; b and 4 le; b	*Antitrichia curtipendula*
1 st; mm; b	*Bryum* sp
2 st; 10 mm; b	*Climacium dendroides*
1 st; 5 mm; b	*Dichodontium pellucidum*
1 le; g	*Fissidens* sp
1 st; 4 mm; g with capsule	*Grimmia apocarpa*
num st; 14 mm; g	*Hylocomium splendens*
2 le; g	*Plagiomnium rugicum*
2 st; 8 mm; vb	*Rhacomitrium* sp
num st; 10 mm; g	*R. lanuginosum*
3 st; 22 mm; vg	*Rhytidiadelphus squarrosus*
3 st; 4 mm; g	*Solenostoma triste* (t)

Angiosperms

Betula sp
B. nana × *pubescens*
B. pubescens

Table 40 *Assemblages from upper silts*

		Depth in m below O.D.							
		0-0.75	0.75-1.00	1.00-1.25	1.25-1.50	1.50-1.75	1.75-2.00	2.00-2.25	2.25-2.50
1 st; 13 mm; vb	*Aulacomnium palustre*		+						
1 st; 3 mm; b and num le; b	*A. turgidum*	+	+	+		+			
1 st; 5 mm; b	*Brachythecium* sp		+						
1 st; 3 mm; b	*Bryum* sp	+	+						
1 st; 10 mm; g	*Campylium stellatum*		+						
num st; 7 mm; g	*Cratoneuron filicinum*			+					
1 st; 5 mm; g and 2 le; g	*Dicranum scoparium*	+					+		+
2 st; 3 mm; g	*Distichium capillaceum* (t)	+							
3 st; 5 mm; g	*Ditrichum flexicaule*		+	+					
1 st; 5 mm; g	*Philonotis* sp	+							
1 st; 7 mm; b	*Pleurozium schreberi*		+						
4 st; 3 mm; b	*Pohlia* sp	+	+			+		+	
1 st; 3 mm; b and 1 le; b	*Polytrichum alpinum*						+	+	
1 st; 6 mm; b	*P. juniperinum*		+						
1 le; b	*P. urnigerum*	+							
7 st; 10 mm; vb	*Rhacomitrium* sp	+	+	+			+		
1 st; 3 mm; b	*R. lanuginosum*		+						
2 st; 8 mm; b	*Tortula* sp		+						
	Angiosperms								
	Betula pubescens		+						
	Carex sp	+							
	Montia fontana subsp *variabilis*	+							
	Oxyria digyna			+					
	Rubus fruticosus	+							
	Salix sp				+				
	S. herbacea			+					
	S. cf. *reticulata*						+		+
	Spergula arvensis	+							

Newgrange, Co. Meath. Fl VII*a*
Professor G. F. Mitchell extracted seeds and mosses from a turf horizon below the 1970 north cutting of the famous Neolithic burial chamber at Newgrange. Two mosses only were present, *Brachythecium rutabulum* (100 st; 30 mm; vg) and *Plagiomnium undulatum* (14 st; 25 mm; g).

North Ferriby, Yorkshire. Fl zone VII*b*
The large, incomplete Late Bronze Age boat found on the north bank of the Humber during 1937 consisted of oak planks stitched together with yew withe (Wright and Wright 1947). A mixture of *Neckera complanata* and *Eurhynchium striatum* filled the seams of the craft which has a radiocarbon age of 2700 ± 150 B.P. (Barker and Mackay 1960).

Very close to the location of the original discovery (SE 990252) part of a boat of similar construction was excavated during 1963 by Mr J. Bartlett, Director of the Hull Museums. In this case the caulking consisted almost entirely of *Neckera complanata*, with very little *Eurhynchium striatum.* A three-strand rope found in one of the seams proved to be made of *Polytrichum commune.*

Seathwaite Tarn, Lake District. Fl zone VIII
Seathwaite Tarn (SD 253988) lies at an altitude of 1250 ft in a narrow valley close to the Cumberland-Lancashire boundary. About half a mile long and oblong in shape, it is surrounded by steep montane grassland.

Mrs W. Tutin of the Botany Department, the University of Leicester, has investigated the sediments of the tarn and analysed the pollen of a core some 6 m deep, covering the period from zone II to zone VIII. At a depth of 2.40 to 2.44 m there is a well-defined stratum consisting of a few layers of plant debris, made up almost entirely of moss fragments. This wash-in layer can perhaps be interpreted as a consequence of forest clearance in the first few centuries A.D., the Brigantian clearance of Pennington (1970). The mosses were extracted by Mrs Tutin and identified by the author. At the level of 250 cm only one species was found, namely *Thamnobryum alopecurum* (1 st; 8 mm; g). Nineteen taxa were recognised from the layer at 2.40 to 2.44 m.

5 st; 10 mm; g	*Breutelia chrysocoma*
1 st; 7 mm; b and 1 le; b	*Bryum* sp
1 st; 6 mm; b	*Calliergonella cuspidata*
3 st; 7 mm; b	*Campylopus flexuosus* (t)
1 st; 4 mm; b	*Dichodontium pellucidum*
2 st; 20 mm; b	*Eurhynchium riparioides*
1 st; 2 mm; b	*Fontinalis* sp
6 st; 20 mm; vg	*Hylocomium splendens*
1 st; 6 mm; g	*H. umbratum*
2 st; 4 mm; g	*Mnium hornum*
1 st; 10 mm; vb	*Plagiomnium rugicum*
2 st; 5 mm; vb	*Polytrichum* sp
1 le; b	*P. commune*
4 st; 13 mm; g	*Rhacomitrium fasciculare*
2 st; 15 mm; b	*Rhytidiadelphus loreus*
2 st; 11 mm; b	*R. squarrosus*
1 le; b	*Sphagnum palustre*
1 le; b	*S. papillosum*
6 st; 4 mm; g	*S.* subg *Litophloea*

Shippea Hill, Cambridgeshire. Fl zone VII*b*
The 1934 excavation at Peacock's Farm (TL 642848), near Shippea Hill, revealed Mesolithic and Neolithic horizons in fen and wood peat overlain by the Fen Clay. During 1960 the site was re-opened in order to radiocarbon date these layers (Clark and Godwin 1962).

The upper part of the peat (wood peat of Fl zone VII*b*) yielded the following mosses:

1 st; 9 mm; vg	*Cratoneuron filicinum*
num st; 36 mm; vg	*Eurhynchium speciosum*
2 st; 6 mm; g	*Homalothecium sericeum*
1 st; 3 mm; b	*Hypnum cupressiforme*
3 st; 5 mm; g	*Neckera complanata*

As a result of the first excavation Dr M. H. Clifford recognised *Aulacomnium androgynum* (now re-identified as *Thamnobryum alopecurum*), *Eurhynchium swartzii* and *Neckera complanata* from peat just below the Fen Clay (zone VII*b*) (Godwin and Clifford 1938).

Sidgwick Avenue, Cambridge. EDe
Lambert, Pearson and Sparks (1963) described the angiosperm flora, beetles and molluscs from temporarily exposed gravels (TL 442579) of the Intermediate Terrace of the River Cam. Mrs Dickson recovered some ninety taxa of angiosperms as well as moss fragments from closely bedded fine sand and detritus, the lowermost 0.27 m of a section some 4.80 m deep. The greater part of the exposure showed gravel and calcareous marl. The mosses listed in table 41 have been recognised from this exposure, labelled *B*. A composite sample partly covering the four measured horizons is indicated *X*. Sample *C* is another bulk gathering from a continuation of the beds in exposure *B*.

Soham Lode, Cambridgeshire. Zone VII*b*
During a brief examination of fen wood peat (zone VII*b*) extracted from just below the Fen Clay close to the outfall of Soham Lode (TL 531767) Dr D. M. Churchill recovered fragments of seven species of moss. They are as follows:

1 st; 3 mm; b	*Bryum* sp
2 st; 8 mm; b	*Calliergonella cuspidata*
5 st; 18 mm; b	*Cratoneuron filicinum*
1 st; 6 mm; g	*Drepanocladus aduncus* (t)
num st; 19 mm; g	*Eurhynchium speciosum* or *swartzii*
4 st; 4 mm; g	*Homalothecium sericeum*
5 st; 10 mm; g	*Neckera complanata*

Stanford, Kent. Fl zone VIII
Samples of peat obtained by Dr D. M. Churchill from Stanford Bog (TR 115384), near Folkestone, proved to be of zone VIII age. The following mosses were recovered by

Table 41

			B				*C*
		X	23-40 cm	40-50 cm	50-60 cm	60-70 cm	Bulk
2 st; 14 mm; g	*Amblystegium kochii* (t)	–	–	–	+	–	–
2 st; 24 mm; b	*Calliergon giganteum*	+	–	–	–	–	–
1 st; 5 mm; g	*C. turgescens*	+	–	–	–	–	–
8 st; 23 mm; g	*Campylium stellatum*	+	–	+	–	+	+
1 le; g	*Cratoneuron commutatum* var *falcatum* (t)	–	–	–	–	–	+
num st; 33 m; vg	*Drepanocladus exannulatus* var *rotae* (t)	+	+	+	+	+	+
1 st; 5 mm; g	*D. revolvens*	–	–	–	–	–	+
2 st; 7 mm; g	*Homalothecium lutescens*	+	–	–	–	–	–
2 st; 14 mm; g	*Hypnum cupressiforme*	–	–	+	–	+	+
4 st; 5 mm; b	*Scorpidium scorpioides*	–	+	+	–	–	+
	Capsule	+	+	–	–	–	–

Dr Churchill and identified by the author:

		3.00-3.25 m	3.25-3.50 m	3.50-3.75 m
8 st; 6 mm; g	*Calliergon giganteum*	+	+	+
1 st; 4 mm; b	*Calliergonella cuspidata*	+	–	–
2 st; 7 mm; b	*Campylium stellatum*	–	–	+
1 st; 3 m; g	*Eurhynchium* sp	+	–	–
num st; 8 mm; g	*Homalothecium nitens*	+	+	+
num st and le	*Sphagnum* spp	+	+	+

Stannon, Cornwall. LDe

In 1950 Conolly, Godwin and Megaw published an account of the flora of four deposits situated on the granitic upland of Bodmin Moor, Cornwall. The principal deposit, Hawks Tor china-clay (kaolin) pit, showed cryoturbated peat and mud, referred to zone II, overlain by solifluction gravels of zone III and Flandrian peat. Such typical Late Devensian plants as *Armeria, Artemisia* and *Helianthemum* (pollen) and *Betula nana, Salix herbacea, Thalictrum alpinum* and *Polytrichum alpinum* (macroscopic fossils) were recovered from the lower peat and mud. The lower peat has been radiocarbon-dated in Cambridge as Q-211,071 ± 180 B.P. However, samples of the same material assayed in two other laboratories gave results widely divergent from the above figure (Godwin and Willis 1958).

During 1959 and 1960 the author paid two visits to the Stannon china-clay pit, one of the minor sites studied by Conolly *et al.* (1950), and discovered a Late Devensian deposit with stratigraphy of a different nature to that of Hawks Tor.

Stannon pit, situated about 4¼ miles north-northeast of Hawks Tor, lies in a shallow valley at about 750 ft. An extension of the pit (SX 134813) was carried out during 1959 and a long exposure of peat was revealed. At the northwestern end of the extension, the peat covered sediments lying in a shallow depression some 3 to 4 m wide, in the kaolin gravel. The sediments contained remains of such plants as *Salix herbacea* and *Polytrichum* spp in sufficient abundance to make recognition in the field easy.

After the exposure had been excavated, the following stratigraphy was recorded:

0 to 0.60 m. *Molinia* peat.

0.0 to 0.40 m. Light brown very fibrous peat with abundant rootlets, stem and leaf fragments. Some *Sphagnum*, mostly *S. papillosum.*

0.40 to 0.60 m. As above but with more *Sphagnum.* Seeds of *Menyanthes* and *Potamogeton polygonifolius* at 0.50 and 0.57 m respectively.

0.60 to 0.80 m. Black peat.

0.60 to 0.72 m. Very humified peat. *Sphagnum* capsule at 0.65 m.

0.72 to 0.80 m. Lighter in colour than above, more humified. Seed of *Viola* at 0.75 m.

0.80 to 0.84 m. Transition. Rhizomes of *Phragmites.*

0.84 to 1.20 m. Silt. Light brown, unstratified silt with little organic material.

1.20 to 1.40 m. Gravel. Angular quartz particles, mostly around 0.5 cm diameter. Transition with above diffuse; with underlying silt sharp and unconformable.

1.40 to 2.10 m. Silt. Markedly stratified silt with laminae of plant detritus, including *Salix herbacea* and *Polytrichum* spp.

2.10 m downward. Kaolin gravel.

The lamination of the lower silt is clearly shown in plate 4.

Two monoliths of the section were taken for pollen and macroscopic fossil analysis. No attempt was made at radiocarbon dating because the deposit was penetrated by living roots.

The long exposure revealed in the pit showed no lower peat as occurred at Hawks Tor. However, at one locality on the southwestern side of the exposure towards the slopes of Rough Tor, large granite boulders were seen in a position suggestive of movement by solifluction as described by Conolly *et al.* (1950).

It appears that at Stannon Late Devensian deposits are much less extensive than at Hawks Tor (where the lower peat stretched for over 200 m) and are restricted to small depressions which supported shallow lakes or ponds. Conolly *et al.* (1950) describe a section from Stannon different from that given above except for the *Molinia* peat. They state that the lowermost layers, assigned to the Late Devensian on the basis of pollen, consisted of nekron mud. No laminated silt was found.

Of the thirty-four angiosperms recovered from the laminated silt some twenty-two are shared with the macroscopic flora of the Hawks Tor lower peat and mud.

Some seventeen species of mosses were recovered from the laminated silt. Numerous small fragments of hypnoid mosses remain unidentified.

leaf fragments on pollen slides	*Aulacomnium palustre* (t)
num st; 12 mm; g	*Calliergon stramineum*
num st; 6 mm; g	*Ceratodon purpureus*
6 st; 15 mm; g	*Drepanocladus exannulatus*
2 st; 17 mm; g	*D. fluitans*
3 st; 9 mm; g	*Hygrohypnum luridum* (t)
15 st; 22 mm; g	*Philonotis fontana* (t)

num st; 7 mm; g	*Pohlia* sp
7 st; 17 mm; g	*Polytrichum commune*
num st; 14 mm; g	*P. juniperinum*
num st; 20 mm; g	*P. alpinum*
8 st; 5 mm; g	*P. norvegicum*
1 st; 6 mm; b	*P. urnigerum*
num st; 15 mm; vb	*Rhacomitrium* sp
2 st; 2 mm; vb	*Sphagnum* sp or spp
2 st; 3 mm; g	*S. imbricatum*
1 st; 3 mm; g	*S. papillosum*

Conolly *et al.* (1950) recovered twelve mosses and one liverwort from the Late Devensian layers at Hawks Tor.

Amblystegium serpens
Antitrichia curtipendula
Aulacomnium palustre
Blindia acuta
Climacium dendroides
Cratoneuron filicinum (t)
Dicranum bonjeani (t)
Drepanocladus revolvens
Fontinalis antipyretica
Mnium affine
Plagiochila asplenoides
Polytrichum alpinum var *septentrionale* (t)
Sphagnum sp

Only three taxa, namely *Aulacomnium palustre, Polytrichum alpinum* and *Sphagnum*, are shared with the seventeen mosses from Stannon.

In samples from throughout the section, the pollen proved to be sparse and badly preserved and the data presented in fig. 96 are quite insufficient for detailed zonation purposes. The counts are very low, being based on 150 total land pollen, inclusive of Gramineae and Cyperaceae.

The upper silt lacks all the open ground, 'cold' indicators of the laminated silts, e.g. *Betula nana, Salix herbacea, Polytrichum alpinum* and *P. norvegicum* and is referable to the early Flandrian.

On the basis of the macroscopic remains and pollen and also the similarity of the flora to that of the Late Devensian layers of Hawks Tor, the laminated silt can be referred to the Late Devensian. However, greater precision is very difficult because of the inadequate pollen data. It is possible that the gravel layer which overlays the laminated silt represented material moved by solifluction during zone III. However, such a correlation must remain doubtful because the full extent of the gravel, which may only have been a lens, was not determined before the deposit was destroyed.

In chapter 5, the moss records from the laminated silt are given merely as 'Late Devensian only'.

The lowermost 50 cm of the laminated silt was sampled as 5 cm blocks and the macroscopic subfossils were counted; no significant changes were revealed. The following taxa of angiosperms were identified with Mrs C. A. Dickson's help.

Alisma plantago-aquatica
Armeria maritima
Betula nana
Bidens cernua
Carduus or *Cirsium* sp
Carex spp
C. aquatilis or *bigelowii*
C. rostrata
Eleocharis palustris
Empetrum sp
Hippuris vulgaris
Luzula sp
Menyanthes trifoliata
Montia fontana sub sp *fontana*
Myriophyllum alterniflorum
Potamogeton alpinus
P. berchtoldii
P. berchtoldii or *pusillus*
P. gramineus (t)
P. natans
P. obtusifolius
P. perfolatus
Potentilla palustris
Ranunculus subg *Batrachium*
R. flammula
Rumex sp
R. acetosa
R. acetosella
Saxifraga herbacea
S. stellaris
Scirpus tabernaemontani (t)
Sparganium angustifolium
Valeriana officinalis
Veronica scutellata
Viola palustris (t)

Upper Brook Street, Winchester. Fl zone VIII*
During 1959 excavations carried out by Mr F. Cottrill, Curator of the Winchester City Museums, revealed an organic layer at Upper Brook Street (SU 483295). On archaeological grounds the deposit can be no older than the late tenth century. The following mosses have been identified:

1 st; 14 mm; vb	*Anomodon* sp
3 st; 8 mm; g	*Antitrichia curtipendula*
1 st; 12 mm; g	*Calliergonella cuspidata*
1 st; 12 mm; g	*Ctenidium molluscum*
2 st; 9 mm; g	*Eurhynchium striatum*
3 st; 13 mm; g	*Homalothecium lutescens* or *sericeum*
2 st; 20 mm; vg	*Isothecium myurum*
6 st; 25 mm; vg	*Neckera complanata*
1 st; 10 mm; g	*Thamnobryum alopecurum*

Upton Warren, Worcestershire. MDe. C^{14} 41,900 ± 800 B.P. GRO-1245
New ground was broken when Professor Shotton and his colleagues published their account of the beetle fauna and vascular plant flora of the Worcestershire of some 42,000 years ago. The richly fossiliferous layers were discovered in a gravel pit (SO 935073) at Upton Warren (Coope *et al.* 1961). The moss remains listed below and those from Fladbury help to fill the gap which existed from the time of Chelford to that of the Lea Valley Arctic Bed.

9 st; 7 mm; b	*Abietinella abietina*
1 st; 10 mm; g	*Brachythecium* sect *rutabula* or *salebrosa*
7 st; 9 mm; g	*Bryum* spp
2 st; 10 mm; g	*B. pseudotriquetrum* (t)
8 st; 14 mm; g	*Calliergon giganteum*
3 st; 8 mm; g	*Campylium polygamum*
6 st; 8 mm; g	*C. stellatum*
3 st; 4 mm; g	*Ceratodon purpureus*
1 st; 6 mm; b	*Climacium dendroides*
5 st; 15 mm; g	*Cratoneuron filicinum*
1 st; 5 mm; b	*Dicranum scoparium* (t)
1 st; 5 mm; g	*Distichium capillaceum* (t)
5 st; 7 mm; g	*Ditrichum flexicaule*
25 st; 33 mm; g	*Drepanocladus* spp
2 st; 6 mm; g	*D. revolvens*
1 st; 2 mm; g	*Encalypta* sp
3 st; 6 mm; g	*Homalothecium lutescens* (t)

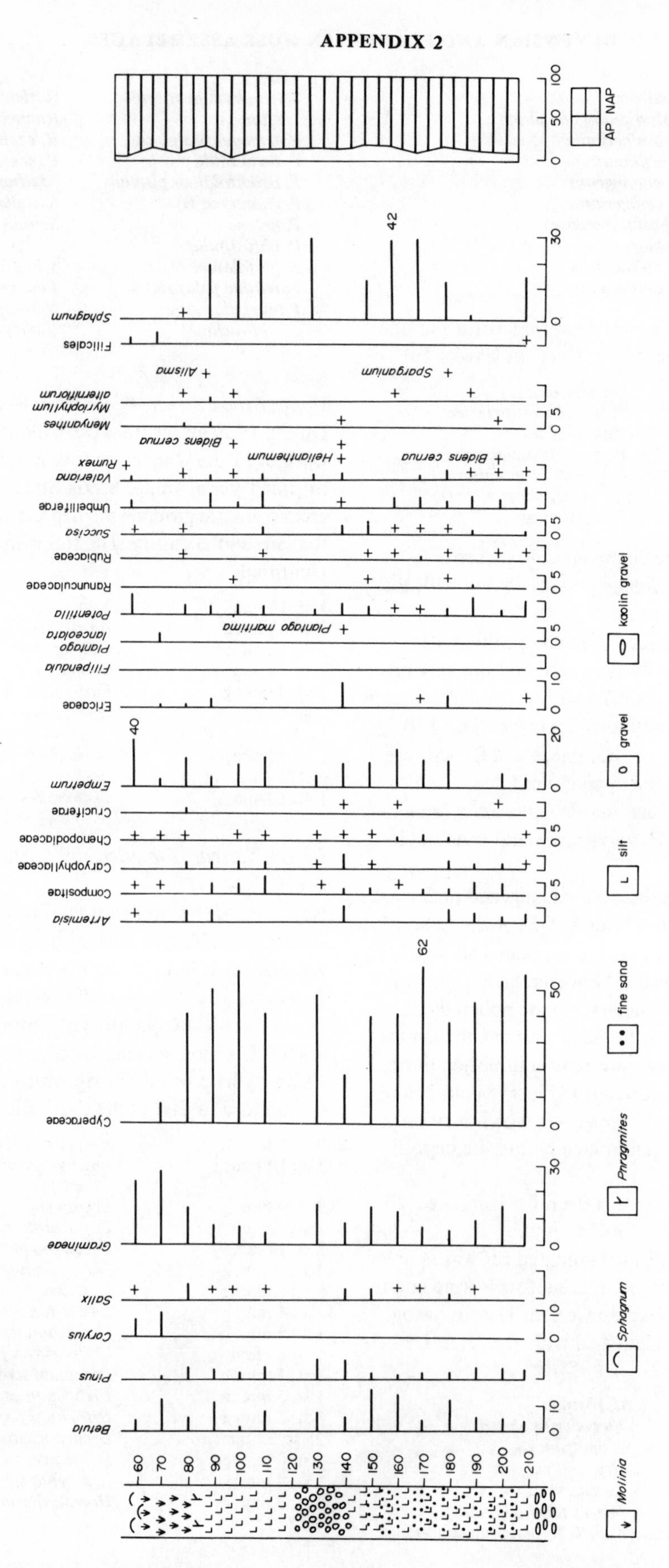

Fig. 96. Pollen diagram from the Late Devensian layers at Stannon, Cornwall. Counts based on 150 total land pollen.

10 st; 14 mm; g	*Philonotis fontana*
12 st; 9 mm; g	*Plagiomnium rugicum*
2 st; 9 mm; g	*Pohlia wahlenbergii*
90 st; 10 mm; g	*Polytrichum juniperinum*
1 st; 3 mm; b	*Scorpidium scorpioides*
1 branch; b	*Sphagnum* subg *Litophloea*
20 st; 10 mm; b	*Tortula ruralis*
1; g	Capsule (bryoid)

Winetavern Street and High Street, Dublin. Fl VIII*; Medieval ar

Extensive excavations carried out in the vicinity of Christ Church Cathedral in Dublin by Mr B. O'Riordain of the National Museum of Ireland revealed a pit largely full of mosses (Winetavern Street, pit 1/C11-12; eleventh century). Amongst the layers of mosses was human excrement containing remains of food plants such as *Rubus, Fragaria, Vaccinium* and *Avena* (Mitchell 1970 pc).

The bryophytes, beautifully preserved and in large masses except where stated, were as follows:

Calliergonella cuspidata	*Plagiochila asplenioides* 1 st only
Cirriphyllum piliferum 1 st only	*Plagiomnium affine* 1 st only
Eurhynchium praelongum	*Rhytidiadelphus squarrosus*
E. striatum	*Thamnobryum alopecurum*
Neckera complanata	*Thuidium tamariscinum*
N. crispa	

From Winetavern Street pit 8/C13 came *Calliergonella cuspidata* (1 st) and *Plagiomnium affine* (1 st).

Neckera complanata was seen as caulking material between timbers, also in the Winetavern Street site.

From pit 1 of High Street, Professor Mitchell extracted a few scraps:

1 st; 3 mm; g	*Brachythecium velutinum*
1 st; 5 mm; g	*Calliergonella cuspidata*
1 st; 4 mm; g	*Neckera complanata*
1 st; 10 mm; g	*Rhytidiadelphus squarrosus.*

REFERENCES

Abbot, L. (1892). The section exposed in the foundations of the new Admiralty Offices. *Proc. geol. Ass.* **12**, 346-56.

Abramova, I. I. (1965). Pliocene bryoflora of Bashkir and the study of fossilized mosses. *Problemy Bot.* **7**, 117-20. In Russian.

Abramova, A. L. and Abramov, I. I. (1955). Mosses from the Kimmeridge deposits of Duab. *Dokl. Akad. Nauk S.S.S.R.* **103**, 699-700. In Russian.

(1956*a*). *Mosses from the Pliocene of Lower Kama*. Symposium dedicated to the 75th birthday of V. N. Sukatschev. In Russian.

(1956*b*). Mosses from the Lower Quaternary deposits near the village Fat'yanovka on the River Oka. *Vest. leningr. gos. Univ. ser. geol. geogr.* **24**, 4. In Russian.

(1958). On certain genetic relations among the fossil mosses of the flora of Duab (Abkhazia). *Bot.Zh. S.S.S.R.* **43**, 1018-24. In Russian.

(1959*a*). Musci Kimmeriensis apud flumen Duab (Abchasia). *Trudy Inst. Bot. V.L. Komarovii Akad. Nauk S.S.S.R.* **12**, 301-59. In Russian.

(1959*b*). Mosses from the Chaudinsk layers of the southwestern Caucasus. *Vest. leningr. gos. Univ. ser. geol. geogr.* **6**, 144-7. In Russian.

(1960). On the Chaduinsk bryoflora of the southwestern Caucasus. *Bot. mater. Otd. spor. rast. BIN AN S.S.S.R.* **13**, 305-12. In Russian.

(1962). The Upper Tertiary and Early Quaternary moss flora of the Central Volga Region. *Problemy Bot.* **4**, 7-17. In Russian.

(1963). *Actinothuidium hookeri* (Mitt.) Broth. in the Kinelian deposits of Bashkiriya. *Bot. mater. Otd. spor. rast. BIN AN S.S.S.R.* **16**, 95. In Russian.

(1964). On the bryoflora of Chaudinsk epoch in the Caucasus. *Bot. Zh. S.S.S.R.* **49**, 1480-7. In Russian.

(1965*a*). Fossil mosses from the deposits at Mamontova Mountain on Aldan. *Nov. Sist. nizch. Rast. Akad. Nauk S.S.S.R.* (1965) 278-96. In Russian.

(1965*b*). Mosses from Sarmat deposits of Precaucasia. *Nov. Sist. nizch. Rast. Akad. Nauk S.S.S.R.* (1965) 280-4. In Russian.

(1967*a*). A new genus and species of fossil mosses from the family Amblystegiaceae. *Nov. Sist. nizch. Rast. Akad. Nauk S.S.S.R.* (1967) 322-36. In Russian.

(1967*b*). Mosses from the Brown Coals of the Ukraine. *Nov. Sist. nizch. Rast. Akad. Nauk S.S.S.R.* (1967) 337-41. In Russian.

(1967*c*). Pleistocene mosses of Western Belorussia. *Nov. Sist. nizch. Rast. Akad. Nauk S.S.S.R.* (1967) 341-50. In Russian.

(1969). Eastern-Asiatic affinities of the Caucasian bryoflora. *J. Hattori Bot. Lab.* **32**, 151-4.

Abramova, A. L. and Abramov, I. I. and Kipiani, M. G. (1965). Sporogonia of mosses from the Quaternary deposits of the U.S.S.R. *Nov. Sist. nizch. Rast. Akad. Nauk S.S.S.R.* (1965) 285-98. In Russian.

Acock, A. M. (1940). Vegetation of a calcareous inner fjord region in Spitsbergen. *J. Ecol.* **28**, 81-106.

Ahti, T. and Isoviita, P. (1962). *Dicranum leioneuron* Kindb. and the other *Dicranum* mosses inhabiting raised bogs in Finland. *Arch. zool. bot. Soc. Vanamo*, **17**, 68-79.

Ahti, T., Isoviita, P. and Maass, W. S. G. (1965). *Dicranum leioneuron* Kindb. new to the British Isles and Labrador, with a description of the sporophyte. *Bryologist* **68**, 197-201.

Albertson, N. (1940*a*). *Scorpidium turgescens* (Th. Jen.) Moenkem. En senglacial relikt i nordisk alvar-vegetation. *Acta phytogeogr. suec.* **13**, 7-26.

(1940*b*). *Rhytidium rugosum* (Hedw.) Lindb. i Fennoscandia. *Svensk bot. Tidskr.* **34**, 77-100.

Aletsee, L. (1959). Zur Geschichte der Moore und Wälder des nordlichen Holsteins. *Nova Acta Leopoldina* **21**, 1-51.

Allison, J., Godwin, H. and Warren, S. H. (1952). Late-glacial deposits at Nazeing in the Lea Valley, North London. *Phil. Trans. R. Soc. Ser. B.* **236**, 169-240.

Allorge, P. (1947). Essai de Bryogéographie de la Péninsule Ibérique. *Encyclopédie Biogéographique et Ecologique* I. Paris; Paul Lechevalier.

Amann, J. (1928). *Bryogéographie de la Suisse. Matériaux pour la Flora Cryptogamique Suisse,* VI, 2. Zurich; Fretz Frères S.A.

Andersen, S. T. (1961). Vegetation and its environment in Denmark in the Early Weichselian Glacial. *Danm. geol. Unders. Raekke 2,* **75**, 1-175.

Andersen, S. T., Vries, H. de and Zagwijn, W. H. (1960). Climatic change and radiocarbon dating in the Weichselian Glacial of Denmark and the Netherlands. *Geol. Mijnb.* **39**, 38-42.

Andersson, G. (1896). Die Geschichte der Vegetation Schwedens. *Bot. Jb.* **22**, 433-550.

Ando, H. (1957). Notes on useful bryophytes. *J. Biol. Soc. Hiroshima Univ.* **7**, 23-6.

Argus, G. W. and Davis, B. (1962). Macrofossils from a Late-Glacial deposit at Cambridge, Massachusetts. *Am. Midl. Nat.* **67**, 106-17.

Arnell, S. (1956). *Illustrated Moss Flora of Fennoscandia. I. Hepaticae.* Lund; Gleerup.

Arnell, S. and Mårtensson, O. (1959). A contribution to the knowledge of the bryophyte flora of W. Spitsbergen and Kongsfjorden (King's Bay 79° N) in particular. *Ark. Bot.* **4**, 105-64.

Ashbee, P. (1963). The Wilsford shaft. *Antiquity* **37**, 116-20.

Atkinson, A. (1887). Notes on an ancient boat found at Brigg. *Archaeologia* **50**, 361-370.

Augier, J. (1966). *Flore des Bryophytes*. Paris; Paul Lechevalier.

Baker, J. G. (1906). *North Yorkshire: Studies of its Botany, Geology, Climate and Physical Geography.* 2nd edition. London; Longmans.

Bardunov, L. V. (1965). *Mosses of Eastern Sayan*. Moscow; Academy of Sciences of the S.S.S.R. In Russian

Barker, H. and Mackay, C. J. (1960). British Museum natural radiocarbon measurements II. *Am. J. Sci. Radiocarbon Suppl.* **2**, 26-30.

Barker, P. A. (1961). Excavations on the Town Wall, Roushill, Shrewsbury. *Medieval Archaeology* **5**, 181-210.

Bartley, D. D. (1962). The stratigraphy and pollen analysis of lake deposits near Tadcaster, Yorkshire. *New Phytol.* **61**, 277-87.

(1964). Pollen analysis of organic deposits in the Halifax region. *Naturalist, Hull.* 1964, 1-11.

REFERENCES

(1966). Pollen analysis of some lake deposits near Bamburgh in Northumberland. *New Phytol.* **65**, 141-56.

Bartosh, T. D. (1963). Concerning the distribution of deposits of calcareous freshwater sediments in the non-chernozem region of the S.S.S.R. Pp. 11-77 in *Material Relating to the Study of Calcareous Freshwater Sediments* II. Riga. In Russian.

Beaumont, P., Turner, J. and Ward, P. F. (1969). An Ipswichian peat raft in glacial till at Hutton Henry, Co. Durham. *New Phytol.* **68**, 779-805.

Behre, K.-E. (1970). Die Flora des Helgoländer Süsswasser-'Tocks', eines Eem-Interglazials unter der Nordsee. *Flora* **159**, 133-46.

Bell, A. M. (1904). Implementiferous sections at Wolvercote. *Q. Jl geol. Soc. Lond.* **60**, 120-32.

Bell, F. G. (1969). The occurrence of southern, steppe and halophyte elements in Weichselian floras of southern Britain. *New Phytol.* **68**, 913-22.

(1970). Late Pleistocene floras from Earith, Huntingdonshire. *Phil. Trans. R. Soc. Ser. B.* **258**, 347-78.

Bell, P. R. and Lodge, E. (1963). The reliability of *Cratoneuron commutatum* (Hedw.) Roth as an 'indicator moss'. *J.Ecol.* **51**, 113-1122.

Bellamy, D. J. and Bellamy, R. (1966). An ecological approach to the classification of the lowland mires of Ireland. *Proc. R. Ir. Acad.* **65**, 237-51.

Bellamy, D. J., Bradshaw, M. E., Millington, G. R. and Simmons, I. G. (1966). Two Quaternary deposits in the Lower Tees Basin. *New Phytol.* **65**, 429-42.

Bellamy, D. J. and Rose, F. (1960). The Waveney-Ouse Valley Fens of the Suffolk-Norfolk Border. *Trans. Suffolk Nat. Soc.* **11**, 367-85.

Benda, V. L. and Schneekloth, H. (1965). Das Eem-Interglazial von Kohlen, Krs. Wesermünde. *Geol. Jb.* **83**, 699-716.

Bennie, J. (1891). On things new and old from the ancient lake of Cowdenglen, Renfrewshire. *Trans. geol. Soc. Glasg.* **9**, 213-25.

(1894*a*). Arctic plants in the old lake deposits of Scotland. *Ann. Scot. Nat. Hist.* **9**, 46-52.

(1894*b*). On the occurrence of peat with Arctic plants in boulder clay at Faskine, near Airdrie, Lanarkshire. *Trans. geol. Soc. Glasgow* **10**, 148-52.

(1896). Arctic plant-beds in Scotland. *Ann. Scot. Nat. Hist.* **17**, 53-6.

Bergeron, T. (1944). On some meteorological conditions for the dissemination of spores, pollen, etc., and a supposed wind transport of *Aloina* spores from the region of the Lower Yenisey to southwestern Finland in July 1936. *Svensk. bot. Tidskr.* **38**, 269-92.

Berglund, B. E. (1966). Late-Quaternary Vegetation in Eastern Blekinge, Southeastern Sweden. A pollen-analytical study. I. Late-glacial time. *Op. bot. Soc. bot. Lund* **12**, 1, 1-180; II. Post-glacial time. *Op. bot. Soc. bot. Lund* **12**, 2, 1-190.

Berry, E. W. (1952). The Pleistocene plant remains of the coastal plain of eastern North America. *Palaeobotanist* **7**, 80-98.

Bertsch, K. (1924). Palaeobotanische Untersuchungen im Reichermoos. *Jahresh. Ver. vaterl. Naturk. Wurttemb.* **80**.

Beug, H.-J. (1957). Untersuchungen zur spatglazialen und frühpostglazialen Floren-und Vegetations-geschichte einiger Mittelgebirge. *Flora* **145**, 167-211.

Billings, W. D. and Anderson, L. E. (1966). Some microclimatic characteristics of habitats of endemics and disjunct bryophytes in the Southern Blue Ridge. *Bryologist* **69**, 76-94.

Birkenmajer, K. and Srodon, A. (1960). Aurignacian Interstadial in the Carpathians. *Biul. Inst. Geol.* **150**, 9-70.

Birks, H. H. and Birks, H. J. B. (1967). *Grimmia agassizii* (Sull. & Lesq.) Jaeg. in Britain. *Trans. Br. bryol. Soc.* **5**, 215-17.

Birks, H. J. B. (1964). Chat Moss, Lancashire. *Mem. Proc. Manchester Lit. Phil. Soc.* **106**, 1-24.

(1965*a*). Late-glacial Deposits at Bag Mere, Cheshire and Chat Moss, Lancashire. *New Phytol.* **64**, 270-85.

(1965*b*). Pollen analytic investigations at Holcroft Moss, Lancashire and Lindow Moss, Cheshire. *J. Ecol.* **53**, 299-314.

Birks, H. J. B. and Dransfield, J. (1970). A note on the habitat of *Scorpidium turgescens* (T. Jens.) Loeske in Scotland. *Trans. Br. bryol. Soc.* **6**, 129-32.

Birks, H. J. B. and Ransom, M. E. (1969). An interglacial peat at Fugla Ness, Shetland. *New Phytol.* **68**, 777-96.

Bisat, W. S. (1948). Interglacial moss at Dimlington, Yorkshire. *Naturalist, Hull.* 1948, 1.

Bishop, W. W. (1958). The Pleistocene geology and geomorphology of three gaps in the Midland Jurassic escarpment. *Phil. Trans. R. Soc. Ser. B.* **241**, 255-306.

Bishop, W. W. and Dickson, J. H. (1970). Radiocarbon dates relating to the Scottish Late-glacial Sea in the Firth of Clyde. *Nature, Lond.* **227**, 480-2.

Blackburn, K. B. (1946). On a peat from the island of Barra, Outer Hebrides. Data for the study of post-glacial history, X. *New Phytol.* **45**, 44-9.

(1952). The dating of a deposit containing an Elk skeleton at Neasham near Darlington, County Durham. *New Phytol.* **51**, 364-77.

Blake, J. H. (1890). *Geology of the Country near Yarmouth and Lowestoft.* Memoirs of the Geological Survey.

Boros, A. (1952). Diluviale Moosfunde in Ungarn. *Foldtani Közlöny* **82**, 294-301.

(1962). Über die Steppenmoose. *Die Pyramide* **3**, 120.

(1968). *Bryogeographie und Bryoflora Ungarns.* Budapest; Akademiai Kiado.

Boulay, N. (1892). *Flore Pliocène du Mont Dore.* Paris.

Boulter, M. C. (1971). A palynological study of two of the Neogene plant beds in Derbyshire. *Bull. Br. Mus. nat. Hist. Geol.* **19**, 361-410.

Braithwaite, R. (1888 to 1895). *The British Moss-Flora.* London; privately published.

Brassard, G. R. and Steere, W. C. (1968). The mosses of Bathurst Island, N.W.T., Canada. *Can. J. Bot.* **46**, 377-83.

Brett, D. W. and Norton, T. A. (1969). Late glacial marine algae from Greenock and Renfrew. *Scott. Jl geol.* **5**, 42-9.

Briggs, D. (1965). The ecology of four British *Dicranum* species. *J. Ecol.* **53**, 69-96.

Burrell, W. H. (1924). Pennine peat. *Naturalist, Hull.* 1924, 145-50.

Camus, F. (1915). Sur les mousses trouvées dans le contenu de l'estomac d'un mammoth. *C.R. Acad. Sci., Paris.* **160**, 842.

Candler, C. (1889). Observations on some undescribed deposits at Saint Cross, South Elmham, Suffolk. *Q. Jl geol. Soc. Lond.* **45**, 504-10.

Cardot, J. (1930). Le peuplement bryologique des Îles Britanniques. In *Contribution à l'étude du peuplement des Îles Britanniques.* Mémoires de la Société de Biogéographie III. Paris; Paul Lechevalier.

Casparie, W. A. (1969). Bult- und Schlenkenbildung in Hochmoortorf. *Vegetatio* **19**, 146-80.

Casparie, W. A. and Van Ziest, W. (1960). A Late-glacial lake deposit near Waskemeer (Prov. of Friesland). *Acta bot. neerl.* **9**, 191-6.

Chandler, M. E. J. (1921). The Arctic flora of the Cam valley at Barnwell. *Q. Jl geol. Soc. Lond.* **77**, 4-22.

Chesters, C. G. C. (1931). On the peat deposits of Moine Mohr. *J. Ecol.* **19**, 46-59.

Clapham, A. R. (1940). The role of bryophytes in the calcareous fens of the Oxford District. *J. Ecol.* **28**, 71-80.

Clapham, A. R. and Clapham, B. N. (1939). The valley fen at Cothill, Berkshire. *New Phytol.* **38**, 167-74.

Clapham, A. R. and Godwin, H. (1948). Studies of the post-glacial history of British vegetation. VIII. Swamping surfaces in peats of the Somerset Levels. IX. Pre-historic trackways in the Somerset Levels. *Phil. Trans. R. Soc. Ser. B.* **233**, 233-86.

REFERENCES

Clark, J. G. D. (1954). *Excavations at Star Carr.* Cambridge University Press.

Clark, J. G. D. and Godwin, H. (1962). The Neolithic in the Cambridgeshire fens. *Antiquity* **36**, 10-23.

Clark, J. G. D., Godwin, H., Godwin, M. E. and Clifford, M. H. (1934). Report on recent excavations at Peacock's Farm, Shippea Hill, Cambridgeshire. *Antiq. Jour.* **15**, 284-319.

Clerici, E. (1892). Illustrazioni della flora nelle fondazioni de ponte in ferro sul tevere a Ripetta. *Boll. Soc. geol. ital.* **11**, 335-68.

Clifford, M. H. (1936). A Mesolithic flora in the Isle of Wight. *Proc. Isle Wight nat. Hist. Archaeol. Soc.* **11**, 582-5.

Clough, T. H. McK. (1971). A hoard of Late Bronze Age metalwork from Aylsham, Norfolk. *Norfolk Archaeology* **35**, 159-69.

Clymo, R. S. (1965). Experiments on breakdown of *Sphagnum* in two bogs. *J. Ecol.* **53**, 747-58.

Coker, P. D. (1967). The effects of sulphur dioxide on bark epiphytes. *Trans. Br. bryol. Soc.* **5**, 341-7.

Conolly, A. P. (1941). A report on plant remains at Minnis Bay, Kent. *New Phytol.* **40**, 299-303.

Conolly, A. P. and Dickson, J. H. (1969). A note on a Late-Weichselian *Splachnum* capsule from Scotland. *New Phytol.* **68**, 197-9.

Conolly, A. P., Godwin, H. and Megaw, E. M. (1950). Studies in the Post-glacial history of British vegetation. XI. Late-glacial deposits in Cornwall. *Phil. Trans. R. Soc. Ser. B.* **234**, 397-469.

Conway, V. (1954). Stratigraphy and pollen analysis of Southern Pennine blanket peats. *J. Ecol.* **42**, 117-47.

Coope, G. R. (1961). A Pleistocene coleopterous fauna with Arctic affinities from Fladbury, Worcestershire. *Q. Jl geol. Soc. Lond.* **118**, 103-23.

Coope, G. R., Shotton, F. W. and Strachan, I. (1961). A Late Pleistocene fauna and flora from Upton Warren, Worcestershire. *Phil. Trans. R. Soc. Ser. B.* **244**, 379-421.

Cöster, I. and Pankow, H. (1968). Illustriester Schüssel zur Bestimmung einiger Mitteleuropaischer *Sphagnum*-Arten. *Wiss. Z. Univ. Rostock.* **415**, 285-323.

Courtney, F. M. and Hardy, D. A. (1969). Wistman's Wood Forest Nature Reserve. *J. Devon Trust Nature Conserv.* **13**, 533-40.

Crum, H. (1965). *Barbula johansenii*, an Arctic disjunct in the Canadian Rocky Mountains. *Bryologist* **68**, 344-5.

(1966). Evolutionary and phytogeographic patterns in the Canadian moss flora. Pp. 28-42 in *The Evolution of Canada's Flora*, edited by R. L. Taylor and R. A. Ludwig. University of Toronto Press.

Crundwell, A. C. (1957). Some neglected British moss records. *Trans. Br. bryol. Soc.* **3**, 174-9.

(1959). A revision of the British material of *Brachythecium glaciale* and *B. starkei*. *Trans. Br. bryol. Soc.* **3**, 565-7.

(1965). *Fossombronia incurva* and *Aongstroemia longipes* in Perthshire, new to the British Isles. *Trans. Br. bryol. Soc.* **4**, 767-74.

(1971). Notes on nomenclature of British mosses. II. *Trans. Br. bryol. Soc.* **6**, 323-6.

Crundwell, A. C. and Nyholm, E. (1964*a*). *Amblystegium saxatile* Schimp. in Cornwall, new to the British Isles. *Trans. Br. bryol. Soc.* **4**, 638-41.

(1964*b*). The European species of the *Bryum erythrocarpum* complex. *Trans. Br. bryol. Soc.* **4**, 597-637.

Crundwell, A. C. and Warburg, E. F. (1963). *Seligeria oelandica* in Ireland, new to the British Isles. *Trans. Br. bryol. Soc.* **4**, 426-8.

Culberson, W. L. (1955). The fossil mosses of the Two Creeks Forest Bed of Wisconsin. *Am. Midl. Nat.* **54**, 452-9.

Curle, J. (1911). *A Roman Frontier Post and its People. The Fort of Newstead.* Glasgow; Maclehose.

Czeczott, H. (1961). The flora of the Baltic amber and its age. Part I. *Prace Muz. ziemi* **4**, 119-45.

Delvosalle, L., Demaret, F., Lambinon, J. and Lawalrée, A. (1969). Plantes rares, disparues ou menacées de disparition en Belgique: L'appauvrissement de la flore indigène. *Trav. serv. Rés. Nat.* **4**, 1-128.

De Vries, B. and Bird, C. D. (1965). Bryophyte subfossils of a Late-glacial deposit from the Missouri Coteau, Saskatchewan. *Can. J. Bot.* **43**, 947-53.

Dewar, H. S. L. and Godwin, H. (1963). Archaeological discoveries in the raised bogs of the Somerset Levels, England. *Proc. prehist. Soc.* **29**, 17-49.

Dickson, C. A., Dickson, J. H. and Mitchell, G. F. (1970). The Late-Weichselian flora of the Isle of Man. *Phil. Trans. R. Soc. Ser. B.* **258**, 31-79.

Dickson, J. H. (1963). Appendix Bryophyta. *Proc. Linn. Soc. Lond.* **174**, 26-7.

(1964*a*). Appendix 2. The Moss flora. *Proc. R. Ir. Acad.* **63**, 182-5.

(1964*b*). Appendix 2. Bryophyta from Ilford. *Phil. Trans. R. Soc. Ser. B.* **247**, 209-10.

(1967). *Pseudoscleropodium purum* (Limpr.) Fleisch. on St Helena and its arrival on Tristan da Cunha. *Bryologist* **70**, 267-8.

Dickson, J. H. and Brown, P. D. (1966). Late Post-glacial *Meesia longiseta* Hedw. in South-eastern England. *Trans. Br. bryol. Soc.* **5**, 100-2.

Dickson, J. H. and Ransom, M. (1968). Report on the caulking from Holme Pierrepont Canoe No. 2. *Trans. Thoroton Soc. Nottingham* **72**, 29.

Dixon, H. N. (1895). Plant remains in peat. *J. Bot. Lond.* **33**, 216.

(1907). List of sub-fossil mosses in 'Summary of Progress for 1907'. *Memoirs of the Geological Survey.*

(1910). Some 'Neolithic' moss remains from Fort William. *Ann. Scot. nat. Hist.* **75**, 103-11.

(1911). Mosses determined from strata in the peat. *Trans. R. Soc. Edinb.* **47**, 829-33.

(1924). *The Student's Handbook of British Mosses*. 3rd edition. Eastbourne; Sumfield and Day.

(1925). Moss remains in Russian peat. *J. Bot. Lond.* **63**, 370.

(1927). *Fossilium Catalogus II: Plantae.* Part 13: *Muscineae.* Berlin; W. Junk.

Doignon, P. (1954). De l'utilisation des mousses dans la construction des châlets valaisons. *Revue bryol. lichen.* **23**, 326-7.

Dombrovskaja, A. V., Koreneva, M. M. and Tyuremnov, S. N. (1959). *Atlas of Plant Remains in Peat*. Moscow, State Press of the Ministry of Power. In Russian.

Donner, J. J. (1957). The geology and vegetation of Late-glacial retreat stages in Scotland. *Trans. R. Soc. Edinb.* **63**, 221-64.

(1959). The Late- and Post-glacial History of the Grampain Highlands of Scotland. *Comment. biol. Helsinf.* **24**, 1-29.

(1962). The Late- and Post-glacial raised beaches in Scotland. II. *Ann. Acad. Sci. Fenn. A* **68**, 1-12.

Douin, R. (1923). Les mousses et les hépatiques fossiles des tufs du Lautaret (Hautes-Alpes). *Revue gén. Bot.* **35**, 113-26.

(1927). Nouvelles recherches sur les mousses et les hépatiques fossiles des tufs du Lautaret (Hautes-Alpes). *Revue gén. Bot.* **39**, 213-17.

Drake, H. C. and Sheppard, T. (1909). Classified list of organic remains from the rocks of the East Riding of Yorkshire. *Proc. Yorks. geol. Soc.* **17**, 4-82.

Duckett, J. G. and Little, E. R. B. (1968). *Mnium medium* B. & S. in Britain. *Trans. Br. bryol. Soc.* **5**, 452-9.

Duigan, S. L. (1955). Plant remains from the gravels of the Summerton-Radley terrace near Dorchester, Oxfordshire. *Q. Jl geol. Soc. Lond.* **111**, 225-38.

(1963). Pollen analyses of the Cromer Forest Bed Series in East Anglia. *Phil. Trans. R. Soc. Ser. B.* **729**, 149-202.

REFERENCES

Duncan, U. K. (1962). Illustrated key to *Sphagnum* mosses. *Trans. Proc. bot. Soc. Edinb.* **39**, 290-301.

(1966). A bryophyte flora of Angus. *Trans. Br. bryol. Soc.* **5**, 1-82.

Du Rietz, G. E. (1940). Problems of bipolar plant distribution. *Acta phytogeogr. suec.* **13**, 215-82.

(1949). Huvudenheter och huvudgränser i svensk myrvegetation. *Svensk bot. Tidskr.* **43**, 274-304.

Durno, S. E. (1957). Certain aspects of vegetational history in North-east Scotland. *Scott. geogr. Mag.* **73**, 176-84.

Dylik, J. and Maarleveld, G. C. (1967). Frost cracks, frost fissures and related polygons. *Meded. geol. Sticht.* **18**, 7-21.

Eggeling, W. J. (1965). Check list of the plants of Rhum, Inner Hebrides. Part II. Lichens, liverworts and mosses. *Trans. bot. Soc. Edinb.* **40**, 60-99.

Erdtman, G. (1957 and 1965). *Pollen and Spore Morphology and Plant Taxonomy. Gymnosperms, Pteridophyta, Bryophyta.* Stockholm; Almquist and Wiksell.

Faegri, K. (1933). Über die Längvariationen einiger Gletscher des Jostedalsbre und die dadurch bedingten Pflanzensukzessionen. *Bergens Mus. Arb.* **7**, 1-255.

Farrand, W. R. (1961). Frozen mammoths and modern geology. *Science* **133**, 729-35.

Fearnsides, M. (1938). Graphic keys for the identification of *Sphagnum. New Phytol.* **37**, 409-24.

Firbas, F., Grünig, G., Weichedel, I. and Werzel, G. (1948). Beitrage zur spat- und Nacheiszeitlichen Vegetationsgeschichte der Vegesen. *Biblthca bot.* **121**, 1-76.

Flenley, J. R. and Pearson, M. C. (1967). Pollen analysis of a peat from the Island of Canna (Inner Hebrides). *New Phytol.* **66**, 299-306.

Franks, J. W. (1960). Interglacial deposits at Trafalgar Square, London. *New Phytol.* **59**, 145-52.

Franks, J. W. and Pennington, W. (Mrs T. G. Tutin) (1961). The Late-glacial and Post-glacial deposits of the Esthwaite Basin, North Lancashire. *New Phytol.* **60**, 27-42.

Fredskild, B. (1967). Palaeobotanical investigations at Sermermuit, Jakobshavn, West Greenland. *Meddr Grønland* **178**, 1-53.

Fries, M., Wright, H. E. and Rubin, M. (1961). A Late Wisconsin buried peat at North Branch, Minnesota. *Am. J. Sci.* **259**, 679-93.

Fulford, M. (1951). Distribution patterns of the genera of leafy hepaticae of South America. *Evolution, Lancaster, Pa.* **5**, 243-65.

(1963). Continental drift and distribution patterns in the leafy hepaticae. Pp. 140-5 in *Polar Wandering and Continental Drift.* American Association of Petroleum Geologists.

Funnell, B. M. and West, R. G. (1962). The Early Pleistocene of Easton Bavents. *Q. Jl geol. Soc. Lond.* **118**, 125-41.

Gadeceau, E. (1919). Les forêts submergées de Belle-Île-en-Mer. *Bull. biol. France* **53**, 276-307.

Gams, H. (1932). Quaternary distribution. Pp. 297-322 in Verdoorn, F. *Manual of Bryology*. The Hague, Martinus Nijhoff.

(1934). Die Moose von Starunia als Vegetations-und-Klimazengen. *Starunia* **2**, 1-6.

(1952). Beiträge sur Verbreitungsgeschichte und Vergesellchaftung der ozeanischen Archegoniaten in Europa. Die Pflanzenwelt Irlands. *Veröff. geobot. Inst. Zürich* **25**, 147-76.

Gepp, A. (1895). Plant remains in peat. *J. Bot. Lond.* **33**, 180-2.

Gilbert, O. L. (1968). Bryophytes as indicators of air pollution in the Tyne Valley. *New Phytol.* **67**, 15-30.

Gimingham, C. H. (1964). Maritime and sub-maritime communities. Pp. 67-142 in *The Vegetation of Scotland*, edited by J. H. Burnett, Edinburgh; Oliver and Boyd.

Gjaerevoll, O. (1956). The plant communities of the Scandinavian alpine snow-beds. *K. norske Vidensk. Selsk. Skr.* **1**, 1-405.

Godwin, H. (1938). *Botany of Cambridgeshire.* Pp. 35-76 in *Victoria County History of Cambridgeshire and the Isle of Ely*, Volume I.

(1941). Studies of the Post-glacial history of British vegetation. VI. Correlations in the Somerset Levels. *New Phytol.* **40**, 108-132.

(1955). Vegetational history at Cwm Idwal: a Welsh plant refuge. *Svensk. bot. Tidskr.* **49**, 36-43.

(1956). *The History of the British Flora.* Cambridge University Press.

(1959). Plant remains from Hartford, Hunts. *New Phytol.* **58**, 85-91.

(1960). The Croonian Lecture. Radiocarbon dating and Quaternary history in Britain. *Proc. R. Soc. Ser. B.* **153**, 287-320.

(1962). Vegetational history of the Kentish Chalk Downs as seen at Wingham and Frogholt. *Veröff. geobot. Inst. Zürich* **37**, 83-99.

(1964). Late-Weichselian conditions in south eastern Britain; organic deposits at Colney Heath, Herts. *Proc. R. Soc. Ser. B.* **160**, 258-75.

Godwin, H. and Clapham, A. R. (1951). Peat deposits on Cross Fell, Cumberland. *New Phytol.* **50**, 167-71.

Godwin, H. and Clifford, M. H. (1938). Studies of the Post-glacial history of British vegetation. I. Origin and Stratigraphy of Fenland deposits near Woodwalton, Hunts. II. Origin and stratigraphy of deposits in southern Fenland. *Phil. Trans. R. Soc. Ser. B.* **229**, 323-406.

Godwin, H. and Conway, V. (1939). The ecology of a raised bog near Tregaron, Cardiganshire. *J. Ecol.* **27**, 313-63.

Godwin, H., Godwin, M. E., Clark, J. G. D. and Clifford, M. A. (1934). A Bronze Age spear head found in Methwold Fen, Norfolk. *Proc. prehist. Soc. East Anglia* **7**, 395-8.

Godwin, H. and Richards, P. W. (1946). Note on the occurrence of *Meesia triquetra* (Hook. & Tayl.) Aongstr. in Post-glacial peat in Somerset (England). *Revue bryol. lichen.* **15**, 123-30.

Godwin, H. and Tallantire, P. A. (1951). Studies of the Post-glacial History of British vegetation. XII. Hockham Mere, Norfolk. *New Phytol.* **39**, 285-307.

Godwin, H. and Willis, E. H. (1958). Radiocarbon dating of the Late-glacial period in Britain. *Proc. R. Soc. Ser. B.* **150**, 199-215.

(1959). Cambridge University natural radiocarbon measurements I. *Am. J. Sci. Radiocarbon Suppl.* **1**, 63-75.

(1960). Cambridge University natural radiocarbon measurements II. *Am. J. Sci. Radiocarbon Suppl.* **2**, 62-72.

(1961). Cambridge University natural radiocarbon measurements III. *Radiocarbon* **3**, 60-76.

(1962). Cambridge University natural radiocarbon measurements V. *Radiocarbon* **4**, 57-70.

Green, B. H. (1968). Factors influencing the spatial and temporal distribution of *Sphagnum imbricatum.* **56**, 47-58.

Greig-Smith, P. (1950). Evidence from hepatics on the history of the British flora. *J. Ecol.* **38**, 320-44.

Grolle, R. (1964). *Jamesoniella carringtonii–eine Plagriochila* in Nepal mit Perianth. *Trans. Br. bryol. Soc.* **4**, 653-64.

Grosse-Brauckmann, G. (1962). Moorstratigraphie Untersuchungen im Niederwesergebiet (Über Moorbildungen an Geestrand und ihre Torfe). *Veröff. geobot. Inst. Zürich* **37**, 100-19.

(1963). Über die Artenzusammensetzung von Torfen aus dem Nordwestdeutschen Marschen-Randgebiet. *Vegetatio* **11**, 325-41.

(1968). Einige Ergebnisse einer Vegetationskundlichen Auswertung Botanischer Torfuntersuchungen, besonders im Hinblick auf Sukzessionsfragen. *Acta bot. neerl.* **17**, 59-69.

(1969). Zur Zonierung und Sukzession im Randgebiet Eines Hochmoores. *Vegetatio* **17**, 33-49.

REFERENCES

Grüger, E. (1968*a*). Vegetationsgeschichliche Untersuchungen an cromerzeitlichen Ablagerungen im nördlichen Randgebiet der deutschen Mittelgebirge. *Eiszeitalter Gegenw.* **18**, 204-35.

(1968*b*). Untersuchungen zur spätglazialen und fruhpostglazialen Vegetationsentwicklung der Südalpen im Umkreis des Gardsees. *Bot. Jb.* **88**, 163-99.

Hallberg, H. (1959). *Rhytidium rugosum* (L. ex Hedw) Kindb. in Bohuslan, western Sweden. *Svensk bot. Tidskr.* **53**, 49-63.

Hallik, R. (1952). Ein 'Weichsel-Fruglazial' Profil in Luneburg. *Eiszeitalter Gegenw.* **2**, 168-72.

(1955). Über eine Verlandungsfolge Weichselinterstadialen Alters in Harksheide bei Hamburg. *Eiszeitalter Gegenw.* **6**, 116-24.

Hamilton, J. R. C. (1968). Appendix II. Wood and Charcoal. Pp. 167-8 in *Excavations at Clikhimin, Shetland.* Ministry of Works Archaeological Report 6.

Hammond, R. F. (1968). Studies into the peat stratigraphy and underlying mineral 'soils' of a raised bog in Ireland. M.Sc. Thesis, University of Dublin.

Hansen, B. (1966). The raised bog Draved Kongsmose. *Bot. Tidsskr.* **62**, 146-85.

Hansen, S. (1965). The Quaternary of Denmark. Pp. 1-90 in *The Quaternary*, volume I, edited by K. Rankama. New York; Interscience Publishers.

Hardy, E. M. (1939). Studies in the Post-glacial history of British vegetation. V. The Shropshire and Flint Maelor mosses. *New Phytol.* **38**, 364-96.

Hartz, N. (1902). Danmarks Senglaciale Flora. *Danm. geol. Unders. Raekke 2* **11**, 1-80.

(1909). Bidrag til Danmarks Tertiaere og Diluviale Flora. *Danm. geol. Unders. Raekke 2* **20**, 1-292.

Hawkesworth, E. (1912). A pre-glacial lake-bed near Northallerton. *Naturalist, Hull.* 1912, 204-5.

Henshall, A. S. (1964). A dagger-grave and other cist burials at Ashgrove, Methilhill, Fife. *Proc. Soc. Ant. Scott.* **97**, 166-79.

Herzog, T. (1926). *Geographie der Moose*. Jena; Gustav Fischer.

Heslop Harrison, J. W. (1948). The passing of the ice age and its effect upon the plant and animal life of the Scottish Western Isles. *New Naturalist* **1**, 83-90.

Hesselbo, A. (1910). Mosrester fra Diluviet ved Skaerumhede. In 'Boring operations through the Quaternary Deposits at Skaerumhede'. *Danm. geol. Unders. Raekke 2* **25**, 101-9.

Holmen, K. (1957). Three West Arctic moss species in Greenland. *Meddr Grønland* **156**, 1-15.

(1959). The distribution of the bryophytes in Denmark. *Bot. Tidsskr.* **55**, 79-154.

(1960). The mosses of Peary Land, North Greenland. *Meddr Grønland.* **163**, 1-96.

Hooker, W. J. and Taylor, T. (1827). *Muscologia Britannia*. 2nd edition. Longman, Rees, Orme, Brown and Green.

Horn af Rantzien, H. (1951). Macrophyte vegetation in lakes and temporary pools of the Alvar of Öland, South Sweden. *Svensk bot. Tidskir.* **45**,

Hulten, E. (1962). Plants of the floating ice-island 'Arliss II'. *Svensk bot. Tidskr.* **56**, 362-4.

Husnot, T. (1884-90). *Muscologia Gallica.* Orne; privately published.

Hutchinson, T. C. (1966). The occurrence of living and sub-fossil remains of *Betula nana* L. in Upper Teesdale. *New Phytol.* **65**, 351-7.

Irmscher, E. (1929). Pflazenverbreitung und Entwicklung der Kontinente. II Teil. Weitere Beitrage zur genetischen Pflanzengeographie unter besonder Berucksichtigung der Laubmoose. *Mitt. Inst. allg. bot. Hamb.* **8**, 169-374.

Isoviita, P. (1966). Studies on *Sphagnum* L. I. Nomenclatural revision of the European taxa. *Annls bot. Fenn.* **3**, 199-264.

Iversen, J. (1939). Plantrester fremdragne i tre Høje i Haderslev Amt. Pp. 18-21 in Broholm, H. C. and Hald, M., Skrydstrupfundet. *Nordiske Fortidsminder* **3**, 2, 1-115.

Jalas, J. (1955). *Rhacomitrium lanuginosum* (Hedw.) Brid. als Klima indikater in Ostfennoskandien. *Arch. Soc. zool. bot. Soc. Vanamo* **9**, 73-88.

Jasnowski, M. (1957*a*). *Calliergon trifarium* Kindb. in der Stratigraphie und Flora del holozänen Niedermoore Polens. *Acta Soc. Bot. Pol.* **26**, 701-18. In Polish with German summary.

(1957*b*). Moosflora quartärer Flachmoorablagerungen. *Acta Soc. Bot. Pol.* **26**, 597-629. In Polish with German summary.

(1959). Origin and classification of Quaternary moss peats. *Acta Soc. Bot. Pol.* **28**, 319-64. In Polish with German summary.

Jessen, A. and Nordmann, A. (1915). Ferskvandslagene ved Nørre Lyngby. *Danm. geol. Unders. Raekke 2* **29**, 1-66.

Jessen, K. (1949). Studies in Late Quaternary deposits and flora-history of Ireland. *Proc. R. Ir. Acad.* **52**, 85-290.

Jessen, K., Andersen, S. T. and Farrington, A. (1959). The interglacial deposit near Gort, Co. Galway. *Proc. R. Ir. Acad.* **60**, 1-77.

Jessen, K. and Milthers, V. (1928). Interglacial fresh-water deposits in Jutland and Northwest Germany. *Danm. geol. Unders. Raekke 2*, **48**, 1-380.

Johnson, G. A. L. and Dunham, K. C. (1963). *The Geology of Moorhouse*. Monographs of the Nature Conservancy 2.

Johnson, T. (1951). Fossil plants from Washing Bay, Co. Tyrone. II. Sphagnaceae. *Ir. Nat. J.* **10**, 150.

Jones, E. W. (1953). A bryophyte flora of Berkshire and Oxfordshire. II. Musci. *Trans. Br. bryol. Soc.* **2**, 220-77.

Jovet-Ast, S. (1967). Bryophyta. Pp. 17-186 in *Traité de Paléobotanique*, Volume II. edited by E. Boureau. Paris; Masson et Cie

Jovet-Ast, S. and Huard, J. (1966). Mousses de la flore néogène d'Arjuzanx (Landes). *Revue bryol. Lichen.* **34**, 807-15.

Kapp, R. O. and Gooding, A. M. (1964). Pleistocene vegetational studies in the Whitewater basin, southeastern Indiana. *J. Geol.* **72**, 307-26.

Karczmarz, K. (1962). Distribution of *Cinclidium stygium* in Poland. *Ann. Univ. M. Curie-Sklodowska Lublin.* **17**, 427.

(1963). The Mosses of the Leczna and Wlodawa Lake District, Part I. *Ann. Univ. M. Curie-Sklodowska* **18**, 367-410. In Polish with English summary.

(1969). *Calliergon turgescens* (Th. Jens.) Kindb. in the stratigraphic profiles of the Quaternary deposits in Poland. *Folia soc. scien. Lublin.* **9**, 111-114.

Karczmarz, K. and Kuc, M. (1966). Notes on *Calliergon orbioulari-cordatum* from Spitzbergen. *Bryologist* **69**, 373-6.

Karenlampi, L. (1968). The first find of the moss *Diphyscium foliosum* (Hedw.) Mohr in Southern Finland. *Memor. Soc. Fauna Flora fenn.* **44**, 21-3.

Kedves, M. (1969). *Palynological Studies on Hungarian Early Tertiary Deposits.* Budapest; Alkademi Kiado.

Keller, F. (1878). *The Lake Dwellings of Switzerland and Other Parts of Europe.* Translated by J. E. Lee. London; Longmans.

Kelly, M. R. (1968). Floras of Middle and Upper Pleistocene Age from Brandon, Warwickshire. *Phil. Trans. R. Soc. Ser. B.* **254**, 401-15.

Kerney, M. P., Brown, E. H. and Chandler, T. J. (1964). The Late-glacial and post-glacial history of the Chalk escarpment near Brook, Kent. *Phil. Trans. R. Soc. Ser. B.* **246**, 136-204.

King, A. L. K. and Morrison, M. E. S. (1956). *Sphagnum imbricatum, Ir. Nat. J.* **12**, 105-7.

King, A. L. K. and Scannell, M. J. (1960). Notes on the vegetation of a mineral flush in Co. Mayo. *Ir. Nat. J.* **13**, 137-40.

Kirk, W. and Godwin, H. (1963). A Late-glacial site at Loch Droma, Ross and Cromarty. *Trans. R. Soc. Edinb.* **65**, 225-249.

Kleiven, M. (1959). Studies on the xerophile vegetation in Northern Gudbransdalen, Norway. *Nytt Mag. Bot.* **7**, 1-60.

Knowles, W. J. (1912). Prehistoric stone implements from the River Bann and Lough Neagh. *Proc. R. Ir. Acad.* **30**, 195-222.

REFERENCES

Knox, E. M. (1939). The spores of the Bryophyta compared with those of Carboniferous Age. *Trans. bot. Soc. Edinb.* **32**, 477-87.

Koelbloed, K. K. and Kroeze, J. M. (1965). *Anthoceros* species as indicators of cultivation. *Boor spade* **14**, 104-9.

Koperawa, W. (1958). A Late-glacial pollen diagram at the north foot of the Tatra Mountains. *Monogr. bot.* **7**, 107-33.

Koponen, T. (1967). *Eurhynchium angustirete* (Broth.) Kop. comb. n. (= *E. zetterstedtii* Storm.) and its distribution pattern. *Memor. Soc. Fauna Flora fenn.* **43**, 53-9.

(1968). Generic revision of the Mniaceae. *Annls bot. Fenn.* **5**, 117-51.

(1969*a*). The moss genus *Cinclidium* (Mniaceae) in Finland. *Annls bot. Fenn.* **6**, 112-18.

(1969*b*). The taxonomic status and typification of *Mnium subpunctatum. Bryologist* **72**, 61-2.

Koponen, T. and Oittinen, V. (1967). *Aloina brevirostris* (Hook. et Grev.) Kindb. found in Finland. *Memor. Soc. Fauna Flora fenn.* **43**, 5-7.

Körber-Grohne, U. (1967). *Geobotanische Untersuchungen auf der Feddersen Wierde.* Text volume 356 pp. Plate volume 84 pp. Wiesbaden; Franz Steiner Verlag GMBH.

Kotilainen, M. J. (1929). Über das boreale Laubmooselement in Ladoga-Karelien. Eine kausal-okologische und floristische studie. *Annls bot. Soc. Zool.-bot. fenn. Vanamo* **11**, 1-142.

Krausel, R. and Weyland, H. (1942). Tertiare und quartare Pflanzenreste aus den vulkanischen Tuffen der Eifel. *Abh. senckenberg. naturf. ges.* **463**, 1-62.

Krusentsjerna, E. V. (1945). Bladmossvegetation och Bladmossflora i Uppsalstrakten. *Acta phytogeogr. suec.* **19**, 1-250.

Krutzsch, W. (1963*a*). *Atlas der Mittel-und jung tertiären dispersen Sporen- und Pollen-sowie der Mikroplanktonformen des nordlichen Mitteleuropas.* Lieferung II. *Die sporen der Anthocerotaceae und der Lycopodiaceae.* Berlin; Deutscher Verlag der Wissenschaften.

(1963*b*). *Ibid.* Lieferung III. *Sphagnacoide und Selaginellacoide sporenformen.*

Kuc, M. (1956). *Scorpidium turgescens* Mnkm. nowy relikt glacjalny we florjemchow Polski. *Kosmos* **4**, 620-1.

(1963). Flora of mosses and their distribution on the north coast of Hornsund (S.W. Svalbard). *Fragm. flor. geobot.* **9**, 291-366.

(1964). Bryogeography of the southern Uplands of Poland. *Monogr. bot.* **17**, 1-211.

(1969). Additions to the Arctic moss flora - I. *Revue bryol. lichen.* **36**, 635-42.

Lacey, W. S. (1969). Fossil bryophytes. *Biol. Rev.* **44**, 189-205.

Lambert, C. A. (1964). Appendix IV. The plant remains from Cist I. Pp. 178-9 in Henshall, A. S., *A dagger-grave and other cist burials at Ashgrove, Methilhill, Fife. Proc. Soc. Ant. Scott.* **97**, 166-79.

Lambert, C. A., Pearson, R. G. and Sparks, B. W. (1963). A flora and fauna from Late Pleistocene deposits at Sidgwick Avenue, Cambridge. *Proc. Linn. Soc. Lond.* **174**, 13-29.

Lambert, J. M., Jennings, J. N., Smith, C. T., Green, C. and Hutchinson, J. N. (1960). *The Making of the Broads.* Royal Geographical Society Research Series 3.

Landwehr, J. (1949). Bladzijden int een oud boek. *Natura* **46**, 142-5.

(1951). Bryologisch onderzoek van subfossiel veen bij Amstelveen. *Buxbaumia* **5**, 26-30.

Lange, B. (1963). Studies in the *Sphagnum* flora of Iceland and the Faroes. *Bot. Tidsskr.* **59**, 220-43.

(1969). The distribution of *Sphagnum* in northernmost Scandinavia. *Bot. Tidsskr.* **65**, 1-43.

Lazarenko, A. S. (1958). Remote dispersal of spores and its importance in the formation of moss ranges. *Ukr. bot. Zhr.* **15**, 71-7. In Russian.

Leach, W. (1931*a*). Note on the effect of growing mosses in a moisture-saturated atmosphere, and under conditions of darkness. *New Phytol.* **30**, 276-84.

(1931*b*). On the importance of some mosses as pioneers on unstable soils. *J. Ecol.* **19**, 98-102.

Lee, L. W. (1944). Some bryophytes from Greenland. *Bryologist* **47**, 114-18.

Lesemann, B. (1969). Pollenanalytische Untersuchungen zur Vegetationsgeschichte des Hannoverschen Wendlandes. *Flora* **158**, 480-519.

Lett, H. W. (1912). Clare Island Survey 11-12. Musci and Hepaticae. *Proc. R. Ir. Acad.* **31**, 1-18.

Lewis, F. J. (1905). The plant remains in the Scottish peat mosses. I. The Scottish Southern Uplands. *Trans. R. Soc. Edinb.* **41**, 699-723.

(1906). The plant remains in the Scottish peat mosses. II. The Scottish Highlands. *Trans. R. Soc. Edinb.* **45**, 335-59.

(1907). The plant remains in the Scottish peat mosses. III. The Scottish Highlands and the Shetland Islands. *Trans. R. Soc. Edinb.* **46**, 33-70.

(1911). The plant remains in the Scottish peat mosses. IV. The Scottish Highlands and Shetland Islands with an appendix on the Icelandic peat deposits. *Trans. R. Soc. Edinb.* **47**, 793-833.

Lid, J. (1925). An account of the cymbifolia group of Sphagna in Norway. *Nyt Mag. Naturvid.* **63**, 224-59.

Lodge, E. (1959). Effects of certain cultivation treatments on the morphology of some British species of *Drepanocladus. J. Linn. Soc. Bot.* **56**, 218-24.

(1960). Studies of variation in British material of *Drepanocladus fluitans* and *Drepanocladus exannulatus*. I. An analysis of the variation. *Svensk bot. Tidskr.* **54**, 369-86; II. An experimental study of the variation. *Svensk bot. Tidskr.* **54**, 387-93.

(1963). The bryophytes of the Small Isles parish of Invernessshire I. *Nova Hedwigia* **5**, 117-48.

Lohammar, G. (1954). The distribution and ecology of *Fissidens julianus* in northern Europe. *Svensk bot. Tidskr.* **48**, 162-73.

Longton, R. E. (1967). Vegetation in the Maritime Antarctic. *Phil. Trans. R. Soc. Ser. B.* **252**, 213-37.

Lundquist, G. (1964). Interglaciala Avlagringar i Sverige. *Sver. geol. Unders. Afh.* **600**, 3-60.

Lye, K. A. (1967). A new classification of Norwegian plant-geographical elements. *Blyttia* **25**, 88-123.

Maass, W. S. G. (1965). *Sphagnum dusenii* and *Sphagnum balticum* in Britain. *Bryologist* **68**, 211-17.

McClymont, J. W. (1955). Spore studies in the Musci, with special reference to the genus *Bruchia. Bryologist* **58**, 287-306.

McClymont, J. W. and Larson, D. A. (1964). An electron-microscopic study of spore wall structure in the Musci. *Am. J. Bot.* **51**, 195-200.

MacCormick, A. G. (1968). Three dug-out canoes and a wheel from Holme Pierrepont, Nottinghamshire. *Trans. Thoroton Soc. Nottingham.* **72**, 14-31.

McMillan, N. F. (1957). Quaternary deposits around Lough Foyle, North Ireland. *Proc. R. Ir. Acad.* **58**, 185-205.

McVean, D. N. and Ratcliffe, D. A. (1962). *Plant Communities of the Scottish Highlands.* Monographs of the Nature Conservancy 1.

MacVicar, S. M. (1926). *The Student's Handbook of British Hepatics.* 2nd edition. Eastbourne; Sumfield.

Mahony, J. A. (1868*a*). Notes on the botany of the Windmillcroft Beds. *Proc. nat. Hist. Soc. Glas.* **1**, 159-65.

(1868*b*). On the occurrence of seaweeds in the Paisley Clay Beds. *Proc. nat. Hist. Soc. Glas.* **1**, 199-202.

(1869). On the organic remains found in clay near Crofthead, Renfrewshire. *Geol. Mag.* **6**, 390-3.

Mai, D. H., Majarski, J. and Unger, K. P. (1963). Pliozän und Altpleistozan von Rippersroda in Thüringen. *Geologie* **12**, 765-815.

Mamakowa, K. (1962). The vegetation of the basin of Sandomierz in the Late-Glacial and Holocene. *Acta Palaeobot.* **3**, no. 2, 1-57. In Polish with English summary.

(1968). Flora from the Paudorf Interstadial at Lazek near Zaklikow. *Acta Palaeobot.* **9**, no. 1, 29-44. In Polish with English summary.

Marr, J. E. and Gardner, E. W. (1916). On some deposits containing an Arctic flora in the Pleistocene Beds of Barnwell, Cambridge. *Geol. Mag.* **3**, 339-43.

Mårtensson, O. (1955 and 1956). Bryophytes of the Torneträsk Area, northern Swedish Lappland. I, II, and III. *Kungl. Svenska Vetens. Avh. i Naturv.* **12**, 1-107, **14**, 1-321 and **15**, 1-94.

Miller, A. G. (1947). Bronze Age graves at Ferniegair, Hamilton. *Trans. Glas. Archaeol. Soc.* **11**, 17-21.

Miller, H. (1854). *My Schools and Schoolmasters; or, The Story of my Education.* Edinburgh; Johnstone and Hunter.

Miller, N. G. and Benninghoff, W. S. (1966). Plant fossils from a Cary-Port Huron Interstade deposit and their paleoecological interpretation. Geological Society of America,

Miller, N. G. and Benninghoff, W. S. (1966). *Plant Fossils from a Cary-Port Huron Interstade deposit and Their Paleoecological Interpretation.* Geological Society of America, Inc. Special Paper 123. INQUA Volume, pp. 225-48.

Mitchell, G. F. (1940). Studies in Irish Quaternary deposits near Dunshaughlin, County Meath. *Proc. R. Ir. Acad.* **46**, 13-37.

(1948). Late-glacial deposits in Berwickshire. *New Phytol.* **47**, 262-4.

(1951). Studies in Irish Quaternary Deposits No. 7. *Proc. R. Ir. Acad.* **53**, 113-206.

(1953). Further identifications of macroscopic plant fossils from Irish Quaternary deposits, especially a Late-glacial deposit at Mapastown, Co. Louth. *Proc. R. Ir. Acad.* **55**, 225-81.

(1956). Post-Boreal pollen diagrams from Irish raised bogs. *Proc. R. Ir. Acad.* **57**, 185-251.

(1967). The Pleistocene deposits of the Isles of Scilly. *Q. Jl geol. Soc. Lond.* **123**, 59-92.

Mitchell, G. F. and Watts, W. A. (1970). The history of the Ericaceae in Ireland during the Quaternary Epoch. Pp. 13-21 in *Studies in the Vegetational History of the British Isles,* edited by D. Walker and R. G. West. Cambridge University Press.

Moar, N. T. (1969*a*). Late Weichselian and Flandrian pollen diagrams from south-west Scotland. *New Phytol.* **68**, 433-67.

(1969*b*). Two pollen diagrams from the Mainland, Orkney Islands. *New Phytol.* **68**, 201-8.

Monkemeyer, W. (1927). Die Laubmoose Europas. IV. Andreales-Bryales. Rabenhorst's *Kryptogamenflora von Deutschland, Osterreich und der Schweiz.* Leipzig; Akademische Verlagsgescellschaft.

Moore, J. J. (1955). The distribution and ecology of *Scheuchzeria palustris* on a raised bog in Offaly. *Ir. Nat. J.* **11**, 320-9.

Moore, P. D. and Chater, E. A. (1969). Studies in the vegetational history of Mid-Wales. I. The Post-glacial Period in Cardiganshire. *New Phytol.* **68**, 183-96.

Morgan, A. (1969). A Pleistocene fauna and flora from Great Billing, Northamptonshire, England. *Opusc. ent.* **34**, 109-29.

Mornsjö, T. (1969). Studies on vegetation and development of a peatland in Scania, south Sweden. *Op. bot. Soc. bot. Lund.* **24**, 1-187.

Morrison, M. E. S. (1959). The ecology of a raised bog in Co. Tyrone, Northern Ireland. *Proc. R. Ir. Acad.* **60**, 291-308.

Morrison, M. E. S. and Stephens, N. (1965). A submerged Late-Quaternary deposit at Roddans Port on the north-east coast of Ireland. *Phil. Trans. R. Soc. Ser. B.* **249**, 221-55.

Muller, E. H. (1964). Quaternary sections at Otto, New York. *Am. J. Sci.* **262**, 461-78.

Munro, R. (1879). Notice of the excavation of a crannog at Lochlee, Tarbolton, Ayrshire. *Proc. Soc. Ant. Scott.* **13**, 175-252.

(1882). *Ancient Scottish Lake Dwellings.* Edinburgh; D. Doylan.

Nagy, E. (1968). Moss spores in Hungarian Neogene strata. *Acta. bot. hung.* **14**, 113-32.

Nathorst, A. (1873). On the distribution of arctic plants during the Post-glacial Epoch. *J. Bot. Lond.* **2**, 225-8.

Neujstadt, M. I. (1957). *History of Forests and Palaeogeography of the S.S.S.R. in the Quaternary.* Moscow; Institute of Geography of the Academy of Sciences of the S.S.S.R. In Russian.

Neuweiler, E. (1905). Die prähistorischen Pflanzenreste Mitteleuropas mit besonderer Berucksichtigung der schweizenschen Funde. *Vjschr. naturf. Ges. Zürich* **50**, 23-134.

Newall, R. S. (1931). Barrow 85, Amesbury. *Wiltsh. Archaeol. nat. Hist. Mag.* **45**, 432-4.

Newbould, P. J. (1953). Bryological note. *Sphagnum squarrosum* Pers. ex Crome. *Trans. Br. bryol. Soc.* **2**, 296.

Nichols, H. (1967). The Post-glacial history of vegetation and climate at Ennadai Lake, Keewatin, and Lynn Lake, Manitoba (Canada). *Eiszeitalter Gegenw.* **18**, 178-97.

(1969). The late Quaternary history of vegetation, and climate at Porcupine Mountain and Clearwater Bog, Manitoba. *Arctic Alpine Research* **1**, 155-67.

Nogouchi, A. (1964). A revision of the genus *Claopodium. J. Hattori Bot. Lab.* **27**, 20-46.

Nyholm, E. (1954-69). *Moss Flora of Fennoscandia. II. Musci.* Volumes 1 to 6. Lund; Gleerup.

Oakley, K. P. (1939). *Geology and Palaeolithic Studies. A Survey of the Prehistory of the Farnham District (Surrey).* Guildford; Surrey Archaeological Society.

Paepe, R. and Vanhoorne, R. (1967). The stratigraphy and palaeobotany of the Late Pleistocene in Belgium. *Mem. cartes geol. Min. Belg.* **8**, 1-95.

Pankow, H. (1966). Die Verbreitung einiger pflanzengeographischen interessanter Moosarten in Mecklenburg und den Grenzegebieten. *Feddes Rep.* **73**, 59-77.

Paton, J. A. (1962). The genus *Calypogeia* Raddi in Britain. *Trans. Br. bryol. Soc.* **4**, 221-229.

Paton, J. (1965*a*). A new British moss, *Fissidens celticus* sp. nov. *Trans. Br. bryol. Soc.* **4**, 779-84.

(1965*b*). *Census Catalogue of British Hepatics.* British Bryological Society.

Pavletic, Z. (1955). *Prodromus Flore Briofita Jugoslavije.* Zagreb; Jugoslavenska Akedemija.

Pearsall, W. H. (1921). The aquatic vegetation of the English Lakes. *J. Ecol.* **8**, 163-201.

(1956). Two blanket-bogs in Sutherland. *J. Ecol.* **44**, 493-516.

Pennington, W. (Mrs T. G. Tutin) (1943). Lake sediments: the bottom deposits of the north basin of Windermere, with special reference to the diatom succession. *New Phytol.* **42**, 1-27.

(1947). Studies of the Post-glacial history of British vegetation. VII. Lake sediments: pollen diagrams from the bottom deposits of the north basin of Windermere. *Phil. Trans. R. Soc. Ser. B.* **233**, 137-75.

(1962). Late-glacial moss records from the English Lake District. Data for the study of Post-glacial history. *New Phytol.* **61**, 28-31.

(1964). Pollen analyses from the deposits of six upland tarns in the Lake District. *Phil. Trans. R. Soc. Ser. B.* **248**, 205-44.

(1970). Vegetation history in the north-west of England: a regional synthesis. Pp. 41-80 in *Studies in the Vegetational History of the British Isles*, edited by D. Walker and R. G. West. Cambridge University Press.

Penny, L. F., Coope, G. R. and Catt, J. A. (1969). Age and insect fauna of the Dimlington silts, East Yorkshire. *Nature, Lond.* **224**, 65-7.

Perry, A. R. and Dransfield, J. (1967). *Orthotrichum gymnostomum* in Scotland. *Trans. Br. bryol. Soc.* **5**, 218-21.

Perry, A. R. and Fitzgerald, R. D. (1963). *Hypnum vaucheri* Lesq. in Perthshire - new to the British Isles. *Trans. Br. bryol. Soc.* **4**, 418-21.

Persson, A, (1964). The vegetation at the margin of the receding glacier Skaftafellsjökull, southeastern Iceland. *Bot. Notiser* **117**, 321-54.

Persson, H. (1942). Bryophytes from the bottom of some lakes in north Sweden. *Bot. Notiser* 1942, 308-24.

(1944). On some species of *Aloina* with special reference to their dispersal by the wind. *Svensk bot. Tidskr.* **38**, 260-68.

(1949). Studies in the bryophyte flora of Alaska-Yukon. *Svensk bot. Tidskr.* **43**, 491-533.

(1960). Bryological examination. In Fromm, E. An interglacial peat at Ale, near Lulea, northern Sweden. *Sver. geol. Unders Afh.* **54**, 1-14.

(1968). Bryophytes from the Aleutian Islands, Alaska, collected mainly by Hansford T. Shacklette. *Svensk. bot. Tidskr.* **62**, 309-87.

Persson, H. and Sjörs, H. (1960). Some bryophytes from the Hudson Bay Lowland of Ontario. *Svensk. bot. Tidskr.* **54**, 247-68.

Petch, C. P. and Swann, E. L. (1968). *Flora of Norfolk*. Norwich; Jarrold and Sons.

Pettersson, B. (1940). Experimentalle Untersuchungen über die euanemochore der Verbreitung der Sporenpflanzen. *Annls bot. Fenn.* **25**, 1-103.

Pettifer, A. J. (1968). A bryophyte flora of Essex. *Essex Nat.* **32**, 83-155.

Péwé, J. L. (1966). Paleoclimatic significance of fossil ice wedges. *Biul. Peryglacjalny* **15**, 65-73.

Pigott, C. D. (1956). The vegetation of Upper Teesdale in the North Pennines. *J. Ecol.* **44**, 545-86.

Pigott, C. D. and Pigott, M. E. (1963). Late-glacial and Post-glacial deposits at Malham, Yorkshire. *New Phytol.* **62**, 317-34.

Pigott, C. D. and Walters, S. M. (1954). On the interpretation of the discontinuous distributions shown by certain British species of open habitats. *J. Ecol.* **42**, 96-116.

Pilous, Z. (1968). Pleistozäne und holozäne Moose im Gebiete von Ostrava (Ostravsko). *Preslia* **40**, 68-75.

Podpera, J. (1954). *Conspectus Muscorum Europaeorum*. Prague; Czechoslovakian Academy.

Polak, B. (1963). A buried Allerød pine-forest. *Acta bot. neerl.* **12**, 533-8.

Polak, B. and Hamming, C. (1963). A peat-layer of early Wurm glacial age. *Geol. Mijnb.* **42**, 202-5.

Polunin, N. (1958). Botany of ice island, T3. *J. Ecol.* **46**, 323-47.

Poore, M. E. D. and Walker, D. (1958). Wybunbury Moss, Cheshire. *Mem. Proc. Manchester lit. phil. Soc.* **101**, 1-24.

Pospisil, V. (1967). Über die Variabilität und Verbreitung der Moosart *Thuidium abietinum* Br. Eur. incl. subsp. *hystricosum* (Mitt.) Kindb. in Tschechoslowakei. *Cas. Morav. Mus.* **52**, 169-96.

(1968). Können die Moose *Camptothecium lutescens* (Hedw.) B.S.G., *Entodon orthocarpus* (Brid.) Lindb., *Rhytidium rugosum* (Hedw.) Kindb. und *Thuidium abietinum* (Hedw.) B.S.G., auf dem gebiet der Tschechoslowakei Präglaciale Relickte Sein? *Cas. Morav. Mus.* **53**, 179-238.

Proctor, M. C. F. (1955). A key to British species of *Sphagnum*. *Trans. Br. bryol. Soc.* **2**, 552-60.

(1959*a*). Appendix in Sparks and West 1959.

(1959*b*). A note on *Acrocladium trifarium* (W. & M.) Richards and Wallace in Ireland. *Trans. Br. bryol. Soc.* **3**, 571-4.

(1960). Mosses and liverworts of the Malham district. *Field Std.* **1**, 61-86.

(1964). The phytogeography of Dartmoor bryophytes. Pp. 141-71 in *Dartmoor Essays*. Devonshire Association for the Advancement of Science, Literature and Art.

(1967). The distribution of British liverworts: a statistical analysis. *J. Ecol.* **55**, 119-36.

Proskauer, J. (1958). Nachtrag zur Familie Anthocerotaceae. Rabenhorst's *Kryptogamenflora* VI, 1304-1319.

Raistrick, A. and Blackburn, K. B. (1932). Linton Mires, Wharfedale. Glacial and Post-glacial history. *Proc. Univ. Durham phil. Soc.* **10**, 24-37.

Raistrick, A. and Woodhead, T. W. (1930). Plant remains in post-glacial gravels near Leeds. *Naturalist, Hull* 1930, 39-44.

Ralska-Jasiewiczowa, M. (1966). Bottom sediments of the Mikolajki Lake (Masurian Lake District) in the light of Palaeobotanical Investigations. *Acta Palaeobot.* **7**, 1-118. In Polish with English summary.

Ratcliffe, D. A. (1960). The mountain flora of Lakeland. *Proc. bot. Soc. Br. Isl.* **4**, 1-25.

(1968). An ecological account of Atlantic bryophytes in the British Isles. *New Phytol.* **67**, 365-439.

Ratcliffe, D. A. and Walker, D. (1958). The Silver Flowe, Galloway, Scotland. *J. Ecol.* **46**, 407-45.

Reese, W. D. (1967). The discovery of *Tortula vectensis* in North America. *Bryologist* **70**, 112-14.

Reich, H. (1953). Die Vegetationsenwicklung der Interglaziale von Grossweil-Ohlstadt und Pfeffbichl im Bayerischen Alpenvorland. *Flora* **140**, 386-443.

Reid, C. (1882). *The Geology of the Country around Cromer.* Memoirs of the Geological Survey.

(1884). On Norfolk amber. *Trans. Norfolk Norwich Nat. Soc.* **3**, 601-3.

(1892). The Pleistocene deposits of the Sussex coast, and their equivalents in other districts. *Q. Jl geol. Soc. Lond.* **48**, 344-64.

(1896). The relation of Palaeolithic Man to the Glacial epoch. Report of the 66th Meeting of the British Association for the Advancement of Science. Pp. 400-16.

(1897). The Palaeolithic deposits at Hitchin and their relation to the Glacial epoch. *Proc. R. Soc.* **61**, 40-9.

(1899). *The Origin of the British Flora*. London; Dulau and Co.

(1904). East Norfolk geology - wells at Mundesley, North Walsham and Metton. *Trans. Norfolk Norwich Nat. Soc.* **7**, 290-8.

Reid, C. and Reid, E. M. (1915). *The Pliocene Floras of the Dutch-Prussian Border.* Gravenhage; Martinus Nijhoff.

Reid, E. M. (1920). On two Preglacial floras from Castle Eden, County Durham. *Q. Jl geol. Soc. Lond.* **76**, 104-44.

(1949). The Late-glacial flora of the Lea Valley. *New Phytol.* **48**, 245-52.

Richards, P. W. (1954). The history of the British oceanic bryophytes. Huitième Congrès International de Botanique. Section 16, Bryologie. Pp. 63-8.

Robertson, D. (1881). On the Post-Tertiary beds of Garvel Park, Greenock. *Trans. geol. Soc. Glasg.* **7**, 1-37.

Robinson, H. (1959). The status of *Calliergon subsarmentosum. Bryologist* **62**, 186-8.

Roeder, E. (1899). Recent Roman discoveries in Deansgate and on Hunt's Bank, and Roman Manchester re-studied (1897-1900). *Trans. Lancs. Ches. Antiq. Soc.* **17**, 87-212.

Romans, J. C. C., Stevens, J. H. and Robertson, L. (1966). Alpine soils of North-East Scotland. *J. Soil Sci.* **17**, 184-199.

Rönning, O. I. (1961). The *Sphagnum* flora of Svalbard. *Blyttia* **19**, 14-53.

Rose, F. (1949). A Bryophyte flora of Kent. I. *Trans. Br. bryol. Soc.* **1**, 202-10.

(1951). A bryophyte flora of Kent. III. Musci. *Trans. Br. bryol. Soc.* **1**, 427-64.

(1953). A survey of the ecology of the British lowland bogs. *Proc. Linn. Soc. Lond.* **164**, 186-211.

(1957). The importance of the study of disjunct distributions to progress in understanding the British flora. Pp. 61-78 in *Progress in the Study of the British Flora*, edited by J. E. Lousley. Botanical Society of the British Isles.

Rosendahl, C. O. (1948). A contribution to the knowledge of the Pleistocene flora of Minnesota. *Ecology* **29**, 284-315.

Rune, D. (1953). Plant life on serpentines and related rocks in the North of Sweden. *Acta phytogeogr. suec.* **31**, 1-139.

REFERENCES

Ruuhijarvi, R. (1963). Zur Entwicklungsgeschichte der Nordfinnischen Hochmoore. *Annls. bot. Soc. Zool. bot. fenn. Vanamo* **34**, 1-40.

Rybnicek, K. (1966). Glacial relics in the bryoflora of the Highlands Ceskomoravska vrchovina. (Bohemian-Moravian Highlands); their habitat and cenotaxonomic value. *Folia geobot. phytotax.* **1**, 101-19.

Rybnicek, K. and Rybnicekova, E. (1968). The history of flora and vegetation on the Bláto Mire in southeastern Bohemia, Czechoslovakia (palaeoecological study). *Folia geobot. phytotax.* **3**, 117-42.

Savicz, L. (1928). Sur la fructification de *Rhytidium rugosum* (Ehrh.) Kindb. en Russe. *Annls bryol.* **1**, 140-3.

Savicz-Ljubitzkaja, L. I. and Abramov, I. I. (1959). The geological annals of the Bryophyta. *Revue bryol. lichen.* **28**, 330-42.

Savicz-Ljubitzkaja, L. I. and Abramova, A. L. (1954). Fossil mosses from the district of the excavations of the Taimyr Mammoth. *Bot. Zh. S.S.S.R.* **39**, 594-603. In Russian.

Savidge, J. P. (1963). The composition of the flora. In *Travis's Flora of South Lancashire* edited by J. P. Savidge. Liverpool Botanical Society.

Schmitz, H. (1967). Pflanzenreste aus dem Wandsbeker Interstadial Saale-(Riss)-Kaltzeit. Pp. 196-202 in *Früke Menschheit und Umwelt*, Volume II, edited by K. Gripp, R. Schutrumpf and H. Schwabedissen.

Schofield, W. B. (1966). The identity of *Polytrichum sphaerothecium* (Besch.) Broth. *Miscnea bryol. lichen.* **4**, 33-5.

(1969*a*). Some common mosses of British Columbia. *British Columbia Provincial Museum handbook*, No. 28.

(1969*b*). Phytogeography of northwestern North America; bryophytes and vascular plants. *Madroño* **20**, 155-207.

Schumacher, A. (1958). Über westdeutsche Standorte von *Sphagnum imbricatum* (Hornsch.) Russ. *Abh. naturw. Verein Bremen* **35**, 335-50.

Schuster, R. M. (1969). Problems of Antipodal distribution in lower land plants. *Taxon* **18**, 46-91.

Schuster, R. M. Steere, W. C. and Thomson, J. W. (1959). The terrestrial cryptogams of northern Ellesmere Island. *Bull. natn. Mus. Can.* **164**, 1-132.

Schütrumpf, R. (1937). Die Paläobotanische-Pollenanalytische Untersuchung. In *Das Altsteinzeitiche Rentierjägerlager Meiendorf.* Neumunster; Karl Wachholtz Verlag.

Scott, T. and Steel, J. (1883). Notes on the occurrence of *Leda Arctica* (Gray); *Lyonsin Arenosa* (Moller), and other organic remains, in the Post-pliocene clays of Garvel Park, Greenock. *Trans. geol. Soc. Glasg.* **7**, 279-83.

Seagrief, S. C. (1955). A pollen analytic investigation of the Quaternary Period in Britain. Ph.D. Thesis, Cambridge University.

(1960). Pollen diagrams from southern England: Cranes Moor, Hampshire. *New Phytol.* **59**, 73-83.

Seagrief, S. C. and Godwin, H. (1960). Pollen diagrams from southern England: Elstead, Surrey. *New Phytol.* **59**, 84-91.

Seddon, B. (1962). Late-glacial deposits at Llyn Dwythwch and Nant Ffrancon, Caernarvonshire. *Phil. Trans. R. Soc. Ser. B.* **244**, 459-81.

Selle, W. (1955). Die vegetationsentwicklung des Interglazials vom Typ Ober-Ohe. *Abh. Naturw. Verein Bremen* **34**, 33-46.

Shackleton, N. J. (1969). The last interglacial in the Marine and Terrestrial records. *Proc. R. Soc. B* **174**, 135-54.

Shackleton, N. J. and Turner, C. (1967). Correlation between Marine and Terrestrial Pleistocene successions. *Nature, Lond.* **216**, 1079-82.

Sheppard, T. (1910). The prehistoric boat from Brigg. *Trans. E. Riding Antiq. Soc.* **17**, 33-54.

Shotton, F. W. (1966). The problems and contributions of methods of absolute dating within the Pleistocene period. *Q. Jl geol. Soc. Lond.* **122**, 356-83.

Shotton, F. W., Sutcliffe, A. J. and West, R. G. (1962). The fauna and flora from the brick pit at Lexden, Essex. *Essex Nat.* **31**, 15-22.

Shotton, F. W. and West, R. G. (1969). Report on the Quaternary Era Sub-Committee.

Simpson, F. G. and Richmond, I. A. (1937). The fort on Hadrian's Wall at Halton. *Archaeologia* **14**, 151-71.

Simpson, I. G. and West, R. G. (1958). On the stratigraphy and palaeobotany of a Late-Pleistocene deposit at Chelford, Cheshire. *New Phytol.* **57**, 239-50.

Simpson, J. R. (1924). Some moss records for Selkirk. *Trans. bot. Soc. Edinb.* **29**, 73-82.

Sinker, C. A. (1962). The North Shropshire meres and mosses: a background for ecologists. *Field Stud.* **1**, 101-38.

Sissons, J. B. (1967). *The Evolution of Scotland's Scenery.* Edinburgh; Oliver and Boyd.

Sjögren, E. (1964). Epilithische und epigäische Moosvegetation in Laubwaldern der Insel Öland (Schweden). *Acta Phytogeogr. suec.* **48**, 1-84.

Sjörs, H. (1949). Om *Sphagnum lindbergii* i Jödra delen ar Sverige. *Svensk bot. Tidskr.* **43**, 568-85.

(1950). Regional studies in north Swedish mire vegetation. *Bot. Notiser* 1950, 175-222.

Smith, A. G. (1958*a*). Pollen analytical investigations of the mire at Fallahogy, Co. Derry. *Proc. R. Ir. Acad.* **59**, 329-43.

(1958*b*). Two lacustrine deposits in the south of the English Lake District. *New Phytol.* **57**, 363-86.

(1958*c*). The content of some Late Bronze Age and Early Iron Age remains from Lincolnshire. *Proc. prehist. Soc.* **24**, 78-84.

(1959). The mires of southwestern Westmorland: stratigraphy and pollen analysis. *New Phytol.* **58**, 105-7.

(1961). Cannons Lough, Kilrea, Co. Derry: stratigraphy and pollen analysis. *Proc. R. Ir. Acad.* **61**, 369-83.

(1970). The influence of Mesolithic and Neolithic man on British vegetation: a discussion. Pp. 81-96 in *Studies in the Vegetational History of the British Isles*, edited by D. Walker and R. G. West. Cambridge University Press.

Smith, R. A. (1911). Lake dwellings in Holderness, Yorks., discovered by Thos. Boynton Esq., F.S.A., 1880-1. *Archaeologia* **62**, 593-610.

Sobolewska, M. and Srodon, A. (1961). Late-Pleistocene deposits at Bialka Tatrzanaka (West Carpathians). *Folia quatern.* **7**, 1-16.

Sobolewska, M., Starkel, L. and Srodon, A. (1964). Late-Pleistocene deposits with fossil flora at Wadowice (West Carpathians). *Folia quatern.* **16**, 1-64. In Polish with English summary.

Solonevicz, K. I. (1935). Sur la flore postglaciaire de Leningrad d'après les matériaux fossiles de Pestchanka. *Sov. Bot.* **6**, 67-77. In Russian.

Sparks, B. W. and West, R. G. (1959). The Palaeo-ecology of the Interglacial deposits at Histon Road, Cambridge. *Eiseitalter Gegenw.* **10**, 123-43.

(1968). Interglacial deposits at Wortwell, Norfolk. *Geol. Mag.* **105**, 471-81

(1970). Late Pleistocene deposits at Wretton, Norfolk. I. Ipswichian interglacial deposits. *Phil. Trans. R. Soc. Ser. B.* **258**, 1-30.

Sparks, B. W., West, R. G., Williams, R. B. G. and Ransom, M. (1969). Hoxnian Interglacial deposits near Hatfield, Herts. *Proc. geol. Assoc.* **80**, 243-67.

Spence, D. H. N. (1967). Factors controlling the distribution of freshwater macrophytes with particular reference to the lochs of Scotland. *J. Ecol.* **55**, 147-70.

Srodon, A. (1952). Last glacial and Post-glacial in the Carpathians. *Biul. Panst. Inst. Geol.* **67**, 27-75. In Polish with English summary.

(1960). Pollen spectra from Spitsbergen. *Folia quatern.* **3**, 1-17.

REFERENCES

(1965). On fossil floras in the terraces of Carpathian valleys. *Folia quatern.* **21**, 1-27.

(1968). On the vegetation of the Paudorf Interstadial in the Western Carpathians. *Acta Palaeobot.* **9**, nr. 1, 1-27. In Polish with English summary.

Stark, P. (1912). Beiträge zur Kenntnis der eiszeitlichen Flora und Fauna Badens. *Ber. Naturf. Ges. Freiburg; Br.* **19**, 153-272.

Steere, W. C. (1938). Pleistocene mosses from Louisiana. *Louisiana Geol. Surv. Dept. Conserv. Bull.* **13**, 97-101.

(1942). Pleistocene mosses from the Aftonian interglacial deposits of Iowa. *Pap. Mich. Acad. Sci.* **27**, 75-104.

(1946). Cenozoic and Mesozoic bryophytes of North America. *Am. Midl. Nat.* **36**, 298-324.

(1948). Musci. In Polunin, N., Botany of the Canadian Eastern Arctic. *Bull. natn. Mus. Can.* **97**, 370-490.

(1953). On the geographical distribution of Arctic bryophytes. *Stanford Univ. Publs. biol. Sci.* **11**, 30-47.

(1965). The Boreal bryophyte flora as affected by Quaternary Glaciation. Pp. 485-95 in *The Quaternary of the United States*, edited by H. E. Wright and D. G. Frey. Princeton University Press.

Stefureac, T. I. (1956). Cercetari asupra speciei *Helodium lanatum* (Stroem.) Broth., in R.P.R. *Bul. Stiint. sect. biol. si agric.* **8**, 237-70. In Roumanian with French summary.

(1962). Relictes subarctiques dans bryoflore du marais eutrophe de Dragoiasa Carpathes orientales. *Revue bryol. lichen.* **31**, 68-73.

Stockmans, F. and Vanhoorne, R. (1954). Étude botanique Gisement de Tourbe de la Région de Pervijze (Plaine Maritime Belge). *Mem. Inst. r. Sci. nat. Belge.* **130**, 1-144.

Störmer, P. (1939). *Brynhia novae-angliae* in Scandinavia. *Annls bryol.* **12**, 154-7.

(1940). Bryophytes from Franz Josef Land and Eastern Svalbard. *Norg. Svalbard Ishavs Unders.* **47**, 1-16.

(1949). Moser funnet i Raknehaugen ved utgravingen 1939-40. *Blyttia* **7**, 92-5.

(1955). *Isothecuim striatulum*, a moss new to Norway. *Nytt Mg. Bot.* **4**, 87-94.

(1965). Mosses from deposits of the Boreal Period in Norway. *Revue bryol. lichen.* **33**,

(1969). *Mosses with a Western and Southern Distribution in Norway.* Oslo; Universitetsforlaget.

Straus, A. (1952). Beitrage zur Pliocänflora von Hillershausen III. *Palaeontographica* **93**, 1-44.

Stuchlik, L. (1964). Pollen analysis of the Miocene deposits at Rypin. *Acta Palaeobot.* **5**, 1-111.

Suggate, R. P. and West, R. G. (1959). The extent of the glaciation in Eastern England. *Proc. R. Soc. B.* **150**, 163-83.

Swinscow, T. D. V. (1959). A Bryophyte flora of Hertfordshire. *Trans. Brit. bryol. Soc.* **3**, 509-57.

Szafer, W. (1954). Pliocene flora from the vicinity of Czorsztyn (West Carpathians) and its relationship to the Pleistocene. *Inst. Geol. Prace* **11**, 1-238. In Polish with English summary.

Szafran, B. (1934). Diluvial mosses from Starunia. *Starunia* **1**, 1-17.

(1948*a*). Mosses from the Pliocene deposit of Kroscienko in Poland. *Acad. Pol. Sc. Lett. Bull. Intern. B.*

(1948*b*). Relicts of past epochs in the moss flora of Poland and adjacent eastern regions *Ochr. Pryr.* **18**, 41-65. In Polish with English summary.

(1949*a*). *Trachycystis Szaferi*, a new species of moss from the Miocene of Poland. *Acta. Soc. Bot. Pol.* **20**, 247-50.

(1949*b*). Pliocene Moss flora of Kroscienko. *Rozpr. Wudz. mat.-przyr. Polsk. Akad. Um. B.* **73**, 9-14. In Polish.

(1949*c*). Pochodzenie torfowcow. The origin of the Sphagna. *Acta Soc. Bot. Pol.* **17**, 219-37.

(1952). Pleistocene mosses from Poland and the adjacent eastern territories. *Panstw. Inst. Geol.* **68**, 5-38. In Polish with English summary.

(1957 and 1961). *Musci*, volumes I and II, *Flora Polska.* Warsaw, Polish Academy. In Polish.

(1958). Tortonian mosses from Stare Gliwice in Silesia. *Mongr. Bot.* **7**, 61-8. In Polish with English summary.

(1964). Tortonian Mosses from 'Zatoka Gdowska' (environs of Cracow). *Acta. Soc. Bot. Pol.* **33**, 557-61.

Szczepanek, K. (1960). Flora Dryasowa z Mokoszyna Kolo Sandomierza. *Inst. Geol. Biul.* **150**, 131-5. In Polish with English summary.

Szweykowski, J. (1958). *Prodromus Florae Hepaticerum Poloniae.* Poznan Society of Friends of Science, volume 17.

Tallis, J. H. (1958). Studies in the biology and ecology of *Rhacomitrium lanuginosum* Brid. I. Distribution and ecology. *J. Ecol.* **46**, 271-88.

(1961). The distributions of *Thuidium recognitum* Lindb., T. *philiberti* Limpr. and T. *delicatulum* Mitt. in Britain. *Trans. Br. bryol. Soc.* **4**, 102-6.

(1962). The identification of *Sphagnum* spores. *Trans. Br. bryol. Soc.* **4**, 209-13.

(1964). The pre-peat vegetation of the southern Pennines. *New Phytol.* **63**, 363-73.

Tansley, A. G. (1939). *The British Islands and Their Vegetation.* Cambridge University Press.

Taylor, J. (1951). Some Irish Bryophyta records. *Ir. Nat. J.* **10**, 131-3.

Terasmae, J. (1955). On the spore morphology of some *Sphagnum* species. *Bryologist* **58**, 306-11.

Thomson, P. W. and Pflug, H. (1953). Pollen und Sporen des mittel-europäischen Tertiars. *Palaeontographica* **95**, 1-138.

Thorpe, J. (1887). A description of an ancient raft recently found by Messrs Ludge and Cole, in a field adjoining the brickyard, in their occupation, belonging to the Right Honourable the Earl of Yarborough, situated at Brigg in the County of Lincoln. *Assoc. Arch. Soc. Re. Pap.* **19**, 95-7.

Tolpa, S. (1952). Interglacial flora at Kalisz. *Buil. panst. Inst. geol.* **68**, 73-117. In Polish with English summary.

(1961). Flora Interglacjalna ze Slawna Kolo Radomia. *Buil. panst. Inst. geol.* **169**, 15-56. In Polish with English summary.

Travis, C. B. (1913). Geological notes on recent dock excavations at Liverpool and Birkenhead. *Proc. L'pool geol. Soc.* **11**, 267-70.

Travis, C. B. and Travis, W. G. (1913). On plant remains in postglacial gravels at Seaforth, Liverpool. *Lancs. Nat.* **6**, 49-51.

Travis, W. G. (1909). Plant remains in peat of the Shirdley Hill Sands at Aintree, South Lancashire. *Trans. L'pool bot. Soc.* **1**, 47-52.

(1922). On peaty bands in the Wallasey sandhills. *Proc. L'pool geol. Soc.* **13**, 207-14.

Tuomikoski, R. (1939). Materialen zu einer Laubmoosflora des Kuusamo-gebietes. *Annls bot. Soc. Zool.-bot. fenn. Vanamo* **12**, 1-124.

(1940). *Calliergon megallophyllum* Mikut. und *Drepanocladus capillifolius* (Warnst.) in Finland. *Annls bot. Soc. Zool.-bot. fenn. Vanamo* **15**, 1-28.

(1946). On the mosses of the cliff Komonkallio in Hollola. *Luonon Ystävä* **50**, 94.

Turner, C. (1968). A Lowestoftian Late-glacial flora from the Pleistocene deposits at Hoxne, Suffolk. *New Phytol.* **67**, 327-32.

(1970). The Middle Pleistocene deposits at Marks Tey, Essex. *Phil. Trans. R. Soc. Ser. B.* **257**, 363-440.

Turner, J. (1970). Post-Neolithic disturbance of British vegetation. Pp. 97-116 in *Studies in the Vegetational History of the British Isles*, edited by D. Walker and R. G. West. Cambridge University Press.

Tyuremnov, S. N. (1963). On the distribution of *Sphagnum*

imbricatum. Byull. Mosk. Obshch. Isp. Prirody. **68**, 98-109. In Russian.
Udar, R. (1964). Palynology of bryophytes. Pp. 79-100 in *Advances in Palynology*, edited by P. K. K. Nair. Lucknow; National Botanic Gardens.
Vaarama, A. (1938). Observations on the locality of *Rhytidium rugosum* in Hollola. *Luonon Ystävä* **42**, 38-9.
(1967). A find of *Pogonatum capillare* (Michx.) Brid. in southern Finland and reflections on its bryo-geographical significance. *Aquilo* **6**, 209-18.
Vanden Berghen, C. (1950). Présence de la mousse *Dicranum bergeri* Bland, à l'état subfossile en Belgique. *Bull. Inst. r. Sci. nat. Belg.* **26**, 1-7.
(1951). Note sur des bryophytes du Pleistocene de Belgique. *Bull. Inst. r. Sci. nat. Belg.* **27**, 1-7.
Van der Hammen, T., Maarleveld, G. C., Vogel, J. C. and Zagwin, W. H. (1967). Stratigraphy, climatic succession and radiocarbon dating of the Last Glacial in the Netherlands. *Geol. Mijnb.* **46**, 79-95.
Van der Flerk, I. M. and Florschütz, F. (1950). Nederland in Het Ijstijdvak. Utrecht: W. de Haan.
Vanhoorne, R. (1954). L'Oscillation d'Alleröd en Belgique. Volume jubilaire V. Van Straelen. *Inst. r. Sci. nat. Belg.* **1**, 141-7.
Varley, W. J. (1968). Barmston and the Holderness Crannogs. *East Riding Archaeologist* **1**, 12-25.
Vasari, Y. and Vasari, A. (1968). Late- and Post-glacial macrophytic vegetation in the lochs of Northern Scotland. *Acta. Bot. Fenn.* **80**, 1-120.
Vevers, H. G. (1936). The land vegetation of Ailsa Craig. *J. Ecol.* **24**, 424-45.
Viereck, L. A. (1966). Plant succession and soil development on gravel outwash of the Muldrow Glacier. *Ecol. Monogr.* **36**, 181-99.
Vishnu-Mittre (1959). Post-glacial history of the Whittlesey Mere region of East Anglian Fenland. Ph.D. Thesis, Cambridge University.
Vogel, J. C. and Zagwijn, W. H. (1967). Groningen radiocarbon dates VI. *Radiocarbon* **9**, 63-106.
Walker, D. (1948). *Tilletia (?) sphagni* Nawaschin: A new British record. *Naturalist, Hull* 1948, 153.
(1953). The interglacial deposits at Histon Road, Cambridge. *Q. Jl geol. Soc. Lond.* **108**, 273-82.
(1955). Studies in the Post-Glacial history of British vegetation. XIV. Skelsmergh Tarn and Kentmere, Westmorland. *New Phytol.* **54**, 222-54.
(1966). The Late Quaternary history of the Cumberland Lowland. *Phil. Trans. R. Soc. Ser. B.* **251**, 1-210.
Walker, D. and Lambert, C. A. (1955). Boreal deposits at Kirkby Thore, Westmorland. *New Phytol.* **54**, 210-15.
Walker, D. and Walker, P. M. (1961). Stratigraphic evidence of regeneration in some Irish bogs. *J. Ecol.* **49**, 169-85.
Wallace, E. C. (1941). The mosses of Surrey. *J. Bot. Lond.* **79**, 1-11, 17-25.
(1956). *Hylocomium splendens* (Hedw.) B. & S. at Verulamium, Hertfordshire. *Trans. Br. bryol. Soc.* **3**, 127.
Warburg, E. F. (1958). *Meesia tristicha* Bruch & Schimp., in the British Isles. *Trans. Br. bryol. Soc.* **3**, 378-81.
(1963). *Census Catalogue of British Mosses.* 3rd edition. The British Bryological Society.
(1965*a*). *Grimmia borealis* in Britain. *Trans. Br. bryol. Soc.* **4**, 757-9.
(1965*b*). *Pohlia pulchella* in Britain. *Trans. Br. bryol. Soc.* **4**, 760-2.
Warburg, E. F. and Crundwell, A. C. (1965). *Tortula vectensis*, a new species from the Isle of Wight. *Trans. Br. bryol. Soc.* **4**, 763-6.
Warren, S. H. (1912). On a Late-glacial stage in the Valley of the River Lea. *Q. Jl geol. Soc. Lond.* **68**, 213-51.
Watson, E. V. (1960). A quantitative study of the bryophytes of chalk grassland. *J. Ecol.* **48**, 397-414.
(1964). An annotated list of the bryophytes of Jan Mayen Island. *Nytt Mag. Bot.* **11**, 151-212.
(1968). *British Mosses and Liverworts.* 2nd edition. Cambridge University Press.
Watson, W. (1925). The bryophytes and lichens of arctic-alpine vegetation. *J. Ecol.* **13**, 2-26.
Watts, W. A. (1959*a*). Interglacial deposits at Kilbeg and Newton, Co. Waterford. *Proc. R. Ir. Acad.* **60**, 79-134.
(1959*b*). Pollen spectra from the interglacial deposits at Kirmington, Lincolnshire. *Proc. Yorks. geol. Soc.* **32**, 145-51.
(1962). Early Tertiary pollen deposits in Ireland. *Nature, Lond.* **193**, 600.
(1964). Interglacial deposits at Baggotstown, near Bruff, County Limerick. *Proc. R. Ir. Acad.* **63**, 167-89.
(1967). Interglacial deposits in Kildrommin Townland, near Herbertstown, Co. Limerick. *Proc. R. Ir. Acad.* **65**, 339-48.
(1970). Tertiary and interglacial floras in Ireland. Pp. 17-33 in *Irish Geographical Studies*.
Webster, G. (1967). Excavations at the Romano-British villa in Barnsley Park, Cirencester 1961-66. *Trans. Bristol Glouc. arch. Soc.* **86**, 74-87.
Welch, W. H. (1948). Mosses and their uses. *Proc. Indiana Acad. Sci.* **58**, 31-46.
Welch, H. (1960). *A Monograph of the Fontinalaceae.* Gravenhage; Martinus Nijhoff.
West, R. G. (1956). Quaternary deposits at Hoxne, Suffolk. *Phil. Trans. R. Soc. Ser. B* **239**, 265-356.
(1961). Vegetational history of the Early Pleistocene of the Royal Society borehole at Ludham, Norfolk. *Proc. R. Soc. B* **155**, 437-53.
(1968*a*). *Pleistocene Geology and Biology*. London; Longmans.
(1968*b*). Evidence for Pre-Cromerian permafrost in East Anglia. *Biul. Peryglacjalny* **17**, 303-4.
(1969). Pollen analyses from interglacial deposits at Avelly and Grays, Essex. *Proc. geol. Ass.* **80**, 271-82.
(1970). Pleistocene history of the British flora. Pp. 1-12 in *Studies in the Vegetational History of the British Isles*, edited by D. Walker and R. G. West. Cambridge University Press.
West, R. G., Lambert, C. A. and Sparks, B. W. (1964). Interglacial Deposits at Ilford, Essex. *Phil. Trans. R. Soc. Ser. B* **247**, 185-212.
West, R. G. and Sparks, B. W. (1960). Coastal interglacial deposits of the English Channel. *Phil. Trans. R. Soc. Ser. B* **243**, 95-133.
West, R. G. and Wilson, D. G. (1968). Plant remains from the Corton Beds at Lowestoft, Suffolk. *Geol. Mag.* **105**, 116-23.
Weyland, E. (1925). Beiträge zur Kenntnis Fossiler I. Die Moose der ober Pliocänen Flora des Frankfurter Klärbechens. *Senckenbergiana* **7**, 3-16.
Whitehead, H. (1921). More about 'Moorlog', a peaty deposit from the Dogger Bank in the North Sea. *Essex Nat.* **19**, 242-52.
Whitehead, H. and Goodchild, H. H. (1909). Some notes on 'Moorlog' - a peaty deposit from the Dogger Bank in the North Sea. *Essex Nat.* **16**, 51-60.
Whitehouse, H. L. K. (1964). Bryophyta. Pp. 281-328. In Perring, F. H., Sell, P. D., Walters, S. M. *A Flora of Cambridgeshire*. Cambridge University Press.
(1966). The occurrence of tubers in European mosses. *Trans. Br. bryol. Soc.* **5**, 103-16.
Williams, R. S. (1930*a*). Pleistocene mosses from Minneapolis, Minnesota. *Bryologist* **33**, 33.
(1930*b*). Notes on some Pleistocene mosses recently discovered. *Jl N. Y. Bot. Gard.* **31**, 154.
Wilson, D. G. (1968). Plant remains from a Roman well at Bunny, Nottinghamshire. *Trans. Thoroton Soc. Nottingham* **72**, 42-9.

REFERENCES

Wilson, L. R. (1932). The Two Creeks Forest Bed, Manitowoc County, Wisconsin. *Trans. Wis. Acad. Sci.* **27**, 31-46.

Wilson, W. (1885). *Bryologica Britannica*. London; Longman, Brown, Green and Longmans.

Woodhead, N. and Hodgson, L. M. (1935). A preliminary study of some Snowdonian Peats. *New Phytol.* **34**, 263-82.

Wright, E. V. and Churchill, D. M. (1965). The boats from North Ferriby, Yorkshire, England. *Proc. prehist. Soc.* **31**, 1-24.

Wright, E. V. and Wright, C. W. (1947). Prehistoric boats from North Ferriby, East Yorkshire. *Proc. prehist. Soc.* **13**, 114-38.

Zagwijn, W. H. (1960). *Aspects of the Pliocene and Early Pleistocene vegetation in the Netherlands*. Maastricht; Uitgevers-Mij. 'Ernest van Aelst'.

(1961). Vegetation, climate and radiocarbon datings in the Late Pleistocene of the Netherlands. Part I. Eemian and Early Weichselian. *Meded. Geol. Stichting* **14**, 15-45.

(1963a). Pollen-analytic investigations in the Tiglian of the Netherlands. *Meded. Geol. Stichting* **16**, 49-71.

(1963*b*). Pleistocene stratigraphy in the Netherlands, based on changes in vegetation and climate. *Verh. Kon. Ned. Geol. Mijnb. Gen. Ser. 21-22* **2**, 173-96.

Zmuda, A. J. (1914). Fossile flora des Krakauer Diluviums. *Bull. Acad. Sci. Cracovie* (1914), 209-352.

INDEX

The British Pleistocene records of chapter 5 are distinguished from other references to bryophytes by numbers in bold type.

Chapter 5 has not been indexed for places or geological periods.

The appendices have been indexed for places but not for bryophytes.

Authors have not been indexed.

Reference to text figures and plates is indicated by a page number and an asterisk.

Synonyms are in *italics*.

INDEX

INDEX

INDEX

INDEX